D1246890

Damage Mechanisms and Life Assessment of High-Temperature Components

Damage Mechanisms and Life Assessment of High-Temperature Components

R. Viswanathan

Technical Advisor
Generation and Storage Division
Electric Power Research Institute

ASM INTERNATIONAL®
Metals Park, Ohio 44073

Library of Congress Catalog Card Number: 89-83684
ISBN: 0-87170-358-0
SAN: 204-7586

Editorial and production coordination by
Carnes Publication Services, Inc.

PRINTED IN THE UNITED STATES OF AMERICA

Preface

A large percentage of power, petroleum, and chemical plants the world over have been in operation for such long durations that the critical components in these plants have been used beyond the "design life" of 30 to 40 years. This percentage is likely to become even higher during the next decade. There are strong economic reasons and technical justifications for continued operation of these plants. In order to realize this in practice, however, techniques and methodologies are needed to assess the current condition of plant components and to project their remaining useful lives. This technology is also of value with respect to younger plants in the contexts of safety, availability, and reliability, and operation, maintenance, and inspection practices. There has been a flurry of research and development activities worldwide during the last few years relating to life-assessment technology for high-temperature components.

With respect to the theory of damage at high temperatures, many books and conference proceedings dealing with creep, fatigue, creep-fatigue, thermal fatigue, hot corrosion, and hydrogen attack have been published. These publications deal extensively with the mechanistic aspects of the various damage phenomena but fail to place them in the framework of an engineering approach. Furthermore, they address each damage phenomenon as a single issue, and

fail to delineate the relationships between phenomena in the over-all context of component integrity. On the other hand, conference proceedings dealing with the engineering aspects of life assessment are component-specific and do not provide sufficient theory to enable a nonspecialist to fully comprehend the methodologies being discussed. A major objective of this book is to bring together in one place the theory and practice of damage assessment of high-temperature components. In accordance with this objective, the first four chapters lay the theoretical groundwork pertaining to damage phenomena, and are followed by illustrations of the practical application of the theory on a component-specific basis in the later chapters. Other objectives of this book are: (1) to provide a complete bibliography to methods, data, and case histories; (2) to provide relevant data with examples; (3) to provide empirical correlations and techniques that enable one to estimate those properties which are often difficult to determine, on the basis of others which are more readily obtained; and (4) to document recent advances in materials technology leading to increased reliability and longevity of components.

The published literature relating to the various damage phenomena described in this book is extensive and vast. It is impossible to provide "in depth" coverage of these phenomena in a single volume. Only a broad coverage of the various issues is therefore provided. Only those damage phenomena which result from high-temperature exposure are discussed. Most of the discussions center around low-alloy steels because these steels represent the largest tonnage of material used in plant construction.

Selection of the contents of this book has been based mainly on my close acquaintance with the concerns that preoccupy utility engineers. In particular, my interaction with the metallurgy and piping task force of the Edison Electric Institute was very useful in identifying the practical needs. It is hoped that this book will be useful to practicing metallurgists and mechanical engineers. In addition, it can also serve to expose college metallurgy students to the industrial aspects of the high-temperature-metallurgy curriculum. Because the features of many types of high-temperature failures are described, failure-analysis consultants may also find the book to be of some use.

This book was written during a sabbatical year that I spent as a visiting professor of metallurgy at the Indian Institute of Technology in Madras, India. I wish to thank Professor L.S. Srinath, the director of IIT Madras, and Professor V.M. Radhakrishnan, past chairman of the department of metallurgy, for allowing me this opportunity. I am also very grateful to the senior management at EPRI, particularly Dr. John Stringer, for granting me this sabbatical leave, for providing constant encouragement, and for allowing me the use of material from EPRI reports. Without the support of the Electric Power Research Institute and the extensive knowledge base developed through its research projects, this book would not have been possible. The various sections of the book were reviewed for technical accuracy by J.M. Allen, F. Ammirato, W.T. Bakker, E. Creamer, R.B. Dooley, F. Ellis, J. Foulds, G. Ibarra, R.I. Jaffee, C. Jaske, S.R. Paterson, M. Prager, A. Saxena, and V.P. Swaminathan. I am indebted to them for the many useful comments and suggestions for improvement that they provided. Several chapters were typed by Perky Perkins and were proofread by Mr. Nallathambi Kandaswamy. The thorough review of the manuscript by the copyeditor C.W. Kirkpatrick is greatly appreciated. I am also thankful to the publisher, ASM International. Timothy Gall of ASM was particularly helpful in streamlining the publication process. Last but not least, I wish to acknowledge the help of my wife Vatsala for her understanding and support in freeing me from many of the household chores during the writing of this book. This book is a humble dedication to Bhagawan Sri Ramana Maharshi.

Contents

1

Introduction and Overview

Need for Remaining-Life-Assessment Technology

A large percentage of power, petroleum, and chemical plants the world over have been in operation for such long durations that the critical components of these plants have been used beyond the design life of 30 to 40 years. This percentage is likely to become even more significant during the next decade because of the hiatus in new plant construction over the last several years. There is a strong desire on the part of many plant owners to continue to operate their plants for another 20 to 40 years. The factors that have led to this situation include:

1. The escalating costs of new construction and diminishing capital resources
2. Excess capacity, although precarious and derived from aging plants
3. Extended lead times in plant construction
4. Uncertainties in projected demand growth rates
5. Limited availability of suitable sites for new construction
6. Increasingly stringent environmental, safety, and other regulations
7. Increasing awareness of the technolog-

ical feasibility of extending component life.

Several preliminary studies have shown that the cost of life extension of a typical fossil power plant may be only 20 to 30% of the cost of constructing a new plant and that the benefit-to-cost ratios are very high (Ref 1). Similar estimates for other types of plants are not available.

The term "life extension" has often been misunderstood. The purpose of life-extension activities is not to continue the operation of a plant beyond its useful life, but merely to ensure full utilization up to its useful life. The idea is to avoid premature retirement of plants and plant components, on the basis of the so-called design life, because actual useful lives could often be well in excess of the design life.

Extension of plant life may cease to be a desirable objective if it results in reduced availability or plant efficiency. Many improvements in material quality and design of critical components have been made over the last two decades. Selective replacement of components with more-modern designs should be part of the life-extension process. In this manner, the availability and efficiency of the life-extended plant may actu-

1

ally be improved in comparison with the initial conditions.

A key ingredient in plant life extension is the remaining-life-assessment technology. If such assessments indicate the need for extensive replacements and refurbishments, life extension may not prove to be a viable option. Above and beyond this objective, life-assessment technology serves many other purposes. It helps in setting up proper inspection schedules, maintenance procedures, and operating procedures. The data and methodologies needed for life extension are the same as those needed for optimizing these factors. It should therefore be recognized at the outset that development of techniques for life assessment is more enduring in value and broader in purpose than simply the extension of plant life. For instance, it has been possible to extend the inspection intervals from six to ten years for modern rotors on the basis of assessments based on fracture mechanics, resulting in considerable savings (Ref 2). Many fossil power plants originally designed 30 years ago for base-load operation are now being pressed into cyclic duty for economic reasons. Life-assessment techniques can quantify the penalty in terms of reduced plant life resulting from the changed operating mode. The start-up and shutdown procedures for plant components can be optimized, resulting in increased efficiency, reliability, and life. In view of the manifold benefits from life-assessment technology, considerable research has been carried out in this area during the last five to ten years. The remainder of this chapter will present a broad overview of the materials problems in fossil power, petroleum, and chemical plants, the failure criteria employed, and their relevance in the context of remaining-life assessment.

Fossil Power Plants

The availability of electrical power and the development of the millions of devices that use it have made electricity the energy of choice in contemporary industrial societies. This convenient energy form, which is available to us at the flip of a switch, is used in countless ways. Practically every economic sector—industrial, commercial, residential, and transportation—depends on the availability of electricity.

Types of Electric Power Plants

The principal route for producing electricity is the conversion of mechanical energy of rotation into electrical energy using a generator. The large generators used by electric utilities employ a shaft comprising the magnetic field (rotor) which rotates inside a stationary electric field containing conducting wires (stator), as shown schematically in Fig. 1.1 (Ref 3). Rotation of the shaft is achieved by coupling it to a turbine in which the kinetic energy of a moving fluid is converted into mechanical energy of rotation. The working fluid can be wind, water, steam, or combustion gases—leading to the resulting classification of turbines as wind turbines, hydroelectric turbines, steam turbines, and combustion turbines. Of these, the most common is the steam turbine, which employs steam produced from burning fossil fuels in a boiler or from the heat produced by atomic reactions inside a nuclear reactor. It is estimated that in the United States approximately 70% of the electricity is produced in fossil power plants, 15% in nuclear power plants, 12% in hydroelectric power plants, and the remainder from other types of sources (Ref 3). This mix may be somewhat different in other countries. In any event, the fossil-fuel power plant is and will continue to be the mainstay of electric power production. The fossil fuel can be employed to make steam to drive a steam turbine, or, alternatively, the combustion gases under high pressure can be used to drive a combustion turbine. Combined-cycle plants in which both combustion and steam turbines are employed in tandem result in greater efficiency and are becoming increasingly common. In these plants, the exhaust gases from a combustion turbine are used to make steam to drive a steam turbine.

The fossil fuel employed in a steam turbine plant can be pulverized coal (PC), oil,

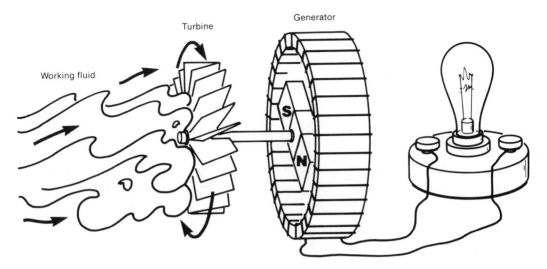

Fig. 1.1. Schematic illustration of the principle of the turbine-generator combination (Ref 3).

or natural gas. Of these, coal is the most abundant and hence the most commonly used fuel for steam turbine plants, while gas turbine plants generally employ oil and natural gas. To cope with increasingly stringent environmental standards, particularly in terms of sulfur- and nitrogen-containing compounds, and to improve the efficiency of combustion, a variety of alternative processes in which coal is combusted in a fluidized bed containing lime or is converted into coal-derived gas or liquid and then combusted are being vigorously pursued worldwide. Jaffee has described ten different plant configurations for generating electricity from coal, as shown in Fig. 1.2 (Ref 4).

Historical Evolution of Fossil Plants

The historical evolution of fossil-fired steam power plants is shown in Fig. 1.3 (Ref 4). In this illustration, the shaft output, maximum steam pressure, steam temperature, and plant thermal efficiency are plotted using different scales. The discontinuous variation in each plot results from the data points being plant-specific; the over-all trends, however, are quite clear. During the period beginning in 1920, plant capacity has increased from less than 0.5 MW to almost 1300 MW; steam throttle pressures have increased from 690 to 1380

kPa (100 to 200 psi) to more than 24.8 MPa (3600 psi). Steam temperature has increased from 230 °C (450 °F) to temperatures in excess of 565 °C (1050 °F). To improve the efficiency of the steam turbines, additional reheat cycles in which the working fluid is reheated again and expanded once again through intermediate pressure turbines have been added. This phenomenal increase in plant capacity and operating conditions has been possible only through corresponding improvements in materials technology.

Typical PC Fossil Plant and Component Damage Mechanisms

Figure 1.4 shows the arrangement of the various elements of a PC fossil plant (Ref 4). Here, water is first preheated to a relatively low temperature in feedwater heaters and pumped into tubes contained in a boiler. The water is heated to steam by the heat of combustion of pulverized coal in the boiler and then superheated. Superheated and pressurized steam is then allowed to expand in a high-pressure (HP) steam turbine and cause rotation of the turbine shaft. The outlet steam from the HP turbine may once again be reheated and made to expand through an intermediate-pressure (IP) turbine and then through a low-pressure (LP) turbine. The turbine

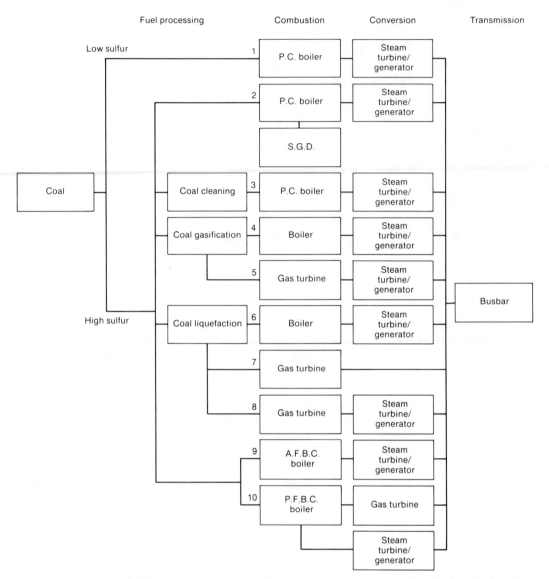

P.C. = pulverized coal. S.G.D. = secondary gas desulfurization. A.F.B.C. = atmospheric fluidized-bed combustion. P.F.B.C. = pressurized fluidized-bed combustion.

Fig. 1.2. Ten routes for generating electricity from coal (Ref 4).

shafts are all connected to one or more generator shafts which in turn rotate and convert the mechanical energy of rotation into electrical energy in the generator. The exit steam from the LP turbine is condensed in the condenser and is once again fed back to the boiler through the feedwater heaters and pumps. A closed loop of the water and steam is thus maintained. A second water loop through a cooling tower provides the cooling water needed to condense the steam exiting from the LP turbine. Combusted gases from the boiler are passed over more heat exchangers to preheat the incoming air to the boiler, are cleaned in scrubbers, and then are allowed to escape into the environment through the stacks. The pressure, temperature, and specific volume of the steam at various stages are illustrated in Fig. 1.5. It is thus clear that a variety of materials of construction are needed to withstand a wide range of these conditions in the plant,

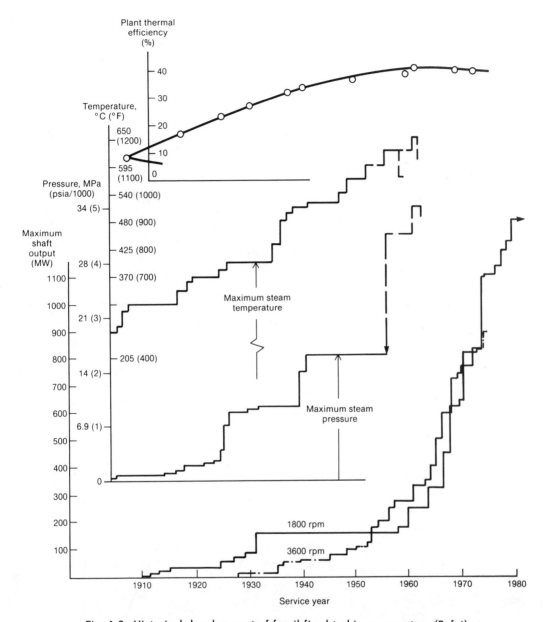

Fig. 1.3. Historical development of fossil-fired turbine-generators (Ref 4).

depending upon the local conditions of pressure, temperature, and chemical environment. The capacity, reliability, efficiency, availability, and safety of plants depend critically on the integrity of the components and materials employed. A number of damage phenomena, such as embrittlement, creep, thermal fatigue, hot corrosion, oxidation, and erosion, can impair plant integrity at elevated temperatures. At lower temperatures, corrosion, erosion, pitting, corrosion fatigue, stress corrosion, hydrogen embrittlement, and fatigue can play major roles. A list of key components, property requirements, and materials of construction for steam power plants is presented in Table 1.1 (Ref 5). As can be seen from the table, low-alloy ferritic steels containing carbon, molybdenum, and/or vanadium constitute the bulk of the materials used in steam power plants. For highly stressed components operating at high temperatures

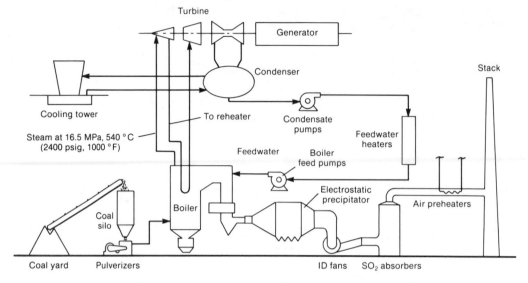

Fig. 1.4. Schematic diagram of a coal-fired steam power plant (Ref 4).

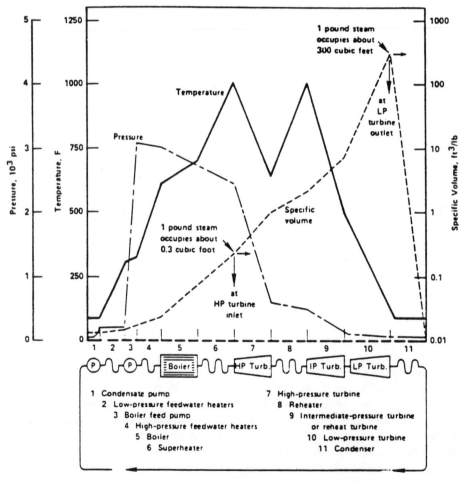

Fig. 1.5. Relationships among steam pressure, temperature, and specific volume in the various components of a large steam power plant (Ref 4).

Table 1.1. Property requirements and materials of construction for fossil steam plant components (Ref 5)

Component	Major property requirements	Typical materials
Boiler		
Waterwall tubes	Tensile strength, corrosion resistance, weldability	C and C-Mo steels
Drum	Tensile strength, corrosion resistance, weldability, corrosion-fatigue strength	C, C-Mo, and C-Mn steels
Headers	Tensile strength, weldability, creep strength	C, C-Mo, C-Mn, and Cr-Mo steels
Superheater/reheater tubes	Weldability, creep strength, oxidation resistance, low coefficient of thermal expansion	Cr-Mo steels; austenitic stainless steels
Steam pipe	Same as above	Same as above
Turbine		
HP-IP rotors/disks	Creep strength, corrosion resistance, thermal-fatigue strength, toughness	Cr-Mo-V steels
LP rotors/disks	Toughness, stress-corrosion resistance, fatigue strength	Ni-Cr-Mo-V steels
HP-IP blading	Creep strength, fatigue strength, corrosion and oxidation resistance	12% Cr steels
LP blading	Fatigue strength, corrosion-fatigue pitting resistance	12% Cr steels; 17-4 PH stainless steel; Ti-6Al-4V
Inner casings, steam chests, valves	Creep strength, thermal-fatigue strength, toughness, yield strength	Cr-Mo steels
Bolts	Proof stress, creep strength, stress-relaxation resistance, toughness, notch ductility	Cr-Mo-V and 12Cr-Mo-V steels
Generators		
Rotor	Yield strength, toughness, fatigue strength, magnetic permeability	Ni-Cr-Mo-V steels
Retaining rings	High yield strength, hydrogen- and stress-corrosion resistance, nonmagnetic	18Mn-5Cr and 18Mn-18Cr steels
Condensers		
Condensers	Corrosion and erosion resistance	Cupronickels; titanium; brass; stainless steels

(for example, turbine blades and bolts), higher-alloy tempered martensitic steels generally containing 12% Cr are used. In combustion turbines, metal temperatures often exceed 760 °C (1400 °F) in the combuster sections and in the early stages of blades and vanes.

Nickel- and cobalt-base alloys known as superalloys are the preferred candidates for the higher-temperature end of the spectrum, whereas components operating near 540 °C (1000 °F), such as the turbine shafts and disks, are made of low-alloy ferritic steels.

Reactor Pressure Vessels for Petroleum Refining

The refining or manufacturing of petroleum products and of chemicals in a refinery involves both physical changes, or separation operations, and chemical changes, or conversion processes. An illustration of an over-all refinery including both types of operations is shown in Fig. 1.6. Physical separation primarily involves distillation of the crude oil into various fractions according to their volatility. The system involves heating the crude by pumping it through

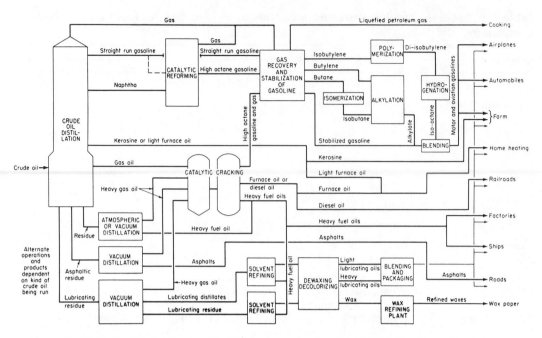

Fig. 1.6. Generalized over-all refinery from crude oil to salable products (American Petroleum Institute, cited in Ref 6).

tubes placed inside a furnace and allowing it to vaporize in a fractionating column, which is tapped at several points to allow side draw of the various boiling fractions. The residues withdrawn from the bottom of the column may be subjected to vacuum or steam distillation and then to further refining operations to convert the higher-boiling-point oils into lighter fuels by chemical conversion processes, as shown in Fig. 1.6 (Ref 6).

The chemical conversion processes in oil refining essentially involve increasing the H/C ratio and the removal of objectionable contaminants such as sulfur, nitrogen, asphaltenes, etc. The two obvious approaches to increasing the H/C ratio are removal of carbon or addition of hydrogen. In the refining industry, these two upgrading approaches are called, respectively, thermal conversion and hydroprocessing. Coal gasification is a drastic example of a thermal conversion process. Thermal conversion includes such well-known processes as coking, thermal cracking, visbreaking, and catalytic cracking. Hydroprocessing includes hydrocracking, hydrodesulfurizing, and catalytic

reforming. Examples of the reactions involved in the various refining processes may be found in Ref 6.

Modern petroleum refining and petrochemical processing may involve operating conditions for the ferritic steel pressure vessels extending to metal temperatures up to 565 °C (1050 °F) and pressures up to 28 MPa (4 ksi). The operating conditions in the pressure vessels of some typical refining processes are listed in Table 1.2 (Ref 7 and 8). For a given application, material selection must consider not only operating conditions but also conditions during start-stop transients. The mechanical behavior considered includes such properties as fracture toughness, creep rupture, and thermal fatigue. In addition, the corrosion and environmental behavior of the materials for normal operation, process upset, and shutdown conditions have to be taken into account. Since fabrication involves extensive welding, the properties of the weldments are of great importance. It is common practice to design, fabricate, and inspect pressure vessels according to the ASME Boiler and Pressure Vessel Code. The code calls

Table 1.2. Operating conditions for petroleum-refining pressure vessels (Ref 7 and 8)

Operation	Temperature range		Pressure range	
	°C	°F	MPa	psi
Catalytic cracking . . . . 449 to 565		840 to 1050	0.05 to 3.1	7 to 450
Hydrocracking . . . . 205 to 482		400 to 900	0.7 to 28	100 to 4000
Catalytic reforming . . . . 427 to 538		800 to 1000	0.35 to 5.2	50 to 750
Hydrogen treating . . . . 205 to 427		400 to 800	0.1 to 10	15 to 1500

for the following material properties: (1) strength necessary for the guarantee of the allowable stress, including room- and design-temperature tensile, creep, and fatigue properties; (2) notch toughness at the lowest operating temperature; and (3) weldability. In addition to the minimum code requirements for the fabricated condition, steels for high-temperature, high-pressure hydrogenation service are required to withstand such environmental degradation processes as temper embrittlement, hydrogen embrittlement, hydrogen attack, and creep embrittlement.

The selection of materials of construction has always been a major concern in the petroleum industry. In the past, the concern has been based largely on safety and economic incentives, which dictate against unexpected equipment failures that could result in hazardous exposures or extended plant shutdowns. Forced shutdown for extensive repair or replacement of reactor pressure vessels would require months or years. It is estimated that unscheduled downtime of a petroleum refinery pressure vessel can cost in excess of $50,000 per hour (Ref 9). More recently, however, the need for extending the lives of current plants well beyond their originally anticipated lives has gained considerable attention. The factors driving this need are essentially the same as those for fossil power plants, as described earlier.

Most steels for refinery and petrochemical applications fall within the following categories: (1) carbon steels; (2) carbon-molybdenum steels; (3) low- and intermediate-alloy chromium-molybdenum steels; and (4) martensitic and ferritic stainless steels. The applicable ASTM specifications covering tubular products, plates, castings, and forgings are listed in Table 1.3 (Ref 10). Allowable stresses used in the manufacture of pressure vessels are designed in accordance with Divisions 1 and 2 of Section VIII of the ASME Boiler and Pressure Vessel Code. The code defines allowable stresses for carbon and carbon-molybdenum steels up to 540 °C (1000 °F). Chromium-molybdenum and ferritic stainless steels are rated up to 650 °C (1200 °F). The only steels for which allowable stresses are given above 650 °C are the austenitic steels.

Selection of materials and upper limits of operating temperature are governed both by the ASME Boiler and Pressure Vessel Code (Section VIII) from a creep point of view and by the Nelson diagrams (described in Chapter 7) from a hydrogen-attack point of view. For reactors used in high-pressure hydrogeneration service, where hydrogen pressure can be as high as 28 MPa (4 ksi), the upper limit of temperature set by the Nelson diagrams is below the limits set by the creep considerations. These vessels (e.g., hydrocrackers, hydrodesulfurizers) are usually made of 2¼Cr-1Mo steels and are permitted to operate only up to about 455 °C (850 °F). Creep *per se* is not a major concern in these vessels. Potential failure mechanisms for these vessels involve brittle fracture at low temperatures under transient conditions aided by embrittlement and environmentally assisted phenomena. In the other extreme, in reactor vessels where refining is carried out purely by thermal processing (e.g., catalytic crackers), hydrogen attack is not an issue; these vessels are made of carbon and carbon-molybde-

Table 1.3. ASTM specifications for steels commonly used in refinery and petrochemical equipment(a) (Ref 10)

Material	Pipes and tubes	Plates	Castings	Forgings
Carbon steel	A53,A106,A134, A135,A139,A155, A178,A179,A192, A210,A214,A226, A333,A334,A381, A524,A587	A283,A285,A299, A433,A443,A455, A515,A516,A537, A573	A27,A216,A352	A105,A181,A266, A350,A372,A465, A508,A541
C-½Mo	A161,A209,A250, A335,A369,A426	A204,A302,A533	A217,A352,A487	A182,A336,A541
1Cr-½Mo	A213,A335,A369, A426	A387	. . .	A182,A336
1¼Cr-½Mo	A199,A200,A213, A335,A369,A426	A387,A389	A217,A389	A182,A541
2Cr-½Mo	A199,A200,A213, A335,A369	. . .	. . .	. . .
2¼Cr-1Mo	A199,A213,A335, A369,A426	A387,A542	A217,A487	A182,A336,A357, A541
3Cr-1Mo	A199,A200,A213, A335,A369,A426	A387	. . .	A182,A336
5Cr-½Mo	A199,A200,A213, A335,A369,A426	A357	A217	A182,A336
7Cr-½Mo	A199,A200,A213, A335,A369,A426	. . .	. . .	A182
9Cr-1Mo	A199,A200,A213, A335,A369,A426	. . .	A217	A182
Ferritic and austenitic stainless steels	A213,A249,A268, A269,A271,A312, A358,A362,A376, A409,A430,A451, A452,A511	A167,A176,A240, A412,A457	A296,A297,A351, A447,A448	A182,A336,A473

(a) Carbon and alloy steel bolts and nuts are covered by specifications A193, A194, A320, A354, A432, A449, A453, A540, and A563.

num steels and operate at temperatures up to about 510 °C (950 °F) in the case of carbon steels and 540 °C (1000 °F) in the case of carbon-molybdenum steels. The potential failure mechanism for these vessels involves primarily creep rupture. In between the above two cases fall reactor vessels which operate under moderate hydrogen pressures and in the creep regime (e.g., catalytic reformers). These vessels are generally made of 1Cr-½Mo and 1¼Cr-½Mo type steels and operate at temperatures up to 540 °C (1000 °F). Failure scenarios here involve creep at high temperatures as well as brittle fracture at low temperatures aided by embrittlement phenomena.

Design Life of Components

Components which operate at low temperatures below the creep regime are generally designed on the basis of yield strength, tensile strength, and fatigue strength by applying suitable safety factors to these values. Because deformation and fracture are not time-dependent under these circumstances, there is no specific value of "design life" associated with them. In principle, as long as the applied stresses do not exceed the design stresses, these components should last indefinitely, although in practice various factors cause reductions in life. In the case of high-temperature components operating in the creep regime, both deformation and fracture are time-dependent. They are therefore designed with respect to a target life usually based on a specified amount of allowable strain or rupture in 100,000 h. A further factor of safety is applied in selecting the stress, which translates into an expected life of 30 to 40 years, leading to the notion of a 30-to-40-year design life for the component. Many metallurgical and operational factors can extend the actual component life beyond the design life. Alternatively, if these factors are adverse, actual life can be reduced. Some of the many favorable and unfavorable factors that hold the balance between design life and actual life are illustrated schematically in Fig. 1.7 (Ref 11).

Built-in safety factors in design with

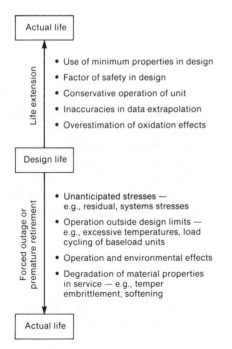

Fig. 1.7. Favorable and adverse factors affecting the useful lives of components (Ref 11).

respect to stress and temperature are intended to ensure that the minimum design life is met. Material-property data are invariably subject to scatter, resulting in a broad band or spectrum of behavior. Designs are generally based on minimum or mean values of mechanical properties, after further corrections for safety have been applied. If the actual materials of construction exceed these expectations, the actual life can then far exceed the design life. This uncertainty in material behavior is illustrated in Fig. 1.8 for a Cr-Mo-V steam turbine rotor steel, with respect to its creep-rupture life. At a stress of 83 MPa (12 ksi) and a temperature of 540 °C (1000 °F), the design curve yields an expected life of 11.4 years. Use of the minimum curve or the mean curve can yield an expected life of 55 or 266 years, respectively. Similar scatter is also encountered with respect to other material properties, leading to major uncertainties in expected life. Units operated conservatively can be operated beyond their design lives. For instance, if a combustion turbine is operated more sparingly than originally intended, it can operate for many years beyond its design life. In the early days, design of components was often based on linear extrapolation of short-time creep and fatigue data to approximate the long-term behavior. Long-term data are now available for many standard materials as a result of international efforts to gather and analyze long-term test data. In some instances it has been found that the original linear extrapolations may have been overly conservative and that the actual expected lives may exceed the design lives. Most of the creep and stress-rupture data used in designing high-temperature components are based on small samples tested in air in the laboratory. For heavy-section components, the oxidation effects may be less pronounced than for small specimens, resulting in an added margin of safety. All of the above factors contribute to extended component life.

In contrast to the above factors, a number of other factors can lead to premature failure of components. Stresses in compo-

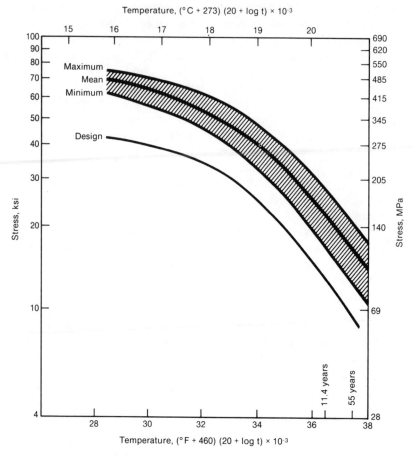

Fig. 1.8. Uncertainty in creep-rupture life assessment due to scatter in the properties of a Cr-Mo-V steel.

nents frequently exceed the design stresses as a result of hidden residual stresses, system stresses, and local stress concentrations. For instance, in many piping systems, bending stresses arise due to failure of supports. Operating temperatures in boilers invariably exceed design temperatures at least over short durations, reducing component life. Unanticipated start-up and shutdown cycles lead to fatigue damage not originally provided for. Conversion of base load fossil power plants into cycling plants is a clear case in point. Unanticipated environmental effects leading to corrosion, pitting, and stress corrosion are major factors in life reduction. Corrosion fatigue of steam turbine blades, stress corrosion of disks, and hot corrosion of combustion turbine components are some examples of this. Pre-existing fabrication defects may cause crack

initiation and growth of cracks during service and lead to premature failures. Inclusions, segregation streaks, reheat cracking, slag inclusions in welds, lack of fusion, incomplete penetration, and numerous other defects such as these have been known to cause catastrophic failures. These defects have been of particular concern with respect to components fabricated in the 1950's when fabrication procedures as well as non-destructive qualification procedures were far inferior to those available today. Extrapolation of short-time data to predict long-time behavior has also sometimes led to overly optimistic expectations, the actual behavior being worse than the expected behavior. The creep-rupture strength of $1\frac{1}{4}$ Cr-$\frac{1}{2}$ Mo steels is a case in point, where a downward revision was made in the allowable stresses specified by the ASME

Boiler and Pressure Vessel Code based on long-term data that became available in the mid-1960's. The last but not the least important factor adversely affecting component life is the in-service degradation of components due to various microstructural changes and embrittlement phenomena such as temper embrittlement and creep embrittlement.

Definitions of Failure

Various definitions of failure are employed in industry, as shown in Table 1.4 (Ref 11). While complete breakage or rupture may be the ultimate and self-evident criterion of failure, more conservative definitions are invariably employed to retire a component prior to such unforeseen and catastrophic failure. Failure of a component may generally be defined as the inability of the component to perform its intended function reliably, economically, and safely. Component-retirement decisions are often based

Table 1.4. Failure criteria and definitions of component life (Ref 11)

History-based criteria

30 to 40 years have elapsed
Statistics of prior failures indicate impending failure
Frequency of repair renders continued operation uneconomical
Calculations indicate life exhaustion

Performance-based criteria

Severe loss of efficiency indicating component degradation
Large crack manifested by leakage, severe vibration, or other malfunction
Catastrophic burst

Inspection-based criteria

Dimensional changes have occurred, leading to distortions and changes in clearances
Inspection shows microscopic damage
Inspection shows crack initiation
Inspection shows large crack approaching critical size

Criteria based on destructive evaluation

Metallographic or mechanical testing indicates life exhaustion

on economic justification rather than on technical need. A logical and technically based decision may, for instance, involve a sequence of steps such as remaining-life calculations based on operating history, inspections, material testing, assessment of remaining life, and final disposition of the component in terms of continued service, repair, or replacement. Unfortunately, there are major cost factors associated with each of these steps. The cost of the component itself is usually a small fraction of the cost of disassembling the unit as necessary and performing all of the above operations. If, after all of this, a wrong decision is made and the component fails in service, the economic penalties are severe. One day of forced outage of a typical 500-MW power plant in the United States is estimated to cost $500,000 to $750,000. Failure of major components may lead to outages as long as one year. The owner of the plant has to weigh all of these economic factors carefully and not make decisions purely on a technical basis. A conservative but not uncommon approach has therefore been simply to replace critical components in a plant after 30 to 40 years, regardless of the technical merits of such action.

In the case of components which are many in number and perform identical functions, it is often possible to make an estimate of future failure rate based on statistics of past failure rates. By defining an acceptable failure rate, the time for total replacement of all similar components can be reasonably anticipated. Boiler-tube failures are a case in point. It is important to ensure that within a given statistical base only similar materials of similar history whose failure mechanisms are identical be included.

Calculational procedures are often employed to determine the expended lives of components under creep, fatigue, and creep-fatigue conditions. From plant records, information on temperature and cycling history is gathered. By use of standard material properties and damage rules, the fractional life expended up to a given point in time can be estimated. Unfortunately, histories of plant operation are usually not available in sufficient detail. Errors of as little as 14 °C (25 °F) in assumed temperature and small errors in assumed stress can lead to errors of 200 to 400% in calculated life. Uncertainties in assumed material behavior can lead to uncertainties of one to two orders of magnitude in estimated life. Damage rules used to calculate cumulative life expenditure contain inherent deficiencies. Due to all these uncertainties, calculation procedures often lead to inaccurate results. These procedures are most valuable when they are used essentially for screening purposes. They are used in determining when and at what locations further and more detailed evaluations may be needed. In many instances where access to NDE equipment is limited, they may provide the only means of estimating expended life.

Dimensional changes can often provide clues to expended life. Gross dimensional changes and distortions can be readily detected and measured. The original dimensions often are not known with sufficient precision to permit accurate calculations of accumulated strain. Many brittle failures occur without manifest dimensional changes. The failure criterion—i.e., the failure strain—can vary with material and operating conditions. Creep damage at elevated temperatures, such as cavitation and crack growth, may occur at localized regions such as weld heat-affected zones and stress concentrations. Such damage cannot be measured in terms of dimensional changes. With certain limitations, the procedure is useful in specific instances such as swelling of headers, rotor-bore expansion, blade elongation, and casing distortion.

The expenditure of creep and fatigue life can sometimes be estimated by removing samples from the component and conducting accelerated tests in the laboratory. Degradation in toughness during service can be determined by conducting impact or fracture-toughness tests. The major difficulty of this approach is that it cannot be used for monitoring on a continuous basis. The available material is often too limited

to conduct a comprehensive test program. Miniature specimens and insufficient numbers of specimens are tested, resulting in inaccuracies. It is essential that the specimens removed be representative of the condition of the material at the critical locations of interest. Often this cannot be ensured, because selection of critical locations has to be done nondestructively. Even if the critical locations can be pinpointed, samples are sometimes removed away from these locations for reasons of safety, convenience, or lack of access. The behavior of the material at the critical locations then has to be estimated based on data from other locations. Plant operators generally are reluctant to permit removal of samples from operating components, because sufficient experience and confidence have not yet been gained in repairing the excavated areas and ensuring their continued safety. In spite of some of these limitations, sample removal and testing is becoming increasingly common because it eliminates a major uncertainty in life-assessment procedures—i.e., the uncertainty due to assumed material behavior. It must be recognized, however, that calculations of expended life as well as the use of mechanical tests are generally based on crack-initiation-based analysis with respect to heavy-section components. In such cases, additional remaining life under crack-growth conditions needs to be taken into account.

Under certain circumstances, any defect or flaw observable in a component by visual or other nondestructive examination constitutes grounds for retirement. Under other circumstances, NDE observations are combined with crack-growth analysis to determine remaining life. These specific circumstances are discussed in detail in the next section. Suffice it to say here that the conservative approach in the old days was to replace components based on crack initiation. With increasing awareness of fracture-mechanics considerations, equipment owners are increasingly eager to take advantage of the crack tolerance of components.

Leaks in tubes and pipes and vibration of rotating components provide forewarnings of more large-scale and massive failures. If they can be detected in time, remedial actions can be taken. In many instances, local repairs can be performed and continued operation of the plant restored. If the frequency or extent of repair is uneconomical, component replacement is indicated. Surface-cracked rotors, cracked casings, cracked gas turbine and steam turbine vanes and diaphragms, welded pipes, and pressure vessels and boiler tubes are some examples to which this scenario might apply.

The ultimate and self-evident criterion of failure is the catastrophic failure of a component. The consequences of such an event are often devastating, both financially and in terms of human life. All the aforementioned failure criteria are employed to forestall such catastrophic events.

Among the many failure definitions given in Table 1.4, the choice of a particular definition will depend on numerous circumstances. Some of the important factors are: (1) the consequences of failure, (2) costs associated with the various levels of remaining-life assessment versus the costs of replacement, (3) outage schedules and procurement lead times, and (4) management philosophy and conservatism. These factors can vary from one organization to another and from country to country.

Appropriateness of Crack Initiation Versus Crack Growth

The general ingredients of a remaining-life-assessment procedure for a commonly encountered failure scenario can be illustrated with the help of Fig. 1.9. Region I corresponds to incipient, microscopic damage events leading up to the initiation of a macroscopic crack. These events include dislocation rearrangements, coarsening of precipitate phases, and formation of creep cavities and microcracks. NDE techniques that are potentially capable of detecting these events include electron microscopy, replication metallography, x-ray diffraction, electrical resistivity measurements,

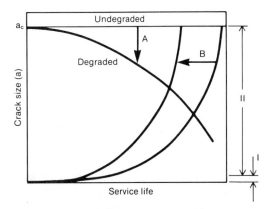

A – embrittlement phenomena. B – unanticipated factors (excess cycling, temperature excursions, corrosion, metallurgical degradation, improper material, excessive stresses). See text for definitions of regions I and II.

Fig. 1.9. Illustration of a remaining-life-assessment procedure for a common failure scenario involving crack initiation and propagation.

positron annihilation measurements, ultrasonic velocity measurements, and hardness measurements. Region II corresponds to propagation of the above-mentioned macrocrack. Conventional NDE techniques such as magnetic-particle inspection, dye-penetrant inspection, ultrasonic examination, x-ray radiography, and eddy-current testing apply to crack detection and sizing in region II. In region II, the crack grows until it reaches critical size, defined as a_c, at which point rapid fracture occurs.

The critical crack size can be defined in a number of ways based on fracture toughness, ligament size, crack-growth-rate transitions, or other considerations as appropriate. A common definition for many heavy-section components is based on the fracture toughness of the material. The critical crack size, a_c, is often not a constant value but decreases with service exposure due to embrittlement phenomena. Similarly, many adverse factors can accelerate the stage II crack-growth behavior, so that the failure point is shifted to the left — to shorter times. To perform a remaining-life analysis, information is needed regarding crack initiation, the rate of crack growth, and the failure point, as defined by the critical crack size a_c, specifically applicable to the component of interest. Conventional ultrasonic, dye-penetrant, and magnetic-particle inspection techniques apply to stage II, and are based on the premise that a detectable crack will form and grow slowly enough to permit periodic inspections and retirement of the component prior to final failure. There are many instances in which crack initiation alone constitutes component failure and conventional NDE techniques and fracture-mechanics analyses serve no useful purpose.

Table 1.5 presents examples of the various circumstances that might dictate whether component failure is governed by crack initiation or crack growth. In the case of very brittle materials such as the heavily segregated bore of a 30-to-40-year-old rotor, a_c may be so small that it is below the limit of detection by conventional NDE techniques. Severely embrittled pressure vessels, bolts, and blades may be other examples of this. High stresses once again have the effect of reducing a_c, sometimes below levels of detection. If a component has a thin cross section (e.g., a blade or a tube), the remaining ligament can be so small that crack propagation is not of importance. In some instances, a_c may be large but the rate of crack growth may be so high that once a crack initiates, it reaches critical size rapidly. Many environmentally induced failures in highly stressed components exhibit this behavior. For instance, in generator retaining rings and in steam turbine blades where crack growth under corrosive conditions is encountered, the presence of a pit or pitlike defect is cause for retirement. Considerations different from those above apply to failures governed by crack propagation. In components where failure is governed by crack initiation, the detection of any defect during an inspection, or, more conservatively, the suspected initiation of a crack based on calculations, can be used to retire the component. Many advanced NDE techniques which can detect incipient damage evolution prior to crack initiation are under development industry wide.

Techniques that use crack initiation as a failure criterion include calculations based on history, extrapolations of failure statistics, strain measurements, accelerated mechanical testing, microstructural evaluations, oxide scale growth, hardness measurements, and advanced NDE techniques. For crack-growth-based analysis, the NDE information, results from stress analysis, and crack-growth data are integrated and evaluated with reference to a failure criterion. The various techniques and their limitations are described in detail in later chapters of this book.

Table 1.5. Examples of circumstances governing crack-initiation- and crack-propagation-controlled failures

Component	Circumstances
Initiation-controlled failures	
HP/IP rotor bores (1950's)	Very brittle (small a_c); highly stressed
Rotor dovetails	Highly stressed; thin section
Rotor grooves	Rapid crack growth (large da/dn)
Blades	Highly stressed; thin section; rapid crack growth (large da/dn)
Bolts	Very brittle (small a_c); highly stressed; thin section; rapid crack growth (large da/dt)
Propagation-controlled failures	
Rotor bores (modern)	Ductile; slow crack growth
Inner casings	Ductile; lower stresses; thick section
Nozzle blocks	Ductile; lower stresses
Valves	Ductile; lower stresses

Implementation of Remaining-Life-Assessment Procedures

In implementing life-assessment procedures, the appropriate failure definition applicable to a given situation must be determined at the outset, and the purpose for which the assessment is being carried out must be kept in mind. While determining the feasibility of extending plant life may be one objective, a more common objective is the setting of appropriate intervals for inspection, repair, and maintenance. In this context, life-assessment procedures are used only to ascertain that failures will not occur between such intervals. It should never be assumed that, having performed a life-assessment study for a 20-year life extension, one can then wait 20 years with no interim monitoring. Periodic checks to ensure the validity of the initial approach are essential. In this sense, life extension should be viewed as an ongoing rather than a one-time activity. The various tools and techniques available should be used in a complementary and cost-effective way rather than as competing techniques. A phased approach in which the initial level includes nonincursive techniques, followed by other levels of actual plant monitoring, then followed by nondestructive inspections and destructive tests would be the most logical and cost-effective approach, as illustrated in Table 1.6 (Ref 12). In level I, assessments are performed using plant records, design stresses and temperatures, and minimum values of material properties from the literature. Level II involves actual measurements of dimensions and temper-

atures, simplified stress calculations, and inspections, coupled with the use of the minimum material properties from the literature. Level III involves in-depth inspection, stress analysis, plant monitoring, and generation of actual material data from samples removed from the component. The level of detail and the accuracy of the results increase from level I to level III, but at the same time the cost of life assessment also increases. Depending on the extent of information available, and the results obtained, the analysis may stop at any level or proceed to the next level as necessary. This iterative process is illustrated in Fig. 1.10 (Ref 12). Depending on the required accuracy of and confidence in the results, each plant owner might make his or her own cost/benefit trade-off and pursue the analysis up to the level of interest to suit his or her circumstances.

The many problems associated with crack-initiation-based life-assessment procedures have already been discussed. These include lack of data on operating history, lack of documentation of prior experience, scatter in material properties, difficulties associated with removal and testing of samples, and unavailability of well-proven nondestructive techniques. Some of the unique problems associated with crack-growth-based approaches and problems of weldments are described below.

The implementation of crack-growth-based life-assessment procedures is limited by uncertainties in material behavior and in NDE results. With respect to material behavior, the major problem is the unavailability of data pertaining to crack growth

Table 1.6. Data requirements for life assessment using a three-level approach (Ref 12)

Item	Level I	Level II	Level III
Feature	Least detail	More detail	Most detail
Failure history	Plant records	Plant records	Plant records
Dimensions	Design or nominal	Measured or nominal	Measured
Condition	Records or nominal	Inspection	Detailed inspection
Temperature and pressure	Design or operational	Operational or measured	Measured
Stresses	Design or operational	Simple calculation	Refined analysis
Material properties	Minimum	Minimum	Actual material
Material samples required?	No	No	Yes

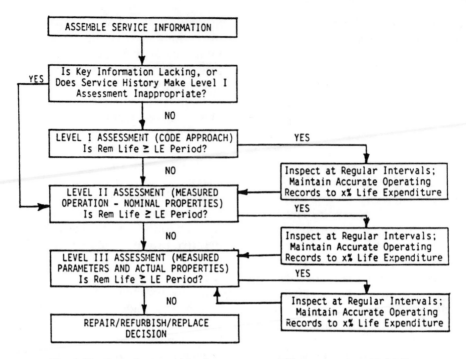

Fig. 1.10. Generic procedure for component life assessment (Ref 12).

and toughness in the service-degraded condition, specific to the component. While considerable data may be available on materials in the virgin condition, the data bank on service-exposed materials is very small. Nondestructive methods are needed to determine those properties with specific reference to a given component. In welded components, the problem is further compounded by the fact that a weldment contains a complex microstructure of many zones with varying material properties. Failure can occur through any of these zones or at the interfaces between them. In cases where crack-growth rates might be rapid, conventional NDE techniques are often inadequate to detect the initial crack. The uncertainties in interpretation of NDE results can sometimes be overwhelming. Difficulties in distinguishing between innocuous versus harmful flaws and identifying their orientations can lead to uncertainties in life assessment. If there are numerous indications closely spaced, the manner of treating them in terms of a link-up analysis could be very crucial. Geometric discontinuities such as fillets, section transitions, and weld backing rings interfere with NDE

signals and mask flaws. Due to lack of access, NDE inspections cannot always be performed at critical locations without extensive disassembly. Cracking on the inside surfaces of insulated pipes, cracking in retaining rings, and blade-groove wall cracking in turbines are some examples of this problem. Above all, wide variations in instrumentation, calibration, and interpretation procedures and operator training and bias lead to divergent results which are often difficult to reconcile.

Despite the many limitations discussed above, major advances have been made during the last few years in development and implementation of life-assessment techniques for various plant components. These techniques are extensively reviewed in the following chapters.

Scope and Organization of the Book

Damage phenomena in high-temperature components include many mechanisms such as creep, fatigue, creep-fatigue, embrittlement, hydrogen attack, hydrogen embrittlement, and hot corrosion. Life-assessment procedures for these components require a multidisciplinary approach involving a knowledge of design, material behavior, and nondestructive inspection techniques.

Books and conference proceedings of the past had limited scope, usually covering a single issue (creep, hot corrosion, etc.). No attempt has been made to pull together all of the issues that affect the reliability of high-temperature components. Further, the various damage phenomena have been treated from a highly scientific point of view with an overemphasis on theoretical models and little emphasis on application. On the other hand, proceedings from recent conferences on life extension have dealt only with the practical application of specialized techniques without providing sufficient background theory. Many recent advances in materials technology have not yet been sufficiently documented. Books dealing with materials issues have been aimed primarily at metallurgists and have not been taken advantage of by the engineering community at large. As a result of all the above, there is not at present a single, authoritative book that integrates all the relevant metallurgical issues affecting the high-temperature components in plants at a level that can be read and understood by the practicing engineer. Such an exercise would also serve to acquaint graduate-level students with the practical needs of the industry in this important area.

The objective of this book is to identify and discuss the metallurgical issues that impact the reliability of plant components operating at elevated temperatures. Broad perspectives on all the key issues and their interrelationships, for the benefit of the "practicing" engineer, are presented. This book is not meant for the "specialist" in any given field and is not intended to cover research topics in the various fields in detail. It provides sufficient theory to give the reader an understanding of the mechanisms of the various damage phenomena while at the same time discussing the practical implications of these phenomena. Many recent advancements that have been made in terms of materials, coatings, and dam-

age-assessment techniques are described. Extensive reference is made to data, literature, and case histories relating to life assessment. Sufficient material-property data are also provided to serve as a ready source for such information. It is hoped that this book will be used as a graduate-level textbook for introducing students to high-temperature plant metallurgy. For this purpose, many worked-out examples of illustration problems have been included. It is nearly impossible to cover all the materials issues pertaining to high-temperature components in a single book of manageable size. This book therefore deals with the problems that occur as a result of exposure to high temperatures—i.e., creep, temper embrittlement, fatigue, thermal fatigue, and fireside corrosion. Problems relating to low-temperature corrosion phenomena (such as pitting, corrosion fatigue, and stress corrosion) that could occur below the creep regime are not addressed. The emphasis inevitably is on ferritic steels because these alloys constitute the largest tonnage of all materials utilized.

The book is divided into nine chapters. The first chapter is an overview of the issues. In Chapters 2, 3, and 4, damage mechanisms that are common to a number of energy systems are discussed. Damage mechanisms such as toughness degradation, creep, and fatigue are common problems with respect to materials used in power plants and in petroleum and chemical processing equipment. The phenomenology of each of these damage mechanisms is described without going into detailed theories of the mechanisms and the controversies surrounding them, because such theories and controversies are deemed irrelevant to the application-minded engineers for whom this book is meant. Unique problems relating to specific components, such as hot corrosion of gas turbine components and hydrogen attack of refinery reactors, are covered in Chapters 5 to 9 on a system-by-system basis. Material data, damage mechanisms, and life-assessment techniques specific to each system are discussed. This appears to be the best approach to providing sufficient background theory necessary for an understanding of the damage phenomena and at the same time describing their practical applications in context. Much of the data and materials development presented is based on recent research. Materials problems as well as currently available solutions are discussed as appropriate.

References

1. R.B. Dooley and R. Viswanathan, Ed., *Life Extension and Assessment of Fossil Power Plants*, proceedings of conference in Washington, June 1986, EPRI CS 5208, Electric Power Research Institute, Palo Alto, CA, 1987
2. R.C. Schwant and D.P. Timo, Life Assessment of General Electric Large Steam Turbine Rotors, in *Life Assessment and Improvement of Turbine-Generator Rotors*, R. Viswanathan, Ed., Pergamon Press, New York, 1985, p 3.23-3.45
3. "Electricity—Today's Technologies, Tomorrow's Alternatives," Electric Power Research Institute, Palo Alto, CA, 1987
4. R.I. Jaffee, Metallurgical Problems and Opportunities in Coal Fired Steam Power Plants, *Met. Trans.*, Vol 10A, May 1979, p 139-165
5. V.A. Altekar, "Power Crisis and the Metallurgical Engineers," N.P. Gandhi Memorial Lecture, Indian Institute of Metals, Baroda, India, Nov 1980
6. R.N. Shreve and J.A. Brink, Jr., *Chemical Process Industries*, 4th Ed., McGraw-Hill, New York, 1977, p 652-656
7. W.F. Bland and R. Davidson, Ed., *Petroleum Processing Handbook*, McGraw-Hill, New York, 1967
8. T.E. Scott, Pressure Vessels for Coal Liquefaction–An Overview, in *Application of 2¼Cr-1Mo Steel for Thick Wall Pressure Vessels*, STP 755, American Society for Testing and Materials, Philadelphia, 1982, p 7-25
9. T.C. Bauman, *J. of Metals*, Vol 12, Aug 1977, p 8-11
10. R.Q. Barr, "A Review of Factors Affecting the Selection of Steels for Refining and Petrochemical Applications," Climax Molybdenum Co., Greenwich, CT, 1971
11. R. Viswanathan and R.B. Dooley, *Creep Life Assessment Techniques for Fossil Plant Boiler Pressure Parts*, Proceedings of International Conference on Creep, JSME-IME-ASTM-ASME, Tokyo, Apr 1986, p 349-359
12. W.P. McNaughton, R.H. Richman, C.S. Pillar, and L.W. Perry, "Generic Guidelines for the Life Extension of Fossil Fuel Power Plants," EPRI CS 4778, Electric Power Research Institute, Palo Alto, CA, Nov 1986

2

Toughness

The toughness of a material is its ability to absorb energy in the form of plastic deformation without fracturing. The ability of a material to withstand occasional stresses above the yield stress without fracturing is very desirable. Toughness is a commonly used property but one which is difficult to define. In simple terms, it may be defined as the area under the stress-strain curve. This area is a measure of the amount of work per unit volume of the material which can be done on the material without causing it to fracture. It is therefore a parameter which combines both strength and ductility.

Ductile-to-Brittle Transition

Ferritic steels used in power-plant components undergo a ductile-to-brittle fracture transition as the temperature is decreased. At low temperatures, fracture occurs by completely brittle cleavage mechanisms with low levels of absorbed energy. At high temperatures, fracture occurs by ductile dimple mechanisms with absorption of considerable energy. In the transition region, fracture is of a mixed mode. A variety of tests have been used to characterize the ductile-to-brittle transition behavior, and, correspondingly, a variety of definitions of transition temperature have emerged. A comprehensive review of various test techniques may be found in the literature (Ref 1 to 3).

Definitions of Transition Temperature

The Charpy V-notch impact test is undoubtedly the test most commonly used to characterize the ductile-to-brittle transition in steel. Several transition temperatures may be derived from any given set of data, because several criteria may be employed. Some of these are illustrated in Fig. 2.1. T_1 is the transition temperature as determined by some fixed level of impact energy E_1 — e.g., the 20-J (15-ft·lb) transition temperature. The specific energy level is usually determined by correlations with other types of tests or service performance. Occasionally, it is defined by what can be expected from commercially available material. This fixed-energy criterion is used fairly extensively, particularly for quality control evaluations and acceptance tests of structural steels. The fracture-appearance transition temperature (FATT, T_2 in Fig. 2.1) is the temperature at which the fracture is 50% brittle and 50% ductile in appearance. The basis for selecting a 50% mixture of the fracture modes is arbitrary. The midpoint of the impact-energy transition region is another arbitrary choice for T_2 based on the ease of measurement. The temperature T_3, above which the fracture appearance is

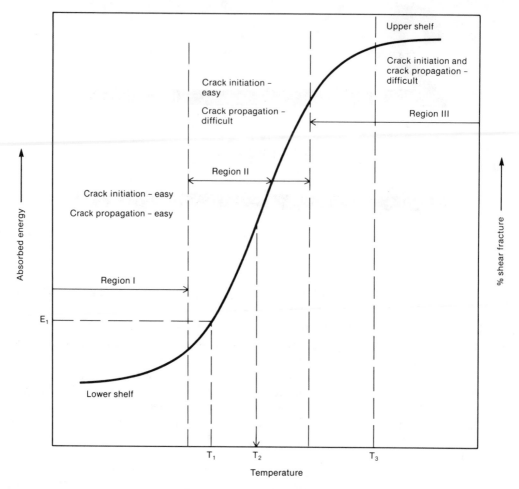

Fig. 2.1. Illustration of a Charpy transition curve.

entirely ductile, represents the most conservative criterion in that it yields the highest transition temperature coupled with the maximum energy for fracture. The low-temperature region where fracture is 100% brittle is referred to as the lower-shelf region, while the region above T_3 is referred to as the upper-shelf region of the curve. The energy levels are correspondingly referred to as the lower-shelf energy and the upper-shelf energy, respectively. In a Charpy test, the lateral expansion which occurs in the specimen on the side opposite to the root of the notch (or lateral contraction of the sides) is a measure of the ability of the material to accommodate plastic deformation in the presence of a notch. An inflection in the temperature dependence of the lateral expansion or some arbitrary fixed

value of expansion (say 1%) may be used to define another kind of transition temperature.

Another generally known type of transition-temperature measurement is the nil-ductility temperature (NDT). This is simply a go/no-go test which defines the temperature below which deformation in the presence of a sharp notch is essentially zero. The NDT is determined from a drop weight test in which a weight is dropped at the midspan of a beaded plate supported at the ends. The test is conducted as a function of temperature, and the highest temperature at which the specimen breaks is the NDT (Ref 4 and 5).

A crack-starter test, known as the explosion bulge test, has sometimes been used primarily to evaluate the crack-propagation characteristics of materials (Ref 5 and 6). A square plate with a brittle weld underbead is subjected to an explosive charge in an open die and the natures of the resulting deformation and cracking are characterized as functions of test temperature. The highest temperature where extensive deformation without brittle cracking occurs is referred to as the fracture-transition plastic (FTP). The temperature below which the cracking begins to extend beyond the deformed material into the elastically loaded region is designated as the fracture-transition elastic (FTE).

Another type of transition temperature, known as the crack-arrest temperature (CAT), has been defined by the work of Robertson (Ref 7 to 9). It is the temperature at which a running cleavage crack will be arrested. In the Robertson test, a relatively large plate is loaded to a uniform tensile stress, with a known temperature gradient across the width of the plate. A cleavage crack is started by an explosive charge on the cold edge of the plate. The crack propagates under the tensile stress until it reaches a zone of the plate where the temperature is sufficiently high to arrest the crack. By varying the stress, the stress/temperature combinations that will cause crack arrest can be determined. Full-thickness tests are necessary to get data representative of the actual application. This approach is particularly attractive to the designer because it provides not only a transition temperature but also information regarding the critical stress for propagation of cracks as a function of temperature.

The confusing mass of transition-temperature approaches can be sorted out by understanding the relationships among them using the Charpy curve as a reference. The relative positions of several of the transition-temperature criteria are shown in Fig. 2.2, superimposed on a Charpy V-notch impact-energy curve for a given material. These data from Pellini and Srawley (Ref 10) are unique, for this is the only example in the literature of such a comprehensive collection of transition temperatures for a specific plate of any given material. This illustration is very useful in understanding the relative degrees of toughness representative of the various transition temperatures. The maximum degree of toughness required in extremely severe applications is depicted by the FTP transition temperature. The FTE (fracture-transition elastic), measured by explosion crack-starter tests, the EBT (Esso brittle temperature) (Ref 11), the CAT (crack-arrest transition temperature), measured by the Robertson test, and the FATT (fracture-appearance transition temperature), measured by impact tests, are all in mutual agreement. All four of them pertain to crack propagation and define the temperature above which crack propagation is difficult. The NDT denotes the minimum level of toughness that is acceptable. At temperatures below the NDT, cleavage-crack initiation and propagation are both easy and can occur at stress levels which are mere fractions of the material's yield strength.

Capabilities and Limitations of the Transition-Temperature Approach

The applicability of the transition-temperature approach to the problem of brittle failure varies considerably depending on the criterion used and the purpose for which it is applied. If a very conservative design protecting against crack initiation and propagation is called for, specifying FTP at a value below the operating temperature might be appropriate. If it is assumed that cracks will be present or will initiate anyway and protection against crack propagation is needed, specifying FTE, CAT, or FATT to be below the expected operating temperature might be appropriate.

The transition-temperature approach is used widely for the purpose of comparing materials. Within a given class of materials of comparable strength, it may generally be true that the material with lower values of transition temperature, regardless of which definition is employed, has greater fracture resistance than another material with higher values of transition temperature. In comparing materials with transition curves of different shapes (for instance, steels at dif-

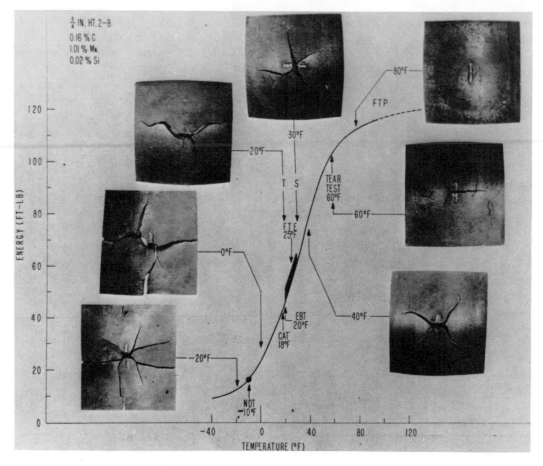

Fig. 2.2. Illustration of the relative positions of various transition temperatures with reference to a Charpy transition curve for steel plate (Ref 10).

ferent strength levels), the transition temperature alone does not tell the whole story. A high-strength steel and a low-strength steel may have the same transition temperatures, but the high-strength steel may have significantly lower energy levels associated with fracture at the upper- and lower-shelf regions.

Specification of acceptable values of transition temperature has been based primarily on empirical correlations with service performance. In one case, for example, statistical studies of steel plates from failed ships indicated that brittle catastrophic failures occurred only in those instances where the 20-J (15-ft·lb) transition temperature was above the minimum service temperature (Ref 12). Hence, specifying the 20-J transition temperature to be below the operating temperature proved to be a suitable

criterion for that particular application. Similar relationships between NDT and service failures contributed to the specification of NDT for certain classes of steels (Ref 13 and 14). Unfortunately, such correlations with service-failure experience often are not available for many applications and specification of values for the transition temperatures can only be arbitrary.

With the exception of the CAT, none of the transition temperatures provides the designer with any information with respect to load-bearing capacity or tolerable stress levels. The CAT provides the stress/temperature combinations that will ensure crack arrest. Unfortunately, the CAT increases with increasing plate thickness. In fact, all of the transition temperatures depend on specimen size, notch geometry, acuity, and a number of test variables. To get data representative of the behavior of cracks in large components, very large specimens with defects typical of those found in components need to be tested. This is a major limitation in using any of the transition temperatures for design purposes.

While the transition-temperature approach may be applicable to low-strength steels for purposes of comparison and for rough design approximations, it is not satisfactory for other materials. High-strength steels do not exhibit abrupt ductile-to-brittle transition behavior, and definition of the various transition temperatures becomes uncertain. Other structural materials such as titanium alloys, aluminum alloys, and other nonferrous alloys, as well as all austenitic stainless steels, do not exhibit characteristic transition behavior at all. The dependence of the fracture stress on the loading and the geometry of the defect (flaw size, shape, acuity) cannot be accommodated in any of the transition-temperature approaches. Lastly, the slow growth of any initially subcritical-size defect to a critical size during the operational lifetime of a statically or cyclically loaded component cannot be accommodated in such an approach. Many of these limitations have paved the way to the more-quantitative modern fracture-mechanics approach for predicting brittle fractures emanating from defects.

Linear-Elastic Fracture Mechanics

The theory of linear-elastic fracture mechanics (LEFM) provides a means of predicting the fracture loads of structures containing sharp flaws of known size and location. An energy-based approach (Ref 15 to 17) and a stress-based approach are both used, the two approaches being mutually related and leading to identical results.

Energy-Based Approach

In this approach, unstable crack propagation is postulated to occur when the energy which could be supplied to the crack tip during an incremental crack extension is greater than or equal to the energy required for the crack to advance. The various en-

ergy components involved in an incremental crack extension are (1) strain energy released, (2) energy supplied to the body by external work, (3) kinetic energy, (4) energy required to form two new fracture surfaces, and (5) energy required to induce plastic deformation of the material at the crack tip. For materials which exhibit some ductility, the fracture-surface energy (item 4) is far outweighed by the other factors and can be neglected (Ref 17). The kinetic-energy term (item 3) is also considered to be small and can be neglected. Therefore, only the other three components (items 1, 2, and 5) are considered in the analysis. Also, if the size of the plastic zone at the tip of the crack is small compared with the total volume of the body, the strain energy released (item 1) is to a good approximation equal to the change in elastic strain energy of the cracked body.

If W_e, W_p, and U represent, respectively, the energy supplied by external work (item 2), the energy required for plastic deformation (item 5), and the elastic strain energy (item 1), and if A represents the crack area, the energy criterion for crack extension can be stated as

$$\frac{\partial W_e}{\partial A} \, dA \geq \frac{\partial U}{\partial A} \, dA + \frac{\partial W_p}{\partial A} \, dA \quad \text{(Eq 2.1)}$$

$$\frac{\partial W_e}{\partial A} - \frac{\partial U}{\partial A} = \frac{\partial W_p}{\partial A} \quad \text{(Eq 2.2)}$$

The left-hand side of the equation is defined as the crack-driving force, G, or the energy available for crack extension. For linear-elastic bodies, G is also equal to the elastic-strain-energy-release rate, $-\partial U/\partial A$, under load or deflection control.

$$G = \frac{\partial W_e}{\partial A} - \frac{\partial U}{\partial A} \quad \text{(Eq 2.3)}$$

The critical energy-release rate at fracture is defined as G_c, where $G_c = \partial W_p/\partial A$. Since this quantity is a function of the material, temperature, strain rate, and state of stress at the crack tip, it is determined from at least one fracture test for which the expression for the energy-release rate is known. G_c is thus a material property determined for specific test conditions.

An expression for G_c can be derived by considering Eq 2.3. When the size of the plastic zone at the crack tip is very small, it has been postulated that the energy-release rate is simply the rate of change of strain energy with respect to crack area when the displacements of the externally applied loads are held constant (Ref 18). For instance, in the case of an infinite plate of unit thickness containing a crack of finite length 2a, perpendicular to a uniaxial stress field, σ, the presence of the crack reduces the elastic energy by the quantity $U = \pi\sigma^2 a^2/E$. By differentiating this with respect to crack size, we get

$$G = \frac{\pi\sigma^2 a}{E} \qquad \text{(Eq 2.4)}$$

Fracture of the plate will occur when the critical stress level is reached:

$$\sigma_c = \sqrt{\frac{G_c E}{\pi a}} \qquad \text{(Eq 2.5)}$$

where G_c can be determined experimentally by performing at least one fracture test on this geometry at some known crack length.

There are two major limitations of this approach. Because the derived equations are restricted to conditions where the size of the crack-tip plastic zone is very small compared with the other dimensions of the body, the approach is limited to materials which are relatively brittle. Secondly, the approach fails to handle adequately the effect of crack-root radius. Thus, it would predict identical fracture stresses for two similarly cracked specimens differing only in crack-tip-root radius, which is contrary to experience (Ref 19). The stress-intensity approach described below suffers from the same limitations, with the only difference being that it is much more general and is not restricted to uniformly loaded infinite bodies.

Stress-Intensity Approach

The stress-intensity approach is a more general approach and yet yields identical results for the specific situation of uniformly loaded infinite bodies. Inherently, it has more appeal to the engineer because it deals with crack-tip stresses and strains rather than energy.

For a through-the-thickness sharp crack in a plate subjected to in-plane loads which are uniform through the thickness and symmetric with respect to the plane of the crack, the elastic stress field in the vicinity of the crack tip (Ref 20 and 21) is given by

$$\sigma_y = \frac{K}{\sqrt{2\pi r}}\, f_1(\theta) \qquad \text{(Eq 2.6)}$$

$$\sigma_x = \frac{K}{\sqrt{2\pi r}}\, f_2(\theta) \qquad \text{(Eq 2.7)}$$

$$\tau_{xy} = \frac{K}{\sqrt{2\pi r}}\, f_3(\theta) \qquad \text{(Eq 2.8a)}$$

$$\sigma_z = \nu(\sigma_x + \sigma_y) \quad \text{[for plane strain]}$$
$$\text{(Eq 2.8b)}$$

$$\sigma_z = 0 \quad \text{[for plane stress]} \qquad \text{(Eq 2.8c)}$$

$$\tau_{xz} = \tau_{yz} = 0 \qquad \text{(Eq 2.8d)}$$

where the coordinate system (r, θ) is as shown in Fig. 2.3. The stress-intensity factor K is a function of plate geometry, applied loads, and the size, location, and orientation of the crack. The crack-tip elastic stress and strain fields, according to the above equations, are uniquely characterized by K with a modification factor that varies with the above variables. The approach is based on the assumption that crack propagation will occur when the stress intensity at the crack tip, K, reaches a critical value K_c. Once the magnitude of K_c has been experimentally determined, the fracture load for a structure containing a crack can be predicted, if the stress-intensity expression for the specific crack geometry is known. Three modes of loading with respect to the crack

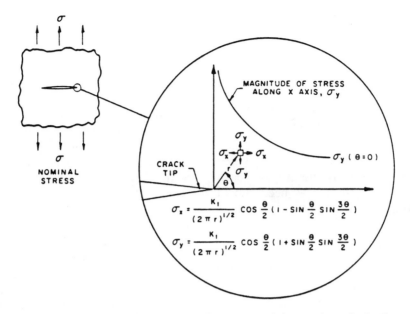

Fig. 2.3. Distribution of stresses in the vicinity of the crack tip (Ref 20).

plane—i.e., tension, shear, and torsion— are possible, and the stress intensities corresponding to these are termed K_I, K_{II}, and K_{III}. The critical stress intensity for mode 1 loading (tension) under plane-strain conditions is termed K_{Ic}.

A general expression for K can be written as

$$K = M\sigma\sqrt{\pi a} \qquad \text{(Eq 2.9)}$$

where M is a constant specific to a given flaw size and geometry. For instance, for the case of a semi-infinite plate containing a finite center crack of length 2a, perpendicular to a uniaxial stress field (see Fig. 2.4), the stress-intensity factor is given by the expression

$$K = \sigma\sqrt{\pi a} \qquad \text{(Eq 2.10)}$$

so that the critical stress for fracture, σ_c, is given by $K_{Ic}/\sqrt{\pi a}$. Other stress-intensity expressions are available for other types of flaws, as illustrated in the example below.

Example:
The K_{Ic} value for a quenched-and-tempered steel is 77 MPa$\sqrt{m}$ (70 ksi$\sqrt{in}$.). Calculate the critical flaw size that would cause failure in a pressure vessel made

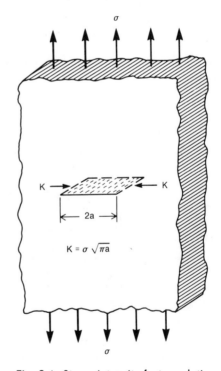

Fig. 2.4. Stress-intensity-factor solution for semi-infinite plate with center crack.

from this steel during a 150% overpressure proof test ($\sigma = 0.75\sigma_{ys} = 1035$ MPa, or 150 ksi). The flaws are assumed to be located in the cylindrical section and oriented normal to the hoop stress.

Four types of defects are envisioned and the stress-intensity expressions are given in Fig. 2.5. When $K = K_{Ic}$, $a = a_c$ and the critical crack sizes a_c can be calculated by substituting the values of K_{Ic} in the various expressions. The a_c values of 1.9, 1.3, 4.1, and 2.3 mm (0.076, 0.050, 0.16, and 0.09 in.) are obtained for the different flaws.

It was stated earlier that the energy-based approach and the stress-intensity approach are mutually related. It has been shown (Ref 21) that

$$K_{Ic} = \sqrt{G_c E} \quad \text{[for plane stress]}$$

(Eq 2.11)

and

$$K_{Ic} = \sqrt{\frac{G_c E}{(1 - \nu^2)}} \quad \text{[for plane strain]}$$

(Eq 2.12)

where ν is Poisson's ratio and E is Young's modulus.

Limitations and Capabilities of the LEFM Approach

Two basic assumptions are inherent in the development of the LEFM approach. Because the stress solutions represented by Eq 2.6 to 2.8 are for elastic stress fields, it is assumed that plastic deformation is restricted to a small region near the crack tip. In other words, the condition of small-scale yielding must be satisfied. Secondly, it is assumed that the defect is cracklike and that the root radius of the crack tip is essentially zero. These two assumptions place a significant restriction on the use of K_{Ic} as a failure criterion.

The size of the plastic zone (r_p) can be estimated from the stress-field equations for elastic–perfectly plastic materials (Ref 22) as follows:

$$r_p = \frac{1}{2\pi} \left(\frac{K_I}{\sigma_y}\right)^2 \quad \text{[for plane stress]}$$

(Eq 2.13)

Type of flaw	Applicable expression for K_I	Φ^2
Type 1 – short, deep surface crack; c = 2a 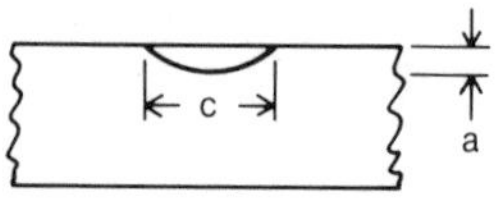	$K_I^2 = \dfrac{1.21\pi\sigma^2 a}{\Phi^2 - 0.212\left(\dfrac{\sigma}{\sigma y}\right)^2}$	1.45
Type 2 – long, shallow surface crack; c > 10a 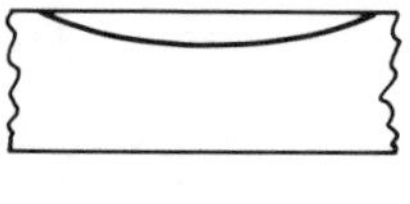	" "	1
Type 3 – internal sphere; c = a 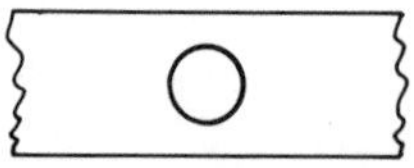	$K_I^2 = \dfrac{\pi\sigma^2 a}{\Phi^2 - 0.212\left(\dfrac{\sigma}{\sigma y}\right)^2}$	2.5
Type 4 – internal elliptical defect; c = 2a 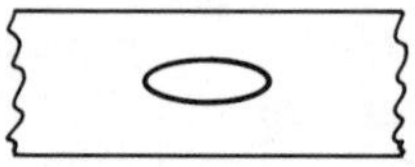	" "	1.45

Fig. 2.5. Stress-intensity data for some typical flaws.

and

$$r_p \simeq \frac{1}{6\pi}\left(\frac{K_I}{\sigma_y}\right)^2 \quad \text{[for plane strain]}$$

(Eq 2.14)

where K_I is the mode 1 applied stress intensity and σ_y is the yield strength.

It has been shown that the elastic stress field distribution in the vicinity of the plastic zone in an elastic–perfectly plastic material is identical to the elastic stress field distribution in the vicinity of a crack (but outside the plastic zone) in a material that is perfectly elastic but whose tip is placed in the center of the plastic zone. Hence, the correction for the presence of a plastic zone can be made by defining an effective crack length (Ref 23) as

$$2a_{eff} = 2a_i + r_p \qquad \text{(Eq 2.15)}$$

The correction is, however, valid only when r_p is small relative to a and the length of the ligament. Hence, in actual determinations of K_{Ic}, the dimensions of the specimen are to be kept as

$$a_i \geq 2.5\left(\frac{K_{Ic}}{\sigma_y}\right)^2 \qquad \text{(Eq 2.16)}$$

$$B \geq 2.5\left(\frac{K_{Ic}}{\sigma_y}\right)^2 \qquad \text{(Eq 2.17)}$$

and

$$b \geq 2.5\left(\frac{K_{Ic}}{\sigma_y}\right)^2 \qquad \text{(Eq 2.18)}$$

where a_i is the initial crack length, B is the thickness of the specimen, and b is the ligament width. The factor 2.5 is sometimes overconservative, and it may be possible to get valid K_{Ic} measurements from smaller specimens. In selecting the appropriate specimen dimensions, one has therefore to guess the K_{Ic} values by trial and error. The only reliable way to get a valid K_{Ic} value for a

material of unknown properties is to measure the toughness values for various specimen thicknesses. K_c at first decreases, reaches a minimum value upon reaching plane-strain conditions, and then remains constant. Above a certain specimen thickness, K_c becomes independent of specimen thickness. This value represents the true K_{Ic} for the material under plane-strain conditions, as shown in Fig. 2.6.

For many materials, K_c decreases in proportion to the decrease in the square root of the root radius until the radius reaches some minimum value (Ref 19). As the radius decreases below this level, K_c approaches its true value, K_{Ic}, and remains constant. To eliminate this effect, K_{Ic} is determined for specimens containing fatigue precracks, and its value can be used for estimating fracture behavior in structures containing either sharp cracks or cracklike discontinuities.

To determine K_{Ic} in the laboratory, a specimen of suitable size and shape, in which a fatigue precrack of known dimensions is present, is loaded monotonically and a load-vs-load-line deflection curve similar to a stress-strain curve is developed. Upon reaching a critical load, P_c, instability sets in, and the rapid crack extension is shown as a sudden change in the slope of the plot. K_{Ic} is then calculated from the critical load by applying known expressions. For instance, for the compact-type specimens (see Fig. 2.7) and single-edge-

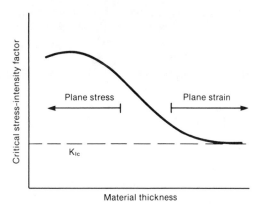

Fig. 2.6. Relationship between specimen thickness and critical stress intensity.

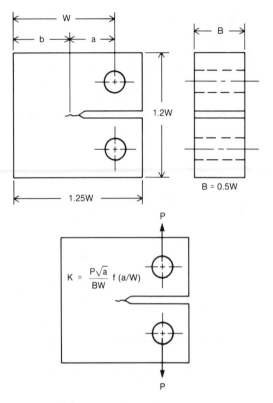

$$K = \frac{P\sqrt{a}}{BW} f(a/W)$$

Fig. 2.7. Schematic illustration of a compact tension specimen.

notched specimens which are commonly used, the following functional form is used:

$$K_{Ic} = \frac{P_c\sqrt{a}}{BW} f(a/W) \qquad \text{(Eq 2.19)}$$

where W is the width of the specimen.

The relationship among the stress-intensity factor, applied load, and specimen dimensions is generally known as K-calibration. Precise calibrations which permit ready calculation of K_{Ic} over a wide range of crack-length-to-specimen-width ratios (a/W) are now available for a variety of specimen geometries (Ref 24 to 28). The standard method for determining fracture toughness is given in ASTM Standard E399 (Ref 29).

In implementing the LEFM approach to a structural component, the following procedure is followed:

1. The K_{Ic} values for the material are estimated, obtained from literature, or

determined experimentally as discussed above.

2. The combination of stress and critical flaw size in the component is then calculated using the K expression appropriate to the flaw geometry and location.

The critical stress intensity for fracture in mode 1 under plane-strain conditions—i.e., K_{Ic}—is now commonly referred to as the fracture toughness of the material. It is quite clear that the higher the value of this parameter, the greater the ability of the material to tolerate cracks. It represents the limiting combination of stress and crack size that would lead to unstable crack propagation. Obviously, service components will rarely contain this limiting combination at the outset, because the first application of load during performance testing will lead to failure. The more applicable scenario is the slow growth of a "subcritical" flaw during service which can eventually reach the critical crack size, resulting in final failure. By properly combining evaluations of crack growth with LEFM, engineering procedures and criteria are established whereby one can answer questions relating to (1) the types and sizes of defects tolerable under design load, (2) the tolerable stress levels corresponding to known defects, (3) the maximum size of flaw that can be accepted initially with the assurance that it will not grow to critical size prior to the next inspection or during the life of the component, (4) the sizes of defects that can be left in and of those that need immediate repair, (5) selection of the appropriate inspection technique and the detection levels needed, and (6) the remaining life of a component which has been in service. An excellent review of the entire subject of various approaches to defining, evaluating, and applying various toughness criteria to assess structural integrity may be found in Wessel, Clark, and Wilson (Ref 30).

Effects of Testing and Material Variables on Fracture Toughness

The two major test variables that have effects on toughness are strain rate and temperature. Decreasing the strain rate results in an increase in K_{Ic}. K_{Ic} increases with increasing temperature in a relationship somewhat analogous in shape to the energy-vs-temperature Charpy curve. Among the metallurgical variables that affect toughness, the significant ones are yield strength, microstructure, grain size, content of inclusions, and impurities. Generally, an increase in yield strength results in a decrease in K_{Ic}. Among the various transformation products, tempered martensite exhibits the highest toughness, followed by bainite, followed by ferrite-pearlite structures.

Increasing the grain size results in a decrease in toughness. Higher levels of sulfide and other inclusions and impurity elements result in reduced toughness. These effects are obtained regardless of whether toughness is defined in terms of the transition-temperature approach or the fracture-toughness approach. For instance, all the factors that decrease the FATT tend to increase K_{Ic} values. The effects of the above variables are illustrated with respect to specific steels and applications in the later chapters and therefore need not be discussed further at this time.

Charpy Test – K_{Ic} Correlations

Although the concepts of fracture mechanics have come into vogue in recent years, many old-time engineers are still familiar with only the impact-transition-curve approach. Furthermore, there is a large body of FATT data available to equipment manufacturers regarding components manufactured in the past for which no K_{Ic} data were generated. Thirdly, K_{Ic} measurements involve use of large specimens which are difficult to excise from operating components. On the other hand, it is easier to generate FATT-type data on the same components. For these reasons, numerous empirical correlations between Charpy test results and K_{Ic} results have been developed that enable the engineer to estimate K_{Ic} values. Roberts and Newton have written an excellent review of the subject (Ref 31). The National Academy of Sciences has also published a review of a variety of techniques for estimating K_{Ic} values from tensile, bend,

and Charpy tests (Ref 32). Because the objective here is to provide K_{Ic} estimation procedures for life assessment and not to outline simplified substitute procedures for K_{Ic} determination in the laboratory, only those correlations dealing with impact-test data are reviewed here. A large body of impact-test data is already available because the impact transition curve has been the most common basis for specifying toughness. Further, the standard Charpy test is a preferred referencing test because the test procedure is described by an ASTM standard.

Correlations in the Upper-Shelf Region. The following correlations apply in the upper-shelf temperature range where the Charpy energy of the steel has reached the maximum or upper-shelf values:

$$\left(\frac{K_{Ic}}{\sigma_y}\right)^2 = 5\left[\left(\frac{CVN}{\sigma_y}\right) - 0.05\right]$$

[Rolfe-Novak (Ref 33)]   (Eq 2.20)

$$\left(\frac{K_{Ic}}{\sigma_y}\right)^2 = 1.37\left(\frac{CVN}{\sigma_y}\right) - 0.045$$

[Ault *et al* (Ref 34)]   (Eq 2.21)

where K_{Ic} is expressed in ksi$\sqrt{in.}$, σ_y is the 0.2% yield strength in ksi, and CVN is the upper-shelf Charpy energy in foot-pounds, and

$$\left(\frac{K_{Ic}}{\sigma_y}\right)^2 = 0.6478\left(\frac{CVN}{\sigma_y} - 0.0098\right)$$

[Iwadate *et al* (Ref 35)]   (Eq 2.22)

where K_{Ic} is in MPa$\sqrt{m}$, σ_y is in MPa, and CVN is in joules. The Iwadate correlation between K_{Ic} and CVN for a variety of low-alloy pressure-vessel and turbine steels is shown in Fig. 2.8 (Ref 36).

The Rolfe-Novak correlation has been developed and applied to steels with σ_y values ranging from 130 to 250 ksi, K_{Ic} values from 87 to 200 ksi$\sqrt{in.}$, and CVN values from 16 to 60 ft·lb. The correlation proposed by Ault is based on higher-strength,

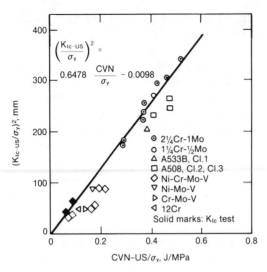

$$\left(\frac{K_{Ic\text{-}US}}{\sigma_y}\right)^2 = 0.6478\,\frac{CVN}{\sigma_y} - 0.0098$$

- ⊚ 2¼Cr-1Mo
- ○ 1¼Cr-½Mo
- △ A533B, Cl.1
- ▢ A508, Cl.2, Cl.3
- ◇ Ni-Cr-Mo-V
- ▽ Ni-Mo-V
- ▷ Cr-Mo-V
- ◁ 12Cr
- Solid marks: K_{Ic} test

y-axis: $(K_{Ic\text{-}US}/\sigma_y)^2$, mm
x-axis: CVN-US/σ_y, J/MPa

Fig. 2.8. Relationship between K_{Ic} and Charpy V-notch energy in the upper-shelf region (Ref 36).

lower-toughness steels, with σ_y values from 234 to 287 ksi, K_{Ic} from 34 to 70 ksi$\sqrt{in.}$, and CVN from 11 to 21 ft·lb. In general, there is a trend of decreasing CVN shelf energy with increasing yield and tensile strengths. Consequently, a plot of CVN vs K_{Ic} often provides a relation nearly as useful as the more refined ones.

Correlations in the Transition-Temperature Region. Several methods and correlations have been proposed and used for estimating K_{Ic} from standard Charpy test data for steels in the transition-temperature range. A variety of proposed correlations among K_{Ic} or K_{Id} (dynamic fracture toughness), CVN, and σ_y have been reviewed by Stahlkopf and Marston (Ref 37) and tabulated by Server and Oldfield (Ref 38) (see Table 2.1).

Table 2.1. Correlations between impact properties and fracture toughness (after Ref 35 to 39 and 41)

Correlation	Comment
Transition-region correlations	
Barsom-Rolfe:	σ_y = 269 to 1696 MPa
$\quad K_{Ic}^2/E = 2(CVN)^{3/2}$	Static tests
$\quad K_{Ic}^2/E = 5(PCVN)$	Precracked Charpy tests
Corten-Sailors:	CVN = 7 to 70 J
$\quad K_{Ic} = 15.5(CVN)^{1/2}$ or $K_{Ic}^2/E = 8(CVN)$	Static tests
$\quad K_{Id} = 15.873(CVN)^{3/8}$	Dynamic (high-strain-rate) tests
Marandet-Sanz:	Static tests
$\quad K_{Ic} = 20(CVN)^{1/2}$	$T_{K_{Ic}}$ at K_{Ic} = 100 MPa$\sqrt{m}$
$\quad T_{K_{Ic}} = 16.2 + 1.37T_{28}$	T_{28} at CVN = 28 J
Begley-Logsdon:	
$\quad K_{Ic}$ at FATT = ½(K_{Ic} from Rolfe-Novak relationship + 0.5σ_y)	σ_y = 269 to 1696 MPa
Iwadate-Watanabe-Tanaka:	
$\quad K_{Ic}/K_{Ic\text{-}US} = 0.0807 + 1.962\exp[0.0287(T - FATT)]$	For $-40\,°C > (T - FATT)$
$\quad K_{Ic}/K_{Ic\text{-}US} = 0.623 + 0.406\exp[-0.00286(T - FATT)]$	For $350\,°C > (T - FATT) > -40\,°C$
Upper-shelf CVN correlations	
Rolfe-Novak:	σ_y = 690 to 1696 MPa
$\quad (K_{Ic}/\sigma_y)^2 = 5[(CVN/\sigma_y) - 0.05]$	Static tests
Wullaert-Server:	σ_y = 345 to 483 MPa
$\quad K_{Jd} = 20(DVN)^{1/2}$	Dynamic J-integral initiation
$\quad K_{Jc} = 2.1(\sigma_y CVN)^{1/2}$ or $(K_{Jc}/\sigma_y)^2 = 4.41(CVN/\sigma_y)$	All loading rates with appropriate σ_y
Lawrence Livermore Laboratory:	
$\quad (K_{Jc}/E)^2 = CVN(9.66 + 0.04\sigma_y)$	$K_{Jc} = (EJ_{Ic})^{1/2}$; $K_{Jc} = (EJ_{Id})^{1/2}$
Ault:	
$\quad (K_{Ic}/\sigma_y)^2 = 1.37(CVN/\sigma_y) - 0.045$	High-strength, low-toughness steels
Iwadate-Karushi-Watanabe:	Pressure-vessel steels
$\quad (K_{Ic}/\sigma_y)^2 = 0.6478(CVN/\sigma_y - 0.0098)$	

A method for defining the K_{Ic}-vs-temperature relationship for steels based on σ_y and CVN has been described by Begley and Logsdon (Ref 39). At the upper-shelf temperature, they estimate K_{Ic} using the Rolfe-Novak correlation given by Eq 2.20. At the lower-shelf temperature, K_{Ic} is estimated as $0.5\sigma_y$. At the 50% ductile-brittle transition temperature, K_{Ic} is taken to be the average of the above two. Good agreement has been shown between the predicted K_{Ic}-vs-T curves and those actually determined from tests for turbine steels as well as for pressure-vessel steels, as illustrated in Fig. 2.9 for the case of turbine-rotor steels.

Based on extensive laboratory tests on Cr-Mo-V, Ni-Cr-Mo-V, and Ni-Mo-V turbine-rotor steels, Greenberg, Wessel, and Pryle (Ref 40) have established a correlation between the excess temperature (test temperature − FATT) and K_{Ic}, as shown in Fig. 2.10. The correlation has considerable scatter, but is nevertheless used extensively in industry for assessments of rotor life.

Iwadate, Watanabe, and Tanaka have reported on a correlation between excess temperature and the value of K_{Ic} at any temperature normalized with respect to the upper-shelf temperature, as shown in Fig. 2.11 (Ref 35 and 36). In this figure, very small scatter is observed and a single master curve could be drawn, despite a variety of chemical compositions of the steels tested, when K_{Ic} is normalized with respect to its value at the upper shelf. The 99% confidence limit curve was reported to result in the following expressions:

$$\frac{K_{Ic}}{K_{Ic\text{-US}}} = 0.0807 + 1.962$$

$$\times \exp[0.0287\,(T - FATT)]$$

$$(Eq\ 2.23)$$

for $-40\ ^\circ C\ (-40\ ^\circ F) > T - FATT$, and

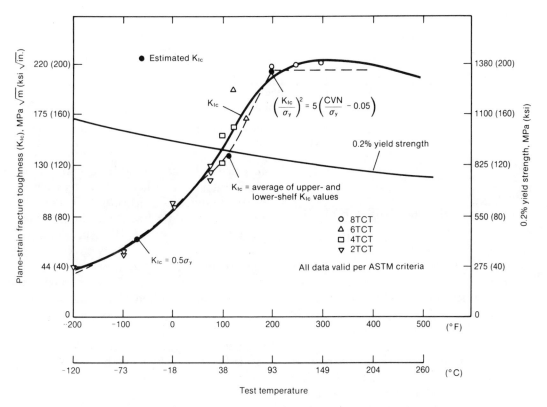

Fig. 2.9. Begley-Logsdon estimation procedure for K_{Ic} and validation of results for a turbine-rotor steel (Ref 39).

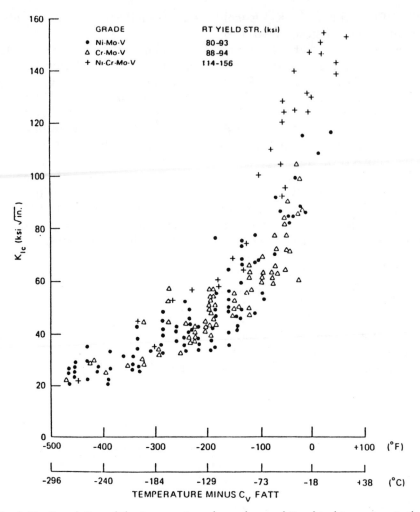

Fig. 2.10. Correlation of the temperature dependence of K_{Ic} of turbine-rotor steels with excess temperature (T − FATT) (Ref 40).

$$\frac{K_{Ic}}{K_{Ic\text{-}US}} = 0.623 + 0.406$$

$$\times \exp[-0.00286(T - FATT)]$$

$$\text{(Eq 2.24)}$$

for 350 °C (660 °F) > (T − FATT) > −40 °C (−40 °F). Using the above correlations, K_{Ic} at any temperature could be estimated as follows. The ratio $K_{Ic}/K_{Ic\text{-}US}$ could first be determined based on Eq 2.23 or 2.24, and $K_{Ic\text{-}US}$ could then be estimated based on σ_y and CVN using the Rolfe-Novak relation (Eq 2.20) or the Iwadate modified relation (Eq 2.22). The value of K_{Ic} at the desired temperature could thus be determined.

It should be noted that the toughness of ferritic steels has a notable dependence on strain rate. Thus, when comparing Charpy results (a dynamic test) with K_{Ic} values obtained from static tests, one should compensate for the strain-rate dependence of K_{Ic} in some fashion. Many investigators have reported correlations between Charpy results and the dynamic fracture toughness K_{Id}. Again, a method of translating these K_{Id} values to K_{Ic} values is needed. Barsom has found that there is an effective lateral temperature shift, ΔT_s, from dynamic to static behavior which can be related to the room-temperature yield strength, σ_y, as follows (Ref 41):

$$\Delta T_s = 215 - 1.5\sigma_y \qquad \text{(Eq 2.25)}$$

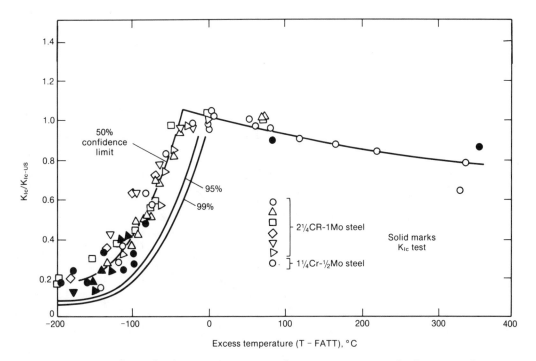

Fig. 2.11. Relationship between K_{Ic}/K_{Ic-US} and excess temperature for ferritic steels (Ref 35 and 36).

where ΔT_s is expressed in °F and σ_y in ksi. Once the K_{Id}-vs-T curve has been established, a lateral leftward (to lower T) shift of the curve by the amount ΔT_s defines the K_{Ic}-vs-T curve.

Values of fracture toughness for steels for nuclear reactor pressure vessels are correlated with the excess temperature using a plot similar to Fig. 2.10 but with two differences. First, the procedure uses the critical stress intensity for crack arrest, K_{Ia}, rather than K_{Ic}. Among K_{Ia}, K_{Id}, and K_{Ic}, K_{Ia} represents the most conservative criterion because it is based on the premise that even if crack extension were initiated, it would be immediately arrested. K_{Ia} is the value of K_I that prevails at the arrest of a rapidly propagating crack and it is utilized exactly in the same manner as K_{Ic}, except to analyze crack arrest. The second major difference is that the excess temperature is not based on FATT but on a reference NDT temperature designated RT_{NDT} and is the higher of: (1) the drop weight NDT (per ASTM E208) and (2) the temperature 33 °C (60 °F) below the temperature at which the

Charpy V-notch impact specimen exhibits an absorbed energy of 68 J (50 ft·lb) and a lateral expansion of 0.9 mm (0.035 in.). The lowest bound of the K_{Ia} and K_{Id} combined data vs excess temperature as defined above is known as the reference toughness curve, or K_{IR} curve. The K_{IR} curve derived in this manner and as adopted for Appendix G of Section III of the ASME Boiler and Pressure Vessel code is shown in Fig. 2.12. Subsequent modifications of this approach to include an analysis of crack initiation and crack arrest have led to the correlation of Appendix A of Section XI of the Code. Specifically, Appendix G procedures are concerned with designing for protection against nonductile fractures, while Appendix A procedures are for evaluating the disposition of flaws detected during in-service inspection. The applicability of the curve shown in Fig. 2.12 is limited to carbon and alloy steels with σ_y at room temperature below 345 MPa (50 ksi). Further details regarding the development of reference toughness curves may be found in Ref 42 to 45.

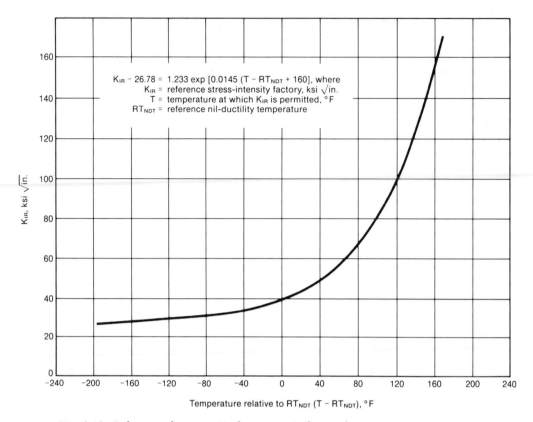

Fig. 2.12. Reference fracture-toughness curve for nuclear-reactor pressure-vessel steels as per ASME Boiler and Pressure Vessel Code, Section III, Appendix G (Ref 45).

Elastic-Plastic Fracture Mechanics (EPFM)

As described earlier, the stress solutions given by Eq 2.6 to 2.8, which form the basis of LEFM, are restricted to cases where the crack-tip stress fields are purely elastic. Accommodation was made for the fact that some amount of yielding at the crack tip is inevitable by postulating that the validity of the elastic stress solutions would continue to hold as long as the size of the plastic zone was small relative to the over-all dimensions of the component, as prescribed by Eq 2.16 to 2.18. This restriction poses major problems with respect to generation of valid K_{Ic} data from laboratory tests and to application of the K_{Ic} concept itself for analyzing the integrity of structures.

Let us consider both of these aspects in relation to the turbine rotor and a steam pipe. For typical rotor steels, yield strength ranges from 620 to 760 MPa (90 to 110 ksi) and K_{Ic} values range from 55 to 110 MPa$\sqrt{m}$ (50 to 100 ksi$\sqrt{in.}$). From Eq 2.16 to 2.18, valid K_{Ic} values can be derived and applied, if the thickness exceeds the range of 13 to 76 mm (0.5 to 3.0 in.). A compact tension specimen 76 mm (3 in.) thick would require a sample measuring 190 by 183 by 76 mm (7.5 by 7.2 by 3 in.)—a size too large and inconvenient to obtain and test. The dimensions of the actual rotor are sufficiently large to justify the use of LEFM. Hence, in the case of a rotor, the problem with LEFM is not in its applicability to the rotor itself but only with respect to data generation in the laboratory. Similar calculations of dimensional considerations will show that for pipe and pressure-vessel steels, which are generally utilized at much lower yield-strength levels and at higher K_{Ic} levels, application of LEFM is severely restricted with respect to both laboratory tests and the actual components. Thus, there is a need for techniques that set the fracture criteria under conditions where the size requirements set forth by Eq 2.16 to 2.18 are no longer satisfied and significant plasticity is associated with crack initiation and propagation. In these instances, application of LEFM analysis leads to overly conservative assessments of component integrity. The J-integral technique and the crack-opening-displacement (COD) technique have been developed as viable crack-initiation parameters, as part of the elastic-plastic fracture-mechanics (EPFM) procedure. To characterize crack growth and final instability, concepts based on tearing modulus have been developed. The entire area of crack initiation, crack growth, and instability is covered by general yielding fracture mechanics (GYFM).

The J-integral concept is essentially an energy criterion, characterizing the plastic stress-strain field at the crack tip (Ref 46). A systematic representation of the stress fields in cracked bodies where the K-fields and J-fields are applicable is shown in Fig. 2.13 (Ref 47 and 48). The J-integral is a two-dimensional energy-line integral (Ref 46) defined as

$$J = \int_{\Gamma} \left(W dy - T \frac{\partial u}{\partial x} ds \right) \quad \text{(Eq 2.26)}$$

where, as shown in Fig. 2.14, Γ is the counterclockwise contour around the crack tip, T is the traction vector defined according to the outward normal n along Γ (i.e., $T_i = \sigma_{ij} n_j$), u is the displacement vector acting along the integration path, ds is an increment of length along the integration path, and W is the strain-energy density defined as

$$W = \int_0^{\epsilon} \sigma_{ij} d\epsilon_{ij} \quad \text{(Eq 2.27)}$$

where σ_{ij} and ϵ_{ij} are the components of stress and strain, respectively.

Rice (Ref 46) has shown that the J-integral is path-independent. The critical value of J at the onset of crack extension provides a fracture-toughness parameter, J_{Ic}, in the elastic-plastic deformation regime. The use of J as a fracture criterion is taken from a model of the fracture process as shown in Fig. 2.15 (Ref 49 to 51). The fracture process starts with a sharp crack when the specimen or structure with the crack is unloaded. For a test specimen, the crack is introduced by fatiguing at a low ΔK level. As the crack undergoes loading, the following sequence of events takes place: (1) sharp precracking is present; (2) blunting of the initial crack occurs; (3) blunting increases with increasing load; (4) crack advance occurs ahead of the blunted crack; and (5) stable growth of the crack occurs until,

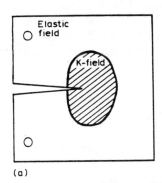 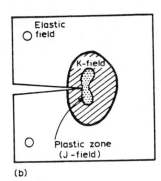 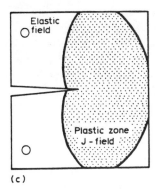

(a) Linear-elastic behavior. (b) Linear-elastic behavior with small-scale yielding. (c) Large-scale yielding.

Fig. 2.13. Schematic illustration of stress fields in cracked bodies (Ref 47 and 48).

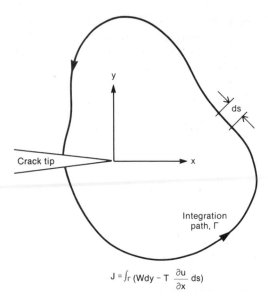

$$J = \int_\Gamma \left(W dy - T \frac{\partial u}{\partial x} \, ds \right)$$

Fig. 2.14. Definition of the J-integral.

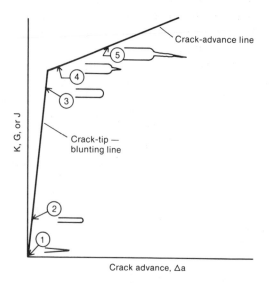

Fig. 2.15. Resistance-curve schematic of the fracture process (Ref 49 to 51).

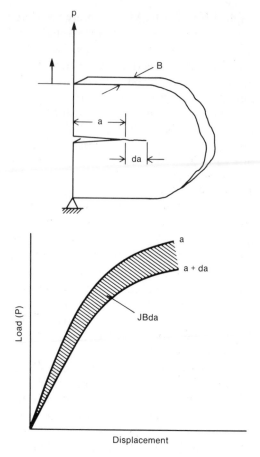

Fig. 2.16. Interpretation of the J-integral (Ref 53).

eventually, ductile instability occurs, terminating the stable crack growth. The cracking process can be related to the characterizing parameter (J for elastic-plastic consideration, and G or K for elastic consideration) by a plot of J vs the crack extension, Δa, as shown in Fig. 2.15. This plot is also known as the R-curve. The point where additional crack growth occurs from the blunted crack is marked by a change in the slope of the R-curve. The intersection

of the crack-blunting line and the crack-advance line essentially defines the critical value of J for crack initiation—i.e., J_{Ic}.

The physical significance of the J-integral can be better appreciated if it is viewed as a potential-energy criterion (Ref 52). The J-parameter is taken as the difference in potential energy of two identically loaded bodies having an incremental difference in crack length—i.e., a, a + da:

$$J = - \frac{1}{B} \frac{dU}{da} \qquad \text{(Eq 2.28)}$$

where B is the specimen thickness and U is the potential energy. This concept is illustrated in Fig. 2.16 (Ref 53).

Using this energy interpretation, Begley and Landes (Ref 54) have proposed a method for measuring J_{Ic}. By measuring the difference in energy at a constant value of displacement for identically loaded specimens of different crack lengths, J can be evaluated as a function of displacement. The various steps involved in this procedure are illustrated in Fig. 2.17. The critical value of J is taken at the point where crack initiation began and is the J_{Ic} value.

An improved method developed for determining J_{Ic} is based on an approximate formulation for calculating J (Ref 53 and 55) by the expression

$$J = 2A/Bb \qquad (Eq\ 2.29)$$

where B is the specimen thickness, b is the remaining uncracked ligament of the specimen, and A is the area under the load-vs-displacement curve. This formulation applies to a specimen with a deep crack ($a/W > 0.6$) under bend-type loading. Compact specimens and bend bars with three- or four-point loading are best suited for these tests. The use of the approximate formula represents a distinct advantage over the energy-rate method in that J can now be calculated from a single specimen. With this capability, the real problem of determining the J_{Ic} measurement point can be

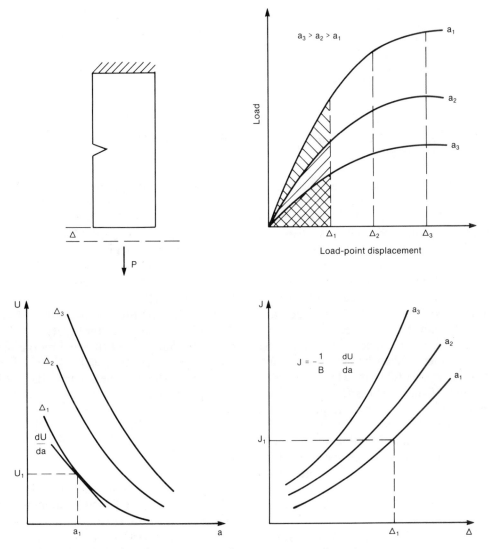

Fig. 2.17. Procedure for experimental J_{Ic} measurement based on energy interpretation of the J-integral (after Ref 54).

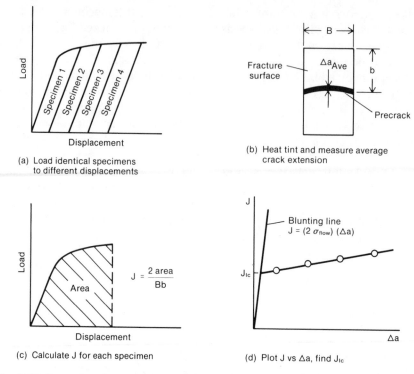

Fig. 2.18. Procedure for experimental J_{Ic} measurement: multiple-specimen R-curve (Ref 49 and 56).

addressed. The method that has been proposed is shown schematically in Fig. 2.18 (Ref 49 and 56). Several identical specimens are loaded to different values of displacement and then unloaded (Fig. 2.18a). These specimens will hopefully exhibit different amounts of crack growth. After unloading, the crack advance is marked and the specimen is broken open so that the crack advance, Δa, can be measured (Fig. 2.18b). Different methods can be used to mark the crack advance, the easiest one being heat tinting by heating to about 315 °C (600 °F) for about 10 min. The specimens subsequently are broken open at liquid-nitrogen temperature. The value of J at the point where the specimen is unloaded is calculated from Eq 2.29 for each specimen (Fig. 2.18c). This value of J is then plotted as a function of Δa for the various specimens to determine the J_{Ic} of the material (Fig. 2.18d).

In the above procedure, several specimens are still needed to define the R-curve, because each specimen is used to get a single J-value. To determine the entire J-vs-Δa curve (hence J_{Ic}) from a single specimen, all that is required is continuous monitoring of crack advance during generation of the load-displacement curves. Elastic-compliance methods and electrical-potential methods have been successfully used for this purpose (Ref 49).

Another quantitative measure of structural stability has been formulated by Paris *et al* (Ref 50) and is expressed by the tearing-modulus (T_j) approach where T_j is related to the slope of the R-curve, dJ/da, by the expression

$$T_j = \frac{dJ}{da} \frac{E}{\sigma_y^2} \qquad \text{(Eq 2.30)}$$

where E is Young's modulus and σ_y is yield strength. An analysis procedure that includes both J_{Ic} and T as the failure criteria considers not only tearing initiation at J_{Ic} but also the resistance of the material to continued crack extension beyond crack initiation. The use of the R-curve to derive

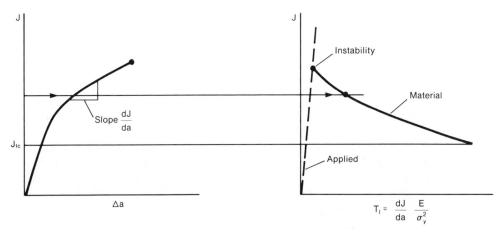

Fig. 2.19. Tearing-modulus concept for stable crack growth.

a J-vs-T stability diagram is illustrated in Fig. 2.19. Instability is predicted at the point where the applied J-T curve intersects the material J-T curve, as shown.

A rational approach to incorporating ductile fracture mechanics into a design philosophy is to use J_{Ic} as a design limit in terms of toughness and to use the ductile stability beyond J_{Ic} as a safety margin (Ref 51). This means that no stable crack advance is permitted by design. However, an added margin of safety is incorporated for unexpectedly high loads (accident conditions) in that ductile crack advance can occur in a stable manner without the risk of catastrophic failure.

Test methods for determining J_{Ic} are described in ASTM E813 (Ref 57). The specimens used are very similar to those used for K_{Ic} except that the size required is much smaller. ASTM E813 requires that

$$B \text{ and } b > \frac{25 J_{Ic}}{\sigma_y} \qquad \text{(Eq 2.31)}$$

where σ_y is an effective yield strength taken as the average of the 0.2% offset yield strength and the ultimate tensile strength, B is the specimen thickness, and b is the uncracked ligament. For a typical rotor steel, this requires a specimen approximately 2.5 mm (0.1 in.) thick or greater. Thus, compared with the LEFM specimen, the size requirements are much smaller. Such small

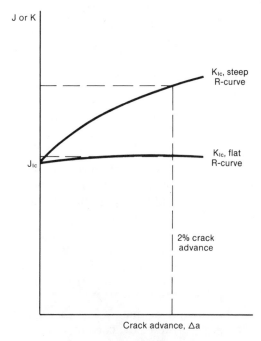

Fig. 2.20. Schematic illustration of R-curve showing differences between J_{Ic} and K_{Ic} measurement points (Ref 49).

specimens can be obtained more readily than the K_{Ic} specimens from rotors, pipes, and pressure vessels.

The relation between J_{Ic} and K_{Ic} can be illustrated with the help of Fig. 2.20. On this R-curve, J_{Ic} is taken as the point of first real crack growth, whereas the K_{Ic} measurement point is taken at 2% crack extension. If the R-curve is relatively flat, such as in the case of a pure brittle fracture, the K_{Ic} value at 2% crack extension would be nearly the same as the K value at the point of first crack growth. In this case the K_{Ic} and J_{Ic} values would be totally compatible. If, on the other hand, the R-curve is fairly steep, such as for failures under purely ductile conditions or under plane-stress conditions, the J_{Ic} value measured from the point of first crack extension would be lower than the K_{Ic} value measured at 2% crack extension.

Based strictly on a linear-elastic definition of J, the equivalence of J_{Ic} and K_{Ic} has been established through the following relationships:

$$J_{Ic} = G_{Ic} = \frac{K_{Ic}^2}{E}\,(1 - \nu^2)$$

[for plane strain]   (Eq 2.32)

and

$$J_c = G_c = \frac{K_c^2}{E}$$

[for plane stress]   (Eq 2.33)

Under elastic-plastic conditions where steep R-curves may obtain and the measurement points for J_{Ic} and K_{Ic} are at different locations on the R-curve, the K_{Ic} values converted from J_{Ic} values using the above equations have been found to be lower than the actually determined K_{Ic} values—i.e., the K_{Ic} equivalents of J_{Ic} represent a lower-bound value. Hence Eq 2.32 and 2.33 could still be used for estimating K_{Ic}, thus ensuring conservatism in the calculations.

To apply the J-integral to analysis of the structural integrity of a component, two alternate procedures are possible: (1) J-expressions developed analytically for the particular component geometry can be directly applied or (2) the equivalence of J and K as expressed by Eq 2.32 and 2.33 can be taken advantage of to perform the analysis using K-solutions alone. A further refinement of this procedure for elastic-plastic and plastic situations is the inclusion of the plastic contribution of J, such that

$$J = J_e + J_P \qquad \text{(Eq 2.34)}$$

where the elastic contribution, J_e, is simply that given by Eq 2.32 or 2.33—i.e., K^2/E or $(1 - \nu^2)K^2/E$—and the plasticity contribution is derived analytically or from handbook solutions. An example of the use of this method for analyzing cracks in casings has been outlined by Saxena, Liaw, and Logsdon (Ref 58).

A parallel and alternative approach to the J-integral approach has been the development of the crack-opening displacement (COD) as a fracture parameter (Ref 47). The crack-opening displacement is defined as the displacement of the two crack surfaces at the tip of the crack, generally measured by clip gages at the specimen outer surfaces near the crack. As a material containing a crack is loaded progressively, crack-opening displacement increases as accommodated by plastic deformation. As the stresses and strains at the plastically deformed region reach a critical value, fracture begins. At the critical load, leading to a critical value of COD, δ_c, the original crack begins to extend in length by either slow growth or rapid propagation. The critical COD, δ_c, is defined as the value of the COD at which the first extension of the crack occurs. The COD-vs-crack-length behavior is very similar to the J-vs-crack-length behavior depicted by the R-curve in Fig. 2.15. By analogy, Shih *et al* have proposed a parameter $\alpha = d\delta/da$ defined as the crack-opening angle, COA (Ref 47). They have defined a tearing modulus, T_δ, very similar to that given in Eq 2.30 in terms of the COA to characterize the stable-crack-growth region.

The J-integral approach and the COD approach are mutually related. It has been shown that $J = M\sigma_y$ (CTOD) where M (= 0.5 to 3.0) is a function of both the state of stress and the strain-hardening exponent (Ref 59). Analytical expressions for computing J for a variety of cracked configurations may be found in Ref 60. The details of the analytical developments pertaining to the J and COD concepts can be found in several references (Ref 47, 50, 61, and 62). It is now clear that the material toughness associated with crack initiation can be characterized by the critical parameter J_{Ic} or δ_{Ic}, while the material toughness associated with crack growth and instability can be characterized by the dimensionless parameter T_j or T_δ. To facilitate the application of general yielding fracture mechanics (GYFM), a ductile fracture handbook approach has been developed, which reduces plastic analysis to simple graphical or semi-analytical procedures (Ref 60 and 63).

Temper Embrittlement of Steels

Temper embrittlement is a major cause of degradation of toughness of ferritic steels. Numerous components otherwise in sound condition become candidates for retirement if they are severely embrittled. The problem is encountered as a result of exposure of a steel in the temperature range 345 to 540 °C (650 to 1000 °F). Tempering, postweld heat treatments, or service exposure in this range must be avoided. The problem may be avoided by heat treating above this range followed by rapid cooling. Unfortunately, in the case of massive components such as rotors, no rate of cooling is fast enough and some residual embrittlement may be inevitable. Subsequent to heat treatment, exposure of the component during service in the critical range can also lead to embrittlement.

Many steel components in a plant invariably are exposed to the critical temperature range during service and hence embrittlement cannot be avoided. Some examples are boiler headers, steam pipes, turbine casings, pressure vessels, blades, HP-IP rotors, and combustion turbine disks. For other components operating at lower temperatures, embrittlement from the heat treatment cycle can play a role. LP rotors of steam turbines, generator rotors, and retaining rings provide some examples of this.

Manifestations and Relevance of Temper Embrittlement

Since embrittlement is related to changes at the grain boundaries, it is always manifested as intergranular fracture. A typical intergranular fracture due to temper em-

Fig. 2.21. Intergranular fracture produced by temper embrittlement in a Ni-Cr-Mo-V steel. (50×; shown here at 80%)

brittlement of Ni-Cr-Mo-V steel is shown in Fig. 2.21. In general, the tensile strength and ductility remain unaffected. Under extremely severe conditions, embrittlement can be detected as a reduction in tensile strength and ductility.

Temper embrittlement manifests itself as a shift of the impact transition curve to the right, as shown in Fig. 2.22 (Ref 64), so that the FATT for the steel is increased. This is sometimes, but not always, accompanied by a lowering of the upper-shelf energy. Since FATT is related to K_{Ic} as described earlier, the increase in FATT is accompanied by a reduction of K_{Ic} in the transition-temperature region—i.e., the K_{Ic}-vs-temperature curve is also shifted to the right, analogous to the FATT curve shown in Fig. 2.22. This means that the tolerable crack size for a given stress level is reduced, thus detrimentally affecting component life and integrity under both base load and cycling conditions. This is illustrated in Fig. 2.23 for the case of a hydrodesulfurizer unit which had failed after 3½ years of service. Temper embrittlement had reduced the tolerable crack size at 10 °C (50 °F) from 86 to 18 mm (3.4 to 0.7 in.) (Ref 65).

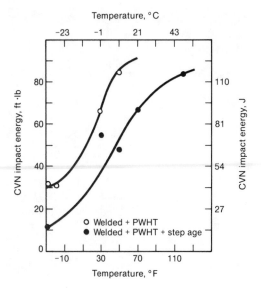

Fig. 2.22. Shift in transition curve due to temper embrittlement in a weld deposit (Ref 64).

The efficiency of a plant is sometimes limited due to temper embrittlement. LP rotors which are made of Ni-Cr-Mo-V steels, and hence are susceptible to embrit-

tlement, are operated below about 370 °C (700 °F), mainly to avoid this problem. If the temperature in the LP turbine could be increased, improved efficiency would result.

In the case of components which are subject to embrittlement such as HP/IP rotors and pressure vessels, restrictions are imposed on the start-up and shutdown procedures. To avoid the risk of brittle failure during these transients, loading is avoided until a certain temperature has been reached. For instance, rotors are prewarmed up to a certain temperature before loading. Similarly, pressure vessels may sometimes need to be depressurized during shutdown prior to reaching a certain temperature. These requirements result in additional operational and maintenance costs and loss of production. Temper-embrittlement phenomena thus adversely affect the longevity, reliability, cycling ability, efficiency, and operating costs of high-temperature equipment. An excellent review of this phenomenon and its characteristics has been published by McMahon (Ref 66).

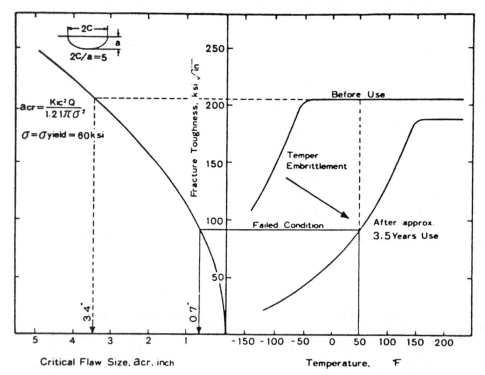

Fig. 2.23. Decrease of critical flaw size for brittle fracture of a 2¼ Cr-1Mo reactor vessel at 10 °C (50 °F) due to temper embrittlement (Ref 65).

Causes of Temper Embrittlement

It is well established at this time that segregation of tramp elements — antimony (Sb), phosphorus (P), tin (Sn), and arsenic (As) — to prior austenite grain boundaries in steel is the principal cause of temper embrittlement. Until the advent of the Auger Electron Spectroscope (AES) in the mid-1960's, no conclusive evidence of such segregation could be obtained. Harris (Ref 67) and Palmberg and Marcus (Ref 68) were the pioneers in demonstrating the segregation of antimony and phosphorus to grain boundaries using the AES technique. Viswanathan (Ref 69) obtained unambiguous evidence of the segregation of phosphorus to grain boundaries in a Ni-Cr steel. Comparison of the grain-boundary AES spectra from the Ni-Cr steel in the embrittled and nonembrittled conditions is shown in Fig. 2.24 (Ref 69). By comparison of the two spectra, evidence for segregation of phosphorus and tin could be seen. Based on these and numerous subsequent investigations (Ref 66 to 78), the following points are now well established:

1. Segregation of both tramp elements and alloying elements occurs (Ref 69 to 75). The concentrations of the former are much higher, sometimes approaching 200 to 300 times their bulk concentrations.
2. Segregation is usually confined to one to two atomic layers and decays exponentially away from the grain boundaries (Ref 69 to 75).
3. Segregation occurs only in ferrite in the critical region from 315 to 540 °C (600 to 1000 °F) and not during austenitizing treatments (Ref 69, 73, and 74).
4. Segregation is reversible and can be reversed at temperatures above the critical range (Ref 69).
5. The extent of segregation is higher in steel with a tempered-martensite structure compared with a tempered-bainite structure and also increases with increasing strength level or decreasing

tempering temperature (Ref 75 and 76).
6. Segregation occurs preferentially and nonuniformly, presumably because of differences in grain-boundary structure (Ref 77 and 78).

Both electronic and elastic misfit interactions of the tramp-element atoms in the host lattice have been investigated as the driving forces for segregation. The Goldschmidt atomic radii for antimony, tin, phosphorus, and arsenic are, respectively, 0.161, 0.158, 0.109, and 0.125 nm (1.61, 1.58, 1.09, and 1.25 Å), compared with a value of 0.128 nm (1.28 Å) for iron. The misfit strain energy in the lattice will therefore be in the decreasing order Sb, Sn, P, As. This will also be expected to reflect in their solubility in iron in the same order. Seah and Hondros (Ref 79) have shown that the theoretically calculated grain-boundary-enrichment ratio is inversely proportional to the atomic solubility of the element in the parent lattice (see Fig. 2.25). Based on this, one could predict that the tendency for segregation in steel would decrease in the order Sb, P, Sn, As, Ni, Cr. This is borne out by experimental results. Hondros (Ref 80 and 81) also found that the solubility in turn was inversely proportional to the term $d\gamma/dC$ in the Gibbs absorption formula

$$\tau_2 = - \frac{C}{kT} \frac{d\gamma}{dC} \qquad \text{(Eq 2.35)}$$

where τ_2 is the grain-boundary concentration of the impurity element in excess of the bulk concentration C, $d\gamma/dC$ is the reduction in grain-boundary energy with the concentration of the impurity at absolute temperature T, and k is Boltzmann's constant. This equation states that any solute which lowers γ tends at equilibrium to be segregated. This tendency increases as the temperature decreases and as $d\gamma/dC$ increases.

Hondros found that a highly segregating element such as phosphorus had large values of $d\gamma/dC$, whereas elements such as

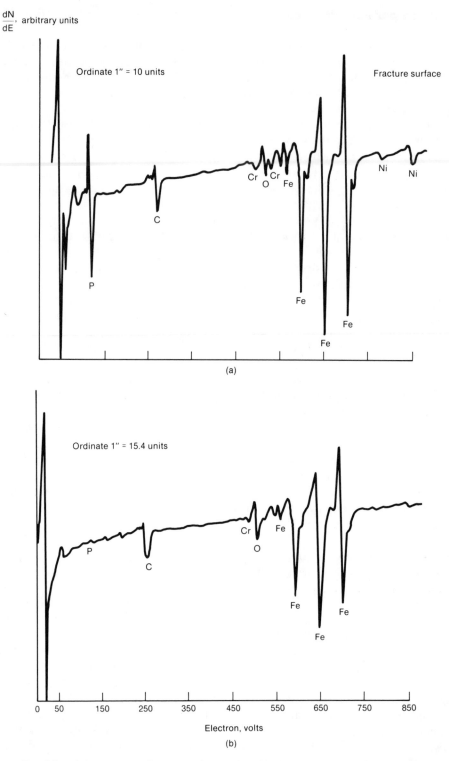

Fig. 2.24. Auger spectra from a Ni-Cr steel in the (a) embrittled and (b) nonembrittled conditions, showing segregation of phosphorus due to embrittlement (Ref 69).

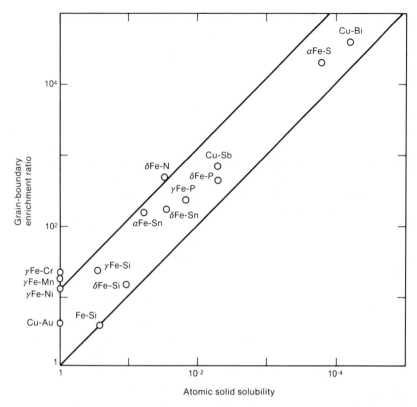

Fig. 2.25. Correlation of predicted grain-boundary-enrichment ratios for various solutes with the inverse of solid solubility (Ref 79).

nickel had much lower values of $d\gamma/dC$. Most experimental results to date indicate that the segregation of phosphorus in steel obeys equilibrium thermodynamics as represented by Eq 2.35 and that it can be explained mainly on the basis of the reduction in grain-boundary energy resulting from segregation.

Example:

(a) A steel contains 300 ppm phosphorus. Given that the grain-boundary energy is lowered by 1190 ergs/cm^2 for each 1 at. % of phosphorus segregated, calculate the Gibbsian segregation:

$$C = 300 \text{ ppm } (0.054 \text{ at. \%})$$

$$k = 1.38 \times 10^{16} \text{ ergs/K per atom}$$

$$d\gamma/dC = 1190 \text{ ergs/cm}^2$$

Substituting these values into Eq 2.35, we get

$$\tau_2 = 0.641 \times 10^{15} \text{ atoms/cm}^2$$

The boundary contains approximately 2.35×10^{15} atoms/cm^2; therefore, τ_2 (phosphorus segregation) = 27.28 at. %.

(b) If segregation of P increases the FATT by 7.3 °F (4.06 °C) per at. %, calculate the ΔFATT due to temper embrittlement:

$$\Delta\text{FATT} = 200 \text{ °F } (111 \text{ °C})$$

A reduced grain-boundary energy implies reduced fracture-surface energy, thus rendering the grain boundaries susceptible to fracture. Whether or not the presence of grain-boundary carbides exacerbates this tendency is not clear. Because commercial steels always contain a network of carbides at the boundaries, the issue is of only academic interest.

Time-Temperature Relationships for Temper Embrittlement

It has been observed by many investigators that temper embrittlement obeys a C-curve behavior in the time-temperature space as indicated in Fig. 2.26 (Ref 82). At high temperatures, the kinetics of impurity diffusion to grain boundaries are rapid, but the tendency to segregate is low because the matrix solubility for the element increases with temperature. Hence, embrittlement occurs rapidly but to a small degree. At low temperatures, the tendency to segregate is high, but the diffusion kinetics are not rapid enough to reach maximum embrittlement. The optimum combination of thermodynamic and kinetic factors favoring embrittlement occurs at some intermediate temperature, called the "knee" of the C-curve. For commercial steels of interest, the knee occurs in the temperature range from 455 to 510 °C (850 to 950 °F) but can be shifted up or down depending on the composition, grain size, and microstructure of the steel.

In addition to the competing thermodynamic and kinetic factors, instability of the microstructure complicates the picture fur-

ther. With increasing exposure, the carbide structures and compositions evolve into more stable configurations with concomitant changes in the ferrite matrix. It is therefore difficult to represent temper-embrittlement kinetics in terms of rate processes with unique activation energies. Attempts to predict long-time behavior based on short-term evaluations have met with little success. For practical purposes, however, it is not uncommon to assume parabolic kinetics, under isothermal conditions.

In the laboratory studies, a step-cooling treatment consisting of 15 h at 540 °C (1000 °F), 24 h at 525 °C (975 °F), 48 h at 495 °C (925 °F), 72 h at 470 °C (875 °F), furnace cooling to 315 °C (600 °F), and air cooling to room temperature is employed to obtain accelerated embrittlement (Ref 83). Other modifications of this treatment have also been employed to maximize the embrittlement in laboratory studies (Ref 84). These treatments are designed to combine the favorable kinetics of embrittlement at high temperatures with the favorable thermodynamics of segregation at lower temperatures. This generally results in ΔFATT values comparable to those obtained isothermally at the knee of the C-curve in about 10,000 to 20,000 h in the case of rotor

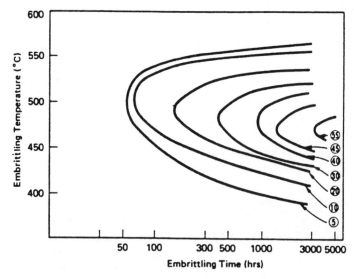

Fig. 2.26. C-curve behavior showing isothermal ΔFATT contours for a 2¼Cr-1Mo steel (Ref 82).

steels. Shaw (Ref 84) employed a slightly modified step-cooling treatment and found the following correlation for 2¼Cr-1Mo pressure-vessel steels:

$$I(t) = 0.67[\log t - 0.91]SCE \quad (Eq\ 2.36)$$

where I(t) is the isothermal embrittlement measured by ΔFATT in °F, SCE is the step-cooled embrittlement measured by ΔFATT, and t is the time in hours. According to this equation, a 30-year isothermal embrittlement could be estimated as three times the step-cooled embrittlement.

Effect of Composition on Temper Embrittlement

As described earlier, the impurity elements antimony, phosphorus, tin, and arsenic are the major contributing causes of temper embrittlement, in decreasing order of efficacy as listed. Antimony is generally not present in large quantities in commercial steels and is therefore neglected from consideration. Arsenic is not a potent embrittler and hence is not very important. Phosphorus and tin are, therefore, the major residual elements of concern.

Among the alloying elements, manganese, silicon, nickel, and chromium are known to exacerbate the effects of impurities. When these elements are present in combination, the effect is further increased. It is well known that nickel and chromium in combination increase embrittlement significantly more than either element alone. For this reason, Ni-Cr-Mo-V steel rotors are considered to be much more susceptible to embrittlement than Cr-Mo-V steel rotors, as shown in Fig. 2.27 (Ref 85). Within a given class of steels, manganese and silicon have the major influence, as may be seen in Fig. 2.28 (Ref 86 and 87). The data show considerable synergism among manganese, silicon, phosphorus, and tin. The maximum embrittlement is observed when all these elements are present together. Because in the old days sulfur levels could not be minimized, the presence of manganese was always necessary for sulfur control. Silicon was generally added for deoxidation. In

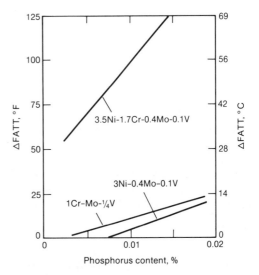

Fig. 2.27. Effect of phosphorus content on the temper embrittlement (ΔFATT) of three step-cooled forging steels (Ref 85).

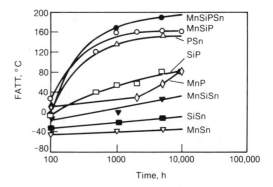

Fig. 2.28. Effects of manganese, silicon, phosphorus, and tin on the kinetics of temper embrittlement at 480 °C (895 °F) for a 2¼Cr-1Mo steel (Ref 86 and 87).

modern practice, silicon can be eliminated by replacing it as a deoxidant by alternate deoxidation processes such as vacuum carbon deoxidation (VCD). Manganese levels can be reduced commensurate with lower sulfur levels. Control of phosphorus and tin to much lower levels can be achieved by careful selection of scrap iron and better steelmaking practices. A combination of all these improvements has brought the temper-embrittlement problem under greater control in recent years.

Various compositional factors for prediction of temper-embrittlement susceptibilities have evolved over the years (Ref 88 and 89). The most commonly used today is the J-factor proposed by Watanabe *et al* (not to be confused with the J-integral discussed earlier). Such correlations between the J-factor—i.e., (Si+Mn) or (P+Sn)—and embrittlement susceptibility for rotor steels and pressure-vessel steels are illustrated in later chapters. It has become common practice to specify an upper limit for the J-factor in purchasing specifications for steels.

The effect of molybdenum in steels has been controversial, with some investigators

claiming beneficial effects and others claiming detrimental effects. Shaw *et al* (Ref 90) have clearly shown that the effect of molybdenum is a function of the Cr/Mo ratio since this ratio determines the extent of molybdenum participation in the carbides. The combined effect of chromium and molybdenum on ΔFATT is shown in Fig. 2.29 for a Ni-Cr-Mo-V steel (Ref 90). Because within a given class of commercial steel, molybdenum levels do not vary widely, information of the type shown in the figure is primarily useful for purposes of alloy

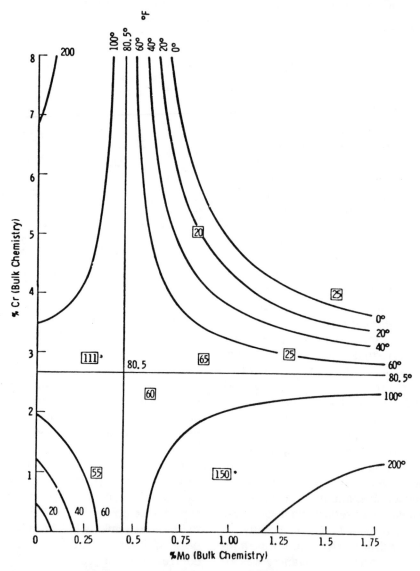

Fig. 2.29. Iso-ΔFATT curves as functions of molybdenum and chromium contents for a Ni-Cr-Mo-V steel doped with 200 ppm of phosphorus and tin (Ref 90).

Table 2.2. Role of composition in temper embrittlement of steels (Ref 91)

Important impurities

Tin, phosphorus . . . . Ni-Cr-base steels (e.g., 3.5Ni-Cr-Mo-V)
Phosphorus. . . . . Cr-Mo-base steels (e.g., 2¼Cr-1Mo; Cr-Mo-V)

Major effects of alloying elements

Nickel . . . . Raises inherent resistance of steel to brittle fracture; promotes segregation of tin and silicon (and antimony, if present)
Chromium . . . . Imparts hardenability; imparts some resistance to softening at elevated temperatures; promotes segregation of phosphorus
Manganese . . . . Imparts hardenability; scavenges sulfur; promotes segregation of phosphorus
Silicon . . . . Deoxidizes; promotes segregation of phosphorus
Molybdenum . . . . Imparts (bainitic) hardenability; imparts resistance to softening; scavenges phosphorus and tin
Vanadium . . . . Imparts resistance to softening; aids in grain refinement
Niobium . . . . Imparts resistance to softening; scavenges phosphorus; aids in grain refinement

development. The effects of various alloying elements and their potential roles in temper embrittlement are summarized in Table 2.2 (Ref 91).

Effects of Microstructural Factors

In heavy-section components such as turbine rotors, casings, and steam chests, inhomogeneities in chemical composition as well as thermal gradients during heat treatment often result in nonuniformities in the microstructure. Hence, the risk of temper embrittlement can vary with location in the component. There have been few systematic studies of these effects.

The effects of strength level and transformation products on the FATT due to temper embrittlement of a 1Cr-1Mo-25V steel are shown in Fig. 2.30 (Ref 75 and 76). The ferrite-pearlite structure showed the least susceptibility to embrittlement, followed by bainite and martensite in increasing order of susceptibility. Increases in strength level (hardness) resulted in increased susceptibility. The ΔFATT results were found to be consistent with the results of Auger analysis, which showed a higher degree of phosphorus and tin segregation in martensite than in bainite, the segregation increasing with increasing strength level. The over-all trend of increasing embrittlement with increasing strength level and with the higher transformation product is borne out by evidence, although fragmentary, on other steels.

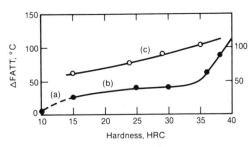

(a) Ferrite-pearlite. (b) Bainite. (c) Martensite.

Fig. 2.30. Correlation between ΔFATT and hardness for Cr-Mo-V steels (Ref 75 and 76).

Another important variable affecting susceptibility to temper embrittlement is grain size. Unfortunately, it is difficult to isolate grain-size effects, because variations in grain size often result in other microstructural changes due to the effect of grain size on hardenability. Over-all results suggest that fine-grain-size steels are less susceptible to embrittlement, as shown in Fig. 2.31 (Ref 85). In a fine-grain steel, there is a larger grain-boundary area per unit volume of the material over which the impurity segregation is distributed. Hence, for a steel of given impurity concentration, the grain-boundary segregation is expected to be less for a fine-grain material. Further, the frequent changes in grain-boundary orientations necessitate more deflections and hence more energy required for a propagating crack.

McMahon *et al* have combined the ef-

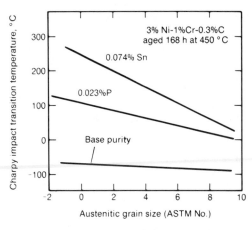

Fig. 2.31. Variation of FATT with austenitic grain size at fixed hardness and impurity levels (Ref 85).

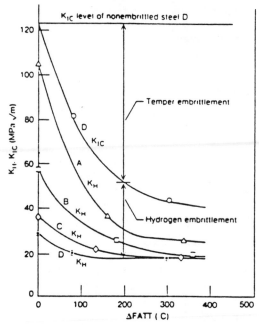

A, B, C, and D denote steels with yield strengths of 862, 986, 1090, and 1175 MPa (125, 143, 158, and 170 ksi), respectively.

Fig. 2.32. Effects of prior temper embrittlement and yield strength on the K_H value of a Ni-Cr-Mo-V steel (Ref 93).

fects of grain size, hardness, and grain-boundary segregation in the form of an embrittlement equation. The embrittlement-equation approach is applicable to laboratory Ni-Cr steels as well as to commercial Ni-Cr-Mo-V rotor steels (Ref 92). Increases in FATT values predicted by the equation agree closely with the experimentally determined values. The approach as applicable to rotor steels is described in a later chapter.

Effects of Temper Embrittlement on Other Properties

Several studies completed in recent years have shown that the problem of hydrogen embrittlement of high-strength low-alloy steels can be exacerbated by temper embrittlement (Ref 93 to 95). Evidence has been obtained in a variety of steels that the threshold stress intensity for cracking in hydrogen can be appreciably reduced if the steel has been subjected to prior temper embrittlement. Figure 2.32 illustrates the combined effect of yield strength and temper embrittlement (FATT) on the K_H value of a Ni-Cr-Mo-V steel (Ref 93). With increasing embrittlement, K_H drops rapidly at first, then levels off. The effect of yield strength is pronounced for small degrees of embrittlement. When severe temper embrittlement has occurred, all K_H values converge to a constant and low value (Ref 93 and 94).

Briant, Feng, and McMahon (Ref 96) investigated crack-growth behavior in high-pressure gaseous hydrogen of HY 130 steel samples aged to various degrees of embrittlement. Their results showed that the stress intensity for crack growth dropped precipitously as impurity concentration and embrittlement increased. Hydrogen-induced intergranular cracking at low levels of stress intensity appeared to be primarily an impurity effect.

Controlled-potential stress-corrosion tests in NaOH using Ni-Cr-Mo-V steels have shown that the threshold stress for cracking is reduced appreciably due to phosphorus (Ref 97). The catastrophic failure of Cr-Mo steel turbine wheels at Hinkley Point was attributed by Kalderon (Ref 98) to stress-corrosion cracking assisted by temper embrittlement.

Hippsley and Bruce (Ref 99 and 100) have shown that during temper embrittlement, general segregation of phosphorus to carbide interfaces can occur, thereby facilitating cavity nucleation. An over-all reduction in the resistance to ductile fracture as measured by upper-shelf energy and the ductile fracture toughness (J_{Ic}) was reported. It has also been reported that temper embrittlement can accelerate near-threshold fatigue-crack growth as well as creep crack growth. The crack-growth effects will be treated separately in later chapters dealing with fatigue and creep.

Failure Analysis of Temper-Embrittled Components

Identification of temper embrittlement as a failure mechanism in a failed component is relatively easy. The presence of large amounts of phosphorus, tin, manganese, and silicon in the steel is an indication that temper embrittlement may be involved. Fractography generally indicates an intergranular fracture with little evidence of ductile dimples on the fracture surface. The extent of the intergranular fracture present is a function of the temperature at which the fracture is produced and the microstructure. It has been observed that a plot of intergranular fracture (%) vs Charpy test temperature for embrittled Cr-Mo-V steels resembles a bell-shape curve with the maximum intergranular fracture being observed at the 50% FATT (Ref 75). For a given degree of embrittlement, more intergranular fracture is observed in martensitic structures than in bainitic structures (Ref 75).

Auger analysis usually is performed on a small sample to look for evidence of phosphorus and tin segregation at the grain boundaries. Frequently, Charpy tests may be conducted from near the failed locations and at other locations which may have operated at lower temperatures. If a higher FATT is encountered in specimens near the failed location, it indicates the occurrence of temper embrittlement. A final test of temper embrittlement consists of comparing the FATT of the as-failed material with the same material after a de-embrittlement heat treatment. If a lower FATT is obtained after de-embrittlement, it proves that a reversible temper-embrittlement mechanism was involved in the failure. Various other techniques for detecting and quantifying temper embrittlement have been explored, including scanning transmission electron microscopy, secondary ion imaging, eddy-current evaluation, electrochemical polarization, and chemical etching (Ref 101 and 102). The more successful among these are described specifically as applicable to steam-turbine rotors in a later chapter.

Control of Temper Embrittlement

It is clear from the discussion so far that reduced strength levels, grain refinement, and avoidance of the temper embrittling temperature range during heat treatment and operation represent some of the means of minimizing embrittlement. The most effective method, however, is reduction of silicon, manganese, phosphorus, and tin contents by improved steelmaking. For many applications, silicon-deoxidized steels have now been replaced by vacuum carbon deoxidation. Phosphorus and tin levels have been brought down to 50 to 100 ppm, compared with the 300 to 500 ppm observed in 1950's-vintage steels. Even further reductions have been shown to be possible, although such low levels have not yet found their way into material specifications for economic reasons.

Nomenclature

a	– Crack length (Eq 2.4)
a_{eff}	– Effective crack length (Eq 2.15)
a_i	– Initial crack length
a_c	– Critical crack length
b	– Uncracked ligament width (Eq 2.18)
k	– Boltzmann's constant
r_p	– Plastic-zone size (Eq 2.13 and 2.14)
u	– Displacement vector (Eq 2.26)
A	– Crack area (Eq 2.1); area under load-displacement curve (Eq 2.29)
B	– Specimen thickness (Eq 2.17)
C	– Concentration of solute (Eq 2.35)
CVN	– Charpy V-notch energy
E	– Young's modulus
FATT	– 50% ductile-brittle fracture-appearance transition temperature
G	– Energy-release rate (Eq 2.3)
G_c	– Critical energy-release rate (Eq 2.5)
J	– J-integral (Eq 2.26)
J_{Ic}	– Critical value of J for crack propagation

J_e, J_p — Elastic and plastic contributions to J

K — Stress-intensity factor (Eq 2.6 to 2.8)

K_{Ic} — Critical value of K in mode 1 loading, also referred to as fracture toughness

$K_{Ic\text{-}US}$ — Value of K_{Ic} at the upper shelf

K_{Id} — Dynamic fracture toughness

K_{Ia} — Crack-arrest fracture toughness

K_{IR} — Reference fracture toughness

M — Parameter related to flaw size and geometry (Eq 2.9)

P — Applied load

T — Traction vector (Eq 2.26)

T_j — Tearing modulus (Eq 2.30)

U — Elastic-strain energy (Eq 2.1)

W — Width of specimen (Eq 2.19); strain-energy density (Eq 2.27)

σ — Stress

σ_y — Yield strength

σ_c — Critical stress for fracture

δ_c — Critical crack-opening displacement

ϵ — Strain

ν — Poisson's ratio

τ_2 — Solute enrichment at grain boundaries (Eq 2.35)

γ — Grain-boundary surface energy (Eq 2.35)

References

1. W.D. Biggs, *The Brittle Fracture of Steel*, Pitman, New York, 1960
2. C.F. Tipper, Testing for Brittleness in Structural Steels, Cambridge Conference on Brittle Fracture, Cambridge, England, Sept 1959
3. E.R. Parker, *Brittle Behavior of Engineering Structures*, John Wiley & Sons, New York, 1957
4. P.P. Puzak, E.W. Eschbacher, and W.S. Pellini, Initiation and Propagation of Brittle Fracture in Structural Steels, *Weld. J. Res. Suppl.*, Vol 31 (No. 12), Dec 1952, p 561 S
5. P.P. Puzak, M.E. Schuster, and W.S. Pellini, Crack Starter Tests of Ship Fracture and Project Steels, *Weld. J. Res. Suppl.*, Vol 33 (No. 10), Oct 1954, p 481 S
6. C.E. Hartbower, Crack Initiation and Propagation in the V-Notch Charpy Impact Specimen, *Weld. J. Res. Suppl.*, Vol 36 (No. 11), Nov 1957, p 494 S
7. T.S. Robertson, Brittle Fracture of Mild Steel, *Engineering* (London), Vol 172, 1951, p 445
8. T.S. Robertson, Propagation of Brittle Fracture in Steel, *J. Iron Steel Inst.*, Vol 175, 1953, p 361
9. T.S. Robertson, D. Hunt, and J.W. Scott, *J. Iron Steel Inst.*, Vol 60, 1953, p 259
10. W.S. Pellini and J.E. Srawley, "Procedures for the Evaluation of Fracture Toughness of Pressure Vessel Materials," U.S. Naval Research Laboratory, Report No. 5609, June 1961
11. F.J. Feeley, D. Hrtko, S.R. Kleppe, and M.S. Northrup, Report on Brittle Fracture Studies, *Weld. J. Res. Suppl.*, Vol 33, 1954, p 99 S
12. M.L. Williams, Analysis of Brittle Behavior in Ship Plate, in *Metallic Materials at Low Temperatures*, STP 158, American Society for Testing and Materials, Philadelphia, 1951, p 11
13. P.P. Puzak, A.J. Babecki, and W.S. Pellini, Correlation of Brittle Fracture Service Failures with Laboratory Nil Ductility Tests, *Weld. J. Res. Suppl.*, Vol 37 (No. 9), Sept 1958, p 391 S
14. W.S. Pellini *et al*, "Review of Concepts and Status of Procedures for Fracture Safe Design of Complex Welded Structures Involving Metals of Low to Ultra High Strength Levels," U.S. Naval Research Laboratory, Report No. 6360, June 1965
15. A.A. Griffith, The Phenomena of Rupture and Flow in Solids, *Phil. Trans. A*, Royal Soc. (London), Vol 221, 1928, p 163-198
16. G.R. Irwin, Fracture Dynamics, in *Fracturing of Metals*, American Society for Metals, Cleveland, 1948, p 147-166
17. E.R. Orowan, Fracture and Strength of Solids, in *Report on Progress in Physics*, Phys. Soc. London, Vol 12, 1949, p 185
18. H.F. Bueckner, The Propagation of Cracks and the Energy of Elastic Deformation, *Trans. ASME, J. Appl. Mech.*, 1958
19. J. Mulherin, D. Armiento, and H. Markus, "The Relation Between Fracture Toughness and Stress Concentration Factors for Several High Strength Aluminum Alloys," Paper No. 63WA306, American Society of Mechanical Engineers, New York, 1963
20. G.R. Irwin, *Hanbuch der Physik*, Vol VI, Springer, Berlin, 1958, p 551
21. P.C. Paris and G. Sih, Stress Analysis of Cracks, in *Fracture Toughness Testing and Its Applications*, STP 381, American Society for Testing and Materials, Philadelphia, 1965
22. G.R. Irwin, Analysis of Stresses and Strains Near the End of a Crack Traversing a Plate, *J. Appl. Mech., Trans. ASME*, 1957
23. G.R. Irwin, Plastic Zone Near a Crack and Fracture Toughness, 1960 Sagamore Ordnance Materials Conference, Syracuse University, 1961
24. W.F. Brown, Jr., and J.E. Srawley (a task group of ASTM Committee E24 SubI), "Plane Strain Fracture Toughness Testing," draft of the 6th committee report, Washington, Feb 1966

25. W.K. Wilson, "Analytic Determination of Stress Intensity Factors for the Manjoine (WOL) Brittle Fracture Test Specimen," AEC Research and Development Report, WERL 0029-3, Westinghouse Research Laboratories, Pittsburgh, Aug 1965

26. G.C. Sih, *Handbook of Stress Intensity Factors for Researchers and Engineers*, Institute of Fracture and Solid Mechanics, Lehigh University, 1973

27. H. Tada, P. Paris, and G. Irwin, *The Stress Analysis of Cracks Handbook*, Del Research Corp., Hellertown, PA, 1971

28. D.P. Rooke and D.J. Cartwright, *Compendium of Stress Intensity Factors*, HMSO (London), 1976

29. Standard Method of Test for Plane Strain Fracture Toughness of Metallic Materials, ASTM E399, American Society for Testing and Materials, Philadelphia, 1985

30. E.T. Wessel, W.G. Clark, and W.K. Wilson, "Engineering Methods for the Design and Selection of Materials Against Brittle Fracture," Final Technical Report, Contract No. DA 30069 AMC602 (T), from U.S. Army Tank and Automotive Center, Warren, MI, report by Westinghouse Research Laboratories, Pittsburgh, June 1966

31. R. Roberts and C. Newton, "Interpretive Report on Small Scale Test Correlations with K_{IC} Data," Bulletin 265, Welding Research Council, New York, Feb 1981

32. *Rapid Inexpensive Tests for Determining Fracture Toughness*, National Academy of Engineering, National Materials Advisory Board, International Standard Book No. 76-39632, National Academy of Sciences, Washington, 1976

33. S.T. Rolfe and S.R. Novak, *Slow Bend K_{IC} Testing Medium Strength, High Toughness Steels*, STP 463, American Society for Testing and Materials, Philadelphia, 1970, p 124

34. R.T. Ault, G.M. Wald, and R.B. Bertola, "Development of an Improved Ultra High Strength Steel for Forged Aircraft Components," AFML TR 7127, Air Force Materials Laboratory, Wright-Patterson Air Force Base, OH, 1971

35. T. Iwadate, T. Karushi, and J. Watanabe, Prediction of Fracture Toughness K_{IC} of 2¼Cr-1Mo Pressure Vessel Steel From Charpy V Notch Test Results, in *Flaw Growth and Fracture*, STP 631, American Society for Testing and Materials, Philadelphia, 1977, p 493-506

36. T. Iwadate, J. Watanabe, and Y. Tanaka, Prediction of the Remaining Life of High Temperature/Pressure Reactors Made of Cr-Mo Steels, *Trans. ASME, J. Pressure Vessel Tech.*, Vol 107, Aug 1985, p 230-238

37. K.F. Stahlkopf and T.U. Marston, A Comprehensive Approach to Radiation Embrittlement Analysis, IAEA Meeting on Irradiation Embrittlement, Thermal Annealing and Surveillance of Reactor Pressure Vessels, Vienna, Austria, Feb 26-28, 1979 (cited in Ref 48)

38. W.L. Server and W. Oldfield, "Nuclear Pressure Vessel Steel Data Base," EPRI-933, Electric Power Research Institute, Palo Alto, CA, Dec 1978

39. J.A. Begley and W.A. Logsdon, "Correlation of Fracture Toughness and Charpy Properties of Rotor Steels," Scientific Paper 71-1E7-MSLRF, Westinghouse Research Laboratories, Pittsburgh, 1971

40. H.D. Greenberg, E.T. Wessel, and W.H. Pryle, Fracture Toughness of Turbine Generator Rotor Forgings, *Engg. Fract. Mech.*, Vol 1, 1970, p 653

41. J.M. Barsom, Development of the AASHTO Fracture Toughness Requirements for Bridge Steels, *Engg. Fract. Mech.*, Vol 7, 1975, p 605-618

42. PVRC Ad Hoc Group on Toughness Requirements, "PVRC Recommendations on Toughness Requirements for Ferritic Materials," Bulletin 175, Welding Research Council, Aug 1972

43. S. Yukawa, ASME Nuclear Code Applications of Structural Integrity and Flaw Evaluation Methodology, in *Structural Integrity Technology*, J.P. Gallagher and T.W. Crooker, Ed., American Society of Mechanical Engineers, New York, 1979

44. T.U. Marston, Ed., "Flaw Evaluation Procedures: ASME Section XI," EPRI NP719-SR, Electric Power Research Institute, Palo Alto, CA, 1978

45. T.R. Mager, Ed., *Reference Fracture Toughness Procedures Applied to Pressure Vessel Materials*, proceedings of a conference held in New Orleans, Dec 9-14, 1984, MPC 24, American Society of Mechanical Engineers, New York, 1984

46. J.R. Rice, A Path Independent Integral and the Approximate Analysis of Strain Concentration by Notches and Cracks, *J. Appl. Mech., Trans. ASME*, Vol 35, 1968, p 379-386

47. T.U. Marston, Ed., "EPRI Ductile Fracture Research Review Document," EPRI NP 701 SR, Electric Power Research Institute, Palo Alto, CA, Feb 1978

48. J.T.A. Roberts, *Structural Materials for Nuclear Power Systems*, Plenum Press, New York, 1981

49. J.D. Landes and J.A. Begley, Recent Developments in J_{Ic} Testing, in *Fracture Mechanics Test Methods Standardization*, STP 632, W.F. Brown, Jr., and J.G. Kaufman, Ed., American Society for Testing and Materials, Philadelphia, 1977, p 57-81

50. P.C. Paris, H. Tada, A. Zahoor, and H. Ernst, in *Elastic Plastic Fracture*, STP 668, J.D.

Landes, J.A. Begley, and G.A. Clarke, Ed., American Society for Testing and Materials, Philadelphia, 1979, p 5-36

51. J.D. Landes, "Future Directions for Ductile Fracture Mechanics," Scientific Paper 81-1D7-JINTF-P4, Westinghouse Research Laboratories, Pittsburgh, Aug 1981

52. J.R. Rice, in *Fracture*, Vol II, Academic Press, New York, 1968

53. R.J. Bucci *et al*, *J Integral Estimation Procedures*, STP 514, American Society for Testing and Materials, Philadelphia, 1972, p 40-69

54. J.A. Begley and J.D. Landes, The J Integral as a Fracture Criterion, in *Fracture Toughness*, Proceedings of the 1971 National Symposium on Fracture Mechanics, Part II, STP 514, American Society for Testing and Materials, Philadelphia, 1972, p 1-20

55. J.R. Rice, P.C. Paris, and J.G. Merkle, Some Further Results on J Integral and Estimates, in *Progress in Flaw Growth and Fracture Toughness Testing*, STP 536, American Society for Testing and Materials, Philadelphia, 1973

56. J.D. Landes and J.A. Begley, *Test Results From J Integral Studies: An Attempt to Establish a J_{Ic} Testing Procedure*, STP 560, American Society for Testing and Materials, Philadelphia, 1974, p 170-186

57. Standard Test Method for J_{Ic}, A Measure of Fracture Toughness, ASTM E813, American Society for Testing and Materials, Philadelphia, 1985

58. A. Saxena, P.K. Liaw, and W.A. Logsdon, Residual Life Prediction and Retirement for Cause Criteria for SSTG Casings–II, Fracture Mechanics Analysis, *Engg. Fract. Mech.*, Vol 25 (No. 3), 1986, p 289-303

59. S.A. Parcinjbe and S. Banerjee, Interrelation of Crack Opening Displacement and J Integral, *Engg. Fract. Mech.*, Vol 11, 1979, p 43-53

60. V. Kumar, M.D. German, and C.F. Shih, "Elastic Plastic Fracture Analysis," EPRI Report NP 1931, Electric Power Research Institute, Palo Alto, CA, 1981

61. J.W. Hutchinson and P.C. Paris, The Theory of Stability Analysis of J Controlled Crack Growth, presented at the 1977 ASTM Symposium on Elastic Plastic Fracture, STP 668, American Society for Testing and Materials, Philadelphia, 1979, p 37-64

62. J.R. Rice, Elastic Plastic Models for Stable Crack Growth, in *Mechanics and Mechanisms of Crack Growth*, M.J. May, Ed., Proceedings of Cambridge Conference, England, Apr 1973, Physical Metallurgy Centre Publication, 1975, p 14-39

63. T.U. Marston, *The EPRI Ductile Fracture Research Program*, Proceedings of the Seminar on Fracture Mechanics, ISPRA, Italy, Apr 2-6, 1979

64. R.F. Bruscato, "High Temperature Embrittlement Phenomena of 2¼Cr-1Mo Weldments," Paper 71-PET 19, ASME Joint Meeting of the Petroleum Mechanical Engineering and Underwater Technology Congress, Houston, Sept 1971

65. J. Watanabe and Y. Murakami, "Prevention of Temper Embrittlement of Chromium-Molybdenum Steel Vessels by Use of Low Silicon Forged Shells," Japan Steel Works, Ltd., Muroran, Japan, Preprint 28-81, 1981

66. C.J. McMahon, Temper Embrittlement – An Interpretive Review, in *Temper Embrittlement in Steel*, STP 407, American Society for Testing and Materials, Philadelphia, 1968, p 127-167

67. L.A. Harris, *J. Appl. Phys.*, Vol 39, 1968, p 1419-1435

68. P.W. Palmberg and H.L. Marcus, *Trans. ASM*, Vol 62, 1969, p 1016

69. R. Viswanathan, Temper Embrittlement in a Ni-Cr Steel Containing Phosphorus as Impurity, *Met. Trans.*, Vol 2, Mar 1971, p 809-815

70. R. Viswanathan and T.P. Sherlock, Long Time Isothermal Temper Embrittlement in NiCrMoV Steels, *Met. Trans.*, Vol 3, Feb 1972, p 459-467

71. D.F. Stein, A. Joshi, and R.P. Laforce, Studies Utilizing Auger Electron Emission Spectroscopy on Temper Embrittlement in Low Alloy Steel, *Trans. ASM*, Vol 62, 1969, p 776-782

72. A.K. Cianelli, H.C. Feng, A.H. Ucsik, and C.J. McMahon, Jr., Temper Embrittlement of NiCr Steel by Sn, *Met. Trans. A*, Vol 8A, July 1977, p 1059-1061

73. C.L. Smith and J.R. Low, Effect of Prior Austenite Grain Boundary Composition on Temper Brittleness in a NiCr Steel, *Met. Trans.*, Vol 5, Jan 1974, p 279-287

74. A. Joshi and D.F. Stein, *Temper Embrittlement of Low Alloy Steels*, STP 499, American Society for Testing and Materials, Philadelphia, 1972

75. R. Viswanathan and A. Joshi, The Effect of Microstructure on the Temper Embrittlement of CrMoV Steels, *Met. Trans. A*, Vol 6A, 1975, p 2289-2297

76. R. Viswanathan, *Influence of Microstructure on the Temper Embrittlement of Some Low Alloy Steels*, STP 672, MICON 1979, American Society for Testing and Materials, Philadelphia, 1979, p 169-183

77. A. Joshi, *Scripta Met.*, Vol 9 (No. 3), 1975, p 251

78. R. Viswanathan, S.M. Breummer, and R.H. Richman, Etching Technique for Assessing Toughness Degradation of In-Service Compo-

nents, submitted to *ASME J. Engg. Mater. Tech.*, 1987

79. M.P. Seah and E.D. Hondros, Grain Boundary Segregation, *Proc. Royal Soc.* (London), Vol A 335, 1973, p 191-212

80. E.D. Hondros and D. McClean, Monograph No. 28, *Soc. Chem. Ind.* (London), 1968, p 39

81. E.D. Hondros, *Proc. Royal Soc.* (London), Vol A 286 (No. 479), 1965

82. I. Masaoka, I. Takase, S. Ikeda, and R. Sasaki, *J. Japan Weld. Soc.*, Vol 46 (No. 11), 1977, p 818

83. D.L. Newhouse *et al*, Temper Embrittlement Study of NiCrMoV Steels–Impurity Effects, Part I, in *Temper Embrittlement of Alloy Steels*, STP 499, American Society for Testing and Materials, Philadelphia, 1972, p 3-36

84. B.J. Shaw, Characterization Study of Temper Embrittlement of Cr-Mo Steels, Preprint 29-81, presented at the 46th Mid-Year Regional Meeting of the API, Chicago, American Petroleum Institute, Washington, May 1981

85. C.J. McMahon, Jr., Problems of Alloy Design in Pressure Vessel Steels, in *Fundamental Aspects of Structural Alloy Design*, R.I. Jaffee and B.A. Wilcox, Ed., McGraw-Hill, New York, 1977, p 295-322

86. C.J. McMahon *et al*, The Effect of Composition and Microstructure on Temper Embrittlement in 2¼Cr-1Mo Steels, *ASME J. Engg. Mater. Tech.*, Vol 102, 1980, p 369

87. R. Viswanathan and R.I. Jaffee, 2¼Cr-1Mo Steels for Coal Conversion Pressure Vessels, *ASME J. Engg. Mater. Tech.*, Vol 104, July 1982, p 220-226

88. J. Watanabe *et al*, Temper Embrittlement of 2¼Cr-1Mo Pressure Vessel Steel, presented at ASME 29th Petroleum Mechanical Engineering Congress, Dallas, Sept 15-18, 1974

89. R. Bruscato, *Weld. Res. Suppl.*, Vol 49, 1973, p 1485

90. B.J. Shaw, "The Effect of Composition on Temper Embrittlement of Low Carbon Rotor Steels," Scientific Paper 77-1D9-GRABO-P1, Westinghouse Research Laboratories, Pittsburgh, 1977

91. C.J. McMahon *et al*, "The Elimination of Impurity Induced Embrittlement in Steels, Part I," EPRI Report NP 1501, Electric Power Research Institute, Palo Alto, CA, Sept 1980

92. C.J. McMahon *et al*, "Impurity Induced Embrittlement of Rotor Steel," EPRI Report CS 3248, Vol 1, Electric Power Research Institute, Palo Alto, CA, Nov 1983

93. R. Viswanathan and S.J. Hudak, Jr., The Effect of Impurities and Strength Level on Hydrogen Induced Cracking in a Low Alloy Turbine Steel, *Met. Trans.*, Vol 8A, Oct 1977, p 1633-1637

94. R. Viswanathan and S.J. Hudak, Jr., The Effect of Impurities and Strength Level on Hydrogen Induced Cracking in 4340 Steels, in *Effect of Hydrogen on Behavior of Materials*, A.W. Thompson and J.M. Bernstein, Ed., The Metallurgical Society of AIME, 1975, p 262

95. K. Yoshino and C.J. McMahon, *Met. Trans.*, Vol 5, 1974, p 363

96. C.L. Briant, H.C. Feng, and C.J. McMahon, Jr., Embrittlement of a 5% Ni High Strength Steel by Impurities and Their Effects on Hydrogen-Induced Cracking, *Met. Trans.*, Vol 9A, May 1978, p 625-633

97. C.L. Briant, unpublished work, General Electric Co., Schenectady, NY, 1982

98. D. Kalderon, Steam Turbine Failure at Hinkley Point–A, *Proc. Inst. of Mechanical Engineers*, England, Vol 186, 1972, p 31-72

99. C.A. Hippsley and S.G. Bruce, The Influence of Phosphorus Segregation of Particle/Matrix Interface on Ductile Fracture in a High Strength Steel, *Acta Met.*, Vol 3 (No. 11), 1983, p 1861-1872

100. C.A. Hippsley and S.G. Bruce, The Influence of Strength and Phosphorus Segregation on the Ductile Fracture Mechanism in a Ni-Cr Steel, *Acta Met.*, Vol 34 (No. 7), 1986, p 1215-1227

101. M. Bruemmer *et al*, "Grain Boundary Composition and Intergranular Fracture of Steels," EPRI Report RD 3859, Electric Power Research Institute, Palo Alto, CA, Jan 1985

102. R. Viswanathan and S.M. Bruemmer, In-Service Degradation of Toughness of Steam Turbine Rotors, *ASME J. Engg. Mater. Tech.*, Vol 107, Oct 1985, p 316-324

3

Creep

One of the most critical factors determining the integrity of elevated-temperature components is their creep behavior. Due to thermal activation, materials can slowly and continuously deform even under constant load (stress) and eventually fail. The time-dependent, thermally assisted deformation of components under load (stress) is known as creep. As a consequence of such deformation, unacceptable dimensional changes and distortions as well as final rupture of the component can occur. Depending on the component, the final failure may be limited either by deformation or by fracture. Local creep processes at the tip of a pre-existing defect or stress concentration can also lead to local crack growth and eventual failure. Numerous texts deal in detail with the extensive body of literature pertaining to the phenomenology, mechanisms, and constitutive relationships for creep and creep fracture (Ref 1 to 3). The intent of this chapter is to introduce only the most necessary concepts from an engineering standpoint.

Creep Curves: Basic Concepts

Creep properties are generally determined by means of a test in which a constant uniaxial load or stress is applied to the specimen and the resulting strain is recorded as a function of time. Typical shapes of creep curves are shown in Fig. 3.1. After the instantaneous strain, ϵ_0, a decelerating strain-rate stage (primary creep) leads to a steady minimum creep rate, $\dot{\epsilon}$ (secondary creep), which is finally followed by an accelerating stage (tertiary creep) that ends in fracture at a rupture time, t_r. The strain at rupture, ϵ_r, represents the rupture ductility.

Creep-Curve Shapes

The shape of the creep curve is determined by several competing reactions, including (1) strain hardening; (2) softening processes such as recovery, recrystallization, strain softening, and precipitate overaging; and (3) damage processes such as cavitation and cracking, and specimen necking. Of these factors, strain hardening tends to decrease

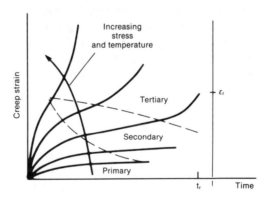

Fig. 3.1. Schematic illustration of creep-curve shapes.

the creep rate whereas the other factors tend to increase the creep rate. The balance among these factors determines the shape of the creep curve. During primary creep, the decreasing slope of the creep curve is attributed to strain hardening. Secondary-stage creep is explained in terms of a balance between strain hardening and the softening and damage processes, resulting in a nearly constant creep rate. The tertiary stage marks the onset of internal- or external-damage processes (item 3), which result in a decrease in the resistance to load or a significant increase in the net section stress. Coupled with the softening processes (item 2), the balance achieved in stage 2 is now offset, and a rapidly increasing tertiary stage of creep is reached.

If we consider a low-temperature creep test in which the recovery and softening processes (item 2) are not operative, the balance between strain hardening (item 1) and the damage processes (item 3) determines the shape of the creep curve. Under constant-load conditions, the primary stage of strain hardening, the secondary stage in which the strain hardening is balanced by higher creep rates due to increased net section stress, and a tertiary accelerating stage will all be present. Under constant-stress conditions, however, the second stage of creep may never be reached. The primary stage may last throughout, resulting in eventual transgranular fracture.

In creep tests in which one or more of the recovery and softening processes become important, all three of the reactions described above determine together the shape of the creep curve. In addition, grain-boundary sliding and intercrystalline fracture may also begin to contribute and accelerate the creep rate and promote the third stage.

In general, the higher the creep strength of a material, the lower its ductility. For instance, if we compare a material of very high creep strength (a brittle material) with the same material in a lower-strength (more-ductile) condition, using a constant-load test, the following differences may be observed. The "ductile" material will be characterized by a higher second-stage creep rate, a shorter time to rupture, and a higher ductility at rupture similar to the high-temperature end of the spectrum of curves shown in Fig. 3.1. A "creep-brittle" material, on the other hand, is usually characterized by higher creep strength, a second stage with a lower creep rate, and a lower-ductility fracture that occurs immediately upon the onset of the tertiary stage. The shape of the curve will be similar to those of the curves at the low-temperature (or low-stress) end of the spectrum in Fig. 3.1. In the latter case, constant-load tests and constant-stress tests lead to nearly identical creep curves.

Variations in the shapes of creep curves are also caused by changes in test temperatures and stresses, as illustrated in Fig. 3.1. Higher temperatures and stresses reduce the extent of the primary stage and practically eliminate the second stage, with the result that the creep rate accelerates almost from the beginning. With decreasing temperatures and stresses, the first two stages become clearly defined, usually at the expense of the tertiary stage. The total elongation at rupture may be found to decrease with decreasing stress and temperature.

In principle, a constant-stress test could be much more meaningful than a constant-load test. A major limitation of the constant-stress test, however, is that in such tests it is difficult to reduce the load commensurate with the progress of straining. Furthermore, in reducing the load, to keep the stress constant at a neck, the stress at other points along the gage length of the specimen is reduced. Under these conditions, the strain measured over the gage length of the specimen is no longer a representative strain. Because the neck can act as a stress concentrator, the strain at the neck is also not simply related to the applied stress. The constant-stress test therefore may be more meaningful when specimen elongation occurs uniformly rather than locally by necking. Constant-stress tests are more difficult to conduct, and hence the database of information from these tests available to the engineer is less extensive than that from constant-load tests. Consequently, for almost all engineering applications, constant-load creep-test data are considered adequate because of the convenience of constant-load testing.

Creep-Curve Descriptions

Various equations, apparently related to each other, describing primary and secondary creep have been published, as follows (Eq 3.1, 3.2, and 3.3 are from, respectively, Ref 4, Ref 1 and 5, and Ref 6 and 7):

$$\epsilon = \epsilon_0 + C_1 t^{1/3} + \dot{\epsilon}t \qquad \text{(Eq 3.1)}$$

$$\epsilon = \epsilon_0 + C_2[1 - \exp(-C_3 t)] + \dot{\epsilon}t \qquad \text{(Eq 3.2)}$$

$$\epsilon = \epsilon_0 + \frac{\dot{\epsilon}}{C_4} \ln\{1 + C_5[1 - \exp(-C_4 t)]\} + \dot{\epsilon}t$$

$$\text{(Eq 3.3)}$$

where the three terms on the right-hand side of each equation denote the contributions due to instantaneous, primary, and secondary strains, respectively, and C_1, C_2, C_3, C_4, and C_5 are empirical constants. For an approximate description of the primary creep strain, these equations are practically equivalent (Ref 8).

The accelerating strain component ϵ_t in the tertiary stage has been proposed (Ref 9) to be represented as

$$\epsilon_t \propto \exp[C_6(t - t_3)] \qquad \text{(Eq 3.4)}$$

and the tertiary creep rate is expressed as (Ref 10)

$$\dot{\epsilon}_t \propto \frac{1}{(1 - D)^{C_7}} \qquad \text{(Eq 3.5)}$$

where C_6 and C_7 are constants, t_3 is the time to the onset of tertiary creep, and D is an arbitrary damage parameter that has values of $D = 0$ at $t = 0$ and $D = 1$ at $t = t_r$.

The initial creep rate, the shape and duration of the primary stage, and the steady creep rate are all interrelated and hence governed by the same mechanisms (Ref 3).

A very general description of the creep curve under constant-stress conditions is given by the "θ" projection concept put forward by Evans, Parker, and Wilshire (Ref 11), in which creep strain, ϵ, is considered to be the sum of two competing processes using the equation

$$\epsilon = \theta_1[1 - \exp(-\theta_2 t)] + \theta_3[\exp(\theta_4 t) - 1]$$

$$\text{(Eq 3.6)}$$

In this expression, θ_1, θ_2, θ_3, and θ_4 are all experimentally determined constants which are functions of stress and temperature: θ_1 and θ_2 define the primary or decaying strain-rate component, and θ_3 and θ_4 describe the tertiary or accelerating strain-rate component. The absence of a steady second-stage creep rate is implied by the model. A wide range of creep-curve shapes can be modeled with various combinations of the constants. Analysis of extensive creep data on ferritic steels by the above authors has shown that the log θ values vary systematically and linearly with stress. Hence, for a given material, if the θ functions can be defined on the basis of short-time tests at high stresses, then the values at lower stresses (longer times) can be obtained by extrapolation and the long-time creep curves under low-stress conditions can be readily predicted by substituting the θ values in Eq 3.6.

Stress and Temperature Dependence

Of all the parameters pertaining to the creep curve, the most important for engineering applications are $\dot{\epsilon}$ and t_r. Specifically, their dependence on temperature and applied stress are of utmost interest to the designer. This dependence varies with the applicable creep mechanism. A variety of mechanisms and equations have been proposed in the literature and have been reviewed elsewhere (Ref 1 to 3). Fortunately, all these mechanisms can be fitted into two basic categories: (1) diffusional creep and (2) dislocation creep.

In diffusional creep, diffusion of single

atoms or ions either by bulk transport (Nebarro-Herring creep) or by grain-boundary transport (Coble creep) leads to a Newtonian viscous type of flow. The flow of atoms is envisaged to occur from regions of local compressive stress toward regions of local tensile stress, balanced by a counterflow of vacancies in the opposite direction. No motion of dislocations is envisaged. In this form of creep, the steady-state creep rates are postulated to vary linearly with stress — i.e., $\dot{\epsilon} \propto \sigma$. At low stresses, diffusional creep is seen only at very high temperatures approaching the melting point of the material and is, therefore, not generally of engineering significance.

The dislocation creep mechanism is operative at intermediate and high stresses and at temperatures above 0.4 of the melting point and is the only mechanism of significance for most engineering materials and applications. The deformation process is controlled by nonconservative motion of dislocations, implying vacancy diffusion or cross slip. Vacancy diffusion can occur by pipe diffusion and grain-boundary diffusion at low temperatures and by bulk diffusion at high temperatures. Various models based on these variations have been proposed. In general, the creep rate varies

nonlinearly with stress, as either a power function or an exponential function of stress, in contrast to the linear stress dependence of diffusional creep. With a knowledge of the stress and temperature dependence of the creep rate for each mechanism, it is possible to construct plots showing the regimes for the various mechanisms in the stress/temperature space (Ref 12 and 13). These plots, which are usually called deformation-mechanism maps or Ashby maps, are constructed by plotting the shear stress, normalized by the shear modulus, against the homologous temperature. Ashby maps for a turbine-blade alloy (MAR-M 200) are shown in Fig. 3.2 (Ref 14). As discussed previously, the diffusional creep mechanism operates at low stresses and high temperatures, whereas at intermediate stresses dislocation creep or power-law creep is operative. By comparing Fig. 3.2(a) and (b) it can be seen that increasing the grain size from 100 μm (3.9 mils) to 1 cm (0.39 in.) expands the power-law creep regime and appreciably decreases the creep rate of the turbine-blade material. These maps are thus very useful in providing insight into alloy design and strengthening mechanisms based on a knowledge of the operative creep mechanism.

At stresses and temperatures of inter-

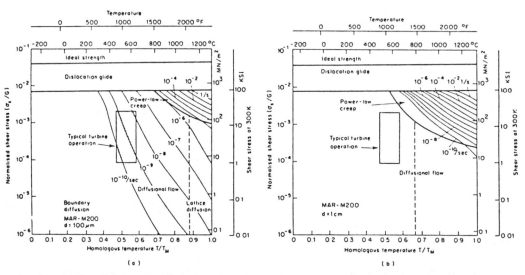

A turbine blade will deform rapidly by boundary diffusion at a grain size of 100 μm (a) but not at a grain size of 1 cm (b).

Fig. 3.2. Ashby deformation maps for MAR-M 200 (Ref 14).

est to the engineer, the following behavior proposed by Norton (Ref 15) and Bailey (Ref 16) is generally obeyed:

$$\dot{\epsilon} = A\sigma^n \qquad \text{(Eq 3.7)}$$

where A and n are stress-independent constants. An exponential relationship, although not generally used, has also been proposed (Ref 17) to explain the behavior at very high stresses, as follows:

$$\dot{\epsilon} = A\exp(C_7\sigma) \qquad \text{(Eq 3.8)}$$

where A and C_7 are stress-independent constants.

Because creep is a thermally activated process, its temperature sensitivity would be expected to obey an Arrhenius-type expression, with a characteristic activation energy "Q" for the rate-controlling mechanism. Equation 3.7 can therefore be rewritten as (Ref 17)

$$\dot{\epsilon} = A_0\sigma^n\exp\left(\frac{-Q}{RT}\right) \qquad \text{(Eq 3.9)}$$

where A_0 and n are constants and R is the universal gas constant.

Although Eq 3.9 suggests constant values for n and Q, experimental results on steels show both of these values to be variable with respect to stress and temperature. An example of the change in the value of n is shown in Fig. 3.3 for a normalized-and-tempered 1¼Cr-½Mo steel (Ref 18). A distinct break in the curve is evident, with n = 4 at low stresses and n = 10 at higher stresses. The breaks in the curves occurred at stresses at which the fracture mode changed from intergranular (I) to transgranular (T) at high stresses [Fig. 3.4 (Ref 18)]. Values of n ranging from n = 1 at low stresses to n = 14 at high stresses have been reported (Ref 18 to 31). Table 3.1 is a sample of studies selected to illustrate this point. Although many investigators report a distinct break in the curve, others view the value of n as continuously

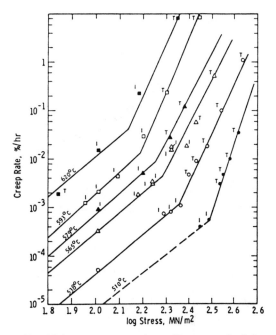

T and I denote transgranular and intergranular failure, respectively.

Fig. 3.3. Variation of minimum creep rate with stress for a normalized-and-tempered 1¼Cr-½Mo steel (Ref 18).

changing with stress and temperature. To account for these changes, it has been suggested (Ref 19, 21, 22, 28, 32, and 33) that the effective stress changes with test conditions due to changes in an "internal back stress σ_0" and that the stress term in Eq 3.7 should be modified to $(\sigma - \sigma_0)^n$. The internal back stress represents a resisting force and is postulated to arise from a variety of microstructural factors including dislocation configurations, precipitate dispersion, solid-solution effects, etc. While discussions continue regarding the natures of n and Q and the reasons for their variations, industrial practice has continued to ignore these controversies and to use a simple power law (Eq 3.9) with discretely chosen values of n and Q. Because variations in n and Q are generally interrelated and self-compensating, no major discrepancies in the end results have been noted.

The behavior of t_r with respect to σ and T is similar to that of $\dot{\epsilon}$, with the differences being that the signs are reversed for the stress exponent and the activation energy

(a)

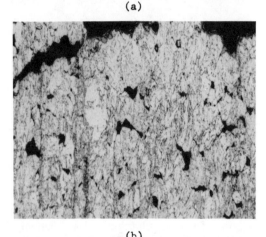

(b)

(c)

(a) Stress, 324 MPa (47 ksi); reduction in area, 84.5%. (b) Stress, 207 MPa (30 ksi); reduction in area, 58.7%. (c) Stress, 152 MPa (22 ksi); reduction in area, 28.8%. Magnification (all), 100×; shown here at 85%.

Fig. 3.4. Progress of intergranular cracking with decreasing stress and ductility at 565 °C (1050 °F) in a 1¼Cr-½Mo steel (Ref 18).

and that the constants A_0 and A in Eq 3.7 and 3.9 have slightly altered values. This behavior occurs when $\dot{\epsilon}$ and t_r are inversely related through Monkman-Grant-type relationships (i.e., $\dot{\epsilon} t_r \simeq$ constant) to be discussed later.

Parametric Extrapolation Techniques

Because components of power plants and process industries are designed to operate for times in excess of 100,000 h, extrapolation of laboratory creep and rupture data to actual service conditions is unavoidable. Even if long-time data are available for selected heats of material, heat-to-heat variations in properties make it necessary to estimate the long-time behavior for other heats. Greater difficulty is encountered in estimating the remaining creep lives of in-service components, where decisions have to be made based on very short-time laboratory tests (usually less than 1000 h). The need for extrapolation techniques that permit estimation of the long-term creep and rupture strengths of materials based on short-duration tests is thus a very real and important one in design, quality control, and plant evaluation.

Basically, parametric techniques incorporate time stress and temperature test data into a single expression. When test data recorded over adequate times and at temperatures above the service temperature are incorporated into a single "master curve," the stress for the service-temperature conditions can be read directly from the master curve.

Extensive reviews of parametric techniques have been given elsewhere (Ref 34 and 35). Their historical evolution up to the present time has been reviewed by Manson and Ensign (Ref 36). It is neither possible nor necessary to discuss all of the various time-temperature-stress parameters that have been proposed, and only some of the more widely accepted parameters will be briefly described here.

Table 3.1. Examples of reported values of stress exponent and activation energy for creep of steels

Reference	System	Temperature, °C	Coefficients in the low-stress region		Coefficients in the high-stress region		Interpretation of coefficients
			n	Q, kJ/mole	n	Q, kJ/mole	
18	1¼Cr-½Mo steel	510-620	4	400	10	625	Grain-boundary sliding at low stresses and matrix deformation at high stresses
19	2¼Cr-1Mo steel	565	2.5	. . .	12	. . .	Deformation governed by matrix deformation
20	1Cr-½Mo steel, heat-affected zone	550-605	3	300	6	300	Diffusive mechanism at low stresses and dislocation mechanism at high stresses
21	1Cr-½Mo steel, base metal	550-605	5.6	. . .	5.6	. . .	. . .
27	Cr-Mo-V steel	550-600	4.9	326	14.3	503	Dislocation climb over particles at low stresses and bowing between particles at high stresses
23	20Cr-25Ni-Nb steel	750	3-4.7	465-532	8-12	440-494	Metallographic measurement showed that $\dot{\epsilon}_{gb} = A\sigma^m$, where m = 3.4. Transition from low-σ to high-σ behavior was attributed to change from grain-boundary sliding to matrix deformation.
24	20Cr-25Ni-Nb steel	750	2-5	250-390; average, 320	. . .	. . .	n ~ corresponds to boundary sliding; n > 3 corresponds to matrix deformation. From scratch displacements it was found that $\dot{\epsilon}_{gbs} = A^{3.7}$ and $Q_{gbs} = 385$ kJ/mole.
22	20Cr-25Ni-Nb steel	700-750	n varied from 6 to 8.4 with increasing T. Q varied with σ. At $\sigma = 79$ MPa, Q = 678 kJ/mole.		. . .	. . .	Values of n and Q are only "apparent" unless a back stress due to NbC precipitate is considered.

Larson-Miller Parameter

Larson and Miller (Ref 37) first introduced the concept of a time-temperature grouping in the form $T(K_1 + \log t)$, based on the earlier Hollomon-Jaffee expression (Ref 38) for tempering of steel. For a given material, a plot of stress vs the above parameter resulted in a single plot, within limits of scatter, regardless of the time-temperature combination employed to derive the parameter, such as that shown in Fig. 3.5 for

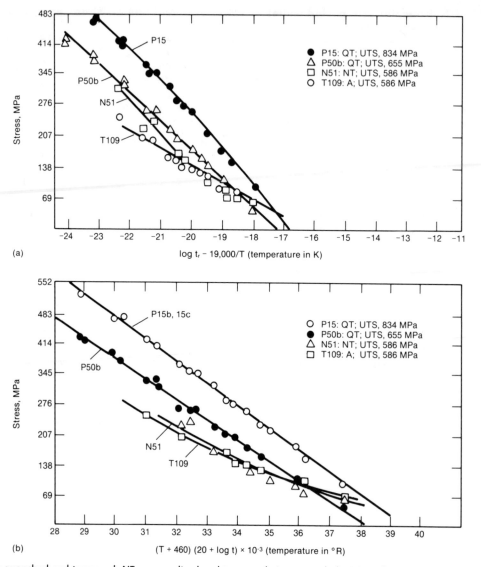

(a)

log t_r − 19,000/T (temperature in K)

(b)

(T + 460) (20 + log t) × 10⁻³ (temperature in °R)

QT = quenched and tempered. NT = normalized and tempered. A = annealed. UTS = ultimate tensile strength.

Fig. 3.5. Variation in stress-rupture strength of 2¼ Cr-1Mo steels under different heat treatment conditions, plotted using (a) the Orr-Sherby-Dorn parameter and (b) the Larson-Miller parameter (Ref 39).

2¼ Cr-1Mo steels subjected to various heat treatment conditions (Ref 39). A value of $K_1 = 20$ was initially proposed, but optimized values between 10 and 40 have subsequently been found to be suitable depending on the material. In common usage, T is taken in absolute units, and t in hours.

It can be shown that the Larson-Miller parameter can be readily derived from the stress and temperature dependence of the creep rate or time to rupture. The rate equations generally can be written as

$$\dot{\epsilon} = A_1 \exp\left(\frac{-B_1}{T}\right) \qquad \text{(Eq 3.10)}$$

or

$$t_r = A_2 \exp\left(\frac{B_2}{T}\right) \qquad \text{(Eq 3.11)}$$

When logarithms are taken, Eq 3.11 becomes

$$\log t_r = \log A_2 + \frac{B_2}{2.3T} \quad \text{(Eq 3.12)}$$

If we assume that $\log A_2$ is a true constant and that only B_2 varies with stress, Eq 3.12 can be rearranged to arrive at

$$\frac{B_2}{2.3} = T(\log t_r - \log A_2) \quad \text{(Eq 3.13)}$$

or

$$P_t = f(\sigma) = T(\log t_r + K_1) \quad \text{(Eq 3.14)}$$

where P_t is the Larson-Miller parameter and K_1 is a constant. Because, in Eq 3.12, $\log A_2$ is assumed to be constant, a plot of $\log t_r$ vs $1/T$ results in straight lines whose intercept is a constant and whose slope, $B_2/2.3$, is a function of stress, as shown in Fig. 3.6(a).

Orr-Sherby-Dorn Parameter

In Eq 3.12, if we assume that B_2 is a true constant but that A_2 is a function of stress, we can rearrange the terms to arrive at

$$\log A_2 = \log t_r - \frac{B_2}{2.3T} \quad \text{(Eq 3.15)}$$

or

$$\theta = f(\sigma) = \log t_r - \frac{Q}{2.3RT} \quad \text{(Eq 3.16)}$$

In this case, a plot of $f(\sigma)$ vs σ yields a straight line. Plots of $\log t_r$ vs $1/T$ result in parallel lines at different stresses, the slope of which is a constant value of $-Q/2.3R$ (where Q is a characteristic activation energy for the process and R is the universal gas constant) (Ref 40). The value of Q for

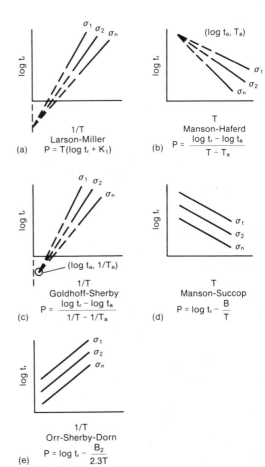

(a) Larson-Miller $P = T(\log t_r + K_1)$

(b) Manson-Haferd $P = \dfrac{\log t_r - \log t_a}{T - T_a}$

(c) Goldhoff-Sherby $P = \dfrac{\log t_r - \log t_a}{1/T - 1/T_a}$

(d) Manson-Succop $P = \log t_r - \dfrac{B}{T}$

(e) Orr-Sherby-Dorn $P = \log t_r - \dfrac{B_2}{2.3T}$

Fig. 3.6. Schematic representations of constant-stress lines for several parametric techniques.

many ferritic steels has been reported to be approximately 380 kJ/mole (90 kcal/mole).

Manson-Haferd Parameter

The Manson-Haferd parameter departs slightly from the Orr-Sherby-Dorn parameter in that the isostress plots of $\log t_r$ vs T (rather than $\log t_r$ vs $1/T$) are assumed to be linear and to intersect the axes at T_a and $\log t_a$, as shown in Fig. 3.6(b). This parameter thus is derived from expressions such as (Ref 41)

$$t_r = A_3 \exp(-B_3 T) \quad \text{(Eq 3.17)}$$

$$\log t_r = \log A_3 - \frac{B_3 T}{2.3} \quad \text{(Eq 3.18)}$$

where A_3 and B_3 are both functions of stress such that all the isostress lines intersect at a point. This gives

$$f(\sigma) = \frac{\log t_r - \log t_a}{T - T_a} \quad \text{(Eq 3.19)}$$

where $f(\sigma)$ is the Manson-Haferd parameter and is plotted against stress to get a linear relationship. The constants t_a and T_a are the coordinates of the point of intercept, as illustrated in Fig. 3.6(b).

Manson-Brown Parameter

Manson and Brown (Ref 42) proposed a generalized parameter in the form

$$f(\sigma) = \frac{\log t_r - \log t_a}{(T - T_a)^q} \quad \text{(Eq 3.20)}$$

Several other parameters could be shown to be special cases of this parameter (Ref 36). For q = 1, its form is equivalent to that of the Manson-Haferd parameter. If q = −1 and $T_a = 0$, it is equivalent to the Larson-Miller parameter. Even the Orr-Sherby-Dorn parameter could be shown to be a special case when log t_a and $1/T_a$ are both taken to be arbitrarily very large numbers with the condition T_a log t_a = Q. The Manson-Succop parameter (Ref 43), defined as (log t + YT), can be regarded as a special case of the Manson-Brown parameter by taking q = 1, while log t_a and T_a are taken as arbitrarily large numbers such that log t_a/T_a = −Y. Following the introduction of the Manson-Brown parameter, efforts continued to develop an even more generalized parameter technique wherein the data would dictate the specific form of the equation to be used, instead of trying to force an equation to fit the data. These efforts led to the development of the minimum-commitment method (MCM).

Minimum-Commitment Method

The basic concept of the minimum-commitment method is to start with a time-temperature-stress relationship sufficiently general to satisfy all the commonly used parameters. Then, the specific functional relationship for a material is numerically established on the basis of experimental data. Not only are the commonly used parameters given an equal chance to emerge as the appropriate one, but additional patterns of behavior not compatible with the common parameters can also be accommodated by use of the MCM. The MEGA (Manson-Ensign Generalized Analysis) computer program has been developed to implement the MCM (Ref 44).

The parameter chosen has the form

$$M \log t + M'X \log t + X = f(\sigma)$$
$$\text{(Eq 3.21)}$$

where M and M′ are temperature-independent constants and X is a function of temperature. It can be shown that when M = 0 and M′ = 0.05, the expression reduces to $0.05 \times (20 + \log t)$, a form compatible with the Larson-Miller parameter. Similarly, if we set M′ = 0 and X = −Q/RT, the expression reduces to (M log t − Q/RT), a form compatible with the Orr-Sherby-Dorn parameter.

The value of M′ is generally set equal to zero, so that Eq 3.21 reduces to

$$M \log t + X = f(\sigma) \quad \text{(Eq 3.22)}$$

X is defined as

$$X = R_1[T - T_{mid}] + R_2[1/T - 1/T_{mid}]$$
$$\text{(Eq 3.23)}$$

where T_{mid} is the mid-value of the temperature range for which data are to be analyzed and R_1 and R_2 are constants. In the usual plot of the MCM, (log t + X) is plotted as a function of log σ.

Considerable refinements designed to take heat-to-heat variations into account have been incorporated more recently into the MCM (Ref 45). The parameter $f(\sigma)$ in

Eq 3.22 is essentially modified by the addition of two "heat terms," as in

$$M \log t + X - R_3 \log \sigma - R_4 = f(\sigma)$$

$$\text{(Eq 3.24)}$$

where R_3 and R_4 are heat-specific terms which can be determined from characterization tests. Addition of these terms considerably reduces scatter in predictions based on multiheat data sets (Ref 45).

Considerable attention has been devoted to an attempt to set an appropriate value for the constant M' in Eq 3.21. It is believed that M' is a measure of the metallurgical stability of the alloy. Manson and Ensign have suggested values of $M' = 0$ for pure metals and aluminum alloys, $M' = -0.05$ for steels and superalloys which are expected to be metallurgically stable, $M' = -0.10$ if moderate instability is suspected, and $M' = -0.15$ for known or suspected cases of serious instability (Ref 36).

Due to the multiplicity of rate processes affecting the creep strengths of complex alloys, it is impossible for a single parameter to successfully describe their behavior over a wide range of stresses and temperatures. At best, the various techniques offer a semiempirical approximation to the trend of data. Among the several parameters, the Larson-Miller parameter enjoys the most widespread use by engineers because it has been used for the longest period of time, is easy to understand and use, and has proved to be at least as accurate as (if not more accurate than) any of the other parameters. The MCM technique is fast emerging as a promising alternative, and its current lack of acceptance is mainly due to a lack of user awareness.

Example:
The purpose of this example is to illustrate the use of the Larson-Miller parameter (LMP). Under design conditions of $\sigma = 7.5$ ksi and T = 1000 °F, a component can be safely used for 40 years (347,520 h). If the plant operates at 1050 °F, calculate the resulting reduction in the life of the component using the Larson-Miller parameter.

Answer:
T = 1000 °F (1460 °R); t = 347,520 h
LMP = T(20 + log t) = 37,289; at
  1050 °F (1510 °R), because σ is
  the same, LMP is the same.
1510(20 + log t) = 37,289
t = 49,573 h (5.7 years)

The life of the component will be reduced to 5.7 years.

Design Rules

For statically loaded components operating at low temperatures, the important "strength" parameters are the ultimate tensile strength (UTS) and the yield strength. Design is based on applying safety factors to these values to avoid failure and gross plastic deformation. Under these conditions, there is no target design life as such, and the component should, in principle, operate indefinitely if no corrosion-related degradation phenomena occur. At elevated temperatures, time-dependent deformation and fracture become operative and hence a target design life can be envisaged based on either the time to rupture or the time to cause a given degree of deformation. All of these factors are taken into account in the design of elevated-temperature components. Paragraph A-150 of Section I, Power Boilers, of the ASME Boiler and Pressure Vessel Code (Ref 46) states that the allowable stresses are to be no higher than the lowest of the following:

1. 1/4 of the specified minimum tensile strength at room temperature
2. 1/4 of the tensile strength at elevated temperature
3. 2/3 of the specified minimum yield strength at room temperature
4. 2/3 of the yield strength at elevated temperature
5. 100% of the stress to produce a creep rate of 0.01% in 1000 h (or 1% in 100,000 h)

6. 67% of the average stress or 80% of the minimum stress to produce creep rupture in 100,000 h as determined from extrapolated data, whichever is lower.

Figure 3.7 illustrates how these criteria are employed to establish the allowable stress for a 2¼Cr-1Mo steel. For this steel, item 1 (room-temperature tensile strength) controls the allowable stress up to about 900 °F. At 900 °F and above, the creep strength and the rupture strength become the dominant factors and the allowable stress takes a sharp downturn. The relative influences of the different criteria vary with material as well as with temperature.

The ASME code criteria listed above are adhered to both for fossil boilers and for pressure-vessel and piping systems in the petroleum and chemical process industries. Design of many other components, such as steam-turbine disks, rotors, and blades, and combustion-turbine disks, blades, vanes, etc., are based on the principle of applying suitable safety factors to stress to cause rupture or a given rate of creep in a specified period. The actual details and the safety factors for these components, however, are proprietary and vary with the equipment manufacturer. Although by application of safety factors and the use of the design criteria, manufacturers strive to ensure the safety of equipment, the safety margins can be reduced in practice by the presence of manufacturing defects, operation outside the design limits, unanticipated system and residual stresses, aggressive environments, and a host of other factors.

Cumulative Damage in Creep

Although it is relatively easy to quantify damage in laboratory creep tests conducted at constant temperature and stress (load), components in service hardly ever operate under constant conditions. Start-stop cycles, reduced power operation, thermal gradients, and other factors result in variations in stresses and temperatures. Procedures are needed that will permit estimation of the cumulative damage under changing exposure conditions.

Damage Rules

The most common approach to calculation of cumulative creep damage is to compute the amount of life expended by using time or strain fractions as measures of damage. When the fractional damages add up to unity, then failure is postulated to occur. The most prominent rules are as follows:

1. Life-fraction rule (Ref 47):

$$\sum \frac{t_i}{t_{ri}} = 1 \qquad \text{(Eq 3.25)}$$

2. Strain-fraction rule (Ref 48):

$$\sum \frac{\epsilon_i}{\epsilon_{ri}} = 1 \qquad \text{(Eq 3.26)}$$

Material: 2¼ Cr-1Mo Steel
Specification: SA-213 Grade T 22

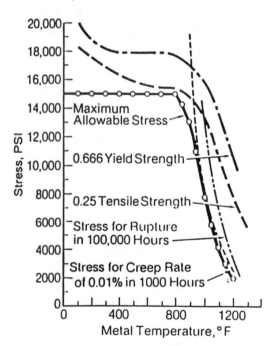

Fig. 3.7. Use of ASME Boiler and Pressure Vessel Code criteria to establish the allowable stress for a 2¼Cr-1Mo steel (Ref 46).

3. Mixed rule (Ref 49):

$$\sum \left(\frac{t_i}{t_{ri}}\right)^{1/2} \left(\frac{\epsilon_i}{\epsilon_{ri}}\right)^{1/2} = 1 \quad \text{(Eq 3.27)}$$

4. Mixed rule (Ref 50):

$$k\sum \left(\frac{t_i}{t_{ri}}\right) + (1 - k)\sum \left(\frac{\epsilon_i}{\epsilon_{ri}}\right) = 1$$

$$\text{(Eq 3.28)}$$

where k is a constant; t_i and ϵ_i are the time spent and strain accrued at condition i; and t_{ri} and ϵ_{ri} are the rupture life and rupture strain under the same conditions.

Example:
The purpose of this example is to illustrate the use of the life-fraction rule. A piping system, made of 1¼Cr-½Mo steel designed for a hoop stress of 7 ksi, was operated at 1000 °F (1460 °R) for 42,500 h and at 1025 °F (1485 °R) for the next 42,500 h. Calculate the life fraction expended using the life-fraction rule. From the minimum curve of LMP for the steel, it is found that, at $\sigma = 7$ ksi,

t_r at 1000 °F = 220,000 h
t_r at 1025 °F = 82,380 h
Life fraction expended, t/t_r,
  at 1000 °F

$$= \frac{42,500}{220,000} = 0.19$$

Life fraction expended, t/t_r,
  at 1025 °F

$$= \frac{42,500}{82,380} = 0.516$$

The total life fraction expended is 0.71.

Validity of Damage Rules

Goldhoff and Woodford (Ref 51) studied the Robinson life-fraction rule and determined that for a Cr-Mo-V rotor steel it worked well for small changes in stress and temperature. Goldhoff (Ref 52) assessed strain-hardening, life-fraction, and strain-fraction rules under unsteady conditions for this steel. While all gave similar results, the strain-fraction rule was found to be the most accurate. A systematic study of the effects of variations in stress and temperature on the creep lives of six steels for test times up to 20,000 h has been carried out by Wiegand *et al* (Ref 53). Figure 3.8 presents an example of the results for a 2¼Cr-1Mo steel. It was concluded that results of variable-temperature tests yielded close agreement between the actual rupture lives and those predicted by use of the life-fraction rule. For stress variations, actual rupture lives were lower than predicted ones.

The effect of stress variation alone on cumulative creep damage, based on the work of Meijers and Etienne (Ref 54), is illustrated in Fig. 3.9. It is evident from this figure that the cumulative life fraction is less than unity if the postcreep exposure stress is larger than the initial creep exposure stress, whereas it is greater than unity if the postcreep exposure stress is smaller than the initial stress.

Hart (Ref 55) performed a series of tests on 1Cr-½Mo steels and concluded that in variable-temperature tests the curve for log t_r vs T at constant stress for predamaged material was shifted in a parallel and proportional manner with respect to the curve for virgin material, confirming the applicability of the life-fraction rule (LFR). On the other hand, in variable-stress tests the curve for log t_r vs σ for predamaged material exhibited a reduced slope compared with the virgin material, indicating that the LFR was not obeyed. Similar findings have been confirmed by Woodford (Ref 56).

Recently, stress-rupture tests for very long times up to 60,000 h have been conducted on two casts of Cr-Mo-V steels by investigators at ERA Technology (Ref 57). Specimens were precrept to life fractions of 0.3, 0.5, 0.7, and 0.9 and subsequently tested to rupture at different temperatures. The cumulative life fractions approached the value of unity for specimens predamaged to life fractions of 0.5, 0.7, and 0.9, but only for a "brittle" cast. For a "ductile" cast of the same steel, however, the total life

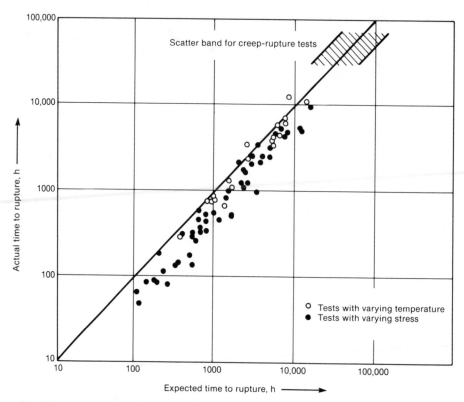

Fig. 3.8. Comparison of actual rupture life with predictions from life-fraction rule for 2¼ Cr-1Mo steel (Ref 53).

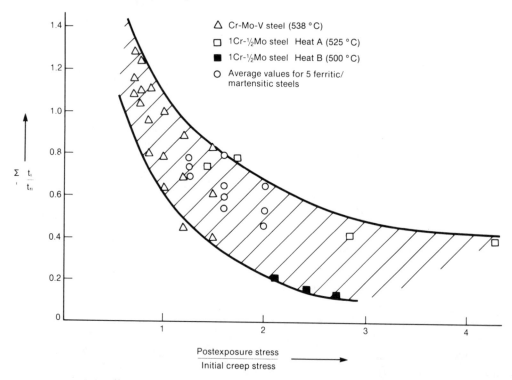

Fig. 3.9. Effect of postexposure stress on cumulative life fraction (Ref 54).

fractions were always in excess of unity. Due to scatter in data, results from this program are still inconclusive and are being further verified.

Hart (Ref 58) has provided an explanation, based on microstructural changes, of why the LFR is not valid for stress-change experiments but is valid for temperature changes. In materials which undergo structural changes at high temperatures in the absence of stress, the kinetics of such changes are governed by time and temperature of exposure. The time and temperature can be "traded" for each other so that equivalent combinations of time and temperature will produce equivalent structural changes. On the other hand, the stress cannot be similarly "traded" against time. For instance, let us say that two identical specimens are tested at a given temperature at two stress levels σ_1 and σ_2, with σ_1 being greater than σ_2. Even if the two specimens are tested to the same life fraction, the larger number of hours of exposure at the lower stress σ_2 will have caused greater structural damage in that sample than in the other sample. If both of these samples now are tested to rupture at a higher stress σ_3, the sample previously exposed to the lower stress will fail in a shorter time. Thus, even though the starting life fractions were the same, the damage levels will be different.

From careful and critical examination of the available results, the following over-all observations can be stated. (1) Although several damage rules have been proposed, none has been demonstrated to have a clear-cut superiority over any of the others. The Robinson life-fraction rule is therefore the most commonly used. (2) The LFR is clearly not valid for stress-change experiments. Under service conditions where stress may be steadily increasing due to corrosion-related wastage (e.g., in boiler tubes), application of the LFR will yield nonconservative life estimates—i.e., the actual life will be less than the predicted life. On the other hand, residual life predictions using postexposure tests at high stresses will yield unduly pessimistic and conservative results. (3) The LFR is generally valid for variable-temperature conditions as long as changing creep mechanisms and environmental interactions do not interfere with test results. Hence, service life under fluctuating temperatures and residual life based on accelerated temperature tests can be predicted reasonably accurately by use of the LFR. (4) The possible effects of material ductility (if any) on the applicability of the LFR need to be investigated. A major limitation in applying the LFR is that the properties of the virgin material must be known or assumed. Postexposure tests using multiple specimens often can obviate the need for assuming any damage rule.

Uniaxial-to-Multiaxial Data Correlation

In design and for remaining-life prediction of elevated-temperature components, extensive use is made of test data generated on laboratory samples under uniaxial stress. Components in service operate under multiaxial stress conditions. It is therefore necessary to establish the effective stress criteria governing creep and rupture under multiaxial stress conditions and to be able to interpret them in terms of uniaxial test data. In addition, the redistribution of stresses occurring with creep must be taken into account. A brief review of the first of these two aspects may be found in the paper by Roberts, Ellis, and Bynum (Ref 59).

Effective Stress for Creep and Rupture

The determination of effective stress and effective strain for creep deformation has been extensively studied. Because the initial inelastic deformation involves a shear process, either the Tresca (maximum shear) definition or the Von Mises (octahedral shear) definition generally has been used. Whereas the Tresca criterion is conservative, the Von Mises relationships given below are more appealing from a continuum calculational and experimental standpoint.

$$\sigma^* = \frac{1}{\sqrt{2}} [(\sigma_1 - \sigma_2)^2 + (\sigma_2 - \sigma_3)^2 + (\sigma_3 - \sigma_1)^2]^{1/2} \qquad \text{(Eq 3.29)}$$

$$\epsilon^* = \frac{\sqrt{2}}{3}[(\epsilon_1 - \epsilon_2)^2 + (\epsilon_2 - \epsilon_3)^2$$

$$+ (\epsilon_3 - \epsilon_1)^2]^{1/2} \qquad \text{(Eq 3.30)}$$

$$\dot{\epsilon}^* = \frac{\sqrt{2}}{3}[(\dot{\epsilon}_1 - \dot{\epsilon}_2)^2 + (\dot{\epsilon}_2 - \dot{\epsilon}_3)^2$$

$$+ (\dot{\epsilon}_3 - \dot{\epsilon}_1)^2]^{1/2} \qquad \text{(Eq 3.31)}$$

where σ^* and ϵ^* are the Von Mises effective stress and strain; σ_1, σ_2, and σ_3 are the principal stresses; and ϵ_1, ϵ_2, and ϵ_3 are the principal strains. The relationship between stress and strain rate is commonly expressed as

$$\frac{\dot{\epsilon}_1 - \dot{\epsilon}_2}{\sigma_1 - \sigma_2} = \frac{\dot{\epsilon}_2 - \dot{\epsilon}_3}{\sigma_2 - \sigma_3} = \frac{\dot{\epsilon}_3 - \dot{\epsilon}_1}{\sigma_3 - \sigma_1} = \frac{3}{2}\frac{\dot{\epsilon}^*}{\sigma^*}$$

$$\text{(Eq 3.32)}$$

The usual assumptions that are made regarding material behavior are as follows: (1) constant volume is maintained, (2) behavior is isotropic, (3) there is no influence of hydrostatic stress, (4) creep rates in the compressive and tensile directions are equivalent, and (5) principal axes of stress and strain are coincident. The isothermal uniaxial tensile data normally are used to infer the functional form of the stress-strain equation, and a wide range of mathematical models incorporating primary, secondary, and tertiary behavior have been used. When either stress or temperature changes as a function of time, a hardening law must be chosen. For simplicity, either time hardening or strain hardening laws have been used, with the latter providing better agreement with results (Ref 59).

Creep rupture under multiaxial stress states with varying stress and temperature histories is of great importance in predicting component performance. However, the most common creep-rupture test is performed on uniaxial samples at constant load or temperature. Most of the life-prediction rules are based on uniaxial tests under variable temperature and load conditions. In uniaxial tests, complete separation of the sample constitutes failure, whereas in multiaxial tests, failure may be defined by another criterion, such as loss of internal pressure in the case of pipes. Other complications also arise in extending the results of uniaxial tests to multiaxial conditions because the effective true stress history under constant load no longer follows the uniaxial true stress history. The appropriate stress criterion under multiaxial stress conditions depends on the nature of the cracking process (Ref 60). Two cases can be distinguished. First, for general and gradual propagation of cracks that are microscopically visible from the outset, the maximum principal stress is the appropriate stress criterion. In contrast, when fracture is not accompanied by microscopically visible cracks until the deformation becomes localized near the end of the component's life, the Von Mises shear stress is the proper fracture criterion. The proportions of life spent in these two phases will control which of these stress criteria in the following equations best applies to the behavior of the component.

$$\sigma = \sigma_{max} \text{ [maximum principal stress]}$$

$$\text{(Eq 3.33)}$$

$$\sigma = \sigma^* \text{ [Von Mises effective stress]}$$

$$\text{(Eq 3.34)}$$

$$\sigma = C_8\sigma_{max} + (1 - C_8)\sigma^*$$

$$\text{[mixed rule (Ref 61)]} \quad \text{(Eq 3.35)}$$

$$\sigma = C_9\sigma_{max} + C_{10}J_1 + C_{11}J_2^{1/2} \text{ [Ref 62]}$$

$$\text{(Eq 3.36)}$$

where C_8, C_9, C_{10}, and C_{11} are constants and where

$$J_1 = \sigma_1 + \sigma_2 + \sigma_3 \qquad \text{(Eq 3.37)}$$

and

$$J_2 = \frac{1}{6}[(\sigma_1 - \sigma_2)^2 + (\sigma_2 - \sigma_3)^2 + (\sigma_3 - \sigma_1)^2]$$

$$\text{(Eq 3.38)}$$

Stresses in Internally Pressurized Components

In the case of internally pressurized tubes and pipes, the initial elastic hoop stress σ_H, axial stress σ_{ax}, and radial stress σ_r at any radial distance can be calculated by the Lame equations (Ref 63):

$$\sigma_H = \frac{Pr_i^2(r_o^2 + r^2)}{r^2(r_o^2 - r_i^2)} \qquad \text{(Eq 3.39)}$$

$$\sigma_{ax} = \frac{Pr_i^2}{r_o^2 - r_i^2} \qquad \text{(Eq 3.40)}$$

$$\sigma_r = \frac{-Pr_i^2(r_o^2 - r^2)}{r^2(r_o^2 - r_i^2)} \qquad \text{(Eq 3.41)}$$

where P is pressure, r is radial distance, and r_i and r_o are the inner and outer radii, respectively. Alternatively, the mean-diameter hoop-stress formula is given as

$$\sigma_H = \frac{Pd}{2x} \qquad \text{(Eq 3.42)}$$

where d is mean diameter and x is wall thickness. If, instead of the mean diameter, the inside diameter is used, the formula is called the thin cylinder formula, the bore formula, or the common stress formula. If the outside diameter is used instead of "d," the formula is known as the Barlowe formula. The hoop stress calculated by use of these formulas is assumed to be a representative stress value, with no regard for any radial stress gradients. The principal stresses given by Eq 3.39 to 3.41 can also be combined using Eq 3.29 to calculate a Von Mises effective stress. Other alternative formulas have also been employed. A total of 31 different formulas for calculating stresses in tubes and pipes have been listed by Burrows, Michel, and Rankin (Ref 64).

Stress Redistribution in Creep

Equations 3.39 to 3.42 describe only the initial elastic stresses in a component. In the creep regime, time-dependent creep deformation will result in a redistribution of the stresses. The rate and extent of this redistribution will depend on (1) the initial stress level and (2) the radius-to-thickness ratio, metal temperature, and creep response of the material. Calculations of the rate and extent of stress redistribution can be made using detailed nonlinear finite-element analysis or can be estimated by simple numerical methods such as those proposed by Bailey (Ref 65). The long-term (i.e., $t \to \infty$) relaxed steady-state stress distributions according to the Bailey solution are given by

$$\sigma_H = \frac{P\{[(2 - n)/n][r_o/r]^{2/n} + 1\}}{(r_o/r_i)^{2/n} - 1} \qquad \text{(Eq 3.43)}$$

$$\sigma_{ax} = \frac{P\{[(1 - n)/n][r_o/r]^{2/n} + 1\}}{(r_o/r_i)^{2/n} - 1} \qquad \text{(Eq 3.44)}$$

$$\sigma_r = \frac{P[(r_o/r)^{2/n} - 1]}{(r_o/r_i)^{2/n} - 1} \qquad \text{(Eq 3.45)}$$

where n is the Norton law exponent; σ_H, σ_{ax}, and σ_r are the hoop, axial, and radial stresses; and r is the radial distance.

Paterson, Rettig, and Clark have shown good agreement between the long-term steady-state stress distributions calculated by the Bailey equations and by finite-element techniques for superheater tubes (Ref 66).

Simplified methods based on a "reference stress" approach have been pursued vigorously in the United Kingdom. This concept had its origin in the findings of Schulte, who pointed out that creep analysis of a rectangular beam under uniform bending moment revealed the existence of a particular location in the beam where the stress remained invariant during stress redistribution by creep (Ref 67). Schulte postulated that this value of the invariant stress could serve as a reference stress for the whole beam and that uniaxial creep and rupture tests conducted at the reference stress would entirely reproduce the behavior of the beam. The validity of this general approach was confirmed by the work of many investigators (Ref 68 to 71). Practical limitations, however, continued to plague evaluations of the reference stress for even relatively simple shapes until the work of Sim (Ref 72) showed that an approximate estimate of the upper boundary for the reference stress could be derived from the simple plasticity/creep relationship

$$\sigma_{ref} = \frac{P\sigma_y}{P_U} \qquad \text{(Eq 3.46)}$$

where P is the current load, P_U is the rigid-plastic collapse load for the component, and σ_y is the yield stress at the temperature of concern. The collapse load is the load at which deformation of the structure becomes excessive due to plastic deformation under short-term loading. It is proportional to the material flow stress and can be determined by testing components to failure at room temperature or from a wide range of theoretical solutions which have been derived. In other words, the term σ_y/P_U is simply a geometric factor that can be obtained analytically or from model tests. Once the reference stress has been established by experiments, creep assessment is straightforward, because the deformation rates and endurance of a complex structure will be very similar to those of simple laboratory specimens. Extensive discussions of the reference-stress approach are available in the literature (Ref 73 to 75). The reference stress has been shown in many cases to give accurate predictions of deformation behavior. Recent experiments on ½Cr-Mo-V steel tubes have shown that rupture data also can be correlated with the reference stress (Ref 76).

Several failure criteria have been applied to stress rupture of tubing at high temperature. Failure is postulated to occur when the "equivalent stress" in the tube under multiaxial conditions becomes equal to the stress for rupture under uniaxial conditions. Several alternative formulas have been examined to determine which one gives the best description of the equivalent stress. If the dominant creep-damage process involves structural coarsening, rather than cavitation, the Von Mises stress is the preferred descriptor of the equivalent stress. If the damage involves early initiation of cracks and cavities and their subsequent evolution to failure, then failure is governed by the maximum principal stress. In this case, hoop-stress formulas based on Eq 3.42 are preferred. It has been reported that in tubes, the mean-diameter hoop-stress formula overestimates the stresses by about 20% and therefore leads to a conservative prediction of life by underestimating it by a factor of two (Ref 76), in comparison with the reference-stress approach. It has also been observed that the thin cylinder formula, using the internal diameter instead of the mean diameter in Eq 3.42, and the Von Mises effective stress bring the uniaxial and tube rupture data much closer together than the mean-diameter formula.

Rupture Ductility

While creep strength and rupture strength have been given considerable attention as design and failure parameters, one of the most important, yet neglected, parameters is rupture ductility. Gross and uniform creep deformation of components is usually the exception rather than the rule. Localized defects and stress concentrations often play decisive roles in failure. Under these circumstances, the growth of cracks and defects is governed by the creep ductility of the material. Because ductility varies inversely with creep (and rupture) strength, both properties have to be optimized for a given application. This point is illustrated in Fig. 3.10, which compares the stress-rupture behavior of a "creep strong" and brittle material (steel A) and a "creep weak" and ductile material (steel B). When smooth test parts of these steels are tested, steel A is clearly superior to steel B. However, when a notch that is simulative of a stress concentration is introduced, the creep strength of steel A plummets whereas that of steel B remains stable. Steel A is known as a notch-sensitive steel, whereas steel B is called a notch-insensitive steel. It is clear from

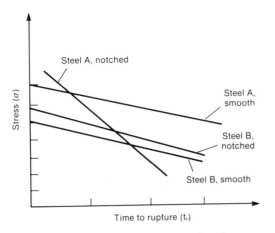

Fig. 3.10. Illustration of notched-bar rupture behavior for a creep-brittle steel (A) and a creep-ductile steel (B).

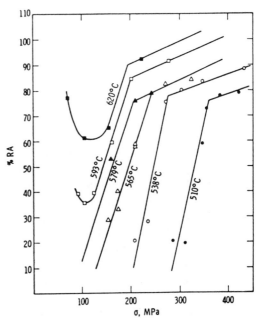

Fig. 3.11. Variation of reduction in area with stress and temperature for 1¼ Cr-½ Mo steels (Ref 18).

Fig. 3.10 that for low-stress, long-term applications, steel B would be preferred to steel A, although it is weaker based on results of smooth-bar rupture tests.

It has been well recognized for many years that notch sensitivity is related to creep ductility. It has been suggested that a minimum smooth-bar creep ductility of about 10% in terms of reduction in area may be desirable for avoidance of notch sensitivity (Ref 52 and 77). For quality control purposes, it is common practice to conduct a rupture test on a combination notched-bar/smooth-bar specimen at a specified stress and temperature, with rupture time usually not to exceed 500 h. If failure occurs at the notch during this test, the material is deemed notch sensitive and is rejected. However, this procedure is inadequate, because notch sensitivity that may develop at lower stresses and longer times may go undetected. Because notch sensitivity is primarily related to ductility, a discussion of ductility is adequate for present purposes.

Time-Temperature Dependence

A typical variation of rupture ductility with stress for a normalized-and-tempered 1Cr-½Mo steel is illustrated in Fig. 3.11 (Ref 18). The same behavior can also be plotted in terms of t_r or $\dot{\epsilon}$. At high stresses (low t_r), ductility is fairly high and constant, and fractures are characterized by transgranular

failure. Below a threshold value of stress, the ductility drops steeply, with fracture modes becoming increasingly intergranular. The value of stress at which ductility drops to about 10% corresponds to the onset of notch sensitivity—i.e., the crossover of smooth- and notched-bar rupture curves, as illustrated in Fig. 3.10. At very low stresses, corresponding to long times to rupture, the ductility starts to recover, especially at the higher temperatures.

The decrease in ductility with decreasing stress is generally attributed to increasing localization of creep strain in a narrow zone adjacent to the grain boundaries. It has been shown that the strain due to grain-boundary sliding as a fraction of the total creep strain increases with decreasing stress (Ref 1 and 18). It has been proposed that strain localization also is assisted by the presence of precipitate-free zones (PFZ's) near the grain boundaries. Evidence for PFZ's in steels has been demonstrated in many investigations. At very high stresses and short times to rupture, the PFZ is absent or is very small, and deformation of the sample takes place uniformly in the matrix. At intermediate stresses and times to rupture, a narrow PFZ develops and facilitates strain localization in the PFZ. The worst case of this occurs at the ductility minimum. At very long times, beyond the ductility minimum, the PFZ widens sufficiently so that the creep strain can be accommodated over a wider PFZ near the boundaries, resulting in recovery of the ductility. The sequence of ductility changes thus has been explained in terms of the evolution of PFZ microstructures. The presence of segregated impurities would also contribute to the creep cavitation independent of the PFZ. Creep-embrittlement theories thus have included one or more of the effects due to grain-boundary sliding, evolution of PFZ's, and impurity effects (Ref 78 to 82).

At high temperatures, a drop in ductility occurs rapidly, but because of early recovery of ductility, very low values of ductility are never reached. At very low temperatures, low values of ductility can be reached, but only after very long times, because the rate at which the ductility minimum is reached is low. The worst combination of ductility minimum and time (stress) dependence occurs at intermediate temperatures. These trends result in a C-curve behavior for curves of isoductility in time-temperature space, as shown for three common varieties of steel in Fig. 3.12 (Ref 83 and 84). Increases in grain size, creep strength, and impurity levels (Ref 39, 78, and 85 to 87) are known to lead to premature decreases in ductility.

Estimation of Long-Term Ductility

Under service conditions, the drop of ductility to critical levels corresponding to the onset of notch-sensitive behavior generally occurs after very long times. It would therefore be very desirable to predict the long-time ductilities of materials based on short-time tests. The earliest attempts to do this were those of Smith (Ref 88), Goldhoff (Ref 89), and Booker (Ref 90).

In the procedure described by Goldhoff (Ref 89), raw data are collected and analyzed

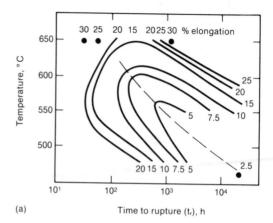

(a)   Time to rupture (t_r), h

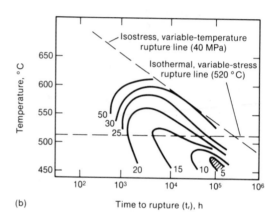

(b)   Time to rupture (t_r), h

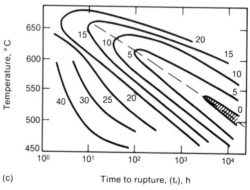

(c)   Time to rupture, (t_r), h

(a) ½ Mo steel (Ref 83). (b) 1Cr-½ Mo steel (Ref 84). (c) 1Cr-1Mo-¼ V steel (Ref 83).

Fig. 3.12. Constant-ductility (% elongation) contours in time-temperature space for three common steels.

using a computer program that correlates stress with a Larson-Miller-type parameter using polynomial equations. These equations have the form

$$\sigma = M_0 + M_1 P_1 + M_2 P_1^2 + M_3 P_1^3 \ldots$$

$$(\text{Eq } 3.47)$$

where $P_1 = (T + 460)(\log t_r + 20)$, and

$$\sigma = M_0' + M_1' P_2 + M_2' P_2^2 + M_3' P_2^3 \ldots$$

$$(\text{Eq } 3.48)$$

where $P_2 = (T + 460)(25 - \log \dot{E})$. M_0, M_1, M_0', M_1', etc., are simply the regression coefficients, and in the definitions of P_1 and P_2, T is temperature in °F, t_r is rupture time in hours, and $\dot{E}$ is the average elongation rate expressed in percent per hour. $\dot{E}$ is obtained by dividing the total elongation at rupture by the time to rupture.

To determine the rupture elongation at any given values of temperature and time to rupture, the stress corresponding to rupture is determined from Eq 3.47. At this stress, the value of $\dot{E}$ is computed using Eq 3.48. Multiplication of $\dot{E}$ by t_r gives the rupture elongation.

Several functional relationships among reduction in area, elongation, temperature, and stress were also explored by Viswanathan and Fardo (Ref 91) using the data on $1\frac{1}{4}$Cr-$\frac{1}{2}$Mo steels shown in Fig. 3.11. Poor correlations were obtained when the entire data set shown in the figure was included. The correlations were improved if the analysis was confined only to the region where percent reduction in area and percent elongation decrease with decreasing stress. An excellent correlation was obtained, however, between the average elongation rate and the time to rupture over the entire data set, as shown in Fig. 3.13. This correlation could be described as

$$\ln \dot{E} = 4.202 - 1.18 \ln t_r \qquad (\text{Eq } 3.49)$$

where t_r is expressed in hours and $\dot{E}$ is expressed in percent per hour. Before this relationship can be used to predict the long-term ductility of a material, a master curve similar to that in Fig. 3.13 will first have to be established on the basis of available data on similar heats. For any heat for which prediction is to be made, the value of $\dot{E}$

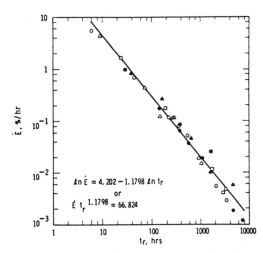

Different symbols denote different temperatures in the range 510 to 620 °C (950 to 1150 °F).

Fig. 3.13. Variation of average elongation rate with time to rupture for $1\frac{1}{4}$Cr-$\frac{1}{2}$Mo steels (Ref 91).

corresponding to the desired t_r and hence the total elongation at rupture ($\dot{E} \times t_r$) can be estimated from the master curve.

Note that the relationship between $\dot{E}$ and t_r depicted in Fig. 3.13 is independent of temperature and stress and is valid for t_r values covering three orders of magnitude. Schlottner and Seeley have recently used a similar approach to develop a ductility-based failure criterion (Ref 92).

Effects of Impurities

Because impurity elements are known to segregate to the prior austenite grain boundaries in steels and adversely affect toughness at low temperatures, their effects on high-temperature creep-rupture ductility have been of considerable interest. Hopkins, Tipler, and Branch showed that commercial-purity Cr-Mo-V-type ferritic steels had much lower rupture ductilities at 550 °C (1025 °F) compared with high-purity steels (Ref 93). The deleterious effects of aluminum and copper in Cr-Mo-V- and Cr-Mo-type steels have been reported by several investigators (Ref 77, 86, and 94 to 96). Deleterious effects of sulfur on rupture ductility also have been confirmed by the work of Middleton (Ref 97) and Pope (Ref 98). The effects of the temper-embrittling impurities antimony, phosphorus, tin, and arsenic seem to be controversial. Viswanathan (Ref 18) showed that the presence of large amounts of these elements in 1Cr-½Mo steels did not affect ductility at 540 °C (1000 °F). Pope *et al* found that in 2¼Cr-1Mo steels, phosphorus-containing steels exhibited higher ductility than a high-purity heat, provided that no prior segregation of phosphorus to grain boundaries had been allowed to take place (Ref 98). If prior segregation by step-cooling treatments had been allowed to occur, then the rupture ductilities were decreased. Gooch observed that in the absence of grain-boundary sulfides, the impurity elements antimony, phosphorus, and arsenic were totally innocuous. On the other hand, when sulfur was present, these impurity elements drastically increased the density of sulfide particles at the grain boundaries and severely impaired rupture ductilities (Ref 99). The most systematic study of the effects of impurities on 1Cr-1Mo-¼V rotor steels was carried out at 595 °C (1100 °F) by Roan and Seth (Ref 100), who showed than antimony, tin, phosphorus, arsenic, and sulfur decreased ductility. On a weight-percent basis, these elements were found to be decreasingly effective as listed. Roan and Seth determined an effective impurity content, which was correlatable to ductility as shown in Fig. 3.14. They defined an effective impurity content I_S, expressed in weight percent, as

$$I_S = 16.1Sb + 13.8Sn + 12.6P$$
$$+ 10.5As + 8.8S \qquad (Eq\ 3.50)$$

Note that antimony, tin, phosphorus, and arsenic rank in decreasing order of efficacy consistent with the misfit strain-energy considerations discussed in Chapter 2.

The discussion above points out that the effects of impurity elements on ductility are rather complex. This result is perhaps caused by synergisms between the impurity elements themselves and with other microstructural and test variables. Pending further studies of these complexities, the

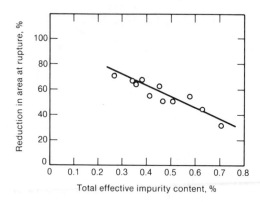

Fig. 3.14. Effect of impurity content (I_S = 16.1Sb + 13.8Sn + 12.6P + 10.5As + 8.8S) on rupture ductility of a Cr-Mo-V steel (Ref 100).

conventional wisdom is to assume that these trace elements are deleterious and to control their contents in steels to levels as low as possible.

Stress-Relief Cracking

Stress-relief cracking (SRC), sometimes called reheat cracking, refers to formation of intergranular cracks in the coarse-grained regions of the heat-affected zone, or occasionally in the weld metal, of a welded assembly when it is reheated to relieve residual stresses or when it is put into service at elevated temperature (Ref 101). Such cracking accompanies the relaxation of the residual stresses by creep and is a manifestation of poor creep ductility. It tends to be found in alloy steels in which some measure of high-temperature strength is imparted by the presence of strong carbide-formers such as chromium, molybdenum, vanadium, niobium, titanium, and tantalum. All the factors which adversely affect creep-rupture ductility also increase susceptibility to SRC.

The tendency for SRC to occur increases with the degree of mechanical constraint imposed on the weld region. It has recently been shown that the likelihood of SRC is related to the purity of the steel (Ref 102). In a study of Mn-Mo-Ni pressure-vessel steels of the A533B type, Brear and King showed that a high-purity heat prepared by vacuum induction melting was not susceptible to SRC (Ref 102). They reported that the cracking tendency increased with the value of the following impurity parameter (compositions in weight percent):

$$I_B = 2.7Sb + 1.9Sn + 1.0P + 1.8As$$
$$+ 0.44S + 0.20Cu \qquad (Eq\ 3.51)$$

This parameter was obtained from experiments on laboratory heats of fixed base composition in which the impurity content was systematically varied. Note that antimony, tin, phosphorus, and arsenic rank approximately in decreasing order of efficacy, consistent with the behavior observed with respect to creep ductility and temper embrittlement, as discussed earlier.

Based on the premise that SRC in steels depends on the presence of strong carbide-formers, two compositional parameters based mainly on such elements have been proposed by Nakamura *et al* (Ref 103) and by Ito and Nakanishi (Ref 104) as follows:

$$I_N = Cr + 3.3Mo + 8.1V - 2 \qquad (Eq\ 3.52)$$

$$I_I = Cr + 2.0Mo + 10V + 7Nb + 5Ti$$
$$+ Cu - 2 \qquad (Eq\ 3.53)$$

There is sound physical basis for the importance of both impurity-based and alloy-content-based parameters. It is expected that strong carbide-formers should facilitate grain-boundary cavitation, because of their strengthening effects on the base metal and because they provide potential sites for cavity nucleation. Impurities can segregate to grain boundaries and promote cavity nucleation by reducing the surface energy for cavitation. Roan and Seth have established clear correlations between impurity content and the extent of creep cavitation in rotor steels.

Pope *et al* investigated the SRC susceptibility of SA533B and SA508, grade 2 pressure-vessel steels (Ref 98). Simulated HAZ microstructures were produced by exposure at 1300 °C (2370 °F) followed by stress relief at 615 °C (1140 °F) for 6 h. Their results showed that the SRC tendency, as measured by notch-opening displacement,

could be rationalized by using a combined parameter which was essentially the one proposed by Brear and King (Eq 3.51) with the addition of a new term for chromium. The correlation between SRC susceptibility and the parameter proposed by Pope *et al* is shown in Fig. 3.15 (Ref 98). At least for their heat treatment conditions, it appeared that the other compositional parameters would have overemphasized the role of strong carbide-formers such as molybdenum and vanadium. Between the pressure-vessel steels SA508-2 (forging grade) and SA533B (plate grade), the former was found to be much more susceptible to SRC than the latter. A forging-grade steel SA508-3, which coupled the low SRC susceptibility of SA533B with the other desirable mechanical properties of SA508-2, was recom-

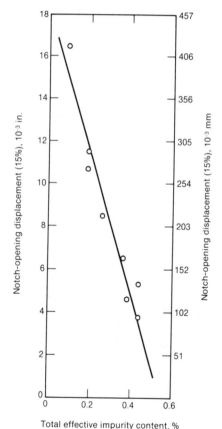

Fig. 3.15. Effect of impurity content (I = 0.20Cu + 0.44S + 1.0P + 1.8As + 1.9Sn + 2.7Sb + 1.0Cr) on stress-relief cracking of Cr-Mo steels as measured by notch-opening displacement (Ref 98).

mended as the best grade for use in nuclear pressure vessels. A reasonably complete, but somewhat outdated, review of the features and mechanisms associated with SRC may be found in the paper by Emmer, Clauser, and Low (Ref 105).

Monkman-Grant Correlation

Monkman and Grant (Ref 106) found that, for many alloy systems, the relation between the minimum creep rate $\dot\epsilon$ and time to rupture t_r can be expressed by the relation

$$\log t_r + m \log \dot\epsilon = \text{constant} \quad (Eq\ 3.54)$$

where m is a constant. For most of the materials the evaluated m had values approaching unity, so that Eq. 3.54 can be rewritten as

$$\dot\epsilon t_r = \text{constant} \quad (Eq\ 3.55)$$

For Cr-Mo-V rotor steels as well as for 2¼Cr-1Mo and other steels (see Fig. 3.16), Eq 3.54 is obeyed over a wide range of strength levels and test conditions. Close examination of such data may reveal a relationship between the Monkman-Grant constant and the rupture ductility of the material under a given set of test conditions. Within the range of scatter, however, such relationships are not readily discernible. Other than the suggestion that the mechanisms controlling deformation and fracture are always interrelated, no fundamental insight has been gained into the significance of the Monkman-Grant constant since its original publication. Currently, its usefulness lies in its ability to estimate rupture life. In the laboratory, a relatively short-time test to determine $\dot\epsilon$ alone may give an estimate of the eventual rupture life. In field components where the creep rate may be known from dimensional measurements, a crude estimate of rupture life can be made, using the Monkman-Grant constant for the steel as the limiting strain.

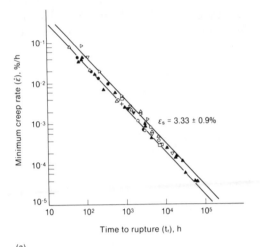

(a)

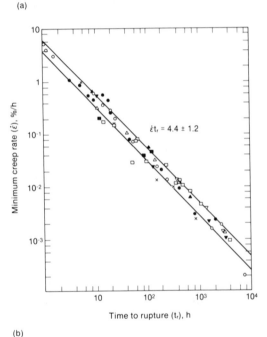

(b)

Fig. 3.16. Monkman-Grant relationships between minimum creep rate and time to rupture for (a) Cr-Mo-V steel (Ref 85) and (b) 2¼Cr-1Mo steel (Ref 39).

Creep Fracture

As with most other types of fracture, creep fractures occur by the nucleation and stable growth of cracks followed by unstable crack growth, leading to final fracture. Specific to creep fracture is the fact that the nucleation and stable growth of cracks are

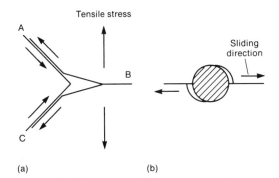

Fig. 3.17. Grain-boundary crack-nucleation mechanisms: (a) triple-junction cracking; (b) cavitation at particles (Ref 87).

time dependent and can occur at constant stress. Cracking can be transgranular at high stresses and intergranular at intermediate and low stresses. Two basic categories of intergranular cracking have been distinguished: (1) cavitation and (2) wedge cracking. Nucleation of both cavities and wedge cracks is generally believed to occur by grain-boundary sliding, as illustrated in Fig. 3.17. The sliding of grains with respect to each other sets up stress concentrations at the grain corners and at various irregularities along the boundaries. These stresses cause decohesion locally and at cavities, and cracks are thereby initiated. The irregularities along grain boundaries responsible for cavity nucleation are believed to be ledges and steps in the boundaries resulting from the interaction of slip traces or subgrain boundaries with the grain boundaries as well as with second-phase particles at the boundaries. The stress concentration, however, may be relaxed by the grain-boundary sliding itself. Cavity-nucleation theories essentially vary in terms of the postulated mechanisms for creating and accommodating the stress concentrations. Growth of cavities has been attributed variously to a purely diffusion-controlled mechanism, a grain-boundary-sliding-controlled mechanism, or a combination of the two. Growth of wedge cracks is attributed primarily to grain-boundary sliding. In general, wedge cracking has been observed at intermediate

stress levels (or strain rates), and cavitation has been observed at low stress levels. These observations are sometimes used to distinguish between stress-relief cracking and in-service creep failures. Various mechanisms for cavity and wedge-crack growth have been reviewed in the literature (Ref 1 to 3 and 107).

Creep-Fracture Maps

In a previous section, the decrease in rupture ductility that occurs with decreasing stress (increasing t_r) and the reasons for this behavior were discussed. The existence of "ductility windows" in time-temperature space for many steels also was described. The changes in ductility also are accompanied by changes in the fracture mode. At high stresses, short-time ruptures tend to be transgranular. With decreasing stresses and increasing times to rupture, the fractures tend to be increasingly intergranular, reaching 100% at the ductility minimum. If ductility recovery occurs in very-long-time tests, then the fracture mode once again becomes transgranular. The time-temperature dependence of fracture-mode transitions thus exhibits a behavior parallel to that of ductility transitions. Using this as a basis, fracture maps have been proposed that are similar to the deformation maps shown in Fig. 3.2. An example of such a map for 2¼Cr-1Mo steel is shown in Fig. 3.18. The stress-temperature regions where different fracture modes operate are delineated in the map (Ref 108).

Cavity Nucleation

Investigations on simulated heat-affected-zone microstructures in 2¼Cr-1Mo steels have shown that grain-boundary particles serve as the sites for cavity nucleation. Incoherent grain-boundary particles such as manganese and other sulfides are effectively nonwetting and provide ready sites for cavity nucleation. Further growth of the cavities would, however, require high stresses. Cavitation associated with such particles is instantaneous, but the nucleation saturates very early in creep life and the rest of the

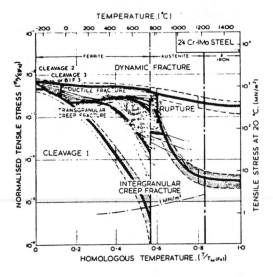

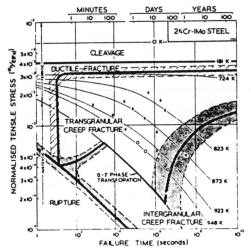

Above: Fracture modes in temperature-stress space; numbers with data points are values of log t_r. Below: Same data plotted differently to show change in fracture mode with t_r.

Fig. 3.18. Fracture-mechanism maps for annealed 2¼Cr-1Mo steel (Ref 108).

life is spent in the growth of these cavities. On the other hand, coherent particles such as grain-boundary carbides may first require decohesion of the particle/matrix interface by local stress-concentration effects. Hence the nucleation of cavities at these sites would have a strong dependence on stress. Cavity nucleation in this case will be a continuous process throughout the creep life. Evidence has been cited for both the instantaneous nucleation associated

with sulfide particles and the continuous nucleation process associated with carbide particles, in 2¼Cr-1Mo steels (Ref 109 to 114). The carbide particles involved in nucleation of cavities are believed to be of the M_2C type (Ref 112) as well as the $M_{23}C_6$ type (Ref 113).

Cavity Growth

The micromechanisms responsible for cavity growth in materials are not yet fully resolved. It seems likely that more than one mechanism pertains, depending on the stress, temperature, and creep duration (Ref 114). Three basic mechanisms have been distinguished and reviewed by Cane: (1) diffusional growth, (2) continuous growth, and (3) constrained growth (Ref 114).

In diffusional cavity growth, the rate of cavity growth, $\dot{V}_D$, is governed by grain-boundary vacancy diffusion such that (Ref 115)

$$\dot{V}_D = K_D\left(\sigma_{max} - \frac{2\gamma}{\omega}\right) \qquad \text{(Eq 3.56)}$$

where $\dot{V}_D$ is the rate of volume growth for a spherical cavity, σ_{max} is the maximum principal stress, ω is the cavity radius, γ is the surface energy, and K_D is a parameter dependent on the grain-boundary-diffusion coefficient, cavity size, and cavity spacing. Additional and special cases arise requiring modifications of Eq. 3.56 depending on the rigidity of the grains and the efficacy of the grain boundary as a vacancy source.

In the case of continuum- or plasticity-controlled cavity growth, cavities are postulated to grow at a rate $\dot{V}_P$ under the action of the hydrostatic stress σ_h, according to the expression (Ref 116)

$$\dot{V}_P = \delta V \dot{\epsilon}^* \sinh\left(\frac{\beta \sigma_h}{\sigma^*}\right) \qquad \text{(Eq 3.57)}$$

where V is the cavity volume, δ and β are parameters dependent on the Norton law exponent (n), $\dot{\epsilon}^*$ is the Von Mises effective creep rate, and σ^* is the corresponding stress.

In each of the above mechanisms, it is implicitly assumed that the dilation due to cavity growth can be readily accommodated by concurrent strain in the adjoining regions. Under conditions where the dilation due to cavity growth occurs at a rate higher than the creep rate of the surrounding matrix, the cavity growth becomes constrained to follow the creep-deformation kinetics, and the stress local to the cavitating regions falls below the applied stress (Ref 117 and 118). The constrained-cavity-growth rate, $\dot{V}_C$, is expressed as

$$\dot{V}_C = \pi c^2 l \dot{\epsilon} = \frac{3\pi}{2} c^2 l A \sigma_{max}^{n-1} J_3 \quad \text{(Eq 3.58)}$$

where c is the cavity half-spacing, l is the grain size, σ_{max} is the maximum principal stress, J_3 is the deviatoric stress in the direction of the principal stress, and A and n are the creep parameters in the equation $\dot{\epsilon}^* = A\sigma^{*n}$ (where σ^* and $\dot{\epsilon}^*$ are the Von Mises effective stress and creep rate, respectively).

The cavity-growth rates at different stress levels, as predicted by Eq 3.56 to 3.58, have been delineated by Cane in the form of a cavity-growth-mechanism map for simulated HAZ material in a 2¼Cr-1Mo steel, as shown in Fig. 3.19 (Ref 110). It is as-

sumed that the grain-boundary-diffusion coefficient is the same as that for pure α-iron (2.1 × 10^{-10} m²/h), the cavity half-spacing is a constant (c = 2 μm), and a continuum-growth-enhancement factor of 10 applies. The predicted growth rates are shown as functions of applied uniaxial stress and cavity radius, ω. The significant feature of this map is that, although the two unconstrained growth mechanisms (diffusional and continuous growth) and their variations may well occur, the constrained-growth law will always provide an upper boundary on the damage-accumulation rate. It also is the most likely mechanism under conditions of low stress, low strain rate, high cavity population (small c), and high creep strength. These are typically the conditions associated with coarse-grain HAZ structures in plant environments.

Failure Prediction Using the Constrained-Growth Model

The constrained-cavity-growth model is most relevant for coarse-grain heat-affected zones in steel weldments operating in plants. It provides an upper-boundary cavitation rate which can form the basis for a predictive failure model. Mathematically, the problem can be modeled using a Kachanov

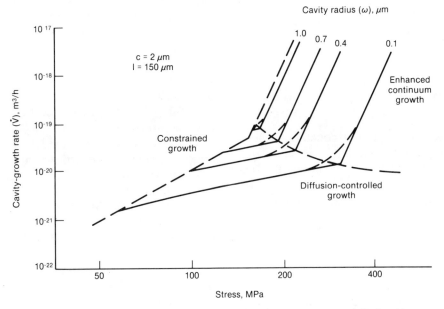

Fig. 3.19. Cavity-growth mechanisms for a 2¼Cr-1Mo steel (Ref 110).

approach with an explicit damage parameter (Ref 10). In the general Kachanov approach, the damage parameter is represented by D (Eq 3.5) and is not defined. In the constrained-cavity-growth model, illustrated by the mechanical analog in Fig. 3.20, there is a physically based damage parameter, A, which represents the number fraction of cavitated boundaries. In this model, two bicrystals are secured between rigid blocks and are subjected to an applied stress. In the region of constrained cavity growth, the growth rate of the cavities in the cavitated material (left-side bicrystal) is constrained by the deformation rate of the right-side bicrystal, which is supporting the applied load. The rate of cavitation is proportional to the strain rate. The steady-state strain rate is assumed to be related to the applied stress by the Norton power law. In the non-cavitated regions of the sample, the applied stress is magnified by the factor $1/(1 - A)$ as a result of the loss of load-bearing area of the cavitated regions in the cross section. Thus, the strain rate–stress relationship is given as

$$\dot{\epsilon} = G \left(\frac{\sigma}{1 - A} \right)^n \qquad \text{(Eq 3.59)}$$

A corresponding power-law relationship has been assumed to describe the time rate of change of the number fraction of cavitating boundaries—i.e., $\dot{A}$ vs stress:

$$\dot{A} = G' \left(\frac{\sigma}{1 - A} \right)^{n'} \qquad \text{(Eq 3.60)}$$

G and G′ are equation constants. These two differential equations can be solved simultaneously between the limits $A = 0$ at $t = 0$ and $A = 1$ at $t = t_r$ to yield the following expression for remaining life fraction in terms of A:

$$\left(1 - \frac{t}{t_r} \right) = (1 - A)^{n\lambda/(\lambda - 1)} \qquad \text{(Eq 3.61)}$$

where t/t_r is the life fraction expended and where $\lambda = \epsilon_r/\epsilon_s$. The remaining life can be

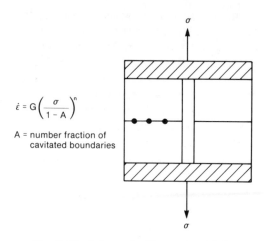

$$\dot{\epsilon} = G \left(\frac{\sigma}{1 - A} \right)^n$$

A = number fraction of cavitated boundaries

Fig. 3.20. Schematic illustration of the constrained-cavity-growth model.

calculated by metallographic measurement of A. Values have to be assumed for n and λ based on a knowledge of the material behavior. Application of this methodology is illustrated in Chapter 5. Detailed derivation of Eq 3.61 is described in Ref 20.

Environmental Effects

Design data for plant components are based on tests conducted in air. In actual operation, however, components may be exposed to steam, ash contaminants, combustion gases, and other aggressive environments. Rupture strength and rupture ductility of the materials in such environments may be reduced by one or more of the following mechanisms: (1) loss of net section and consequent increase in stress, (2) grain-boundary attack leading to notch effects and embrittlement, and (3) local changes in alloy composition.

The effects of simple oxidation on the creep and rupture properties of a Cr-Mo-V steel are shown in Fig. 3.21, which compares the creep behavior of specimens of four different sizes (Ref 119). The specimen with the largest diameter was found to have a rupture life almost three times that of the smallest-diameter specimen. When the thin specimens were tested in argon, however, the effect of section size disappeared, indicating a purely environmental effect (Ref 119).

Because most boiler piping is designed on

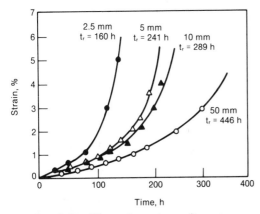

Fig. 3.21. Effect of specimen diameter on creep behavior of a ½ Cr-Mo-V steel tested in air at 675 °C (1245 °F) (Ref 119).

the basis of test data generated in air using conventional ASTM specimens, its design is essentially conservative from this point of view. Application of small-specimen data to large components can thus provide an added safety margin. On the other hand, use of the same data to predict the rupture life of superheater tubes exposed to ash corrosion can lead to nonconservative results. Hence, use of the appropriate data, taking the environmental effect into account, is always the prudent course.

Creep-Crack Growth

Failure due to creep can be classified as resulting either from widespread bulk damage or from localized damage. The structural components that are vulnerable to bulk damage (e.g., boiler tubes) are subjected to uniform loading and uniform temperature distribution during service. If a sample of material from such a component is examined, it will truly represent the state of damage in the material surrounding it. The life of such a component is related to the creep-rupture properties. On the other hand, components which are subjected to stress (strain) and temperature gradients (typical of thick-section components) may not fail by bulk creep rupture. It is likely that at the end of the predicted creep-rupture life, a crack will develop at the critical

location and propagate to cause failure. A similar situation exists where failure originates at a stress concentration or at preexisting defects in the component. In this case, most of the life of the component is spent in crack propagation, and creep-rupture-based criteria are of little value.

Crack-Tip Parameters

Several macroscopic load parameters, such as net section stress, σ_{net}, stress-intensity factor, K, the path-independent integrals J and C*, and other parameters such as C(t) and C_t, have been proposed and applied in the fracture-mechanics literature to describe crack-growth behavior in materials. The idea of a crack-tip parameter is that identical values of the appropriate parameter in differently shaped specimens or structures generate identical conditions of stress and/or deformation near the crack tip, so that the crack-growth rate must be the same provided that the material, the environment, and the temperature at the crack tip are also the same. Thus, such a parameter may be thought of as a transfer function from specimen to structural behavior. It then will be sufficient to measure the crack-growth rate as a function of the load parameter in the laboratory, and to calculate the value of the crack-tip parameter for the crack in the structure. The expected crack-growth rate in the structure then can be estimated.

In the subcreep-temperature regime involving crack growth under elastic or elastic-plastic conditions, the fracture-mechanics approach for predicting crack-growth behavior is well established. In the creep-temperature regime, the crack-tip parameter must take into account time-dependent creep deformation. Depending on the material and on the extent of creep deformation, various parameters mentioned above have been successfully correlated with rates of creep-crack growth.

Three regimes of crack growth—namely, small-scale, transient, and steady-state—can be distinguished for materials exhibiting elastic, power-law creep behavior, depending on the size of the crack-tip creep zone relative to the specimen dimensions, as shown in Fig. 3.22 (Ref 120). In the early stages of crack growth, the creep zone may be very small and localized near the crack tip. This regime is defined as the small-scale creep regime. At the other extreme, cracking may occur under widespread creep conditions where the entire uncracked ligament is subjected to creep deformation, as shown in Fig. 3.22(c). This regime is termed the large-scale or steady-state creep condition. Even in the latter case, creep-crack growth usually begins under small-scale conditions and, as the creep proceeds, the steady-state creep conditions develop. In between, the specimen passes through the transition creep conditions shown in Fig. 3.22(b). The transition time, t_1, from small-scale creep to steady-state creep conditions depends on several factors, including specimen geometry and size, load level, loading rate, temperature, and the kinetics of the creep. During the small-scale and transition creep conditions, the size of the creep zone and the stress at the crack tip change continuously with time (Ref 121 to 123). Under large-scale creep conditions, the crack-tip stress no longer changes with time. Hence, this regime is known as the steady-state regime. The nature (plasticity or creep) and size of the crack-tip deformation zone relative to the size of the specimen determine which of the parameters K, J, C^*, C(t), and C_t might be applicable to a given situation. For creeping materials, description of the phenomenology surrounding C^*, C_t, and C(t) is adequate. The parameters K and J, which do not account for time-dependent strain that occurs in the creep regime, are not applicable here.

The C^* Parameter. This parameter specifically addresses the steady-state (large-scale) creep-crack-growth regime. Under steady-state conditions, the path-independent integral C^* is defined as (Ref 124 and 125)

$$C^* = \int_\Gamma W^* dy - T_i \frac{\partial \dot{u}_i}{\partial x} ds \qquad (Eq \ 3.62)$$

where W^* is the strain-rate-energy density given by

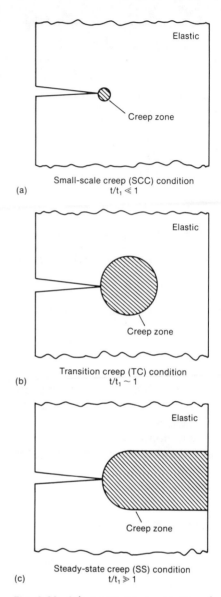

Fig. 3.22. Schematic representation of the levels of creep deformation under which creep-crack growth can occur (Ref 120).

$$W^* = \int_0^{\dot{\epsilon}_{ij}} \sigma_{ij} d\dot{\epsilon}_{ij} \qquad (Eq \ 3.63)$$

In Eq 3.62, T_i is the traction vector along the path τ which originates at a point along the lower crack surface, goes counterclockwise, and ends at a point on the upper crack surface. Thus the contour encloses the crack tip. The terms σ_{ij} and $\dot{\epsilon}_{ij}$ are the stress and strain-rate tensors, $\dot{u}_i$ is the deflection-rate vector along the direction of the traction, and ds is a length element along τ. The C^* parameter is analogous to the J contour integral (see Eq 2.26 and 2.27, and Fig. 2.14) with the difference that strain and strain-energy density are replaced by strain rate and strain-energy-rate density, respectively. When secondary creep dominates, for a given crack length and loading conditions, C^* will be independent of time. C^* will be subject to the same restrictions with respect to path independence as J. The following equations relate C^* to the crack-tip stress and strain-rate fields when $r/a \to 0$ (Ref 126):

$$\sigma_{ij} = \left(\frac{C^*}{AI_n r}\right)^{1/(n+1)} \tilde{\sigma}_{ij}(\theta) \qquad \text{(Eq 3.64)}$$

$$\dot{\epsilon}_{ij} = A \left(\frac{C^*}{AI_n r}\right)^{n/(n+1)} \tilde{\epsilon}_{ij}(\theta) \qquad \text{(Eq 3.65)}$$

where a is the crack length, r is the distance from the crack tip, θ is the angle from the plane of the crack, and $\tilde{\sigma}_{ij}(\theta)$ and $\tilde{\epsilon}_{ij}(\theta)$ are angular functions specified in Ref 127. A is the Norton law coefficient in the relation between stress and steady-state creep rate. I_n is a constant dependent on the steady-state creep exponent n, whose values may be found in tables (Ref 127). For most values of n of practical interest, I_n can be expressed approximately as 3 for plane-stress conditions and 4 for plane-strain conditions. Thus, C^* characterizes the strength of the crack-tip-stress singularity commonly known as the Hutchinson-Rosengren-Rice (HRR) singularity (Ref 128 and 129).

Examination of Eq 3.64 indicates that when n = 1, C^* predicts the same stress distribution ahead of a crack tip as K. Consequently, for n = 1, or where elastic strains dominate, good correlations of crack-growth rate with K would be expected. In the limit of $n \to \infty$, the singularity at the crack tip disappears and C^* and the net section stress or reference stress give equivalent stress distributions. Thus, the C^* parameter is capable of encompassing parameters K and σ_{net} as special cases (Ref 130).

The analytical expression for C^* has the general form

$$C^* = aA\sigma_{net}^{n+1} g_1 \left(\frac{a}{W}, n\right) \qquad \text{(Eq 3.66)}$$

where the dimensionless quantity g_1 is a function of a/W and n. Kumar *et al* have evaluated the values of g_1 using finite-element solutions for various specimen and component geometries and a/W ratios (Ref 131). For example, for ASTM compact tension specimens, they express g_1 as

$$g_1 = \frac{h_1[W/(a-1)]}{(1.455\eta)^{n+1}} \qquad \text{(Eq 3.67)}$$

The function h_1 is tabulated for plane stress and plane strain, whereas η is given analytically. Based on a knowledge of a, A, σ_{net}, n, and g_1 applicable to a specific specimen or component geometry, C^* can be determined analytically.

Experimental determination of C^* in the laboratory involves testing precracked specimens of known geometry under an applied load P and measuring the creep deflection rate $\dot{V}$. The expression for obtaining C^* in test specimens can be written as (Ref 123 and 132)

$$C^* = \frac{P\dot{V}}{BW} \eta \left(\frac{a}{W}\right) \qquad \text{(Eq 3.68)}$$

where a is the crack half-length, B is the specimen thickness, and W is the width of the CT specimen and half-width of the CCT specimen. Specific expressions for $\eta(a/W)$ and the necessary constants for calculating η for various a/W values are available in handbooks (Ref 131). For instance, for the center-cracked tension (CCT) specimen, η is given by (Ref 120)

$$\eta\left(\frac{a}{W}\right) = \left[\frac{1}{2(1-a/W)}\right]\left[\frac{n-1}{n+1}\right]$$

$$\text{(Eq 3.69)}$$

and for the compact type (CT) specimen by

$$\eta\left(\frac{a}{W}\right) = \left[\frac{1}{1 - a/W}\right]\left[\frac{n}{n + 1}\right]\left[\phi_1 - \frac{\phi_2}{n}\right]$$

(Eq 3.70)

where the values of ϕ_1 and ϕ_2 are obtained from the literature (Ref 133). Methods for estimating C^* have been discussed in several papers (Ref 124, 125, and 134). Because C^* is the analog of the J-integral as applied to creep, it may be obtained by making appropriate substitutions in the expression for estimating the J-integral under fully plastic conditions. These expressions are given in Ref 135 and 136 and in many others for several configurations, primarily through the work of Hutchinson, Shih, and Kumar. The EPRI handbook (Ref 131) contains a collection of several such solutions. Recently, Jaske has developed procedures for estimating values of the C^*-integral from previously developed estimates of the J-integral (Ref 137). Simple expressions for calculating C^* have been developed for several commonly used test-specimen configurations. These include the compact-type (CT), center-cracked tension (CCT), three-point bend (TPB), low-cycle fatigue (LCF), and edge-notch tension (ENT) specimens. Two examples that illustrate the simplicity of these expressions are as follows:

$$C^* = \left[\frac{n}{n + 1}\right]\left[\frac{2.3P(d\delta_c/dt)}{BW(1 - a/W)}\right]$$

(Eq 3.71)

for CT specimens, and

$$C^* = \frac{2Pn(d\delta_c/dt)}{B(n + 1)(W - a)}$$

(Eq 3.72)

for TPB specimens, where P is load and δ_c is load-line deflection.

Small-Scale Creep Regime. In view of the fact that C^* is a parameter applicable to large-scale (steady-state) creep conditions, its applicability to small-scale creep conditions, where crack extension occurs under predominantly elastic conditions, is limited (Ref 138). In most real situations, however, crack growth will start in the small-scale creep regime and eventually progress to the large-scale creep regime. Hence, the early crack-growth rates will not correlate with C^*, and an alternative parameter must be sought. The transition from one regime to the other can be characterized in terms of the time it takes for the initial short-term elastic stresses characterized by K at the crack tip to relax to the long-range stresses characterized by C^* beyond the crack tip. Ohji *et al* have reviewed the applicability of K or C^* for a variety of steels (Ref 139).

The relaxation of the stress near the crack tip is analytically described by the time-dependent, HRR-type asymptotic stress field (Ref 140):

$$\sigma_{ij} = \left[\frac{K^2(1 - \nu^2)/E}{(n + 1)I_n Art}\right]^{1/(n+1)} \tilde{\sigma}_{ij}(\theta)$$

(Eq 3.73)

where A is the Norton law coefficient, r is radial distance from the crack tip, and t is time. When this stress level becomes equal to the long-time steady stress level as given by Eq 3.64, the transition will occur from K control to C^* control. By equating Eq 3.64 and Eq 3.73, the transition time can be shown to be (Ref 138 and 141)

$$t_1 = \frac{K^2(1 - \nu^2)}{E(n + 1)C^*}$$

(Eq 3.74)

where t_1 is the time for transition from the initial, elastically dominated to the final, creep-dominated response of the material. It is therefore desirable to calculate the transition time when creep-crack-growth tests are done. For times shorter than t_1, a good correlation can be expected with K, but for times in excess of t_1, better correlations will result with C^*. Based on the above line of reasoning, Ehlers and Riedel have defined a crack-tip parameter C(t)

(Ref 142). This parameter provided a reasonable estimate of the HRR field over a wide range of conditions. The value of C(t) was determined by simply adding its asymptotic limits for t → 0 and t → ∞ as follows:

$$C(t) = \frac{K^2(1 - \nu^2)}{E(n + 1)t} + C^* \qquad (Eq\ 3.75)$$

As t → 0, the first term becomes the controlling parameter, while as t → ∞, C(t) → C*. Further, from Eq 3.73 and 3.75 it can be seen that, under small-scale creep conditions, the crack-tip stress is characterized by K but the relationship is not unique. Hence, K by itself is not a likely creep-crack-growth parameter even under small-scale creep conditions.

The C_t Parameter. The parameter C_t, proposed by Saxena, is another attempt to extend the C* concept into the nonsteady-state crack-growth regime (Ref 123). The connection is made through an energy-rate interpretation of C*.

Consider several identical pairs of cracked specimens. Within each pair, one specimen has a crack length a and the other has an incrementally different crack length (a + Δa). The specimens of each pair are loaded to various load levels P_1, P_2, P_3, etc., at elevated temperatures and the load-line deflection as a function of time is recorded as in Fig. 3.23(a).

The load-line deflection due to creep is V_c. It is assumed that no crack extension occurs in any of the specimens and that the instantaneous response is linearly elastic. We first limit our consideration to small-scale creep conditions. At a fixed time t, load vs $\dot{V}_c$ can be plotted as shown in Fig. 3.23(b). Several such plots can be generated by varying the time.

The area between P-vs-$\dot{V}_c$ curves for specimens of crack lengths a and (a + Δa) is designated ΔU_t^*. It represents the difference in the energy rates (or power) supplied to the two cracked bodies with identical creep-deformation histories. The C_t parameter is defined as (Ref 120 and 123)

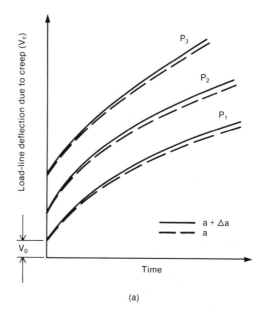

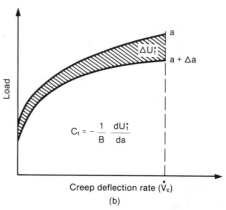

Fig. 3.23. (a) Load-line deflection as a function of time for bodies of crack lengths a and (a + Δa) at various load levels, and (b) definition of the C_t parameter (Ref 120).

$$C_t = \lim_{\Delta a \to 0}\left(-\frac{1}{B}\frac{\Delta U_t^*}{\Delta a}\right) = -\frac{1}{B}\frac{\partial U_t^*}{\partial a}$$

$$(Eq\ 3.76)$$

The generalized expression for calculating C_t from measurements of load vs deflection rate on laboratory samples has been given as (Ref 120)

$$C_t = \frac{P\dot{V}_c}{BW}\frac{F'}{F} - C^*\left(\frac{F'}{F}\frac{1}{\eta} - 1\right) \quad (Eq\ 3.77)$$

Under extensive creep conditions, C_t can be simply calculated from the C^* expression (Eq 3.68).

The analytical expression for calculating C_t has been given as (Ref 120)

$$C_t = \frac{4\alpha\beta(1 - \nu^2)}{E(n - 1)} \frac{K^4}{W}$$

$$\times (EA)^{2/(n-1)} t^{-(n-3)/(n-1)} \frac{F'}{F} + C^*$$

$$\text{(Eq 3.78)}$$

where β has a value of approximately 1/7.5. In Eq 3.77 and 3.78, the first term denotes the contribution from small-scale creep and the second term denotes the contribution from steady-state, large-scale creep. Clearly, the first term is time-variant whereas the second term is time-invariant. In the limit of $t \to 0$, approaching small-scale creep conditions, the first term dominates, implying that K is the controlling parameter in crack growth, with time also explicitly entering the relationship. In the limit $t \to \infty$, the first term becomes zero and C_t becomes identical with C^*. In Eq 3.77, F is a function of (a/W) and F' is given by dF/d(a/W). In Eq 3.78, α is a constant whose value is a function of n, and A and n are the Norton law coefficients.

Equation 3.78 can be used to estimate C_t from an applied load (stress) and from a knowledge of the elastic and creep behavior of the material, the K calibration expression, and the C^* expression for the geometry of interest. The K and C^* expressions can be found in handbooks—at least for selected geometries (Ref 131 and 143). The material properties A and n can be obtained from creep tests. The C^* expressions are not as abundantly available for different geometries as the K expressions. At the present time, this is viewed as a limitation of the technology. More detailed descriptions of the derivations of the C^* and C_t expressions, and the manner of obtaining some of the constants and calculating their values, are presented in the literature

which has been referred to in this section. Procedures for estimating C^* based on the reference-stress approach also have been described by Ainsworth *et al* (Ref 75).

Applicability of the J Parameter. Development of the J-integral concept was discussed in Chapter 2. The applicability of the J-integral as the crack-tip parameter for describing creep-crack growth has been reviewed by Riedel (Ref 138).

In relatively short-time tests, the load level may be so high that extensive plasticity develops at the crack tip immediately upon application of the load. In such a material, the J-integral determines the short-time stress field (instead of K), whereas the C^*-integral determines the long-time stress field at the crack tip. For a strong strain-hardening material, the transition can be treated in a manner similar to that for the elastic nonlinear viscous case described earlier. The characteristic transition time has been found to be (Ref 144)

$$t_2 = \frac{J}{(n + 1)C^*} \qquad \text{(Eq 3.79)}$$

Saxena *et al* have performed creep-crack-growth tests on AISI type 316 stainless steel at 595 °C (1100 °F) and at relatively high load levels so that the specimen became fully plastic upon application of load (Ref 145). The specimens used were single-edge-notched specimens. Calculations using Eq 3.79 for the test conditions showed that the transition time from J-controlled to C^*-controlled crack growth was about 200 h. Because experimental durations were less than the transition time, Saxena *et al* found that the crack-growth rates correlated better with J than with C^*.

Experimental Results

Experimental results usually consist of plots of the creep-crack-growth rate (da/dt or $\dot{a}$) vs the appropriate parameter chosen in the study (K, C^*, or C_t). The effects of such variables as temperature, environment, prior degradation, and impurity content on the crack-growth behavior also have been evaluated in some studies. In reviewing these results, a clear-cut separation exists between the use of K on the one hand, and the use of C^* and C_t on the other hand, as the applicable load parameter. As described earlier, the appropriate use of K is limited to small-scale creep conditions where the creep zone is appreciably smaller than the specimen or component dimensions. Brittle and highly creep-resistant materials, test durations below the transition time as calculated from Eq 3.74, applications to large components such as rotors, and instances where oxidation plays a major role in crack growth justify the use of K as the load parameter. Based on this rough distinction, data on many nickel-base superalloys have generally been characterized in terms of K, whereas the data on many ferritic steels have been characterized in terms of C^* or C_t. Although the C_t parameter, in principle, is capable of characterizing small-scale creep, its applicability to superalloys has not been sufficiently explored. Hence, for convenience, we can separate the results on ferritic steels and those on nickel-base superalloys.

Ferritic Steels. Ferritic steels used for piping, casings, and pressure vessels are generally characterized by relatively low creep strength and good ductility. The transition times from small-scale creep to large-scale creep often have been sufficiently small compared with experimental times to justify the use of C^* or C_t parameters (Ref 138). Several experimental studies have demonstrated the applicability of C^* (Ref 124, 125, 136, and 146) as well as that of C_t (Ref 120 to 122 and 125). A limited number of studies have utilized K as the load parameter (Ref 147 and 148). Airoldi, Bianchi, and D'Angelo concluded that although K was a suitable parameter in the case of brittle weld metal, it was not adequate to describe the behavior of the ductile Cr-Mo-V base metal (Ref 147). Saxena, Han, and Banerji collected, reviewed, and reanalyzed most of the published data relating to crack growth in ferritic piping and pressure-vessel

steels (Ref 149). In order to have a consistent basis for comparison, they correlated all creep-crack-growth rates in terms of C_t, because this parameter is suitable for characterizing crack growth over a wide range of creep conditions. If the data in the original study had been correlated with K, the raw data were completely reanalyzed in terms of C_t. In order to perform the reanalysis, they needed measurements of crack size and load-line deflection as functions of time, and thus only those data which provided these basic measurements could be included in their analysis. If the creep-crack-growth data in the original study had been correlated with C^*, they were used directly in the $\dot{a}$-vs-C_t plot, provided the data were obtained on compact-type (CT) specimens, because the estimated error in doing this would be small. By normalizing all available data in terms of a single parameter, these investigators were able to draw several useful conclusions, as described below.

Effect of Alloy Content and Service Exposure. Creep-crack-growth-rate data for several Cr-Mo-type steels are all plotted together in Fig. 3.24. These data pertain only to base metal and include material in the virgin condition as well as in the service-exposed condition. Regardless of alloy con-

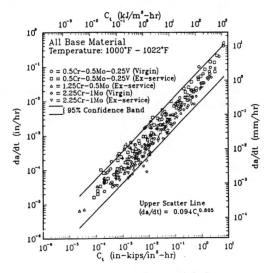

Fig. 3.24. Creep-crack-growth behavior of ferritic steel base metal (Ref 149).

tent and prior degradation, all the data could be plotted within a small band of scatter. The correlation between à and C_t could be expressed in the form

$$\dot{a} = bC_t^m \qquad \text{(Eq 3.80)}$$

Values of the constants b and m for all the materials analyzed by Saxena *et al* are listed in Table 3.2. Based on Eq 3.64 and 3.65, it can be shown that m should have the approximate value $n/(n + 1)$, where n is the creep-rate exponent. The results shown in Fig. 3.24 are not to be interpreted as indicating that alloy content and service exposure do not affect creep-crack-growth rates. For identical test conditions, differences in creep rates, and hence in C_t values, are likely to exist, resulting in appreciably different creep-crack-growth rates in the materials. Figure 3.24 simply shows that the C_t parameter normalizes this behavior so that a single plot describes all the data. Thus, the material with the greatest creep resistance would exhibit the lowest creep-crack-growth rate under comparable loading conditions.

Crack Growth in Weldments. A limited amount of data pertaining to welds and simulated HAZ structures has also been analyzed and plotted similar to Fig. 3.24 by Saxena *et al* (Ref 149). The scatter in these data was found to be greater than the scatter for the base metal. Once again, virgin material and service-exposed material did

not show appreciable differences. Crack-growth rates in the weld metal were generally found to be higher, for a given C_t, by a factor of 4 in the case of 2¼Cr-1Mo steels, but a similar trend could not be confirmed for 1¼Cr-½Mo steels.

In the case of heat-affected-zone/fusion-line (HAZ/FL) crack growth, the over-all scatter band was found to be shifted upward by a factor of 4 to 5 in comparison with that for the base metal, as shown in Fig. 3.25. No difference was observed between ¼Cr-½Mo and 2¼Cr-1Mo steels. Analysis of the data of Konosu and Maeda

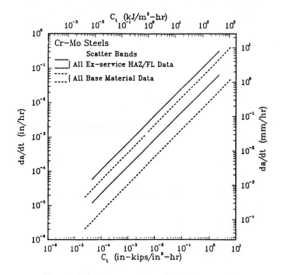

Fig. 3.25. Comparison of crack-growth-rate scatter bands for ex-service heat-affected-zone/fusion-zone material and base material for ferritic steels (Ref 149).

Table 3.2. Summary of creep-crack-growth constants b and m (Eq 3.80) for various ferritic steels (Ref 149)

Material	b Upper scatter line BU(a)	b Upper scatter line SI(b)	b Mean BU(a)	b Mean SI(b)	m Upper scatter	m Mean
All base metal	0.094	0.0373	0.022	0.00874	0.805	0.805
2¼Cr-1Mo weld metal	0.131	0.102	0.017	0.0133	0.674	0.674
1¼Cr-½Mo weld metal	(c)	(c)	(c)	(c)	(c)	(c)
2¼Cr-1Mo and 1¼Cr-½Mo heat-affected-zone/fusion-line material	0.163	0.0692	0.073	0.031	0.792	0.792

(a) BU = British units: da/dt in in./h; C_t in in.·lb/in.·h × 1000. (b) SI = Système Internationale units: da/dt in mm/h; C_t in kJ/m²·h. (c) Insufficient data; creep-crack-growth-rate behavior comparable to that of base metal.

(Ref 150) by Saxena *et al* (Ref 149) showed that postweld heat treatment also may be an important variable affecting the rate of creep-crack growth in the HAZ. As shown in Fig. 3.26, simulated HAZ material postweld heat treated at 1148 °F (620 °C) exhibited higher rates of crack growth than material postweld heat treated at 1328 °F (720 °C) — presumably because of reduced ductility resulting from the lower-temperature heat treatment.

Effect of Temperature. From Eq 3.66, it can be seen that the temperature dependence of C^* or C_t comes from the temperature dependence of A through an expression of the form $A = A_0 \exp(-Q/RT)$. Increasing the temperature would lead to an increase in C_t and hence increased crack growth, as dictated by Eq 3.80. Although this effect may be operative, it may be countered by a trend of increasing ductility at high temperature. The effect of changing ductility with temperature is not currently taken into account. Figure 3.27 is a plot of à vs C^* for 1Cr-½Mo steel at 540, 565, and 590 °C (1000, 1050, and 1095 °F) based on the work of Henshall and Gee (Ref 151). It is clear from this figure that for a given value of C^*, the da/dt lines are shifted to lower values as the temperature is increased. These trends are in agreement with other test observations that as the test temperature is increased, the crack-growth rate for a given value of C_t or C^* decreases (Ref 130 and 152). They are, however, in disagreement with the test results of Riedel and Wagner on a similar steel showing little influence of temperature variation in the range 450 to 600 °C (840 to 1110 °F) on the da/dt-vs-C^* behavior (Ref 153).

Effect of Ductility. Nikbin, Smith, and Webster have developed a model that explicitly takes into account the effects of ductility and constraint on crack-growth rate (Ref 152). According to this model, the proportionality factor b in Eq 3.80 was found to be inversely proportional to the ductility ϵ_f appropriate to the state of stress at the crack tip but insensitive to the size of the process zone. The appropriate modifi-

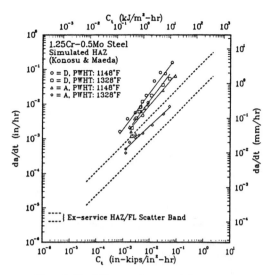

Fig. 3.26. Effects of material composition (steel A had a lower impurity content than steel D) and simulated postweld heat treatment on creep-crack-growth behavior of 1¼Cr-½Mo steels (Ref 149).

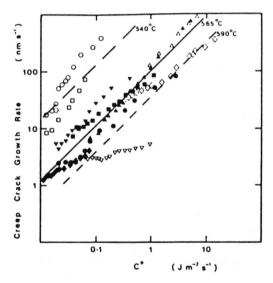

Fig. 3.27. Effect of temperature on crack-growth rate for a 1Cr-½Mo steel (Ref 151).

cations of b led to the following expressions:

$$\dot{a} = \frac{300}{\epsilon_f} C^{*0.85} \qquad \text{(Eq 3.81)}$$

for plane-stress conditions, and

$$\dot{a} = \frac{15,000}{\epsilon_f} C^{*0.85} \qquad (Eq\ 3.82)$$

for plane-strain conditions, where $\dot{a}$ is expressed in mm/h, ϵ_f in %, and C^* in $MJ/m^2 \cdot h$. It was found that crack growth for plane-stress situations can be predicted approximately by substituting the uniaxial creep ductility of the material into the equation. As constraint is increased, higher crack-growth rates are obtained and a reduced ductility more relevant to the increased degree of triaxiality at the crack tip must be used.

Effects of Impurities. The effects of impurities on crack-growth rates are also incorporated in Fig. 3.26 (Ref 149), based on the work of Konosu and Maeda (Ref 150). In this figure, steel D is an impure heat whereas steel A is a high-purity heat of $1\frac{1}{4}$Cr-$\frac{1}{2}$Mo steel. For both postweld heat treatments used, the impure heat exhibits a higher crack-growth rate than the pure heat, for a given value of C_t. This effect has also been confirmed by the work of Lewandowski, Ellis, and Knott (Ref 154 and 155). Between these two studies, there is some disagreement regarding the role of sulfur in causing embrittlement. Konosu and Maeda did not find any segregation of sulfur on the grain-boundary fracture surfaces, whereas Lewandowski *et al* indicate considerable sulfur segregation. The effects of impurities and the role of segregation need to be clarified by further research.

Limitations of Current Approaches. Several limitations of C^* as a crack-tip parameter have been reviewed by Riedel (Ref 138). Subsequent development of C(t) and C_t has served to overcome at least one of these limitations — i.e., extending the parameter to include the elastic, or small-scale, creep regime. Only limited attempts have been made to modify these parameters so as to take into account deformation due to primary creep. It may also be important to take into account variations in creep-rate response of microstructurally unstable materials entering tertiary creep. Crack-tip blunting also may be a limiting factor in application of these parameters and may become particularly important in the context of cyclic loading. Because the effect of grain-boundary cavitation on the stress field is not included in the nonlinear viscous descriptions of the material, the C^*, C(t), and C_t approaches might be invalidated by profuse cavitation of the whole ligament. Although analytical solutions for C^* are available from handbooks for a range of geometries, the list is far from complete. Liaw and Saxena have also raised a cautionary note that the HRR field for stationary cracks assumed in deriving C^* and C_t may not be valid for growing cracks (Ref 120).

In general, crack-growth-rate data relating to various materials are insufficient and have too much scatter to permit optimal use of the crack-growth methodology for plant assessment. The effects of variables such as temperature, recovery processes, ductility, impurity content, and crack-tip constraint have not been sufficiently sorted out. These variables can affect crack-growth rates by changing the value of C^* or C_t or by changing the proportionality constant b in Eq 3.80. Although these effects sometimes may be in the same direction, at other times they may counteract each other. For instance, an increase in temperature may contribute to an increase in C^* but to a decrease in b as a result of increased ductility. Similarly, an increase in creep strength may decrease the value of C^* but increase the value of b by decreasing the ductility of the material. Impurities are expected to have little effect on C^* or C_t but can change crack-growth rates by changing the value of b. Many of these subtleties have yet to be sorted out by experiments.

Recently, the results of a multiorganizational cooperative test program using an ASTM 470, class 8, Cr-Mo-V rotor steel as the reference material have been reported by Saxena and Han (Ref 156). These tests were conducted in five separate laboratories in the United States, West Germany, and Japan. The da/dt behavior in compact-type specimens tested at 595 °C (1100 °F) were found to correlate better with C_t than with K or C^* over a wide range of creep conditions. Differences in crack-growth behavior due to utilization of different types of specimens (CT vs CCT) also were observed. The study group identified the need for developing ASTM standards for creep-crack-growth testing and for developing guidelines for selecting proper crack-tip field parameters as a major one deserving additional attention by the research community.

Nickel-Base Superalloys. Most creep-crack-growth studies conducted on superalloys justify the use of K as the appropriate crack-tip parameter. These materials are generally characterized by high strength and low ductility, and the times for transition from small-scale to large-scale creep are far in excess of the test durations employed.

Crack growth in Nimonic 80A has been studied at 650 °C (1200 °F) by Riedel and Wagner (Ref 153). The data suggest a bi-linear dependence of da/dt on K on a log-log plot, as shown in Fig. 3.28. At high values of K, the slope was found to be between 4 and 5, whereas at low K values the slope was found to be about 13. The existence of a threshold K value below which no crack growth occurred was found. The activation energy for crack growth was determined to be in the range 240 to 300 kJ/mole (60 to 75 kcal/mole). Crack-growth rates in an environment of argon plus 3% hydrogen were found to be ten times higher than those in air, indicating a strong environment sensitivity.

The environment sensitivity of creep-crack growth in Inconel 718, Inconel X-750, and Udimet 700 was investigated by Sadananda and Shahinian (Ref 157). In all cases, the log da/dt–vs–log K behavior was similar to that shown in Fig. 3.28, with a threshold K value and a bilinear relationship above that value. For materials that showed a brittle mode of crack growth characterized by continuous growth of the main crack, such as Inconel 718 and Inconel X-750, the crack-growth rates in air were significantly higher than those in vacuum. In Udimet 700, a ductile crack-growth mode characterized by nucleation of microcracks

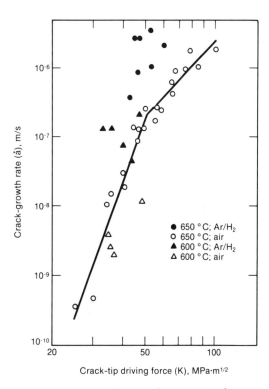

Fig. 3.28. Crack-growth rate vs K for Nimonic 80A (Ref 153).

ahead of the main crack and subsequent joining of the main crack with the microcracks was observed. In this case, crack growth was found to be relatively insensitive to environment, but much more sensitive to changes in microstructure. Some of the results from this work are presented in Fig. 3.29.

The creep-crack-growth behavior of cast Inconel 738 and Inconel 939 at various temperatures around 850 °C (1560 °F) has been investigated by Nazmy and Wuthrich (Ref 158). Figure 3.30 shows the dependence of da/dt on K for these two alloys. Because these data have been generated mainly in the high-K regime, the bilinear variation discussed earlier is not apparent. The stress-intensity exponent m—i.e., the slope of the log da/dt–vs–log K plots in Fig. 3.30 were found to be identical to the creep-rate exponent n, suggesting that growth was governed by the nucleation and growth of cavities at the grain boundaries. Crack-growth rates increased with increasing temperature. Plots of log da/dt vs 1/T at constant stress re-

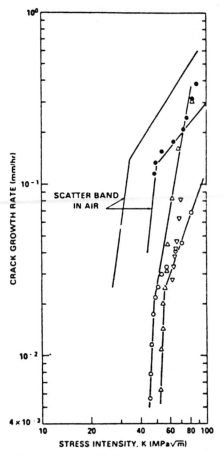

o 20 kN in vacuum. △ 24.4 kN in vacuum. ▽ 27.8 kN in vacuum. ● 20 kN in air.

Fig. 3.29. Effect of environment on creep-crack growth in Inconel 718 at 540 °C (1000 °F) (Ref 157).

sulted in an activation energy of 280 kJ/ mole (70 kcal/mole), which closely corresponds to the activation energy for volume diffusion in Ni-Cr alloys.

Lupinc has suggested that the da/dt – vs – K behavior in a very general sense should be characterized by trilinear dependence. The second-stage crack growth presumably represents the steady-state crack-growth rate, similar to the steady-state creep rate. As with the Monkman-Grant relationship, the equation (Ref 8)

$$\dot{a} t_r = \text{constant} \qquad \text{(Eq 3.83)}$$

where $\dot{a}$ is the second-stage crack-growth rate at constant stress intensity, was found

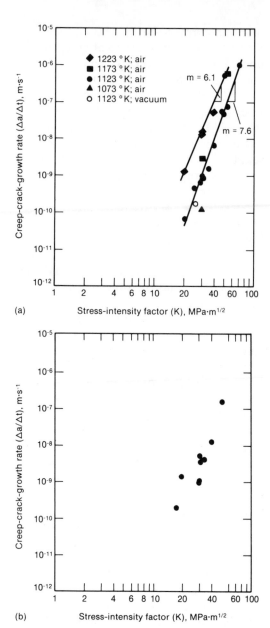

Fig. 3.30. Crack-growth rate vs stress-intensity factor for (a) Inconel 738 and (b) Inconel 939 (Ref 158).

to be valid in many gas-turbine superalloy disk materials, as shown in Fig. 3.31.

Remaining-Life-Assessment Methodology

The contexts in which bulk damage by creep and localized damage by creep-crack growth are applicable have already been discussed.

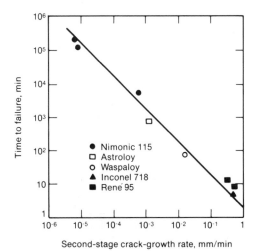

Fig. 3.31. Monkman-Grant-type correlation between time to failure and creep-crack-growth rate for gas-turbine disk alloys (Ref 8).

In this section, remaining-life-assessment methodologies applicable to these two situations are briefly reviewed. More extensive descriptions of these methodologies are presented in later chapters on a component-specific basis.

Assessment of Bulk Creep Damage

The current approaches to creep-damage assessment of components can be classified into two broad categories: (1) history-based methods, in which plant operating history in conjunction with standard material-property data is employed to calculate the fractional creep life that has been expended, using the life-fraction rule or other damage rules described earlier; and (2) methods based on postservice evaluation of the actual component.

In history-based methods, plant records and the time-temperature history of the component are reviewed. The creep-life fraction consumed for each time-temperature segment of the history can then be calculated and summed up using the lower-bound ISO data and the life-fraction rule. This procedure usually is inaccurate because of errors in assumed history, in material properties, and in the life-fraction rule itself. The temperature-history information

may be somewhat refined by supplemental information concerning the current oxide-scale thickness and microstructural details. In spite of such refinements, only gross estimates of creep damage are obtained using the calculation technique.

Direct postservice evaluations represent an improvement over history-based methods, because no assumptions regarding material properties and past history are made. Unfortunately, direct examinations are expensive and time-consuming. The best strategy is to combine the two approaches. A history-based method is used to determine if more detailed evaluations are justified and to identify the critical locations, and this is followed by judicious postservice evaluation. Table 3.3 summarizes the techniques that are in use for life assessment and some of the issues pertaining to each technique (Ref 159).

Current postservice evaluation procedures include conventional NDE methods (e.g., ultrasonics, dye-penetrant inspection, etc.), dimensional (strain) measurements, and creep-life evaluations by means of accelerated creep testing. All of these methods have limitations. Normal NDE methods often fail to detect incipient creep damage and microstructural damage, which can be precursors of rapid, unanticipated failures. Due to unknown variations in the original dimensions, changes in dimensions cannot be determined with confidence. Dimensional measurements fail to provide indications of local creep damage caused by localized strains such as those in heat-affected zones of welds and regions of stress concentrations in the base metal. Cracking can frequently occur without manifest over-all strain. Furthermore, the critical strain accumulation preceding fracture can vary widely with a variety of operational material parameters, and with stress state.

A common method of estimating the remaining creep life is to conduct accelerated rupture tests at temperatures well above the service temperature. The stress is kept as close as possible to the service stress value, since only isostress-varied temperature tests are believed to be in compliance with the

Table 3.3. Life-assessment techniques and their limitations for creep-damage evaluation for crack initiation and crack propagation (Ref 159)

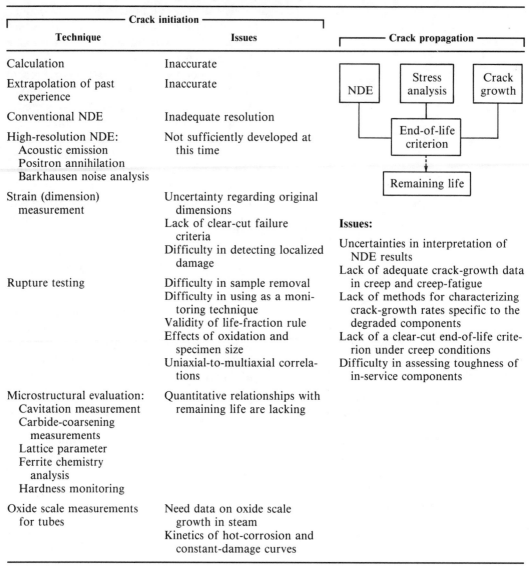

Crack initiation		
Technique	**Issues**	
Calculation	Inaccurate	
Extrapolation of past experience	Inaccurate	
Conventional NDE	Inadequate resolution	
High-resolution NDE: Acoustic emission Positron annihilation Barkhausen noise analysis	Not sufficiently developed at this time	
Strain (dimension) measurement	Uncertainty regarding original dimensions Lack of clear-cut failure criteria Difficulty in detecting localized damage	
Rupture testing	Difficulty in sample removal Difficulty in using as a monitoring technique Validity of life-fraction rule Effects of oxidation and specimen size Uniaxial-to-multiaxial correlations	
Microstructural evaluation: Cavitation measurement Carbide-coarsening measurements Lattice parameter Ferrite chemistry analysis Hardness monitoring	Quantitative relationships with remaining life are lacking	
Oxide scale measurements for tubes	Need data on oxide scale growth in steam Kinetics of hot-corrosion and constant-damage curves	

Issues:

Uncertainties in interpretation of NDE results

Lack of adequate crack-growth data in creep and creep-fatigue

Lack of methods for characterizing crack-growth rates specific to the degraded components

Lack of a clear-cut end-of-life criterion under creep conditions

Difficulty in assessing toughness of in-service components

life-fraction rule as discussed earlier. The results are plotted as shown in Fig. 3.32 (Ref 160). By extrapolating the test results to the service temperature, the remaining life under service conditions is estimated.

Implementation of the above procedure requires a reasonably accurate knowledge of the stresses involved. For cyclic stressing conditions, and in situations involving large stress gradients, selection of the appropriate stress for the isostress tests is clouded in uncertainty. Furthermore, the procedure involves destructive tests requiring removal of large samples from operating components. There are limitations on the number of available samples and the locations from which they can be taken. Periodic assessment of the remaining life is not possible. The costs of cutting out material, machining specimens, and conducting creep tests can add up to a significant expenditure. These costs are further compounded by the plant outage during this extended period of evaluation and decisionmaking. Develop-

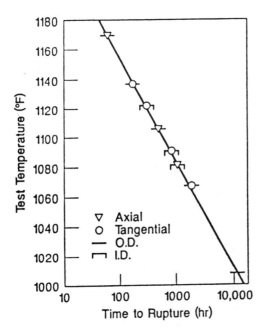

Fig. 3.32. Plot of data from accelerated creep-rupture tests on retired header specimens, illustrating the isostress method (Ref 160).

ment of nondestructive techniques, particularly those based on metallographic and miniature-specimen approaches, has therefore been a major focus of the programs aimed at predicting crack initiation. In addition, the effects of stress state, environment, cyclic operation, and prior damage on life prediction have been investigated so that accelerated laboratory tests currently in use can be refined. Specific life-estimation procedures applicable to individual components are discussed in detail later in this book.

Localized Damage by Crack Growth

For heavy-wall components, the initiation criteria will be combined with crack-growth data to perform a fracture-mechanics analysis of remaining life. In the crack-growth area, both crack-growth data and the methodologies for data analysis are lacking and are being developed. To estimate critical crack sizes for end-of-life under brittle-fracture conditions, methods are needed for characterizing the toughness of the in-service component as nondestructively as possible.

Several methods, including Auger analysis, chemical analysis, miniature-specimen testing, chemical etching, electron microscopy, and others, have been explored for ferritic steels, and the results are described in a subsequent chapter. An excellent overview of remaining-life techniques may be found in Ref 161.

To perform a remaining-life assessment of a component under creep-crack-growth conditions, two principal ingredients are needed: (1) an appropriate expression for relating the driving force K, C^*, or C_t to the nominal stress, crack size, material constants, and geometry of the component being analyzed; and (2) a correlation between this driving force and the crack-growth rate in the material, which has been established on the basis of prior data or by laboratory testing of samples from the component. Once these two ingredients are available, they can be combined to derive the crack size as a function of time. The general methodology for doing this is illustrated below, assuming C_t to be the driving crack-tip parameter.

The general expression for C_t given in Eq 3.78 essentially reduces to the form

$$C_t = \sigma \dot{\epsilon}(A,n)aH \text{ (geometry, n)} \quad \text{(Eq 3.84)}$$

where σ is the stress far from the crack tip, obtained by stress analysis; $\dot{\epsilon}$ is the strain rate far from the crack tip, which is a function of the constants A and n in the Norton relation; a is the crack depth from NDE measurements; and H is a tabulated function of geometry and the creep exponent n. The values of A and n are either assumed from prior data or generated by creep testing of samples. By assembling all the constants needed, the value of C_t can be calculated.

Once C_t is known, it is correlatable to the crack-growth rate through the constants b and m in Eq 3.80. Combining Eq 3.80 and 3.84 provides a first-order differential equation for crack depth, a, as a function of time, t. Theoretically, this equation can be solved by separating variables and integrating. However, the procedure is complicated by the time-dependency of C_t and

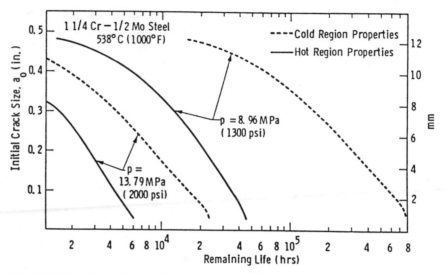

Fig. 3.33. Remaining life as a function of initial crack size for an internally pressurized cylinder, illustrating a typical output from crack-growth analysis (Ref 120).

the a (crack size) dependency of the term H in Eq 3.84. To circumvent this, crack-growth calculations are performed with the current values of a and the corresponding values of da/dt, to determine the time increment required for incrementing the crack size by a small amount Δa—i.e., $\Delta t = \Delta a/\dot{a}$. This provides new values of a, t, and C_t, and the process is then repeated. When the value of a reaches the critical size a_c as defined by K_{Ic}, J_{Ic} wall thickness, remaining ligament thickness, or any other appropriate failure parameter, failure is deemed to have occurred.

Although this procedure appears complex at first sight, the calculations are relatively easy once the principles are understood. Computer programs have been developed which perform the entire analysis on personal computers. The only judgment involved is in selecting proper values for the constants A, n, b, and m, because large scatter in creep and crack-growth data necessitates subjective choices. If actual creep and/or crack-growth tests could be performed, more accurate results could be obtained. Several case histories are available in the literature to acquaint the reader with the procedures involved (Ref 120 and 162 to 164). A sample output may be in the form of a table of crack depth vs time or a

plot of crack size vs remaining life, as illustrated in Fig. 3.33. This plot was generated for a thick-wall cylinder under internal pressure containing a longitudinal crack. The outside radius and wall thickness of the cylinder were assumed to be 45.7 and 7.62 cm (18 and 3 in.), respectively, and the hoop stresses were calculated for internal pressures of 8.96 and 13.79 MPa (1.3 and 2 ksi). Material properties in the degraded condition (hot region) as well as in the undegraded condition were considered. The results show that the remaining life is a function of the stresses as well as of prior degradation. Plots of this type could be used to determine remaining life or to set inspection criteria and inspection intervals. Examples of remaining-life analyses are presented in the chapters on boilers and rotors (Chapters 5 and 6).

Ainsworth *et al* have recently described a unified approach for structures containing defects (Ref 75). This approach incorporates structural failure by rupture, incubation behavior preceding crack growth, and creep-crack growth in a single framework. Service life is governed by a combination of time to rupture, time of incubation, and time of crack growth. All of these quantities are calculated using a reference stress that is specifically applicable to the geometry of the component and is derived analytically or based on scale-model tests. If the desired service life exceeds the calculated rupture time, retirement may be necessary. In the opposite situation, further analysis is carried out to calculate the incubation time during which no crack growth is expected to occur. If the calculation indicates that the incubation time t_i is less than the desired service life, then a crack-growth analysis is performed to calculate the crack-growth life t_g, If the total life, $t_i + t_g$, is less than the desired service life, safe operation beyond that point would be considered undesirable. This approach seems very promising and needs further exploration.

Physical Metallurgy of Creep-Resistant Steels

The key to development of creep-resistant steels is increasing the resistance of the grains and grain boundaries to flow, while at the same time retarding recovery and other softening processes. In the creep temperature and stress range of interest, the flow of material is by the motion of dislocations. Hence the common methods used for impeding dislocation motion—namely, solid-solution strengthening and precipitation strengthening—are applicable to creep strengthening. In solid-solution strengthening, the solute atoms of the alloying elements carbon, chromium, molybdenum, vanadium, nickel, etc., cluster around dislocations and impede their motion. Higher alloy contents thus favor increased creep strength. For instance, in the series of steels C steel, 0.5Cr steel, 0.5Cr-0.5Mo steel, 1.25Cr-0.5Mo steel, and 2.25Cr-1Mo steel, the creep strength increases progressively with increasing contents of Cr and Mo. In addition to the solid-solution effect, alloying elements also form carbide precipitates which impede dislocation motion by a mechanism known as precipitation hardening. The stability of the carbides increases in the following order of alloying elements: chromium, molybdenum, vanadium, and niobium. Fine and closely dispersed precipitates of NbC and VC are the most desirable, followed by the other carbides.

All of the hardening mechanisms become unstable at high temperatures. Their creep-strengthening effects, therefore, are limited to the time-temperature regions in which they are stable. In solid-solution hardening, an increase in temperature increases the diffusion rates of solute atoms in the dislocation atmospheres while at the same time dispersing the atoms of the atmospheres, with both effects making it easier for dislocations to move. In precipitation hardening, heating of the alloy to an excessively high temperature can cause solutionizing of the precipitates. At intermediate temperatures, the precipitates can coarsen and become less-effective impediments to dislocation motion. High stresses and high-strain cyclic loading also can lead to accelerated softening.

Nomenclature

a	– Crack depth (or length)
Δa	– Incremental crack depth
$\dot{a}$	– Rate of crack growth, da/dt
b	– Coefficient in crack-growth rate (Eq 3.80)
c	– Cavity half-spacing (Eq 3.58)
d	– Mean diameter of pipe/tube
l	– Grain size (Eq 3.58)
m	– Exponent in the crack-growth rate (Eq 3.80); also, coefficient in the Monkman-Grant correlation (Eq 3.54)
n	– Norton law creep-rate exponent (Eq 3.7)
q	– Exponent of denominator in Manson-Brown parameter (Eq 3.20)
r	– Radial distance
r_i	– Inside radius of pipe/tube
r_o	– Outside radius of pipe/tube
t	– Time
t_a	– Coordinate of point of intercept (Eq 3.19)
t_i	– Time spent under condition i
t_r	– Time to rupture
t_{ri}	– Time to rupture under condition i

t_1	– Time to transition from K- to C*-controlled crack growth (Eq 3.74)	T_a	– Coordinate of point of intercept (Eq 3.19)
t_2	– Time to transition from J- to C*-controlled crack growth (Eq 3.79)	T_i	– Traction vector (Eq 3.62)
		ΔU_t^*	– Difference in rates of energy supplied to two cracked bodies with identical deformation histories but different crack sizes (Eq 3.76)
t_3	– Time to onset of tertiary creep		
$\dot{u}_i$	– Deflection-rate vector (Eq 3.62)		
x	– Wall thickness of pipe/tube	V	– Cavity volume (Eq 3.57)
A	– Norton law creep-rate coefficient (Eq 3.7); also, number fraction of cavitating boundaries (Eq 3.59)	$\dot{V}_c$	– Creep-deflection rate
		$\dot{V}_D, \dot{V}_P, \dot{V}_C$	– Rates of growth of spherical cavities for different mechanisms (Eq 3.56 to 3.58)
$\dot{A}$	– Time rate of change of number fraction of cavitating boundaries (Eq 3.60)	W	– Width of a crack-growth test specimen
		W − a	– Width of uncracked ligament
B	– Thickness of crack-growth-test specimen	W^*	– Strain-rate-energy density (Eq 3.62)
C*, C_t, C(t)	– Integrals defining crack-tip driving force for creep-crack growth	γ	– Surface energy
		δ_c	– Load-line deflection
D	– Kachanov damage parameter (Eq 3.5)	ϵ	– Creep strain
		ϵ_0	– Instantaneous strain in a creep test
E	– Young's modulus	$\epsilon_1, \epsilon_2, \epsilon_3$	– Principal strains
$\dot{E}$	– Average elongation rate (total elongation divided by t_r)	ϵ_s	– Secondary creep strain
		ϵ_t	– Tertiary creep strain
F	– K-calibration factor, f(a/W) (Eq 3.77)	$\dot{\epsilon}_t$	– Tertiary creep rate
		ϵ_r	– Rupture strain
F'	– dF/d(a/W) (Eq 3.77)	ϵ_f	– Creep ductility chosen as appropriate
H	– A constant crack growth, f(geometry, n) (Eq 3.84)		
		ϵ^*	– Von Mises effective strain
I_n	– A constant dependent on n (Eq 3.64)	$\dot{\epsilon}$	– Secondary creep rate, or minimum creep rate
I_S, I_B, I_N, I_I	– Parameters for expressing effective impurity content	$\dot{\epsilon}^*$	– Von Mises effective creep rate
		ϵ_i	– Strain accumulated under condition i
J	– Integral defining crack-tip driving force for crack growth under elastic-plastic conditions	ϵ_{ri}	– Strain to rupture under condition i
J_1, J_2	– Stress functions defined by Eq 3.37 and 3.38	$\theta_1, \theta_2, \theta_3, \theta_4$	– Constants in the θ projection method (Eq 3.6)
K	– Crack-tip driving force for crack growth under linear-elastic conditions	ν	– Poisson's ratio
		σ	– Applied stress
		σ_0	– Back stress
P	– Applied load, or pressure	σ_{net}	– Net section stress
P_t	– Larson-Miller parameter (Eq 3.14)	$\sigma_1, \sigma_2, \sigma_3$	– Principal stresses
		σ_{max}	– Maximum principal stress
P_U	– Rigid-plastic collapse load	σ^*	– Von Mises effective stress
Q	– Activation energy	σ_H	– Hoop stress
R	– Universal gas constant	σ_r	– Radial stress
T	– Temperature	σ_{ax}	– Axial stress

σ_{ref} – Reference stress
σ_y – Yield stress
σ_h – Hydrostatic stress
ω – Cavity radius (Eq 3.56)

References

1. F. Garofalo, *Fundamentals of Creep and Creep Rupture in Metals*, M. Fine, J. Weertman, and J.R. Weertman, Ed., MacMillan Series in Materials Science, The MacMillan Co., New York, 1965
2. J. Bressers, Ed., *Creep and Fatigue in High Temperature Alloys*, Applied Science Publishers, London, 1981
3. G. Bernasconi and G. Piatti, Ed., *Creep of Engineering Materials and Structures*, Applied Science Publishers, London, 1979
4. E.N. La and C. Andrade, *Proc. Royal Soc. (London)*, Vol A 84, 1910, p 1
5. P.G. McVetty, *Mech. Engg.*, Vol 56, 1934, p 49
6. J.C. Li, *Acta Met.*, Vol 11, 1963, p 1269
7. N.S. Akulov, *Acta Met.*, Vol 12, 1964, p 1195
8. V. Lupinc, Creep: Introduction and Phenomenology, in Ref 2, p 7-39
9. P.W. Davies, W.J. Evans, K.R. Williams, and B. Wilshire, *Scripta Met.*, Vol 3, 1969, p 671
10. L.H. Kachanov, IZV. Akad. Nauk, USSR, Vol 8, 1958, p 26
11. R.W. Evans, J.D. Parker, and B. Wilshire, An Extrapolative Procedure for Long Term Creep-Strain and Creep Life Prediction, in *Recent Advances in Creep and Fracture of Engineering Materials and Structures*, Pineridge Press, 1982, p 135-184
12. M.F. Ashby, A First Report on Deformation Mechanism Maps, *Acta Met.*, Vol 20, 1972, p 887-895
13. M.F. Ashby, Proc. ICSMA 3, sponsored by the Institute of Metals and the Iron and Steel Institute, Cambridge, England, Vol 2, 1973, p 8
14. J. Gittus, *Creep, Viscoelasticity and Creep Fracture in Solids*, Applied Science Publishers, London, 1975, p 473
15. F.H. Norton, *The Creep of Steel at High Temperatures*, McGraw-Hill, New York, 1929
16. R.W. Bailey, Creep of Steel Under Simple and Compound Stress, *Engineering*, Vol 121, 1930, p 265
17. O.D. Sherby and P.M. Burke, Mechanical Behavior of Crystalline Solids at Elevated Temperatures, *Prog. Mater. Sci.*, Vol 13, 1967, p 325
18. R. Viswanathan, The Effect of Stress and Temperature on the Creep and Rupture Behavior of a 1.25 pct. Chromium–0.5 pct. Molybdenum Steel, *Met. Trans. A*, Vol 8A, 1977, p 877-883
19. B.J. Cane, "The Process Controlling Creep and Creep Fracture of 2¼Cr-1Mo Steel," CEGB Report RD/LR 1979, Central Electricity Generating Board Research Laboratories, Leatherhead, England, 1979
20. M.S. Shammas, "Estimating the Remaining Life of Boiler Pressure Parts," EPRI Final Report on RP 2253-1, Vol 4, Electric Power Research Institute, Palo Alto, CA, 1987
21. M.C. Askins *et al*, EPRI Final Report on RP 2253-1, Vol 5, Electric Power Research Institute, Palo Alto, CA, 1987
22. I.R. McLauchlin, in *Creep Strength in Steel and High Temperature Alloys*, The Metals Society, London, 1974, p 86
23. J.M. Adamson and J.W. Martin, Ref 22, p 106
24. J.M. Silcox and G. Willoughby, Ref 22, p 122
25. F.E. Asbury and G. Willoughby, Ref 22, p 122
26. M.J. Collins, Ref 22, p 217
27. V. Foldyna, A. Jakobova, T. Prnka, and J. Sabotka, Ref 22, p 230
28. P.L. Threadgill and B. Wilshire, Ref 22, p 8
29. K.E. Amin and J.E. Dorn, *Acta Met.*, Vol 17, 1969, p 1429
30. F. R. Beckitt, T.M. Bauks, and T. Gladman, *Creep Strength in Steels and High Temperature Alloys*, The Metals Society, London, 1974, p 71
31. V. Foldyna, A. Jakobova, T. Prnka, and J. Sabotka, Ref 30, p 130
32. K.R. Williams and B. Wilshire, *Met. Sci. J.*, Vol 7, 1973, p 176
33. C.N. Aneceuist, R. Gasca-Neri, and W.D. Nix, A Phenomenological Theory of Steady State Creep Based on Average Internal and Effective Stress, *Acta Met.*, Vol 18, June 1970, p 663-671
34. J.B. Conway, *Stress Rupture Parameters: Origin, Calculations and Use*, Gordon and Breach, New York, 1969
35. S.S. Manson, *Time-Temperature Parameters—A Re-Evaluation of Some New Approaches*, ASM Publication D-8-100, American Society for Metals, Metals Park, OH, 1968, p 1-115
36. S.S. Manson and C.R. Ensign, A Quarter Century of Progress in the Development of Correlation and Extrapolation Methods for Creep Rupture Data, *ASME J. Engg. Mater. Tech.*, Vol 101, 1979, p 317-325
37. F.R. Larson and J. Miller, *Trans. ASME*, Vol 74, 1952, p 765
38. J.H. Hollomon and L.D. Jaffee, Time-Temperature Relations in Tempering Steel, *Trans. AIME*, Vol 162, 1945, p 22
39. R. Viswanathan, Strength and Ductility of

2¼Cr-1Mo Steels in Creep, *Met. Tech.*, June 1974, p 284-293

40. R.L. Orr, O.D. Sherby, and J.E. Dorn, *Trans. ASM*, Vol 46, 1954, p 113

41. S.S. Manson and A.M. Haferd, "A Linear Time-Temperature Relation for Extrapolation of Creep and Stress Rupture Data," NASA TN2890, Mar 1952

42. S.S. Manson and W.F. Brown, Time-Temperature Stress Relations for Correlation and Extrapolation of Stress Rupture Data, *Proc. ASTM*, Vol 53, 1953, p 683-719

43. S.S. Manson and G. Succop, *Stress Rupture Properties of Inconel 700 and Correlation on the Basis of Several Time-Temperature Parameters*, STP 174, American Society for Testing and Materials, Philadelphia, 1956

44. S.S. Manson and C.R. Ensign, "A Specialized Model for Analysis of Creep Rupture Data by the Minimum Commitment Method," NASA Tech. Memo TMX 52999, Washington, 1971

45. S.S. Manson and U. Muralidharan, "Analysis of Creep Rupture Data for Five Multi-heat Alloys by the Minimum Commitment Method Using Double Heat Term Centering Technique," EPRI CS 317, Electric Power Research Institute, Palo Alto, CA, 1983

46. ASME Boiler and Pressure Vessel Code, Section I, American Society of Mechanical Engineers, New York

47. E.L. Robinson, Effect of Temperature Variation on the Creep Strength of Steels, *Trans. ASME*, Vol 160, 1938, p 253-259

48. Y. Lieberman, Relaxation, Tensile Strength and Failure of E1 512 and Kh1 F-L Steels, *Metalloved Term Obrabodke Metal*, Vol 4, 1962, p 6-13

49. H.R. Voorhees and F.W. Freeman, "Notch Sensitivity of Aircraft Structural and Engine Alloys," Wright Air Development Center Technical Report, Part II, Jan 1959, p 23

50. M.M. Abo El Ata and I. Finnie, "A Study of Creep Damage Rules," ASME Paper No. 71-WA/Met-1, American Society of Mechanical Engineers, New York, Dec 1971

51. R.M. Goldhoff and D.A. Woodford, *The Evaluation of Creep Damage in a CrMoV Steel*, STP 515, American Society for Testing and Materials, 1982, p 89

52. R.M. Goldhoff, Stress Concentration and Size Effects in a CrMoV Steel at Elevated Temperatures, *Joint International Conference on Creep*, Institute of Mechanical Engineers, London, 1963

53. H. Wiegand, J. Granachar, and M. Sander, Zeitstandbruchverhalten Einiger Warmfester Stahle Unter Rechteckzyklish Veranderter Spannung und Temperatur, *Arch. Eisenhüttenwesen*, Vol 45, 1975, p 533-539

54. P. Meijers and C.F. Etienne, Research Related to Residual Life of Structures in the Creep Range, in *Proceedings of the Conference on Determination of the Life Expectancy of Plant Which Has Operated at High Temperatures*, Institute of Mechanical Engineers, London, Nov 21, 1978

55. R.V. Hart, Assessment of Remaining Creep Life Using Accelerated Stress Rupture Tests, *Met. Tech.*, Vol 13, 1976, p 1

56. D.A. Woodford, Creep Damage and the Remaining Life Concept, *Trans. ASME, J. Engg. Mater. Tech.*, Vol 101, 1979, p 311

57. J.M. Brear, "Residual Life Assessment Methods," Project 2023, Ref. 3A/20, 11th Progress Report, ERA Technology, Apr 1987

58. R.V. Hart, The Effect of Prior Creep on the Rupture Properties of the Power Plant Alloys, Paper C 207/80, in *International Congress on the Engineering Aspects of Creep*, Vol 2, University of Sheffield, Institute of Mechanical Engineers, London, 1980, p 207-211

59. B.W. Roberts, F.V. Ellis, and J.E. Bynum, Remaining Creep or Stress Rupture Life under Non Steady Temperature and Stress, *Trans. ASME, J. Engg. Mater. Tech.*, Vol 101, 1979, p 331-336

60. A.E. Johnson, J. Henderson, and B. Kahn, *Complex Stress Creep, Relaxation and Fracture of Metallic Alloys*, Her Majesty's Stationery Office, Edinburgh, 1962

61. M. Abo El Ata and I. Finnie, On the Prediction of Creep Rupture Life of Components Under Multiaxial Stress, in *Creep in Structures*, IUTAM Symposium, Gothenburg, 1970

62. D.R. Hayhurst, Creep Rupture Under Multiaxial State of Stress, *J. Mech. Phys. Solids*, Vol 20, 1972, p 381-390

63. S.P. Timoshenko and J.N. Goodier, *Theory of Elasticity*, 3rd Ed., McGraw-Hill, New York, 1951

64. W. Burrows, R. Michel, and R. Rankin, A Wall Thickness Formula for High Pressure, High Temperature Piping, *Trans. ASME*, Apr 1954, p 427-444

65. R.W. Bailey, Creep Relationships and Their Application to Pipes, Tubes and Cylindrical Parts Under Internal Pressure, *Proc. Inst. Mech. Eng.*, Vol 164, 1956, p 324

66. S.R. Paterson, T.W. Rettig, and K.J. Clark, Creep Damage and Remaining Life Assessment of Superheater and Reheater Tubes, in *Life Extension and Assessment of Fossil Power Plants*, R.B. Dooley and R. Viswanathan, Ed., EPRI CS5208, Electric Power Research Institute, Palo Alto, CA, 1986, p 455-474

67. C.A. Schulte, Predicting Creep Deflections of Plastic Beams, *Proc. ASTM*, Vol 60, 1960, p 895-904

68. A.E. Johnson, J. Henderson, and B. Kahn, The Behavior of Metallic Thick Walled Tubes Under Internal Pressure at Elevated Temper-

atures, *Proc. Inst. Mech. Eng.*, Vol 175 (No. 25), 1961, p 1043-1069

69. D.L. Marriott and F. A. Leckie, Some Observations on the Deflections of Structures During Creep, in Thermal Loading and Creep in Structures and Components, *Proc. Inst. Mech. Eng.*, Vol 178 (Pt 3L), 1963-64, p 115-125

70. A.E. McKenzie, On the Use of a Single Uniaxial Test to Estimate the Deformation Rates in Some Structures Undergoing Creep, *Int. J. Mech. Sci.*, Vol 10 (No. 5), 1968, p 441-453

71. R.K. Penny and D.L. Marriott, *Design for Creep*, McGraw-Hill, London, 1971

72. R.G. Sim, "Creep in Structures," Ph.D. Thesis, University of Cambridge, 1968

73. A.M. Goodman, Design Aspects of Creep and Fatigue, in *Creep and Fatigue in High Temperature Alloys*, J. Bressers, Ed., Applied Science Publishers, London, 1981, p 145-186

74. I.W. Goodall, F.A. Leckie, A.R.S. Ponter, and C.H.A. Townley, The Development of High Temperature Design Methods Based on Reference Stresses and Bounding Theorems, *Trans. ASME, J. Engg. Mater. Tech.*, Vol 101, 1979, p 349-353

75. R.A. Ainsworth *et al*, CEGB Assessment Procedure for Defects in Plant Operating in the Creep Range, *Fatigue Fract. Engg. Mater. Struct.*, Vol 10 (No. 2), 1987

76. B.J. Cane and R.D. Townsend, "Prediction of Remaining Life in Low Alloy Steels," TPRD/ L/2674/N84, Central Electricity Generating Board, England, Sept 1984

77. R. Viswanathan and C.G. Beck, Effect of Aluminum on the Stress Rupture Properties of CrMoV Steels, *Met. Trans. A*, Vol 6A, Nov 1975, p 1997-2003

78. C.R. Roper, "An Investigation of the Causes of Creep Embrittlement in A 387D Steel," Lukens Steel Co., RDR 69-11, Coatesville, PA, June 1969

79. R.A. Swift, *The Mechanism of Creep Embrittlement in 2¼ Chrome 1 Molybdenum Steel in Pressure Vessels and Piping*, A.O. Schaefer, Ed., American Society of Mechanical Engineers, New York, 1971, p 123-136

80. J. Nutting and J.M. Arrowsmith, in *Structural Processes in Creep*, Special Report 70, *Iron Steel Inst.*, London, 1961, p 246

81. J. Glen, *J. Iron Steel Inst.*, Vol 190, Sept 1958, p 30

82. R.M. Goldhoff and H.J. Beatty, *Trans. AIME*, Vol 233, Sept 1965, p 1743

83. C.F. Etienne, H.C. Van Helst, and P. Meijers, Procedure for the Estimation of Residual Life in High Temperature Installations, in *Creep of Engineering Materials and Structures*, G. Bernasconi and G. Piatti, Ed., Applied Science Publishers, London, 1979, p 149-193

84. B.J. Cane and R.D. Townsend, "Prediction of Remaining Life in Low Alloy Steels," TPRD/

L/2674/N84, Central Electricity Generating Board, England, Sept 1984

85. R. Viswanathan, Metallurgical Factors Affecting the Creep Properties of CrMoV Steels, *ASTM Journal of Testing and Evaluation*, Vol 3, 1975, p 93

86. R. Viswanathan, Effect of Impurities on the Creep Properties of Ferritic Steels, *ASM Metals Engg. Quarterly*, Nov 1975, p 50-55

87. D.A. Woodford and R.M. Goldhoff, An Approach to the Understanding of Brittle Behavior of Steel at Elevated Temperatures, *Mater. Sci. Engg.*, Vol 5, 1969/70, p 303-324

88. G.V. Smith, *Trans. ASME, J. Engg. Mater. Tech.*, Apr 1975, p 188

89. R.M. Goldhoff, A Method for Extrapolating Rupture Ductility, in *Elevated Temperature Testing Problems*, STP 400, American Society for Testing and Materials, Philadelphia, 1971, p 82-90

90. M.K. Booker, C.R. Brinkman, and V.K. Sikka, *Structural Materials for Service at Elevated Temperature in Nuclear Power Generation*, MPC 1, ASME, New York, 1975, p 108

91. R. Viswanathan and R.D. Fardo, Parametric Techniques for Extrapolating Rupture Ductility, in *Ductility and Toughness Considerations in Elevated Temperature Service*, G.V. Smith, Ed., MPC 8, ASME, New York, 1978

92. G. Schlottner and R.E. Seeley, Estimation of Remaining Life of High Temperature Steam Turbine Components, in *Residual Life Assessment, Non Destructive Examination and Nuclear Heat Exchanger Materials*, PVP-Vol 98-1, ASME, New York, 1985, p 35-44

93. B.E. Hopkins, H.R. Tipler, and G.D. Branch, Improvements in the Behavior of CrMoV Steels Through Enhanced Purity, *J. Iron Steel Inst.*, Vol 209, 1971, p 745

94. W.E. Trumpler, R.E. Clark, and E.V. Black, ASME Paper 67 WA/Met-12, 1967

95. J.L. Ratliff and R.M. Brown, *Trans. ASM*, Vol 60, 1967, p 176

96. F. Benek and P. Skvor, *Hutn Listy*, Vol 3, 1972, p 197 (English translation, British Iron and Steel Institute Paper 10480)

97. C.J. Middleton, "Re-Heat Cavity Nucleation in Bainitic Creep Resistant Low Alloy Steels, Part I: The Influence of Sulfides," Report RD/L/N37/79, Central Electricity Generating Board, England, Aug 1979

98. D.P. Pope *et al*, "Elimination of Impurity-Induced Embrittlement in Steel," Part 2, EPRI Report NP 1501, Electric Power Research Institute, Palo Alto, CA, Sept 1980

99. D.J. Gooch, "Effects of Tramp Elements on the Creep Strength and Ductility of Low Carbon 2¼ Cr-1Mo Steels," Report RD/L/N52/ 80, Central Electricity Generating Board Research Laboratories, England, July 1980

100. D.F. Roan and B.B. Seth, Metallographic

and Fractographic Study of the Creep Cavitation and Fracture Behavior of 1Cr1Mo0.25V Rotor Steels with Controlled Residual Impurities, in *Ductility and Toughness Considerations in Elevated Temperature Service*, G.V. Smith, Ed., MPC 8, ASME, New York, 1978

101. C.F. Meitzner, WRC Bulletin No. 211, Welding Research Council, Nov 1975

102. J.M. Brear and B.L. King, *Proceedings of the Conference on Grain Boundaries*, The Metals Society, London, 1976

103. H. Nakamura, T. Naiki, and H. Okabayashi, *Proceedings of the First International Conference on Fracture*, Vol 2, 1965, p 863

104. Y. Ito and M. Nakanishi, ITW Document Y 668-72, 1972

105. L.G. Emmer, C.D. Clauser, and J.R. Low, "Critical Literature Review of Embrittlement in 2¼Cr-1Mo Steel," WRC Bulletin 183, Welding Research Council, May 1973

106. F.C. Monkman and N.J. Grant, *Proc. ASTM*, Vol 56, 1956, p 595

107. R. Logneborg, Creep Fracture Mechanisms, in *Creep of Engineering Materials and Structures*, G. Bernasconi and G. Piatti, Ed., Applied Science Publishers, London, 1979, p 35-46

108. R.J. Fields, T. Weerasurya, and M.F. Ashby, Fracture Mechanisms in Pure Iron, Two Austenitic Steels and One Ferritic Steel, *Met. Trans. A*, Vol 11A, Feb 1980, p 333-347

109. B.J. Cane, Intergranular Creep Cavity Formation in Low Alloy Bainitic Steels, *Met. Sci.*, Vol 15, 1981, p 295-301

110. N.G. Needham and B.J. Cane, Creep Strain and Rupture Predictions by Cavitational Assessment in 2¼Cr-1Mo Steel Weldments, in *Advances in Life Prediction Methods*, D.A. Woodford, Ed., ASME, New York, 1983, p 65-73

111. N.G. Needham and T. Gladman, Intergranular Creep Cavitation in 2¼Cr-1Mo Steels, Conference on Advances in the Physical Metallurgy and Application of Steel, The Metals Society, London, 1981

112. N.G. Needham and T. Gladman, BSC Report SH/PROD/PM/8914/10/82/A, British Steel Corp., 1982

113. B.J. Cane, "The Process Controlling Creep and Creep Fracture in 2¼Cr-1Mo Steel," CEGB Report RD/L/R/1979, Central Electricity Generating Board, England, Apr 1978

114. B.J. Cane, Mechanistic Control Regions for Intergranular Creep Cavity Growth in 2¼Cr-1Mo Steel under Various Stresses and Stress States, *Met. Sci.*, Vol 15, 1981, p 302-310

115. M.V. Speight and W. Beere, Vacancy Potential and Void Growth on Grain Boundaries, *Met. Sci.*, Vol 9, 1975, p 190

116. T.K. Hellan, *Int. J. Mech. Sci.*, Vol 17, 1975, p 369

117. B.F. Dyson, Constraints on Creep Cavity Growth, *Met. Sci.*, Vol 10, 1976, p 349

118. R. Raj and A.K. Ghosh, Stress Rupture, *Met. Trans. A*, Vol 12A, 1981, p 1291-1302

119. R.D. Townsend *et al*, unpublished work, Central Electricity Generating Board, England, 1987

120. P.K. Liaw and A. Saxena, "Remaining Life Estimation of Boiler Pressure Parts—Crack Growth Studies," EPRI Report CS 4688, Electric Power Research Institute, Palo Alto, CA, July 1986

121. H. Riedel and J.R. Rice, in *Fracture Mechanics: Twelfth Conference*, STP 700, American Society for Testing and Materials, Philadelphia, 1980, p 112-130

122. J.L. Bassani and F.A. McClintock, *Int. J. Solids and Structures*, Vol 7, 1981, p 479-492

123. A. Saxena, in *Fracture Mechanics: Seventeenth Volume*, STP 905, American Society for Testing and Materials, Philadelphia, 1986, p 185-201

124. J.D. Landes and J.A. Begley, in *Mechanics of Crack Growth*, STP 590, American Society for Testing and Materials, Philadelphia, 1976, p 128-148

125. A. Saxena, in *Fracture Mechanics: Twelfth Conference*, STP 700, American Society for Testing and Materials, Philadelphia, 1980, p 131-135

126. N.L. Goldman and J.W. Hutchinson, *Int. J. Solids and Structures*, Vol 11, 1975, p 575-591

127. C.F. Shih, "Tables of Hutchinson-Rice-Rosengren Singular Field Quantities," MRLE 147, Materials Research Laboratory, Brown University, 1983

128. J.W. Hutchinson, *J. Mech. Phys. Solids*, Vol 16, 1968, p 13-31

129. J.R. Rice and G.F. Rosengren, *J. Mech. Phys. Solids*, Vol 16, 1968, p 10-12

130. K.M. Nikbin, D.J. Smith, and G.A. Webster, An Engineering Approach to the Prediction of Crack Growth, *Trans. ASME, J. Engg. Mater. Tech.*, Vol 108, 1986, p 186-191

131. V. Kumar, M.D. German, and C.F. Shih, "An Engineering Approach for Elastic-Plastic Fracture Analysis," EPRI Report NP 1931, Electric Power Research Institute, Palo Alto, CA, 1981

132. D.J. Smith and G.A. Webster, *Inelastic Crack Analysis*, Vol 1 of *Elastic Plastic Fracture: Second Symposium*, STP 803, American Society for Testing and Materials, Philadelphia, 1983, p I 654-I 674

133. H.A. Ernst, in *Theory and Analysis*, Vol 1 of *Fracture Mechanics: Fourteenth Sympo-*

sium, STP 791, American Society for Testing and Materials, Philadelphia, 1983, p I 499–I 599

134. K.M. Nikbin, G.A. Webster, and C.E. Turner, in *Cracks and Fracture*, STP 601, American Society for Testing and Materials, Philadelphia, 1976, p 47-62

135. C.F. Shih and J.W. Hutchinson, *Trans. ASME, J. Engg. Mater. Tech.*, Series H, Vol 98, 1976, p 289-295

136. M.Y. He and J.W. Hutchinson, in *Inelastic Crack Analysis*, Vol 1 of *Elastic Plastic Fracture: Second Symposium*, STP 803, American Society for Testing and Materials, Philadelphia, 1983, p I 291–I 305

137. C.E. Jaske, "Estimation of the C* Integral for Creep Crack Growth Test Specimens," ASM Conference, Salt Lake City, Dec 5, 1985

138. H. Riedel, "The Use and the Limitations of C* in Creep Crack Growth Testing," International Symposium on Fracture Mechanics, Beijing, China

139. K. Ohji, K. Ogura, S. Kubo, and Y. Katada, The Application of Modified J Integral to Creep Crack Growth, in *Engineering Aspects of Creep*, Vol 2, Institute of Mechanical Engineers, London, 1980, p 9-16

140. H. Riedel and J.R. Rice, Tensile Cracks in Creeping Solids, in *Fracture Mechanics*, STP 700, American Society for Testing and Materials, Philadelphia, 1980, p 112-140

141. K. Ohji, K. Ogura, and S. Kubo, *J. Soc. Mater. Sci. Japan*, Vol 29 (No. 320), 1980, p 465-471

142. R. Ehlers and H. Riedel, in *Advances in Fracture Research*, ICF 5, Vol 2, D. François *et al*, Ed., Pergamon Press, 1981, p 691-698

143. H. Tada, P. Paris, and G.R. Irwin, *The Stress Analysis of Cracks Handbook*, Del Research Corp., Hellertown, PA, 1971

144. H. Riedel, Creep Deformation at Crack Tips in Elastic-Viscoplastic Solids, *J. Mech. Phys. Solids*, Vol 29, 1981, p 35

145. A. Saxena, H.A. Ernst, and J.D. Landes, "Creep Crack Growth Behavior in 316 Stainless Steel at 594 C," Scientific Paper 82-107, REACT-P1, Westinghouse Research Laboratories, Pittsburgh, 1982

146. S. Taira, R. Ohtani, and T. Kitamura, *Trans. ASME, J. Engg. Mater. Tech.*, Vol 101, 1979, p 156-161

147. G. Airoldi, P. Bianchi, and D. D'Angelo, "Microstructure and Creep Crack Growth Behavior in Low Alloy Steels," International Symposium on Microstructure and Mechanical Behavior of Materials, Xian, China, Oct 21-24, 1985

148. D.J. Gooch, "Effects of Geometric Constraint on Analysis Methods for Creep Crack Growth in 2¼Cr-1Mo Weld Metal," CEGB Report TPRD/L/2384/N82, Central Electricity Generating Board, England, Jan 1983

149. A. Saxena, J. Han, and K. Banerji, "Creep Crack Growth Behavior in Power Plant Boiler and Steam Pipe Steels," EPRI Report on Project 2253-10, Electric Power Research Institute, Palo Alto, CA, Apr 1987

150. S. Konosu and K. Maeda, "Creep Embrittlement Susceptibility and Creep Crack Growth Behavior in Low Alloy Steels," Third International Conference on Non Linear Fracture Mechanics, Oct 1986

151. J.L. Henshall and M.G. Gee, Creep Crack Propagation in Bainitic 1Cr-0.5Mo Steel Between 540 and 590°C, *Mater. Sci. Engg.*, Vol 80, 1986, p 49-57

152. K.M. Nikbin, D.J. Smith, and G.A. Webster, Influence of Creep Ductility and State of Stress on Creep Crack Growth, in *Advances in Life Prediction Methods*, International Conference, Albany, NY, Apr 18-20, 1983, ASME, New York, 1983

153. H. Riedel and W. Wagner, Creep Crack Growth in Nimonic 80A and in a 1Cr-½Mo Steel, *Sixth International Conference on Fracture* (ICF6), New Delhi, S.R. Valluri, J.F. Knott, P. Rama Rao, and D.M.R. Taplin, Ed., Pergamon Press, 1984

154. J.J. Lewandowski, M.B.D. Ellis, and J.F. Knott, Impurity Effects on Sustained Load Cracking of 2¼Cr-1Mo Steel at High Temperature, in *Fracture Control of Engineering Structures*, Vol III, H.C. Van Helst and A. Bakker, Ed., 1986, p 1905-1914

155. J.J. Lewandowski, C.A. Hippsley, M.B.D. Ellis, and J.F. Knott, Effects of Impurity Segregation on Sustained Load Cracking of 2.25Cr-1Mo Steels—I, Crack Initiation, *Acta Met.*, Vol 35 (No. 3), 1987, p 593-608

156. A. Saxena and J. Han, "Evaluation of Crack Tip Parameters for Characterizing Crack Growth Behavior in Creeping Materials," ASTM Task Group Report, Joint Task Group E.24.08.07/E.24.04.08, 1987

157. K. Sadananda and P. Shahinian, The Effects of Environment on the Creep Crack Growth Behavior of Several Structural Alloys, *Mater. Sci. Engg.*, Vol 43, 1980, p 159-168

158. M.Z. Nazmy and C. Wuthrich, Creep Crack Growth in In 738 and In 939 Nickel Base Superalloys, *Mater. Sci. Engg.*, Vol 61, 1983, p 119-125

159. R. Viswanathan and R.B. Dooley, Creep Life Assessment Techniques for Fossil Power Plant Boiler Pressure Parts, in *Proceedings of the International Conference on Creep*, JSME, Tokyo, Apr 14-18, 1986, p 349-359

160. R. Viswanathan, R.B. Dooley, and A. Sax-

ena, A Methodology for Evaluating the Integrity of Longitudinally Seam Welded Steam Pipes, submitted to ASME *Journal of Pressure Vessels Technology*, Vol 110, 1988, p 283-290

161. C.H.A. Townley *et al*, "A Review of the Present State of the Art of Assessing Remanent Life of Pressure Vessels and Pressurised Systems Designed for High Temperature Service," British Standards Institute Report PD 6550, 1983

162. F.L. Becker, S.M. Walker, and R. Viswanathan, "Guidelines for the Evaluation of Seam Welded Steam Pipes," EPRI Report CS4774,

Electric Power Research Institute, Palo Alto, CA, Feb 1987

163. A. Saxena, T.P. Sherlock, and R. Viswanathan, Evaluation of Remaining Life of High Temperature Headers: A Case History, in *Life Extension and Assessment of Fossil Power Plants*, R.B. Dooley and R. Viswanathan, Ed., EPRI CS5208, Electric Power Research Institute, Palo Alto, CA, 1987, p 575-605

164. V.P. Swaminathan, N.S. Cheruvu, and A. Saxena, An Initiation and Propagation Approach for the Life Assessment of an HP-IP Rotor, in Ref 163, p 659-676

4
Fatigue

A metal subjected to repetitive or fluctuating stress will fail at a stress much lower than that required for failure on a single application of load. Failures occurring under cyclic loading are termed fatigue failures. Vibrational stresses on turbine blades, alternating bending loads on blades and shafts, and fluctuating thermal stresses during start-stop cycles and due to power changes are some examples of cyclic loading that can occur in a plant. For convenience, two types of fatigue are distinguished: high-cycle fatigue (HCF) and low-cycle fatigue (LCF). Although phenomenologically there is no distinction and no clear-cut border between these two types of fatigue, the traditional approach is to classify failures occurring above about 10^4 cycles as HCF and those occurring below that value as LCF.

High-Cycle Fatigue and the S-N Curve

Most laboratory fatigue testing is done either with axial loading or in bending, thus producing only tensile and compressive stresses. The stress usually is cycled either between a maximum and a minimum tensile stress or between a maximum tensile stress and a maximum compressive stress. The latter stress is considered a negative tensile stress, is assigned an algebraic minus

sign, and therefore is known as the minimum stress.

The stress ratio is the algebraic ratio of two specified stress values in a stress cycle. Two commonly used stress ratios are the ratio, A, of the alternating stress amplitude to the mean stress ($A = \sigma_a/\sigma_m$) and the ratio, R, of the minimum stress to the maximum stress ($R = \sigma_{min}/\sigma_{max}$).

If the stresses are fully reversed, the stress ratio R becomes -1; if the stresses are partially reversed, R becomes a negative number less than 1. If the stress is cycled between a maximum stress and no load, the ratio R becomes zero. If the stress is cycled between two tensile stresses, R becomes a positive number less than 1. An R value of 1 indicates no variation in stress, making the test a sustained-load creep test rather than a fatigue test.

Applied stresses are described by three parameters. The mean stress, σ_m, is the algebraic average of the maximum and minimum stresses in one cycle, $\sigma_m = (\sigma_{max} + \sigma_{min})/2$. In the completely reversed cycle test, the mean stress is zero. The range of stress, $\Delta\sigma$, is the algebraic difference between the maximum and minimum stresses in one cycle, $\Delta\sigma = \sigma_{max} - \sigma_{min}$. The stress amplitude, σ_a, is one-half the range of stress, $\sigma_a = \Delta\sigma/2 = (\sigma_{max} - \sigma_{min})/2$.

During a fatigue test, the stress cycle usually is maintained constant so that the

111

applied stress conditions can be written $\sigma_m \pm \sigma_a$, where σ_m is the static or mean stress, and σ_a is the alternating stress, which is equal to half the stress range. Nomenclature to describe test parameters involved in cyclic stress testing are shown in Fig. 4.1(a).

The basic method for presenting engineering HCF data is by means of the S-N curve, which plots the number of cycles to failure N vs the stress range, as shown in Fig. 4.1(b). N is usually taken to denote the cycles for complete specimen fracture. For most engineering materials, the S-N curve becomes almost flat at low stresses, indicating a threshold value of stress below which failure will not occur for practical purposes. The threshold value of stress, σ_e, is defined as the "fatigue limit." If the S-N curve is not flat, the stress level corresponding to some arbitrarily chosen value of N, say 10^8 cycles, is defined as the fatigue limit. A safety factor is applied to the fatigue limit in design of components. The S-N curve does not distinguish between crack initiation and crack propagation. The number of cycles corresponding to the fatigue limit (i.e., at low stresses) denotes primarily initiation, whereas at high stresses the fatigue life corresponds primarily to crack propagation. The number of cycles to failure at any arbitrarily chosen stress level is termed the fatigue life for that stress. Similarly, the stress to cause failure at a given value of N is called the fatigue strength.

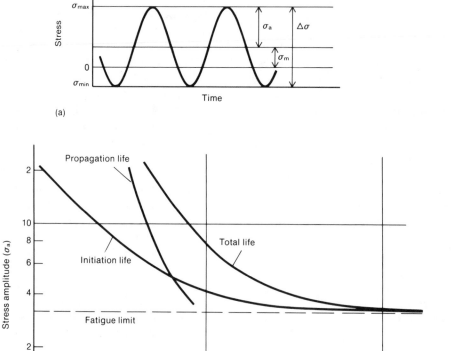

(a)

(b)

Fig. 4.1. (a) Nomenclature for test parameters involved in cyclic stress testing, and (b) typical S-N curve for fatigue.

Effects of Test Variables

Many variables in the test procedure, such as mean stress, temperature, environment, specimen size, specimen surface condition, and stress concentrations, affect the fatigue life.

Mean Stress and Temperature. If the mean stress is assumed to be zero, the stress range denotes a completely reversed tension-compression cycle with the maximum and minimum stresses simply having opposite signs. In most engineering situations, the mean stress is not zero, and the amplitude varies. In such cases, fatigue limit can be quite different, depending on the R ratio ($\sigma_{min}/\sigma_{max}$). As R becomes more positive, which is equivalent to the mean stress becoming greater, the endurance limit becomes greater. Designing components against the specific R ratio to be encountered in service is therefore very critical. For each value of σ_m there is a different value of the alternating stress σ_a (i.e., $\sigma_{max} - \sigma_{min}$) that can be withstood without failure. Two common ways of representing such data are shown in Fig. 4.2(a) and (b) using, respectively, the Goodman diagram and the modified Goodman diagram that includes the Gerber parabola (Ref 1 and 2). In both diagrams, it can be seen that the maximum alternating stress or maximum stress range can be tolerated when σ_m approaches zero. As σ_m approaches the ultimate tensile strength of the material, σ_a approaches zero. Most practical cases lie in between these two extremes. A general relationship among σ_a, σ_e, σ_m, and the ultimate strength σ_u (Ref 3) is

$$\sigma_a = \sigma_e[1 - (\sigma_m/\sigma_u)^x] \qquad \text{(Eq 4.1)}$$

where x = 1 for the Goodman and Soderberg approaches, x = 2 for the Gerber parabola approach, and σ_u is replaced by the yield strength σ_y for the Soderberg approach.

The effects of temperature *per se* are a reduction in the fatigue limit and a lowering of the fatigue strength, as shown in Fig. 4.3 (Ref 4).

Specimen Size and Surface Condition. Specimen size is known to have an effect on fatigue strength (Ref 4). Increasing the specimen size increases the surface area, decreases the stress gradient across the specimen, and increases the probability of fatigue-crack initiation. In general, an increase in specimen size is believed to decrease the fatigue strength, and one must keep this fact in mind when designing large structures based on small-specimen test data.

Fatigue properties are very sensitive to surface condition. Except in special cases involving internal defects or case hardening, all fatigue cracks initiate at the surface. Increasing surface roughness decreases the fatigue strength. Methods which increase the surface strength (e.g., surface hardening) and those which introduce compressive residual stresses improve the fatigue strength significantly (see Fig. 4.4; Ref 4). It is common practice to shot peen steam-turbine blades for this reason. The effects of environment on fatigue strength are complex, although in most cases fatigue strength is reduced by aggressive environments (see Fig. 4.5; Ref 4). In certain circumstances, the filling up of the crack with corrosion products can decrease crack-growth rates. This is discussed later.

Effects of Stress Concentrations. Fatigue strength is reduced appreciably by the presence of stress-concentrating features. Most fatigue failures are caused by poor design which allows stress raisers to exist. Three broad groups of stress concentrations can be recognized (Ref 5), as follows.

Group 1 stress concentrations are those due to changes in the configuration of the part. Examples include steps in shaft diameters; cross-sectional changes in piping; broad integral collars; fillets; abrupt corners; holes; threads in bolts, grooves, and keyways; and undercuts and toes in welds.

Group 2 stress concentrations are those arising from surface discontinuities such as nicks, notches, machining marks, die marks, and corrosion pits.

Group 3 stress concentrations comprise metallurgically inherent discontinuities such as inclusions, microcracks, voids, porosity, and casting defects.

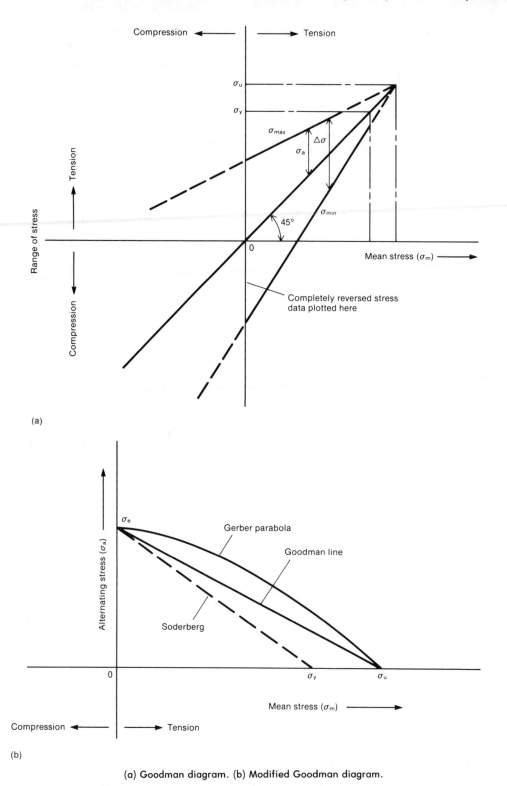

(a) Goodman diagram. (b) Modified Goodman diagram.

Fig. 4.2. Graphical methods for presenting the combined effects of alternating stress and mean stress on fatigue life (Ref 3).

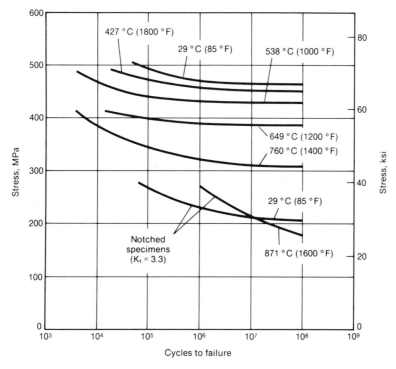

Fig. 4.3. Effect of temperature on S-N curves for Inconel 625 (Ref 4).

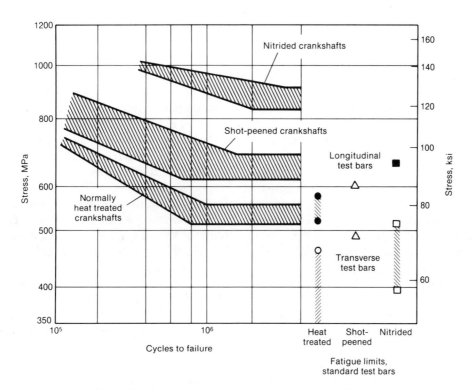

Fig. 4.4. Effect of surface treatment on S-N curves for crankshafts (Ref 4).

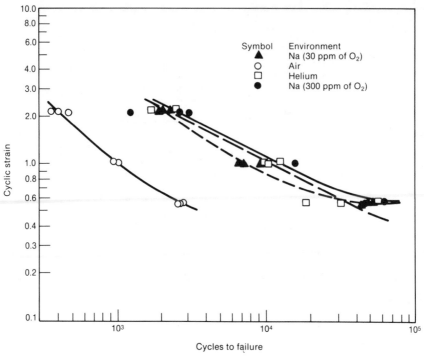

The cycle used was approximately up for 5 s, hold for 5 s, down for 5 s, hold for 5 s.

Fig. 4.5. Influence of environment on fatigue endurance of 2¼Cr-1Mo steel in sodium, air, and helium at 865 K (Ref 4).

Sometimes, several of these stress raisers can occur together, exacerbating the problem. For instance, pitting can occur preferentially at sulfide inclusions in blades or at section transitions in shafts and make the problem worse.

Any geometrical discontinuity such as a notch or a hole results in a nonuniform stress distribution in the vicinity of the discontinuity. The local stress in the vicinity of the discontinuity is higher than the nominal stress at regions in the far field. This stress concentration is expressed as the theoretical elastic stress-concentration factor K_t, which is the ratio of the maximum stress to the nominal stress. In addition to producing a stress concentration, a notch also creates a localized condition of biaxial or triaxial stress. The stress distributions around a circular hole and an elliptical hole in a plate are illustrated in Fig. 4.6 (Ref 3). From elastic analysis (Ref 6), the maximum stress at the ends of the elliptical hole can be shown to be given by

$$\sigma_{max} = \sigma\left[1 + 2\left(\frac{a'}{b'}\right)\right] \qquad \text{(Eq 4.2)}$$

For a circular hole, $a' = b'$, and Eq 4.2 reduces to $\sigma_{max} = 3$. Equation 4.2 shows that stress increases with a'/b'. Hence a very narrow hole, such as a crack normal to the tensile direction, will result in a very high stress concentration.

Due to the mathematical complexities, calculations of K_t have been performed only for simple geometries. Neuber (Ref 7) has compiled many of these expressions. For practical problems, K_t values are determined experimentally. Photoelastic analysis of models was once the most widely used technique (Ref 3). However, with the advent of high-speed digital computers and general-purpose finite-element-analysis programs, stress analysis is now widely used. Most of the available data on stress-concentration factors have been collected by Peterson (Ref 8). McClintock and Argon

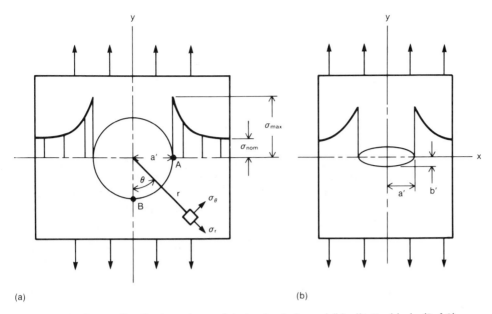

Fig. 4.6. Stress distributions due to (a) circular hole and (b) elliptical hole (Ref 3).

have outlined a simple procedure for estimating elastic-stress-concentration factors for a number of stress-concentration shapes (Ref 9). According to them,

$$K_t = 1 + (0.5 \text{ to } 2)\sqrt{\frac{d}{r}} \qquad \text{(Eq 4.3)}$$

where r is the radius of curvature at the root of the notch and d is the lowest of the relevant dimensions, such as the half-thickness of the remaining ligament, the half-length of a two-sided crack, the length of a single-ended crack, or the height of a shoulder.

In performing calculations of K_t, it is assumed that the elastic limit of the material is not exceeded and that the loading is static. In practice, however, local yielding may be expected to occur and will locally affect the stress distribution. Because the extent of yielding will depend on the nominal stress, higher nominal stresses may render the stress-concentration effects increasingly innocuous.

The effect of notches on fatigue strength is determined by comparing the S-N curves for notched and unnotched specimens. The data for notched specimens are usually plotted in terms of the nominal stress based on the net section of the specimen. The effectiveness of the notch in decreasing the fatigue limit is expressed by the fatigue-strength-reduction factor or fatigue-notch factor K_f. This is simply the ratio of the fatigue limit or fatigue strength of the unnotched specimen to that of the notched specimen. Values of K_f have been found to vary with the severity of the notch, the type of notch, the material, the environment, the type of loading, and the stress level. In general, K_f decreases initially as K_t increases but eventually saturates at large values of K_t. The notch sensitivity of the material in fatigue is usually expressed by a notch-sensitivity index, q:

$$q = \frac{K_f - 1}{K_t - 1} \qquad \text{(Eq 4.4)}$$

For a material that is totally notch-insensitive, q = 0, whereas for a material in which a notch has its full theoretical effect, q = 1. The notch-sensitivity index is not a true material constant but varies with the severity and type of notch, specimen size, and type of loading. It increases with section size and tensile strength.

Fatigue Under Combined Stresses. Components are subjected to complex loadings with both alternating and steady components of stress. Fatigue tests with variable combinations of bending and torsion have shown that for ductile materials the distortion-energy criterion provides the best fit (Ref 3). For brittle materials, the maximum principal stress serves as a better failure criterion. Sines (Ref 10) has suggested a failure criterion that includes the effect of combined stresses and the effect of a static mean stress:

$$\frac{1}{\sqrt{2}}\left[(\sigma_1 - \sigma_2)^2 + (\sigma_2 - \sigma_3)^2 + (\sigma_3 - \sigma_1)^2\right]^{1/2}$$

$$= \sigma_e - C_x(S_x + S_y + S_z) \qquad \text{(Eq 4.5)}$$

where σ_1, σ_2, and σ_3 are the alternating principal stresses, S_x, S_y, and S_z are the static stresses, σ_e is the fatigue strength for completely reversed stress, and C_x is the slope of the Goodman plot of σ_m vs σ_a. Fatigue failure is expected to occur if the left side of Eq 4.5 exceeds the right side.

Effects of Metallurgical Variables

The most significant variables affecting fatigue strength are manufacturing defects. Inclusions in steels, particularly sulfides, serve as initiation sites and as sources of increased crack growth. The effects of surface and subsurface inclusions on the S-N curve for AISI 4340 steel are shown in Fig. 4.7. Fine grain size, which decreases creep strength, is considered to be desirable for fatigue strength (see Fig. 4.8). For alloy steels, tempered martensitic structures usually have better fatigue strengths than bainitic and ferritic-pearlitic structures.

Fatigue properties frequently are correlated with tensile properties. The ratio of the fatigue limit to the ultimate tensile strength is known as the endurance ratio. Endurance ratios for steels generally range from 0.5 to 0.6, whereas those for nonferrous alloys are somewhat lower. Endurance ratios for notched specimens generally range from 0.2 to 0.3.

Low-Cycle Fatigue

Low-cycle fatigue, or high-strain fatigue, is tentatively defined as the fatigue mechanism that controls failures occurring at $N < 10^4$ cycles and typically is of concern when there is significant cyclic plasticity. Skelton (Ref 11) has shown that S-N curves based on LCF tests conducted at constant strain blend nicely with data from HCF tests, indicating that there are no fundamental differences between the mechanisms of the two processes. An important distinction between HCF and LCF is that in HCF most of the fatigue life is spent in crack initiation, whereas in LCF most of the life is spent in crack propagation, because cracks are found to initiate within 3 to 10% of the fatigue life. Traditionally, LCF tests are conducted in the same manner as HCF tests except that the strain range is held constant and the stresses are allowed to vary. This procedure results in a plot of $\Delta\epsilon$ vs N that is similar to the S-N plots discussed previously.

The variation of stresses with strains in LCF tests typically leads to a hysteresis loop such as the one shown in Fig. 4.9. A tension-compression stress range $\Delta\sigma$ is established corresponding to the strain range $\Delta\epsilon$ imposed on the specimen. The total (tip-to-tip) width of the loop corresponds to the total strain range $\Delta\epsilon_t$, which can be broken up into the elastic strain range and the plastic strain range. The height of the hysteresis loop is the stress range. Over a limited

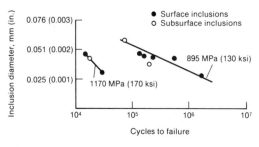

Fig. 4.7. Effects of inclusions on fatigue life of type 4340 steel (Ref 4).

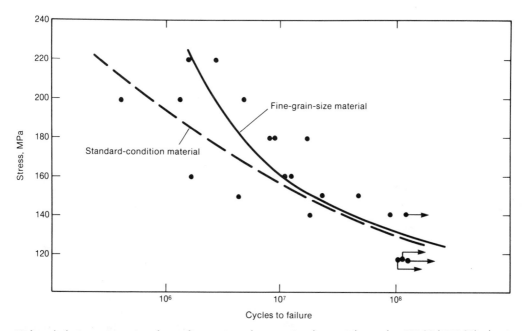

High-cycle fatigue properties of extrafine-grain and conventional material tested at 850 °C (1560 °F), showing the effect of grain size.

Fig. 4.8. S-N curves for IN-738 LC (Ref 4).

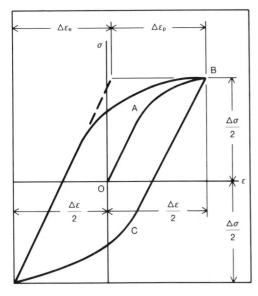

Fig. 4.9. Typical stress-strain hysteresis loop generated in a constant-strain low-cycle fatigue test.

$$\Delta\epsilon_t = B\Delta\epsilon_p^\gamma \qquad \text{(Eq 4.6)}$$

where B and γ are experimentally determined constants. Equation 4.6 is only an approximation of the behavior at intermediate strain ranges and breaks down for small values of $\Delta\epsilon_t$ and large values of $\Delta\epsilon_p$ where $\Delta\epsilon_t \simeq \Delta\epsilon_p$. However, because the two parameters are related, it is common practice to plot the cyclic life N in terms of either $\Delta\epsilon_t$ or $\Delta\epsilon_p$.

Cyclic Stress-Strain Behavior

In the course of a low-cycle fatigue test, the stress range does not remain constant. With increasing N, $\Delta\sigma$ initially either increases or decreases and eventually reaches an approximately steady value. This is known as the saturation or cyclically stable condition. The decrease of $\Delta\sigma$ with N is known as cyclic strain softening, whereas the reverse process is called cyclic strain hardening. Well-annealed materials usually undergo cyclic strain hardening, whereas initially well-hardened materials undergo cyclic

range of strain, the plastic strain range $\Delta\epsilon_p$ is generally correlatable with the total strain range, as illustrated in Fig. 4.10 (Ref 12 and 13), through an expression of the form

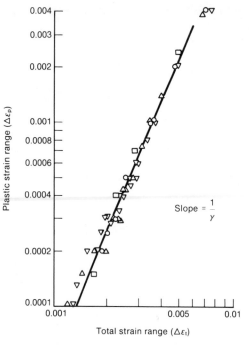

Fig. 4.10. Correlation between total strain range and plastic strain range (Ref 12).

strain softening. Normalized-and-tempered and quenched-and-tempered 2¼Cr-1Mo, 1¼Cr-½Mo, and Cr-Mo-V steels fall in the strain-softening category, whereas annealed steels of the same compositions show light hardening. Once saturation has been achieved and $\Delta\sigma$ has reached a stable value, the cyclic stress-strain relationship can be expresssed in a way similar to the monotonic stress-strain curve:

$$\Delta\sigma = A\Delta\epsilon_p^\beta \qquad \text{(Eq 4.7)}$$

where A is a strength coefficient and β is the cyclic strain-hardening exponent. This is the cyclic analogue of the tensile stress-strain curve. Cyclic stress-strain relationships for several engineering alloys are shown in Fig. 4.11 (Ref 14). The cyclic stress-strain curves for many materials are appreciably different from the monotonic stress-strain curves, and use of the appropriate curves for cyclic loading situations is critical.

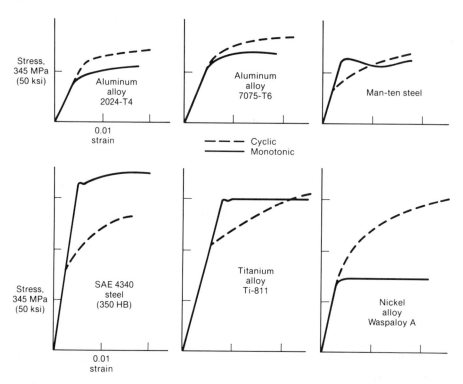

Fig. 4.11. Monotonic and cyclic stress-strain curves for several engineering alloys (Ref 14).

Coffin-Manson Relationships

Curves of strain amplitude (one-half of the strain range) vs cyclic life obtained in LCF tests can be separated into elastic and plastic components of the strain range, as shown in Fig. 4.12. Power-law expressions are found to hold true for both regions, leading to the well-known Basquin (Ref 15) and Coffin-Manson (Ref 16 and 17) relationships

$$\Delta\epsilon_e N_f^{\alpha_1} = C_1 \qquad \text{(Eq 4.8)}$$

and

$$\Delta\epsilon_p N_f^{\alpha_2} = C_2 \qquad \text{(Eq 4.9)}$$

where α_1 and α_2 are material constants related to the slopes, and C_1 and C_2 are material constants related to the fatigue-ductility coefficient ϵ_f' and the fatigue-strength coefficient pertaining to the elastic and plastic regimes, as illustrated in Fig. 4.12. Summing Eq 4.8 and 4.9 results in the expression

$$\Delta\epsilon_t = C_1 N_f^{-\alpha_1} + C_2 N_f^{-\alpha_2} \qquad \text{(Eq 4.10)}$$

Manson (Ref 18) has proposed that

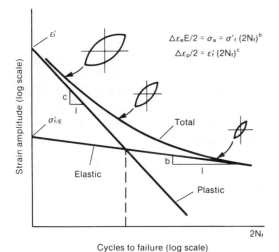

Fig. 4.12. Fatigue life as a function of elastic, plastic, and total strain amplitude.

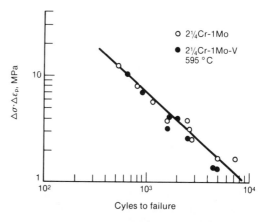

Fig. 4.13. Alternative presentation of data where $\Delta\sigma \cdot \Delta\epsilon_p$ is proportional to hysteresis energy (Ref 19).

$$\Delta\epsilon_t = \frac{3.5\sigma_u}{E} N_f^{-0.12} + D^{0.6} N_f^{-0.6} \qquad \text{(Eq 4.11)}$$

where σ_u and E are expressed in ksi and D is the ductility defined as $D = \ln[100/(100 - RA)]$, where RA is a reduction in area (in %). Equation 4.11 is known as the Method of Universal Slopes. The implication of this equation is that in the elastic region the fatigue strength is governed by the tensile strength whereas at large $\Delta\epsilon$ (small N_f) it is governed by ductility. Considerable laboratory data have been expressed using Eq 4.11, but in regions of practical interest (say $\Delta\epsilon \simeq 0.002$) the plastic strain is too small to be measured accurately. An alternative plot of the product $\Delta\sigma \cdot \Delta\epsilon_p$, which is proportional to the energy dissipated per cycle vs N, has been suggested, as shown in Fig. 4.13 (Ref 19). This method is useful when alloys of widely different strength are to be compared.

Creep-Fatigue Interaction

In components which operate at high temperatures, changes in conditions at the beginning and end of operation or during operation result in transient temperature gradients. If these transients are repeated, the differential thermal expansion during each transient results in a thermally induced cyclic stress. The extent of the resulting fatigue damage depends on the nature and frequency of the transient, the thermal gradient in the component, and the material properties. Components which are subject to thermally induced stresses generally operate within the creep range so that damage due to both fatigue and creep have to be taken into account. Gas-turbine blades and disks are particularly subject to severe thermal gradients during start-ups. In steam-turbine rotors and casings, the large section size of the component results in large temperature gradients even under conditions of long start-up and prewarming of the component. Under fast starting conditions from relatively low temperatures, severe thermal stresses can result. This problem has been exacerbated in recent years by the pressing of base-load units into cyclic operation for economic reasons. Surface craze cracking can occur in steam-turbine pipework and valves due to condensation of steam on cold metal surfaces during certain modes of operation.

The effect of start-stop cycles can be illustrated with respect to a high pressure (HP) rotor of a steam turbine. During a start-up, the surface of the rotor heats up faster than the bulk and attempts to expand, but is held in check by the bulk of the rotor. This action sets up a compressive strain at the surface which eventually relaxes to zero strain but leaves a tensile residual stress when the entire rotor reaches the steady operating temperature. During shutdown, the process is reversed. During steady operation between start-stops, stress relaxation as well as creep processes operate under centrifugal or pressure stresses.

Situations similar to the above apply for many other heavy-section components such as petroleum- and chemical-industry pressure vessels and nuclear-reactor pressure vessels. In view of the importance of combined creep and fatigue damage with respect to component reliability, many attempts have been made to develop damage rules that will help in design as well as in component life prediction under creep-fatigue conditions. Several reviews of this subject are available (Ref 20 to 27). In developing these

damage rules, four types of laboratory tests have been utilized:

1. Strain-controlled tests with hold periods at constant stress or strain
2. Creep tests under cyclic stress or strain
3. Interspersed creep and fatigue tests
4. Strain-controlled tests under athermal conditions.

Type 4 tests, generally known as thermomechanical fatigue tests, are discussed in a separate section. Results from tests of types 2 and 3 are meager and are referred to whenever appropriate. Because the strain-controlled fatigue test (type 1) is the most common, it is described in detail in the next section.

Hold-Time Effects in Strain-Controlled Fatigue

The principal method of studying creep-fatigue interactions has been to conduct strain-controlled fatigue tests with variable frequencies with and without a holding period (hold time) during some portion of the test. The lower frequencies and the hold times can allow creep to take place. In pure fatigue tests, at higher frequencies and short hold times, the fatigue mode dominates and failures start near the surface and propagate transgranularly. As the hold time is increased, or the frequency decreases, the creep component begins to play a role with increasing creep-fatigue interaction. In this region, fractures are of a mixed mode involving both fatigue cracking and creep cavitation. With prolonged hold times with occasional interspersed cycles, creep processes completely dominate and can be treated almost as pure cases of creep. In instances where oxidation effects contribute significantly to the creep-fatigue interaction, the situation is more complex than described above (Ref 28).

Ferritic Steels. The effects of frequency as well as of hold time on the LCF behavior of Cr-Mo-V rotor steels has been studied by a number of investigators (Ref 29 to 34). Leven (Ref 31) reported that the effect of frequency (cycles per minute) at 540 °C

(1000 °F) could be adequately represented by the equation

$$\Delta\epsilon_t = \Delta\epsilon_e + \Delta\epsilon_p$$
$$= 0.0097 N_f^{-0.095} \nu^{0.08}$$
$$+ 2.8 N_f^{-0.831} \nu^{0.162} \quad \text{(Eq 4.12)}$$

A good systematic study of hold-time effects has been carried out by Ellison and Patterson at 565 °C (1050 °F) (Ref 29). Their tests included either constant-strain or constant-load hold periods in tension or compression. The resulting data are shown in Fig. 4.14 (Ref 29). The hold time (in minutes) during the tensile/compressive portion of the cycle is also shown on the figure. These data show that the addition of tensile hold periods dramatically reduced the life of the material compared with results of continuous cycling tests, with the longer hold periods leading to shorter lives and creep-dominated failures. Compressive holds offset the detrimental effect of tensile holds, and, in the limiting case of equal-duration tensile and compressive holds, the fatigue curve approaches that of the pure fatigue case. In contrast to the above results, Kramer *et al* (Ref 32) found that the

effects of hold time and frequency were only marginal at 425 °C (800 °F), presumably because the temperature was too low for creep effects.

Long-term high-strain fatigue data have been obtained for a forged 1Cr-1Mo-¼V rotor steel by Thomas and Dawson (Ref 34). Their results show that the effect of hold time on fatigue life is a function of the type of strain cycle employed, the strain range, and the test temperature. They compared cyclic lives under two types of strain cycles: the laboratory-type cycle, in which the hold time is normally imposed at the maximum strain in the tensile cycle; and a type II cycle in which the hold time is imposed at the zero strain. In the laboratory-type cycle, the maximum tensile stress occurs at the start of the hold period, while in the type II cycle the tensile stress at the start of the hold period depends on the extent of the yielding during the previous compressive part of the cycle. In addition, the hold period is followed by further tensile strain, rather than by a strain reversal. The two types of cycles and the corresponding fatigue data are illustrated in Fig. 4.15. The type II cycle simulates the actual strain cycles expected on the surface of a HP

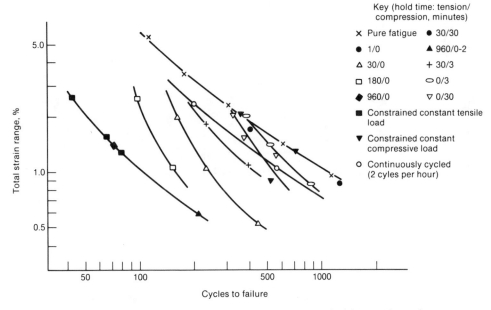

Fig. 4.14. Effects of hold time and tensile vs compressive hold on cyclic endurance life of 1Cr-Mo-V rotor steel at 565 °C (1050 °F) (Ref 29).

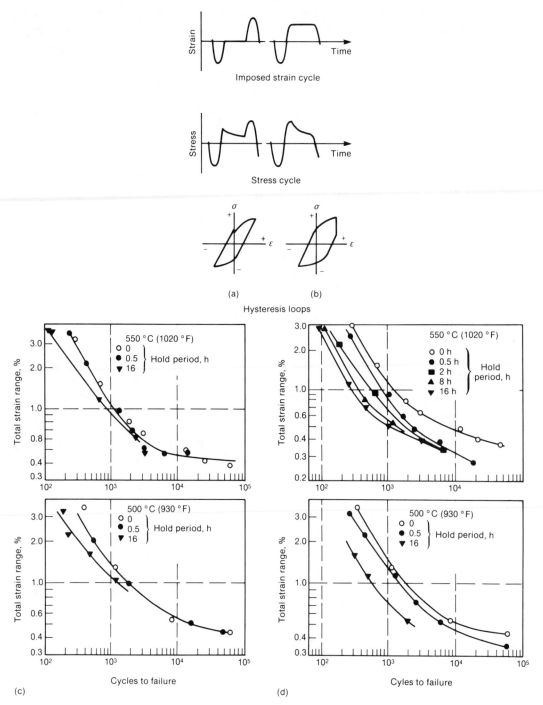

(a) Type II cycle. (b) Laboratory cycle. (c) Endurance data for type II cycle. (d) Endurance data for laboratory cycle.

Fig. 4.15. Types of LCF cycles employed, and corresponding endurance data, for a 1Cr-Mo-V rotor steel (Ref 34).

rotor. The data show that for laboratory cycles, at 550 °C (1020 °F), increasing hold time progressively decreases fatigue life at strain ranges above about 0.4%. At lower strain ranges, all the fatigue curves converge to the same values as those of the 0.5-h hold-time curve. At 500 °C (930 °F), similar trends are apparent, although the

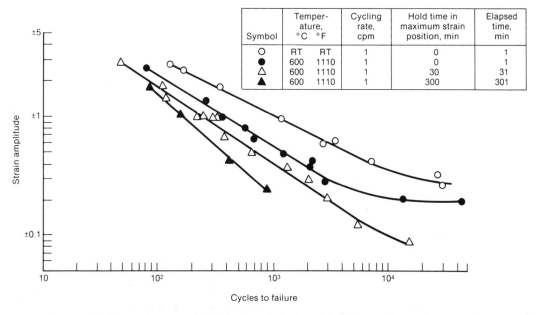

Symbol	Temperature, °C	°F	Cycling rate, cpm	Hold time in maximum strain position, min	Elapsed time, min
○	RT	RT	1	0	1
●	600	1110	1	0	1
△	600	1110	1	30	31
▲	600	1110	1	300	301

Fig. 4.16. Effect of tensile hold time at 600 °C (1110 °F) on cyclic endurance of 2¼Cr-1Mo steel (Ref 38).

actual convergence of the various hold-time fatigue curves is indicated at lower strains. For the type II service cycle, which is representative of the HP rotor, the effect of a 0.5-h hold period on endurance at both 550 and 500 °C is negligible. The 16-h hold period has a very small effect at low strain ranges; at high strain ranges, the effect increases but is still less than for laboratory cycles. These data clearly point out the pitfalls of using unrealistic laboratory tests in life prediction of components and the need for simulating the component strain cycles in the laboratory tests in order to generate the appropriate data.

Creep-fatigue tests on 2¼Cr-1Mo steels have been conducted by several investigators (Ref 35 to 38). Brinkman, Strizak, and Booker (Ref 36) showed that tensile holds are innocuous whereas compressive holds are detrimental to the cyclic life. Challenger *et al* (Ref 37) have attributed this effect to oxide growth during compressive hold, which results in increased tensile strains during the tensile cycle. Miller *et al* (Ref 20) have attributed the above behavior to errors in diametric strain measurements, especially under oxidizing conditions. Furthermore, in the tests of Brinkman *et al*, the hold times were too short to reveal the true creep-

fatigue behavior of the steel. Other data by Edmunds and White (Ref 38) have shown that, as expected, tensile holds are damaging, as shown in Fig. 4.16. It should once again be emphasized that in applying these data, the actual strain cycles in a component may be sufficiently different to warrant modification of the laboratory results. Results on 1Cr-½Mo steels have confirmed the damaging effects of tensile holds, as shown in Fig. 4.17 (Ref 39).

Many fossil plants which were originally designed for base-loaded operation now have been converted to cyclic operation. The effect of prior creep on subsequent creep-fatigue damage is therefore of great interest. Miller and Gladwin (Ref 39) investigated the creep-fatigue behavior of samples which had been previously tested in creep to life fractions of 0.2 and 0.6. Their results (Fig. 4.17) show that in simulated heat-affected-zone material where prior creep had resulted in cavitation damage, the prior creep greatly decreased cyclic life. On the other hand, in the base material of the 1Cr-½Mo steel, where prior creep damage consisted merely of softening, subsequent low-cycle fatigue behavior was actually improved in comparison with the nonprecrept samples. Softening can lead to increased

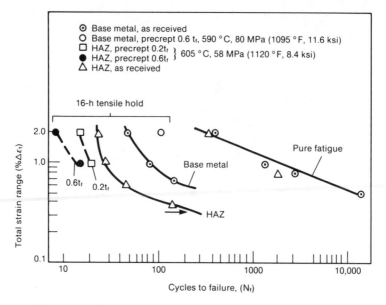

Fig. 4.17. Effects of hold time and prior creep damage (0.2 and 0.6 life fractions) on cyclic endurance of 1Cr-½Mo steel (Ref 39).

ductility and hence to improved low-cycle fatigue life, as reflected in Eq 4.11. The nature of prior creep damage thus plays a key role in subsequent LCF damage.

Austenitic Stainless Steels. Early studies (Ref 40 to 45) on AISI type 316 stainless steel showed that tensile hold periods in the temperature region from 550 to 625 °C (1020 to 1160 °F) were very damaging, as shown in Fig. 4.18 (Ref 41). Because the strain ranges were fairly high and the hold periods were short, failures were dominated by fatigue. More recent results at lower strain ranges and longer hold periods have revealed that creep-dominated failures also occur in stainless steels (Ref 46 to 48). Creep-dominated failures have been observed by Goodall *et al* (Ref 46) for tensile hold times

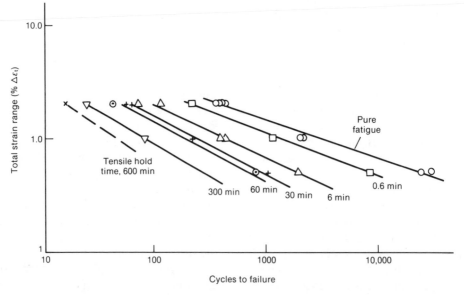

Fig. 4.18. Effect of tensile hold time on fatigue endurance of type 316 stainless steel (Ref 41).

up to 16 h at 600 °C (1110 °F). Similar results have been reported by Wood *et al* (Ref 47) in tests at 625 °C (1160 °F) with tensile hold times up to 48 h. Life reduction due to tensile hold has been observed to be related to stress-rupture ductility, thus leading to heat-to-heat variations (Ref 48 and 49). Some investigators have observed a saturation in the detrimental effects of tensile hold periods — a recovery of the endurance occurring at longer hold periods. This has been attributed to microstructural changes leading to increases in ductility. Aging and concomitant precipitation and growth of large carbides prior to testing have been shown to eliminate creep-fatigue effects altogether in type 316 stainless steel at 650 °C (1200 °F) (Ref 50). Compressive holds have been found to have an effect similar to that in Cr-Mo-V steels — i.e., they nullified the detrimental effects due to tensile holds. The effect of slow-fast cycles, in which the strain increased slowly during the tension cycle but increased rapidly during the compression-going cycle, on the endurance of type 304, type 316, and other stainless steels has been investigated (Ref 51 to 54). In other tests, lower strain rates in the tension cycle were found to reduce the endurance.

Nickel Alloys. There is some evidence based on studies of cast Inconel 738 (Ref 55), René 95 (Ref 56), MAR-M 200 (Ref 57), and cast René 80 (Ref 58 and 59) that compressive holds are more damaging to the endurance of nickel-base alloys than tensile holds. Wells and Sullivan were the first investigators who noted the unexpectedly damaging effect of compressive holds (Ref 60 and 61). They suggested that compressive holds promoted elongated cracks which were more detrimental to endurance than rounded cavities produced by tensile holds. Lord and Coffin also observed that compressive holds often introduce tensile, cyclic mean stresses in some nickel-base alloys (Ref 62).

Pronounced detrimental effects due to tensile holds were observed by Viswanathan, Beck, and Johnson during fatigue testing of Udimet 710 in air (Ref 63). Continuous-cycle tests in which the frequency of the strain rate was held constant were conducted in the total strain range of 0.6 to 2%, at selected temperatures ranging from 540 to 980 °C (1000 to 1800 °F). Cycling tests with hold times at the maximum tensile strain were conducted using 5-h hold times, mostly for tests with a total strain range of 2%. The fracture morphology also was characterized.

The relationship between total strain range and number of cycles to failure from the study of Viswanathan, Beck, and Johnson is shown in Fig. 4.19 (Ref 63). Because data at 790 °C (1450 °F) were the most abundant, curves were drawn only through these data points, although data for other conditions were also included. Data from thermomechanical cycling also are included in this figure, but will be discussed in a later section. The fracture morphology was characterized as transgranular (T), intergranular (I), or mixed (M). The first letter in parentheses denotes the initiation mechanism and the second letter denotes the propagation mechanism. For instance, the designation (I,T) would indicate that the crack initiation was intergranular but that the crack propagation was transgranular. The principal conclusions from this study were as follows. (1) A 5-h tensile hold caused a reduction in the fatigue life at all temperatures. The extent of the reduction was, however, a function of the test temperature, as shown in Fig. 4.20. The maximum reduction in life due to hold time occurred at an intermediate temperature of about 845 °C (1550 °F). (2) Plots of log N_f vs $1/T$ for the pure fatigue tests at the 1% strain-range level resulted in a linear relationship, as shown in Fig. 4.21. The apparent activation energy for failure, Q, using an Arrhenius relationship of the type $N_f = N_0 \exp(Q/RT)$, was found to be 67 kJ/mole (16 kcal/mole). (3) The magnitude of the strain range employed had no effect on the fracture morphology. Increasing the temperature, hold time, and thermal cycling, on the other hand, increased the intergranular component of the fracture, changing first the morphology of the crack initiation and then that of the propagation. The effect of temperature on fracture mor-

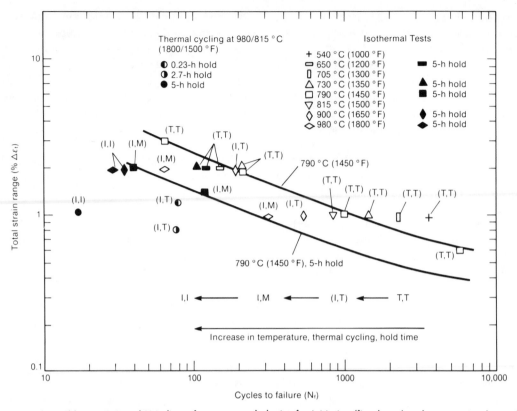

Parenthetical letters T, I, and M indicate fracture morphologies for initiation (first letter) and propagation (second letter): T = transgranular; I = intergranular; M = mixed.

Fig. 4.19. Variation of number of cycles to failure with strain range for Udimet 710 (Ref 63).

phology in a test with 5-h hold times is illustrated in Fig. 4.22.

In contrast to the above results, Whitlow *et al* showed that in isothermal fatigue tests at 730 °C (1350 °F), a hold time of 1 h at maximum tensile strain was relatively innocuous in Udimet 710 and 720 (Ref 64). These results can be rationalized in terms of Fig. 4.19, which shows that the worst effects of hold time occur at 845 °C (1550 °F). The low temperature employed by Whitlow *et al* combined with the shorter hold times were probably responsible for the negligible effect of hold time in their tests.

The effects of waveform, including hold-time effects, on the fatigue endurance of cast Inconel 738 at 850 °C (1560 °F), were investigated by Nazmy (Ref 65 and 66). The results of these studies, depicted in Fig. 4.23, show the detrimental effects of tensile holds.

The CP-type cycles in which the tension-going cycle was at a lower rate or where a tensile hold was imposed at the maximum tensile strain were found to lead to a larger reduction in life compared with the opposite type of cycle, termed the PC cycle.

Low-cycle fatigue behavior of a cast gas-turbine vane alloy, MAR-M 509, at 900 °C (1650 °F) in air has been investigated by Remy *et al* (Ref 67). For continuous saw-tooth-type cycles, reducing the frequency from 20 Hz to 6×10^{-2} Hz resulted in a severe reduction in fatigue life. A further decrease in frequency to 5×10^{-3} Hz did not lead to additional degradation, indicating a saturation effect (see Fig. 4.24). Tensile holds and compressive holds (2 min) were found to be equally damaging and reduced the fatigue life by a factor of 2 compared with the pure fatigue tests.

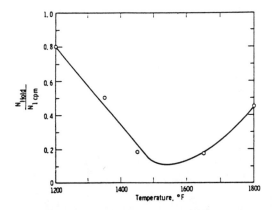

Fig. 4.20. Variation of the effect of 5-h tensile hold time on cyclic life as a function of test temperature ($\Delta\epsilon = 2\%$) for Udimet 710 (Ref 63).

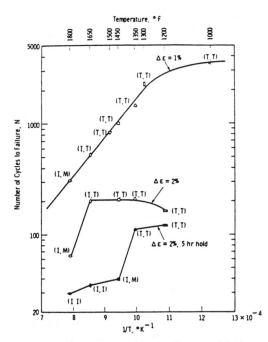

See caption for Fig. 4.19 for identification of fracture morphologies.

Fig. 4.21. Effect of temperature on cyclic life of Udimet 710 (Ref 63).

Effect of Rupture Ductility

There is ample evidence to show that rupture ductility has a major influence on creep-fatigue interaction. Because this effect is believed to be caused by the influence of rupture ductility on the creep-fracture com-

ponent, endurance in continuous-cycle and in high-frequency or short-hold-time fatigue tests (where fracture is fatigue-dominated) will be relatively unaffected. As the frequency is decreased or as the hold time is increased, the effect of rupture ductility becomes more pronounced, as illustrated by the work of Kadoya *et al* (Ref 68). In Fig. 4.25(a), hold-time effects on the fatigue lives of two rotors differing primarily in terms of their rupture-ductility behavior are compared. The fatigue life of the low-ductility rotor steel is much more adversely affected by hold time than that of the high-ductility rotor steel. The rupture ductilities diverge with increasing time to rupture (Fig. 4.25b), which is correspondingly reflected in the long-hold-time tests. Endurance data for several ferritic steels, in relation to the range of rupture ductility exhibited by them, are illustrated in Fig. 4.26, from the work of Miller *et al* (Ref 20). The lower the ductility, the lower the creep-fatigue endurance. In addition, long hold periods, small strain ranges, and low ductility favor creep-dominated failures, whereas short hold periods, intermediate strain ranges, and high creep ductility favor creep-fatigue-interaction failures. Similar results have been presented by Miller *et al* for austenitic stainless steels (Ref 20).

In a tensile hold period, a range of strain rates is encountered. If the strain rates are above the critical strain rate needed to cause constrained cavity growth, and hence a significant drop in ductility, the cyclic endurance is unaffected. This is generally the case for short hold times. With increasing hold times, strain rates drop to sufficiently low levels to cause cavity growth and low-ductility failures. Hence, the differentiation between low-ductility and high-ductility materials is exhibited in long-hold-time tests. For further discussion of this subject, the reader is referred to the paper by Miller *et al* (Ref 20).

Effects of Environment

The possible roles that can be played by environment in affecting fatigue life are too

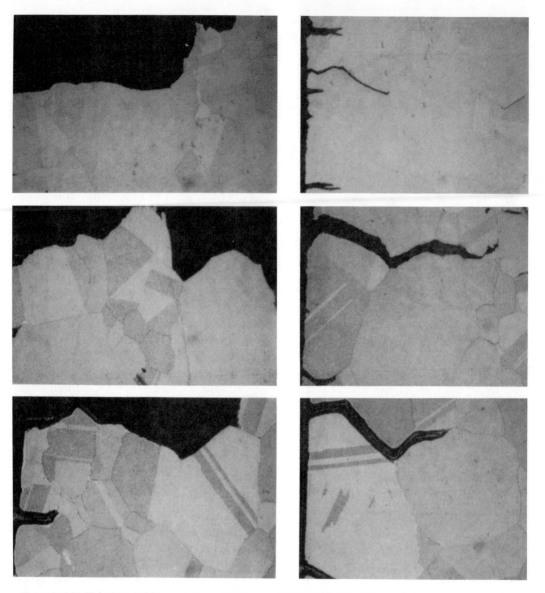

Top pair: 730 °C (1350 °F); T,T morphology. Middle pair: 790 °C (1450 °F); I,M morphology. Bottom pair: 900 °C (1650 °F); I,I morphology. Magnification (all), approximately 100×. See caption for Fig. 4.19 for identification of fracture morphologies.

Fig. 4.22. Morphologies of fractures (at left) and spikes (at right) in low-cycle fatigue specimens of Udimet 710 tested isothermally with 5-h hold times at a strain range of 2% (Ref 63).

numerous to describe. The more important ones include formation of oxide notches, grain-boundary embrittlement, increase of net section stress, corrosion-product wedging, and shielding. The first four are fairly obvious and can cause reductions in fatigue life. The last effect, shielding, is described as a net reduction in the effective strain range or stress-intensity range due to corrosion products which in some instances can result in decreased crack growth and increased life. Where detrimental effects occur, it is to be expected that longer hold times or lower frequencies during test-

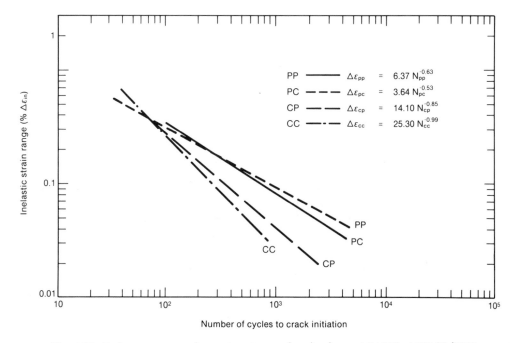

Fig. 4.23. Endurance curves for various types of cycles for cast IN-738 at 850 °C (1560 °F) (Ref 65 and 66).

ing will display the environmental effects more prominently, as shown in Fig. 4.27 (Ref 69). At high frequencies, fracture is fatigue-dominated in air and in vacuum. At intermediate frequencies, the effect of the environment is most significantly felt. At very low frequencies where pure creep processes begin to dominate, the data for air and for vacuum start to converge. In the intermediate region where the creep-fatigue interaction will be most pronounced, the effects due to environment also are most obvious. The presence of the environment promotes the onset of the intergranular fracture mechanisms at higher frequency levels.

Detrimental effects resulting from air oxidation also have been reported for Udimet 500 (Ref 70), type 304 stainless steel (Ref 70), type 316 stainless steels (Ref 71 and 72), type 304 stainless steel and Hastelloy (Ref 73), and wrought IN-738 LC (Ref 74). Reduction in fatigue life due to hot corrosion of Udimet 710 and 720 is well documented (Ref 64).

The effects of environment are not always straightforward. Harrod and Manjoine (Ref 75) claim to have found no difference in fatigue behavior of type 304 stainless steel tested in air and in vacuum at 650 °C (1200 °F). Gell and Leverant (Ref 76) showed air to be actually beneficial to the fatigue life of MAR-M 200 at 910 °C (1670 °F) in comparison with vacuum, presumably because of the oxide shielding effect. Taking credit for beneficial effects of shielding in actual applications can, however, be dangerous, because one cannot ensure that the oxide products will remain intact during service as they do in laboratory tests. The shielding effect also has been found to be beneficial and to lead to increasing threshold stress intensity for crack propagation with decreasing test frequency for a 1Cr-Mo-V steel at 550 °C (1020 °F) (Ref 77). Once the threshold K value was exceeded, however, the air environment caused accelerated crack growth. In cast Inconel 738, an air environment has been found to be beneficial because of crack branching along oxidized dendritic boundaries, thus resulting in reduced crack growth (Ref 74). The effects of environment are thus found to be very complex and to vary with the mate-

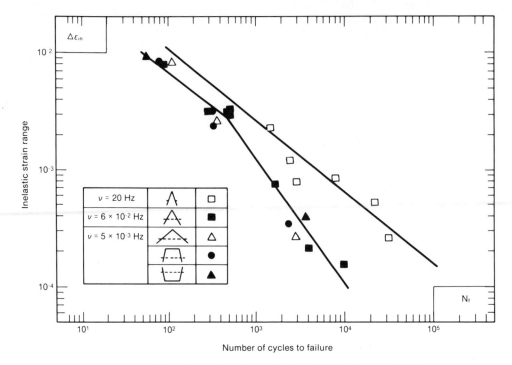

Fig. 4.24. Variation of number of cycles to failure (N_f) in low-cycle fatigue as a function of inelastic strain range and frequency (ν) for MAR-M 509 (Ref 67).

rial, environment, temperature, and test frequency.

Summary of Hold-Time and Frequency Effects

Many studies have been carried out on ferritic steels, austenitic steels, and nickel-base alloys under a variety of test conditions. Hold times and reduced frequencies have been reported to be detrimental or innocuous. For each material, there is controversy regarding whether a tensile hold or a compressive hold is detrimental. These controversies seem to arise mainly because data obtained under different test and/or material conditions have been compared. The effect of hold time seems to depend on a variety of factors including type of cycle, strain range, temperature, environment, and material ductility. Frequently, test conditions employed in the laboratory seem to be irrelevant to actual conditions existing in a component. The importance of conducting tests that are appropriate to a specific com-

ponent under service conditions cannot be overemphasized.

Damage Rules and Life Prediction

In general, creep-fatigue design considerations are intended to prevent crack initiation, where crack initiation may be defined arbitrarily as the presence of cracks which can be detected visually, say 1 mm in size. The difference between crack initiation and failure life in a small specimen is often a small proportion of the total life, and it can be argued that the failure endurance of a small specimen corresponds to the endurance at crack initiation in a large component.

Several damage rules have been enunciated for estimating the cumulative damage under creep-fatigue conditions. Interest has tended to focus on four basic types of approaches:

1. The damage-summation method
2. The frequency-modified strain-range method

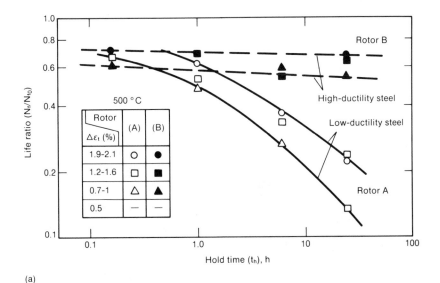

(a)

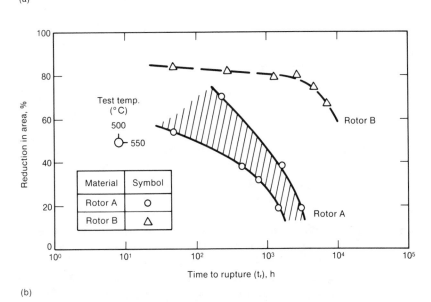

(b)

Fig. 4.25. Effect of rupture ductility on hold-time effects during low-cycle fatigue testing of 1Cr-Mo-V rotor steel (Ref 68).

3. The strain-range-partitioning method
4. The ductility-exhaustion method.

In addition to these, several other approaches have also been selectively applied.

Linear Damage Summation. The most common approach is based on linear superposition of fatigue and creep damage. Indeed, the mainstay of the present design procedures is the linear life-fraction rule, which forms the basis of the ASME Boiler and Pressure Vessel Code, Section III, Code Case N-47 (Ref 78). This approach combines the damage summations of Robinson for creep (Ref 79) and of Miner for fatigue (Ref 80) as follows (Ref 81):

$$\Sigma \frac{N}{N_f} + \Sigma \frac{t}{t_r} = D' \qquad (Eq\ 4.13)$$

where N/N_f is the cyclic portion of the life fraction, in which N is the number of cycles

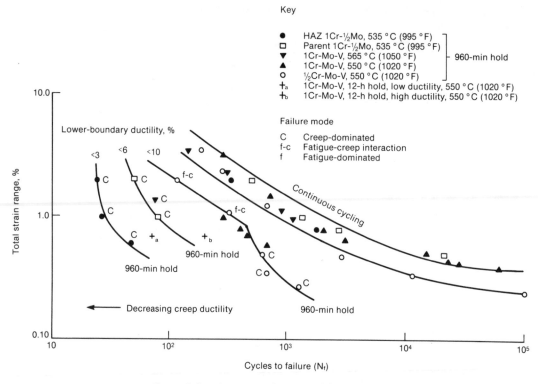

Key

●	HAZ 1Cr-½Mo, 535 °C (995 °F)	⎤
□	Parent 1Cr-½Mo, 535 °C (995 °F)	
▼	1Cr-Mo-V, 565 °C (1050 °F)	⎬ 960-min hold
▲	1Cr-Mo-V, 550 °C (1020 °F)	
○	½Cr-Mo-V, 550 °C (1020 °F)	⎦
+$_a$	1Cr-Mo-V, 12-h hold, low ductility, 550 °C (1020 °F)	
+$_b$	1Cr-Mo-V, 12-h hold, high ductility, 550 °C (1020 °F)	

Failure mode

C — Creep-dominated
f-c — Fatigue-creep interaction
f — Fatigue-dominated

Fig. 4.26. Effect of ductility on endurance of ferritic steels (Ref 20).

at a given strain range and N_f is the pure fatigue life at that strain range. The time-dependent creep-life fraction is t/t_r, where t is the time at a given stress and t_r is the time to rupture at that stress. The stress-relaxation period is divided into time blocks during which an average, constant value of stress prevails, and for each time block t/t_r is computed and summed. D′ is the cumulative damage index. When D′ = 1, failure is presumed to occur.

If Eq 4.13 were obeyed, a straight line of the type shown in Fig. 4.28 between the fatigue- and creep-life fractions would be expected. Results on Cr-Mo-V rotor steels, types 304 and 316 stainless steels, 2¼ Cr-1Mo steels, and Incoloy 800 have shown that the straight-line behavior is not obeyed. The behavior actually observed for 1Cr-Mo-V rotor steels is shown in Fig. 4.28 (Ref 82). Values of D′ are found to be both above and below unity in different regions and are characterized by a bilinear curve. The actual damage curve for the steels shown can be used as the upper boundary

for safe design, or in life assessment, implying a varying value of D′. Unfortunately, several material and test parameters may affect the distribution of D′, and there is no satisfactory way of applying the linear damage rule at present. For types 304 and 316 stainless steels and for Incoloy 800, the damage parameter D′ is also characterized by a bilinear distribution and has been adopted by the ASME Code Case N-47. Design envelopes based on the bilinear distribution of data are shown in Fig. 4.29 (Ref 78). Note, however, that in this figure N_f and t_r have been converted to design-allowable values N_d and t_d by applying safety factors. The code case procedure is described later in this chapter, under "Design Rules for Creep-Fatigue."

The life-fraction rule is purely phenomenological, having no mechanistic basis. Its applicability is, therefore, material-dependent. Contrary to experience, it also assumes that tensile and compressive hold periods are equally damaging. The strain softening behavior encountered in many

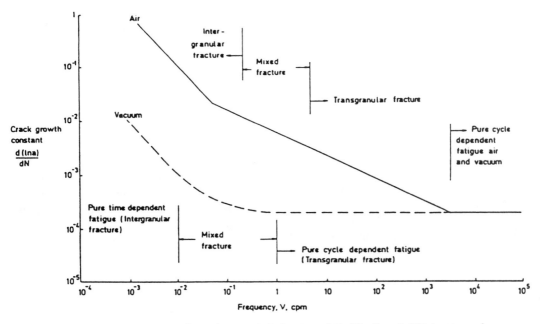

Fig. 4.27. Comparison of crack-growth behavior of Fe-Ni alloy A-286 in air and vacuum at 595 °C (1100 °F) (Ref 69).

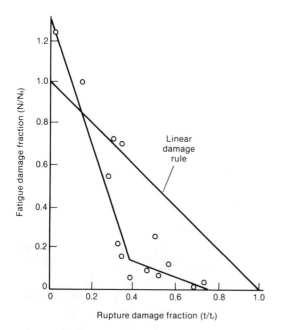

Fig. 4.28. Creep-rupture/low-cycle-fatigue damage interaction curve for 1Cr-Mo-V rotor steel at 540 °C (1000 °F) (after Ref 82).

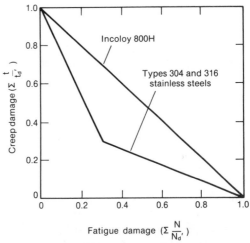

Fig. 4.29. Linear damage fatigue-creep design envelopes (Ref 78).

steels and the effect of prior plasticity on subsequent creep are not taken into account. Use of virgin-material rupture life to compute creep-life fractions is, therefore,

inaccurate. In spite of these limitations, the damage-summation method is very popular because it is easy to use and requires only standard S-N curves and stress-rupture curves. An example of the application of this method to a steam-turbine rotor may be found in Ref 83.

It should be noted that the use of the life-fraction rule is not synonymous with the use of ASME Code Case N-47 because the latter incorporates several assumptions and safety factors that make it unduly conservative for purposes of life assessment.

Frequency Modification. This approach is essentially a modification of the Coffin-Manson relationship for pure fatigue, given earlier as Eq 4.10, by incorporating a frequency term (Ref 84), as follows:

$$\Delta \epsilon_p = C[N_f \nu^{k-1}]^{-\alpha} \qquad \text{(Eq 4.14)}$$

where C, k, and α are constants.

This equation has been rewritten in various forms to indicate the significance of the constant C and the exponents k and α. In terms of total strain range, Eq 4.14 can be written as

$$\Delta \epsilon_t = \Delta \epsilon_e + \Delta \epsilon_p$$
$$= C_3[N_f \nu^{k_1-1}]^{-\alpha_3}$$
$$+ C_4[N_f \nu^{k_2-1}]^{-\alpha_4} \qquad \text{(Eq 4.15)}$$

where C_3, C_4, k_1, k_2, α_3, and α_4 are constants. This form of the equation has been used widely to describe data for several steels. In the early descriptions (Ref 84), the frequency term was related to the total time as

$$\frac{1}{\nu} = t_{cy} + t_h \qquad \text{(Eq 4.16)}$$

where t_{cy} is the cycle time and t_h is the hold time. More recently, however, modifications have been made to include the effects of cycle shape by postulating that the damage is dependent on the rate of straining during tension and in compression (Ref 85 to 88). Equation 4.14 is rewritten in the form

$$N_f = \left(\frac{C}{\Delta \epsilon_p} \right)^{1/\alpha} \nu^{1-k} \qquad \text{(Eq 4.17)}$$

and is altered to give the frequency-separation equation

$$N_f = \left(\frac{C}{\Delta \epsilon_p'} \right)^{1/\alpha} \left(\frac{\nu_t}{2} \right)^{1-k} \qquad \text{(Eq 4.18)}$$

where the term $\Delta \epsilon_p'$ accounts for the imbalance of the loop and the term $(\nu_t/2)^{1-k}$ accounts for the time spent in the tension part of the cycle. $\Delta \epsilon_p'$ is the equivalent plastic strain, defined as

$$\Delta \epsilon_p' = \Delta \epsilon_p \left[\frac{(\nu_c/\nu_t)^{k_1'} + 1}{2} \right]^{\alpha/\alpha'} \qquad \text{(Eq 4.19)}$$

In this form, $\nu_t = 1/t_t$ is the tension-going frequency and $\nu_c = 1/t_c$ is the compression-going frequency. t_t is the time spent in the tension part of the cycle, t_c is the time spent in the compression part of the cycle, and k_1' and α' are equation constants related to each other through the expression

$$k_1' = k_1 + \alpha'(k - 1) \qquad \text{(Eq 4.20)}$$

The total frequency ν is given by

$$\nu = \frac{1}{1/\nu_t + 1/\nu_c} \qquad \text{(Eq 4.21)}$$

In general, the frequency-modified strain-range equations are straightforward in terms of application to laboratory data, provided that a complete description of the stress-strain hysteresis loop is available for the cycles under consideration. The constants C, α, and k are obtained from balanced-loop data, and the constants α' and k_1' are determined from imbalanced-loop data.

Equation 4.18 is the latest in the development of a series of frequency-modified approaches. Coffin has shown good agreement between fatigue life predicted from this equation and actual life measured in experiments for type 316 stainless steel (Ref 87). Such agreement for a 1Cr-Mo-V steel at 565 °C (1050 °F) has been reported to be less convincing (Ref 89). Unfortunately, this method suffers from the need to determine a number of material constants from cyclic hold-time tests. These constants also can vary from material to material, and generalizations cannot be made to other materials and test conditions because there is little mechanistic foundation for the empirical relationships.

Example:
A Cr-Mo-V rotor operating in the creep range has been subjected to daily start-stop cycles (similar) amounting to 730 cycles over a period of about two years. Calculate the fatigue life consumed using the frequency-modified strain-range approach. The total strain range at the critical location has been determined to be 0.0084.

Answer:
A simplified form of the frequency-modified strain range (Eq 4.15) can be written as

$$\Delta\epsilon_t = C_1 N_f \nu^b + C_2 N_f \nu^d$$

From literature, the equation applicable to Cr-Mo-V rotor steels at elevated temperatures has been found to be

$$\Delta\epsilon_t = 0.0094 N_f^{-0.092} \nu^{0.033}$$
$$+ 0.885 N_f^{-0.759} \nu^{0.034}$$

where ν is the frequency in cycles per minute. For the turbine, $\nu = 0.0007$ cpm and $\Delta\epsilon_t = 0.0084$. Substituting these values in the above equation, we get the number of cycles to failure, $N_f = 800$ cycles. Thus,

Fatigue-life consumption
$$= N/N_f = 730/800 = 0.93$$

We conclude that 93% of the fatigue-crack-initiation life has been consumed.

Strain-Range Partitioning. The strain-range-partitioning (SRP) approach involves partitioning of the total inelastic strain range into four possible components depending on the direction of straining (tension or compression) and the type of inelastic strain accumulated (creep or time-independent plasticity) (Ref 90 to 95). Figure 4.30 shows the four generic types of hysteresis loops for the four types of strain ranges. The actual hysteresis loop from a creep-fatigue test (i.e., LCF test with hold time) is broken down into the component strains: $\Delta\epsilon_{pp}$, $\Delta\epsilon_{cc}$, $\Delta\epsilon_{pc}$, and $\Delta\epsilon_{cp}$. The terms $\Delta\epsilon_{pp}$ and $\Delta\epsilon_{cc}$ represent the pure reversed plastic and reversed creep strain ranges, respectively,

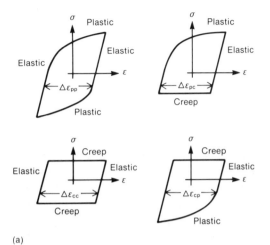

(a)

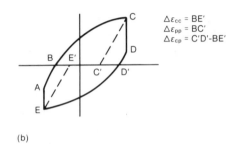

(b)

$$\Delta\epsilon_{cc} = BE'$$
$$\Delta\epsilon_{pp} = BC'$$
$$\Delta\epsilon_{cp} = C'D' - BE'$$

(a) Idealized hysteresis loops for the four basic types of inelastic strain range. (b) Hysteresis loop containing $\Delta\epsilon_{pp}$, $\Delta\epsilon_{cc}$, and $\Delta\epsilon_{cp}$.

Fig. 4.30. Illustration of partitioning of the strain range into component strains.

and the other two terms represent combined creep and plastic strain ranges. For each type of strain range, the Coffin-Manson relationship (Eq 4.10) can be applied. For instance, $N_{pp} = A(\Delta\epsilon_{pp})^\alpha$, etc. The fractional strain for each type of strain with respect to the total inelastic strain can be expressed as

$$F_{pp} = \frac{\Delta\epsilon_{pp}}{\Delta\epsilon_{in}} \qquad F_{cc} = \frac{\Delta\epsilon_{cc}}{\Delta\epsilon_{in}}$$
$$\text{(Eq 4.22)}$$
$$F_{pc} = \frac{\Delta\epsilon_{pc}}{\Delta\epsilon_{in}} \qquad F_{cp} = \frac{\Delta\epsilon_{cp}}{\Delta\epsilon_{in}}$$

By adding up the fractional damage for each type of strain, the total damage is estimated by the expression

$$\frac{1}{N_f} = \frac{F_{pp}}{N_{pp}} + \frac{F_{cc}}{N_{cc}} + \frac{F_{cp}}{N_{cp}} \left[\text{or} \frac{F_{pc}}{N_{pc}} \right]$$

(Eq 4.23)

where N_{pp}, N_{cc}, etc. represent the number of cycles to failure for each type of strain.

Predictions using the SRP approach for a number of materials have met with varying degrees of success. One of the major problems with this approach is the need to generate baseline data based on complex hold-time tests. Extrapolation of predictions to long hold times and small strain ranges also needs further verification. Application of the SRP technique to multiaxial loading conditions has been discussed and illustrated by Manson and Halford (Ref 96). In order to apply Eq 4.23 to life prediction for any arbitrary cycle, the following information is needed.

1. From a stable hysteresis loop of the stress-strain cycle, the partitioned strain-range components $\Delta\epsilon_{pp}$, $\Delta\epsilon_{cc}$, and $\Delta\epsilon_{cp}$ (or $\Delta\epsilon_{pc}$) and the total inelastic strain range $\Delta\epsilon$ are obtained.
2. The fractional strains F_{pp}, F_{cc}, and F_{cp} are then calculated by use of the information obtained in step 1 and by use of Eq 4.22.
3. The number of cycles to failure for each given type of strain (i.e., the relationships $N_{pp} = A\Delta\epsilon_{pp}^{\alpha}$, etc.) must be known from independent laboratory experiments, as illustrated for various materials in Fig. 4.31.
4. Now that the fractional strains F_{pp}, F_{cc}, etc. and the cyclic life for each type of strain, N_{pp}, N_{cc}, etc., are known, the interaction damage rule (Eq 4.23) is used to calculate the cyclic life for the arbitrary cycle.

Example:
In a laboratory test at 540 °C (1000 °F), a sample of a 2¼Cr-1Mo steel is subjected to fatigue cycles similar to that shown in Fig. 4.30(b). From the resulting half-life hysteresis loop, the following information is obtained:

$$\Delta\epsilon_{cp} = 0.00095; \ \Delta\epsilon_{pp} = 0.01192;$$

$$\Delta\epsilon_{cc} = 0.0095$$

From independent laboratory tests, the following relationships are given:

$$\Delta\epsilon_{pp} = 0.559(N_{pp})^{-0.570}$$

$$\Delta\epsilon_{cp} = 0.233(N_{cp})^{-0.515}$$

$$\Delta\epsilon_{cc} = 0.15(N_{cc})^{-0.52}$$

Calculate the total numbers of cycles of the above types at which failure of the specimen will occur.

Answer:

$$\Delta\epsilon_{in} = \Delta\epsilon_{pp} + \Delta\epsilon_{cp} + \Delta\epsilon_{cc} = 0.02237$$

$$F_{pp} = \frac{0.01192}{0.02237} = 0.5329$$

$$F_{cp} = \frac{0.00095}{0.02237} = 0.0425$$

$$F_{cc} = \frac{0.0095}{0.02237} = 0.4246$$

$$N_{pp} = \left(\frac{0.559}{0.01192}\right)^{1/0.57} = 856 \text{ cycles}$$

$$N_{cp} = \left(\frac{0.233}{0.00095}\right)^{1/0.515} = 43,627 \text{ cycles}$$

$$N_{cc} = \left(\frac{0.150}{0.0095}\right)^{1/0.52} = 200 \text{ cycles}$$

Using the interaction rule (Eq 4.23), we get

$$\frac{1}{N_f} = \frac{0.0425}{43,627} + \frac{0.5329}{856} + \frac{0.4246}{200}$$

$$N_f = 364 \text{ cycles}$$

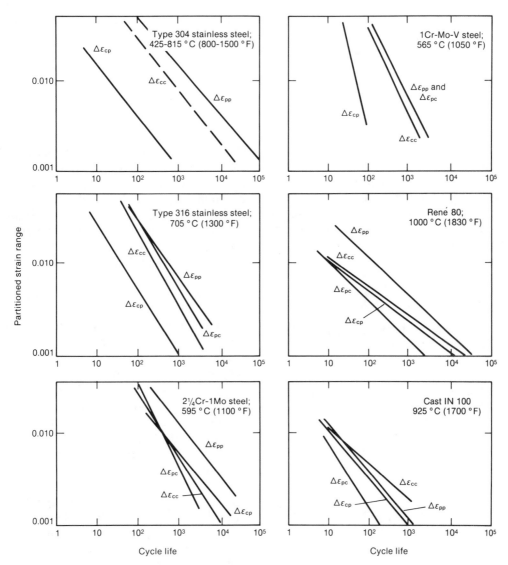

Fig. 4.31. Combined strain-range-partitioning relationships for various alloys (Ref 20).

An illustration of the application of SRP to analysis of the MPC interspersion test cycles (Ref 82) has been provided by Saltsman and Halford (Ref 97).

Ostergren's Damage Function. A damage function has been proposed by Ostergren for predicting low-cycle fatigue at elevated temperatures (Ref 98 and 99). Defined as $\sigma_{max}\Delta\epsilon_{in}$, where σ_{max} is the maximum principal stress in the cycle and $\Delta\epsilon_{in}$ is the inelastic strain range, the damage function was correlated with cyclic life as follows:

$$\sigma_{max}\Delta\epsilon_{in}N_f^\phi = C_5 \qquad \text{(Eq 4.24)}$$

where C_5 and ϕ are constants. Equation 4.24 contains the influence of mean stress in view of the identity

$$\sigma_{max} = \sigma_m + \Delta\sigma/2 \qquad \text{(Eq 4.25)}$$

Where mean stresses do not exist, Eq 4.24 reduces to the well-known Coffin-Manson equation (i.e., Eq 4.10), with the difference being that the product of stress and strain are considered instead of strain alone. The damage function was derived from the postulate that the net tensile hysteretic energy is a measure of low-cycle fatigue damage.

When time-dependent damage mechanisms become important, a frequency-modified damage function was introduced as

$$\sigma_{max} \Delta \epsilon_{in} N_f^{\phi} \nu^{\phi(k-1)} = C \qquad \text{(Eq 4.26)}$$

where k and C are material constants. When k = 1, Eq 4.26 reduces to Eq 4.24. It was postulated that two categories of time-dependent damage exist and that the definition of ν depends on the specific category of the material in question. In the first category, time-dependent damage is independent of waveshape and ν is defined as the frequency of cycling, or

$$\nu = \frac{1}{t_{cy}} = \frac{1}{t_0 + t_t + t_c} \qquad \text{(Eq 4.27)}$$

where t_{cy} is the cycle period, t_0 is the time for the continuous cycle portion, t_t is the tension hold time, and t_c is the compression hold time. In the second category, time-dependent damage is dependent on waveshape and is defined as

$$\nu = \frac{1}{t_0 + t_t - t_c} \quad \text{for } t_t \geq t_c \qquad \text{(Eq 4.28)}$$

and

$$\nu = \frac{1}{t_0} \quad \text{for } t_t < t_c \qquad \text{(Eq 4.29)}$$

The latter category utilizes an effective frequency, which accounts for the greater time-dependent damage associated with un-reversed tensile creep deformation. Tensile hold times which are not reversed by compressive hold times, as denoted by the term $(t_t - t_c)$, are the most damaging.

It was theorized by Ostergren that the difference in the influence of waveshape was attributable to differences in the mechanism of time-dependent damage. In cases where the time-dependent damage arises from environmental damage, waveshape may not be important. In cases where creep damage is important, then, the waveshape can be expected to have an effect on cyclic life.

Based on an analysis of his data, Ostergren concluded that for a Cr-Mo-V steel, René 80, and Inconel 738, time-dependent damage was primarily environment-related and, hence, waveshape had no effect on cyclic life. An example of the correlation obtained between the frequency-modified damage function (see Eq 4.26) and the cyclic life is shown in Fig. 4.32 for a Cr-Mo-V steel. For René 80 and Inconel 738, the complete absence of any time dependence of damage was indicated and the cyclic life could be directly correlated with the damage function using Eq 4.24 without any frequency modifier (i.e., k = 1 in Eq 4.26), as shown in Fig. 4.33. In contrast to the above alloys, time-dependent creep damage and hence the effect of waveshape on cyclic life were found to be important for AISI types 304 and 316 stainless steels and for Incoloy 800.

The frequency-modified damage function proposed by Ostergren thus appears to be capable of correlating strain-rate and hold-time effects for a wide range of materials. The major limitation at the present time is the need to separate the materials into two categories depending on whether the time-dependent damage is related to environmental effects or creep effects. Low-cycle fatigue tests with hold times will need to be carried out on the material under the relevant test conditions to make this determination. Furthermore, the separation of materials into the two categories is at present based purely on the data correlations and not on mechanistic evidence. Despite these limitations, the Ostergren approach offers a very useful and simple technique for life prediction under creep-fatigue conditions.

Bisego's Energy Criterion. Bisego, Fossati, and Ragazzoni conducted low-cycle fatigue tests with strain control at temperatures ranging from 480 to 560 °C (895 to 1040 °F) on Cr-Mo-V rotor steels (Ref 100). A triangular waveform with and without a 20-s hold time was employed. For all the tests, the stress-strain hysteresis loops were analyzed. In spite of continuous material cyclic softening, the hysteretic energy e, defined as the mechanical energy absorbed

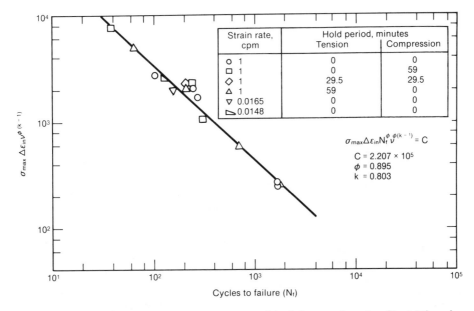

Strain rate, cpm	Hold period, minutes Tension	Compression
○ 1	0	0
□ 1	0	59
◇ 1	29.5	29.5
△ 1	59	0
▽ 0.0165	0	0
◿ 0.0148	0	0

$$\sigma_{max}\Delta\varepsilon_{in}N_f^\phi v^{\phi(k^{-1})} = C$$

$$C = 2.207 \times 10^5$$
$$\phi = 0.895$$
$$k = 0.803$$

Fig. 4.32. Correlation between frequency-modified damage function (Eq 4.26) and cyclic life for a Cr-Mo-V steel (Ref 98 and 99).

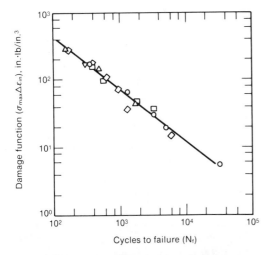

Hold period, minutes Tension	Compression
○ 0	0
◇ 0	2
□ 2	0
△ 0	10
▽ 10	0

Fig. 4.33. Correlation between damage function of Eq 4.24 and cyclic life for IN-738 (Ref 98 and 99).

during one cycle in one mole of material, was approximately constant. The value of e is determined from the area of the stress-strain loop assuming a density of 7.8 g/cm³

(0.28 lb/in.³) and a molar weight of 56 g/mole (1.98 oz/mole). All the results at 20, 480, and 560 °C (68, 895, and 1040 °F) could be fitted to the equation

$$eN_f = Q_1 \qquad \text{(Eq 4.30)}$$

where Q_1 might be regarded as the maximum amount of mechanical work the material could absorb up to failure. Values of Q_1 were found to be 56.5 and about 40 kJ/mole (13.5 and about 9.6 kcal/mole) at 20 and 560 °C (68 and 1040 °F), respectively, for one heat of steel. Different values were obtained for another heat of steel. If the value of Q_1 for a given heat under given test conditions could be established, then the number of cycles to failure for any arbitrary cycle of known stress-strain loop could be predicted. The fact that Q_1 varies with test temperature and from heat to heat might be a serious limitation on the use of this technique. Test data also are too limited to permit evaluation of the general applicability of the correlation proposed. Further tests by Bartoloni and Ragazzoni have shown that the energy Q_1 is also a function of strain rate, as described (Ref 101) by $Q_1 = 89.4\dot\varepsilon^{0.142}$ kJ/mole. The strain-rate dependence of cyclic life could be described by the relationship

$$\Delta\epsilon_t = 7.3 \times 10^{-3} N_f^{-0.0416} \dot{\epsilon}^{0.0339}$$
$$+ 8.1359 N_f^{-0.9069} \dot{\epsilon}^{0.1071} \quad \text{(Eq 4.31)}$$

where $\dot{\epsilon}$ is expressed as strain per second. The authors term their approach the strain-rate-modified strain-range approach. Equation 4.31 could also be used to analyze hold-time effects by postulating that $\dot{\epsilon}_{eq} = 2\Delta\epsilon_t/t_{cy}$, where $\dot{\epsilon}_{eq}$ is an equivalent strain rate and t_{cy} is the cycle time including the hold time.

Ductility Exhaustion. The ductility-exhaustion approach is simply a strain-based life-fraction rule in which the fatigue damage and creep damage are summed up in terms of the fractional strain damage for each category, as follows (Ref 102):

$$\frac{1}{N_f} = \frac{\Delta\epsilon_p}{D_p} + \frac{\Delta\epsilon_c}{D_c} \quad \text{(Eq 4.32)}$$

where $\Delta\epsilon_p$ is the plastic strain-range component at half life, D_p is the fatigue ductility obtained from pure fatigue tests, $\Delta\epsilon_c$ is the true tensile creep-strain component, and D_c is the lower-boundary creep-rupture ductility of the material. The first term in Eq 4.32 denotes the fatigue-damage component and the second term denotes the creep-damage component. Although the first term is fairly easy to understand and obtain from test data, the second term, especially the definitions of $\Delta\epsilon_c$ and D_c, needs to be clarified. The problem arises from two issues. First, not all creep strain is viewed as damaging, and only that strain which accumulates below a critical strain rate necessary to cause constrained cavity growth is viewed as damaging (Ref 20). Secondly, the rupture ductility of a material is not a constant but decreases with decreasing strain rate. Hence, in defining a failure criterion, an appropriate lower-boundary value has to be defined for D_c.

During a low-cycle fatigue test with hold time, stress relaxation occurs during the hold time from some initial stress σ_0 to the relaxed stress σ_r, as illustrated in Fig. 4.34 (Ref 39). The decrease in stress corresponds to a decrease in strain rate. Hence, creep is occurring under a progressively decreasing strain rate. The maximum possible strain that can be tolerated for each strain condition is different because the rupture strain (or ductility) generally decreases with decreasing strain rate (or decreasing stress and increasing time to rupture), as illustrated in Fig. 4.34. Most of the creep damage in terms of cavitation is viewed as occurring only when the strains are accumulated below the critical strain rate (i.e., only in region III). The lower-boundary ductility corresponding to strain rates of $\dot{\epsilon}_c$ and below is defined as D_c. Only those strains occurring below $\dot{\epsilon}_c$ are summed up to give the total damaging strain $\Delta\epsilon_c$. The ratio $\Delta\epsilon_c/D_c$, therefore, denotes the strain-life fraction expended under the "damaging" creep conditions. Application of this method for actual damage calculation involves the following four steps. (1) $\Delta\epsilon_p$ is obtained from the hysteresis loop for the actual creep-fatigue cycle after the loops have stabilized. D_p is obtained from a pure fatigue test conducted to failure under the same strain range. The ratio $\Delta\epsilon_p/D_p$ can be readily calculated. (2) From the literature or from tests, the variation of rupture ductility with strain rate, as shown in Fig. 4.34, is established. From this plot, the critical strain rate for transition from region II to region III, and the corresponding ductility D_c, are established. (3) From the stable hysteresis loop, the tensile creep strain $\Delta\epsilon_c$ is determined. Only that part of the strain which occurs in the damaging region of strain rate is taken into account. The ratio $\Delta\epsilon_c/D_c$ gives the fractional creep damage. (4) The fatigue damage and creep damage are then summed to determine N_f using Eq 4.32.

The ductility-exhaustion approach is simple to use and has some mechanistic basis. It has been applied to the treatment of laboratory data with reasonable success by a number of workers for both ferritic steels (Ref 39 and 103 to 105) and austenitic steels (Ref 102, 106, and 107). Selection of appropriate values for D_c and $\Delta\epsilon_c$ is, however,

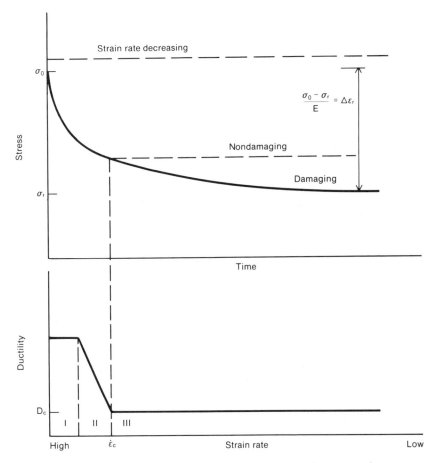

Fig. 4.34. Schematic representation of stress relaxation and associated strain rate, strain, and creep ductility (Ref 39).

somewhat arbitrary and subject to errors. It is not easy to judge which part of the creep strain accumulated during the hold time corresponds to true creep damage and which part is due to relatively undamaging anelastic effects. The values of $\Delta\epsilon_c$ and D_c have been found to be functions of specimen constraint, triaxiality, test temperature, impurity content in the material, and a host of other variables. Hence, selection of $\Delta\epsilon_c$ and D_c based on uniaxial tests on one heat and application of the results to the prediction of component behavior is liable to lead to serious errors. A detailed discussion of rupture ductility is presented in Chapter 3 and is very useful in understanding this section.

An alternative approach to that of Miller *et al* would be to avoid the distinction between the damaging and nondamaging re-

gions of strain rate (or stress) and simply treat the entire region where ductility drop occurs (i.e., regions II and III in Fig. 4.34) as damaging. In this case D_c is not the lower-boundary value but simply the appropriate value of rupture ductility at any given stress level. The ratio $\Delta\epsilon_c/D_c$ can be computed for various stress decrements during the stress-relaxation process to estimate the total creep-life fraction expended. The method employed by Schlottner and Seeley is somewhat analogous to this approach, but differs in the sense that time life fractions spent at various strain rates are employed rather than strain fraction *per se* (Ref 108). Their procedure is described in Chapter 5. This approach avoids the need to make subjective judgments in selecting $\Delta\epsilon_c$ and D_c. However, the need to estimate the long-term rupture ductility at low strain rates and the problems in applying laboratory data to components still remain.

Yamaguchi *et al* have recently reported on an empirical life-prediction technique based on normalizing the inelastic strain with respect to the fatigue ductility or the creep ductility (Ref 109). This method is simply based on a modification of the Coffin-Manson relationship. In a pure fatigue test, the fatigue life N_0 was found to be related to the inelastic strain and tensile fracture ductility D_p through the relationship

$$\frac{\Delta\epsilon_{in}}{D_p} N_0^m = G \qquad \text{(Eq 4.33)}$$

In hold-time tests, where creep-fatigue interaction was found to occur, as evidenced by intergranular fractures, the creep-fatigue life N_h was found to be related to the inelastic strain and rupture ductility D_c through the relationship

$$\frac{\Delta\epsilon_{in}}{D_c} N_h^m = G \qquad \text{(Eq 4.34)}$$

Because the values of m and G were identical in Eq 4.33 and 4.34, both the fatigue data and the creep-fatigue data could be plotted as a unique relationship over a wide range of conditions, independent of materials and hold time, as shown in Fig. 4.35. To use this approach, one must know the variation of rupture ductility with time (or strain rate or stress). The simplicity of this approach and the excellent correlations reported justify further investigation. The model implies that failures are either fatigue-dominated or creep-dominated and that no interaction effects are present.

Damage Rate. A strain-based approach which takes into account the rate of damage accumulation has been proposed by Majumdar and Maiya (Ref 110 to 113). They view the total damage as consisting of crack damage (fatigue) and cavitation damage (creep). If the two damage mechanisms are additive, the damage rate is given by the sum of the equations

$$\frac{1}{a}\frac{da}{dt} = \begin{cases} T|\Delta\epsilon_p|^s|\dot{\epsilon}_p|^{k_2} & \text{[for tension]} \\ C|\Delta\epsilon_p|^s|\dot{\epsilon}_p|^{k_2} & \text{[for compression]} \end{cases}$$

$$\text{(Eq 4.35)}$$

and

$$\frac{1}{c}\frac{dc}{dt} = \begin{cases} G|\Delta\epsilon_p|^s|\dot{\epsilon}_p|^{k_3} & \text{[for tension]} \\ -G|\Delta\epsilon_p|^s|\dot{\epsilon}_p|^{k_3} & \text{[for compression]} \end{cases}$$

$$\text{(Eq 4.36)}$$

Equation 4.35 describes the crack damage due to fatigue, whereas Eq 4.36 describes the cavitation damage due to creep. T, C, G, k_2, k_3, and s are material parameters which are functions of temperature, environment, and the metallurgical state of the material; $\Delta\epsilon_p$ and $\dot{\epsilon}_p$ are current absolute values of plastic strain and strain rate, respectively; and a and c are the crack size and cavity size, respectively, at time t. T and C are included to account for differences in growth rates occurring in tension and compression. The parameter G is given the appropriate sign for the tensile or the compressive stress regime. Final failure is calculated as the reciprocal of the sum of the crack and cavity damage.

For cases where the fatigue and creep damage are interactive (not additive), Majumdar and Maiya have proposed a slightly modified equation for the crack damage components, as follows (Ref 114):

$$\frac{1}{a}\frac{da}{dt} = \left(\frac{T}{C}\right)\left[1 + \alpha\ln\frac{c}{c_0}\right]|\Delta\epsilon_p|^s|\dot{\epsilon}_p|^{k_2}$$

$$\text{(Eq 4.37)}$$

The expression for the cavitation damage remains the same as Eq 4.36.

The damage-rate approach allows one to take into account the effects of various

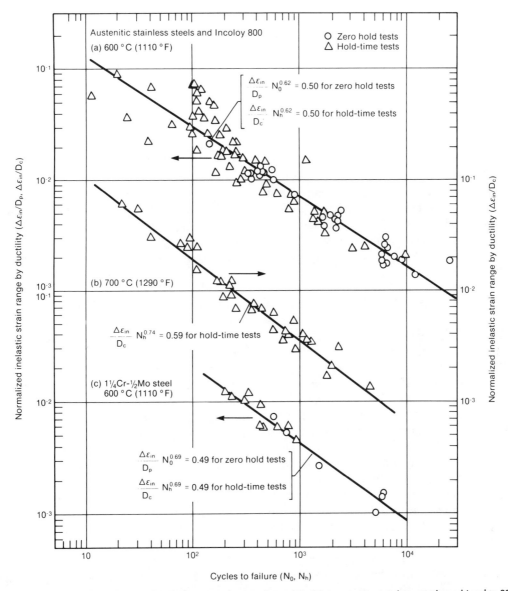

(a) Austenitic stainless steels and Incoloy 800 at 600 °C (1110 °F). (b) Austenitic stainless steels and Incoloy 800 at 700 °C (1290 °F). (c) 1¼ Cr-½ Mo steel at 600 °C (1110 °F). Use appropriate scale as indicated by arrows.

Fig. 4.35. Correlation between normalized inelastic strain range and cycles to failure (Ref 109).

waveshapes on fatigue life. It has been applied successfully to 1Cr-Mo-V steels (Ref 89), austenitic stainless steels (Ref 110 to 114), Incoloy 800, and 2¼ Cr-1Mo steels (Ref 111). Agreement between predicted and actual life has been within a factor of 2. The principal limitation of this approach is the number of coefficients that must be determined from complex cyclic hold tests. Furthermore, its capability of predicting life

within a factor of 2 is not a significant improvement over the other approaches.

Evaluation of Life-Prediction Methods. The relative merits of one or more of the damage rules described so far in predicting the lives of specific materials have been assessed by a number of investigators. With respect to Cr-Mo-V rotor steels, Leven compared the linear-damage (LD), frequency-modified strain-range (FM), and strain-range-partitioning (SRP) approaches and concluded that all of them could predict life within a factor of 2 (Ref 31). Similar conclusions were reached by Kuwabara and Nitta (Ref 115) and by Batte (Ref 27). Batte has claimed, however, that in the low strain ranges the LD approach is better than the others. Curran and Wundt reported that the LD approach gave nonconservative predictions (Ref 82). Melton compared the FM and SRP approaches and the Ostergren damage function and found the data to be best described by the FM approach (Ref 116). Bisego, Fossati, and Ragazzoni claimed better fit of data to the SRP approach than to the LD approach (Ref 100).

Several studies can be cited with respect to 2¼Cr-1Mo and 1Cr-½Mo steels. Ellis *et al* compared the LD and SRP approaches and concluded that neither was adequate to predict life within a factor of 3 (Ref 117). Brinkman *et al* claimed good agreement of data with SRP (Ref 36). Kuwabara, Nitta, and Kitamura claimed prediction capability within a factor of 2 for the SRP method (Ref 35). Saltsman and Halford (Ref 118 and 119) and Majumdar and Maiya (Ref 112) have reported similar experience with SRP. Miller and Gladwin compared the ductility-exhaustion (DE) approach with the LD approach and concluded that the former offered better prediction capability (Ref 39).

For austenitic stainless steels, both the LD and SRP approaches have been used although the relative merits of different approaches have not been compared extensively.

Nazmy and Wuthrich compared the applicability of the SRP, FM, and Ostergren damage approaches to life prediction of Inconel 738 and concluded in favor of SRP (Ref 120). For the same alloy, prediction capability within a factor of 2 has been claimed for SRP (Ref 121). Contrary experience indicating inapplicability of SRP to Inconel 738 and to René 95 also has been documented (Ref 122 and 123). For a cobalt-base vane alloy (MAR-M 509), the SRP method was found to predict life

within a factor of 3 (Ref 67). For Inconel 738 LC, Persson *et al* claim better correlation with the Ostergren approach than with the Coffin-Manson relationship (Ref 124). An alternative approach termed the strain-rate-modified accumulation of time-dependent damage (SRM), in which both the creep and fatigue components are converted to an equivalent creep damage and then summed as a life fraction, has also been advocated (Ref 125).

It is clear from the above review that there are divergent opinions regarding which damage approach provides the best basis for life prediction. It is quite clear that a number of variables, such as test temperature, strain range, frequency, time and type of hold, waveform, ductility of the material, and damage characteristics, affect the fatigue life. The conclusions drawn in any investigation may therefore apply only to the envelope of material and test conditions used in that study. The validity of any damage approach has to be examined with reference to the material and service conditions relevant to a specific application. Broad generalizations based on laboratory tests, which often may have no relevance to actual component conditions, do not appear to be productive. Thus, one should use a tailored, case-specific approach for any given situation.

One of the major problems in evaluating the applicability of different life-prediction methods is that in many cases it is necessary to use all the available data in deriving the life-prediction method and thus it is possible to examine only the accuracy with which a given method describes the data. With a few exceptions (Ref 115, 126, and 127), there also is a scarcity of instances in which service experience has been compared with specific life-prediction methods. In general, the available methods are utilized only to predict the lives of samples tested under laboratory conditions. Validation against component test data in the laboratory and in-service monitoring of actual equipment would lead to more confidence in the use of the various rules.

Results from most studies show that even

the best of the available methods can predict life only to within a factor of 2 to 3. Some of the cited reasons for these inaccuracies are: failure of the methods to model changing stress-relaxation and creep characteristics caused by strain softening or hardening, use of monotonic creep data instead of cyclic creep data, and lack of sufficiently extended-duration test data. None of the damage rules available today is entirely based on sound mechanistic principles. They are all phenomenological in nature, involving empirical constants that are material-dependent and difficult to evaluate. Extrapolation of the rules to materials and conditions outside the envelope covered by the specific investigation often results in unsuccessful life predictions. For application to service components, the stress-strain variation for each type of transient and its time dependence must be known with accuracy. Such calculations are difficult and expensive to perform. Because of these limitations and the simplicity of the linear-damage summation using the life-fraction rule, the latter approach continues to enjoy widespread popularity in engineering applications.

Design Rules for Creep-Fatigue

It is clear from the discussion so far that although a variety of damage rules have been proposed, none of them has proved capable of accurately predicting the creep-fatigue life for all materials and test conditions. The prospect of establishing such an all-encompassing approach seems remote.

Considerable effort has been directed recently toward the development of a creep-fatigue analysis method for incorporation in the ASME Boiler and Pressure Vessel Code. In Section III of the code, which gives design rules appropriate to nuclear power plants, the rules for high-temperature design have been under development over a number of years, leading up to the most recent rules codified under Code Case N-47 (Ref 78).

The approach to creep-fatigue design in Code Case N-47 is essentially based on the linear damage-summation method given by Eq 4.13. It is written for two alternate routes, as follows:

$$\sum \frac{N}{N_d} + \sum \frac{t}{t_r} = D' \quad \text{[for inelastic route]}$$

$$\text{(Eq 4.38a)}$$

$$\sum \frac{N}{N_d} + \sum \frac{t}{t_r} = 1 \quad \text{[for elastic route]}$$

$$\text{(Eq 4.38b)}$$

where N is the number of cycles of the loading condition, N_d is the corresponding number of allowed cycles in design under the same conditions, t is the duration of the loading condition, and t_r is the time to rupture for that condition at 1.1 times the applied stress.

In the inelastic route, N_d is determined from pure fatigue (continuous cycle with no hold time) design curves incorporated in the code and shown in Fig. 4.36. These design curves have a built-in factor of safety and are established by applying a safety factor of 2 with respect to strain range or a factor of 20 with respect to the number of cycles, whichever gives the lower value. The creep-life fraction is determined on the basis of time life fraction per cycle using assumed stresses 1.1 times the applied stress and the minimum stress-rupture curves incorporated in the code. The total damage D' must not exceed the envelope defined by the bilinear damage curve shown in Fig. 4.29.

In the elastic route, creep-life fraction is calculated in the same way as in the inelastic route, but the fatigue-life fraction is computed using alternative design curves shown in Fig. 4.37. These curves are a more conservative set of curves compared with those of Fig. 4.36. They incorporate the effect of creep damage by applying a fatigue-life-reduction factor which includes hold-time effects as well as the factor of safety (2 in strength and 20 in cycles, whichever gives the lower N_d). The total damage D' is not allowed to exceed unity in this case.

With respect to the inelastic route, con-

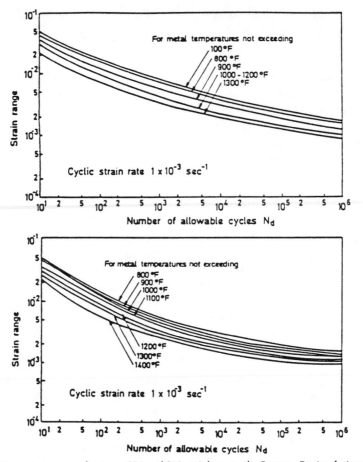

Top: Design fatigue strain range for types 304 and 316 stainless steels. Bottom: Design fatigue strain range for Ni-Fe-Cr alloy Incoloy 800H.

Fig. 4.36. Design curves from ASME Code Case N-47 for inelastic route for austenitic stainless steels and Incoloy 800H (Ref 78).

servatism is built into the procedure through the safety factor applied to N_f (i.e., N_d) and by assuming an effective stress 1.1 times the applied stress in calculating the creep-life fraction. With respect to the elastic route, conservatism is built in by considering hold-time effects in addition to the safety factor contained in N_d used in developing the design curves and through the safety factor of 1.1 in stress during calculation of the creep-life fraction. In its current form, ASME Code Case N-47 is conservative and reflects the numerous uncertainties in creep-fatigue life prediction. It is also purely empirical, but is the best that is currently available. Use of Code Case N-47 for remaining life assessment of components represents an extremely conservative approach that will lead to premature and unwarranted retirements. The philosophy and development of Code Case N-47 have been described in detail elsewhere in the literature (Ref 128 to 130).

Strain-Concentration Effects

Many components contain stress-raising notches, grooves, or defects, and it is required to know the local stress-strain conditions at these regions in terms of those remote from it. In the case of elastic loading, the elastic stress-concentration factor K_t adequately describes the concentrated stress at the tip. Because $\epsilon = \sigma/E$, the strain-concentration factor is identical to the stress-concentration factor.

In LCF and thermal-fatigue situations,

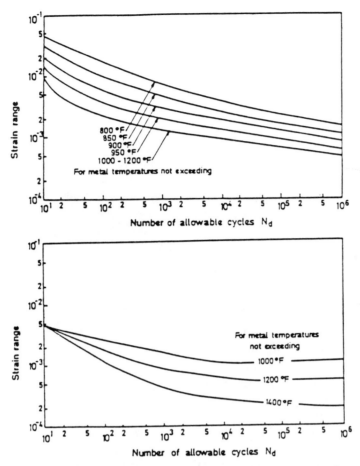

Top: Design fatigue strain range for types 304 and 316 stainless steels. Bottom: Design fatigue strain range for Ni-Fe-Cr alloy Incoloy 800 H.

Fig. 4.37. Design curves from ASME Code Case N-47 for elastic route for austenitic stainless steels and Incoloy 800H (Ref 78).

the yield strain is generally exceeded and the strain-concentration factor increases above the elastic stress-concentration factor. Thus, the use of K_t to estimate concentrated strains would significantly underestimate their magnitudes. In these cases, the strain-concentration factor K_ϵ and the actual (not elastic) stress-concentration factor K_σ are approximately related to the elastically calculated K_t through the Neuber expression (Ref 131):

$$K_\epsilon K_\sigma = K_t^2 \qquad \text{(Eq 4.39)}$$

It can be shown (Ref 13) that

$$K_\sigma = K_t^{2\beta/(\gamma+\beta)} \qquad \text{(Eq 4.40)}$$

$$K_\epsilon = K_t^{2\gamma/(\gamma+\beta)} \qquad \text{(Eq 4.41)}$$

$$K_{\epsilon p} = K_t^{2/(\gamma+\beta)} \qquad \text{(Eq 4.42)}$$

where β is the strain-hardening exponent from the cyclic stress-strain curve at saturation, as defined by Eq 4.7; γ is the exponent in the relationship between the total strain range and the plastic strain range, as defined by Eq 4.6; K_ϵ is the total strain-concentration factor; and $K_{\epsilon p}$ is the plastic strain-concentration factor. For instance, assuming values of $\beta = 0.39$ and $\gamma = 0.64$ for an austenitic steel, Eq 4.41 leads to $K_\epsilon = K_t^{1.2}$. If a typical value of K_t is assumed to be 2.8, K_ϵ turns out to be equal to 3.5.

Equation 4.41 applies only to cases where total yielding of the specimen occurs, such as during push-pull testing of laboratory specimens. The situation encountered more often in practice is where yielding is localized to the notch tip but elastic conditions prevail elsewhere. A modified equation for this case (Ref 13) has been given as

$$K_\epsilon = K_t^{2\gamma/(\gamma+\beta)} \left[\frac{(EB)^\beta \Delta\sigma^{(\gamma-\beta)}}{A^\gamma} \right]^{1/(\gamma+\beta)}$$

(Eq 4.43)

where A and B are coefficients from Eq 4.7 and 4.6, respectively. By multiplying the nominal strain by K_ϵ, the concentrated strain at the tip of a notch, groove, or other strain concentration can be calculated. The concentrated strain then is used in conjunction with the LCF data on smooth specimens to compute the life of a notched specimen or component. This procedure avoids the need to generate notched-bar fatigue data for damage calculations in components containing stress concentrations. In applying laboratory LCF data generated from uniaxial tests to notched bars and to components, additional corrections for multiaxial loading, in terms of "equivalent strain," also need to be made. The true stress-strain curves of the material must be used in calculating strain-concentration factors. Because many materials strain harden or soften, the cyclic true stress-strain curves must be used. Some procedures for calculating K_ϵ are described elsewhere (Ref 132 and 133). A modified form of the Neuber equation is incorporated in ASME Code Case N-47 to take into account the effects of strain concentrations.

Illustration of Linear Damage Summation

Service components are subject to different types of starts and stops. The strain ranges corresponding to these events (i.e., the severity of the transients) vary. The maximum strains and stresses may peak at intermediate temperatures or near the operating temperature. The components also contain stress raisers, so that the calculated nominal strains have to be converted to the actual concentrated strains at the roots of the stress raisers. Typically, damage calculations may involve the following steps:

1. From operating records, determine the steady operating temperature and the duration of the steady operation, the types of transients and the number of each, and the temperature changes associated with the different types of transients.
2. Perform a simple elastic or elastic-plastic creep finite-element analysis to determine the spatial and timewise distributions of nominal strains and stresses.
3. Calculate the theoretical elastic stress-concentration factor and determine the strain-concentration factor K_ϵ from data derived from cyclic stress-strain curves. Using the K_ϵ values for each type of transient, calculate the concentrated strains at the stress raisers, from the nominal strain.
4. Assemble the strain-range-vs-N_f curves from literature. For each type of transient, the appropriate curve or expression corresponding to the peak temperature must be available. If the peak strains occur below the creep regime, hold-time and frequency effects may be neglected. If the peak strains occur in the creep regime, fatigue curves incorporating these effects should be used.
5. For each type of transient, calculate N_f and, hence, N/N_f.
6. Sum up the total damage due to the various transients to arrive at the total fatigue damage.

An illustration of the methodology is provided below, based on the calculations of Kramer *et al*, performed in connection with a rotor failure analysis (Ref 32).

Example:
The high-pressure rotor of a steam turbine which had been in operation for 17 years failed catastrophically during a cold start, the fracture being ascribed to a radial-axial crack at the bore. Estimate the possibility that crack initiation could have occurred at MnS inclusions as a result of low-cycle fatigue. The following information is given:

Operational data:

- Total operational hours, 106,000
- Maximum temperature at failure location during steady operation, 800 °F
- Number of hot-start cycles, 183
- Number of cold-start cycles, 105

A hot-start cycle is a cycle in which the rotor has been allowed to cool from the operating temperature for less than 72 h, so that the rotor is still hot at the time of the next start-up. Thermal stresses during such a start-up are assumed to be minimal. A cold-start cycle is a cycle in which the rotor has been allowed to cool for more than 72 h prior to start-up, so that appreciable thermal stresses are set up during the cold start.

Stress-analysis data: From an elastic finite-element stress analysis, the following information was calculated:

- Peak tangential stress at the bore = tangential stress at steady state at the bore = 49,600 psi (i.e., no stress relaxation occurred even after 106,000 h)
- For hot starts, total stress = tangential stress (at 800 °F)
- For cold starts, total stress = tangential stress + thermal stress = 78,200 psi, which peaked at 230 °F.

Material-property data: The following $\Delta\epsilon$-vs-N_f LCF curves can be obtained from literature:

- Pure fatigue LCF curve at 230 °F (cold start)
- Pure fatigue LCF curve at 800 °F (hot start)
- Fatigue curves as modified for frequency and long-hold-time effects at 800 °F.

Answer:

1. *Pure Fatigue Case – No Creep Effects*
Nominal strain at the bore is calculated as

$$\epsilon = \frac{\sigma}{E} = \frac{78,200}{29.5 \times 10^6}$$

$$= 0.00265 \text{ (cold start)}$$

$$\epsilon = \frac{\sigma}{E} = \frac{49,600}{29.5 \times 10^6}$$

$$= 0.0017 \text{ (hot start)}$$

For a large elliptical MnS inclusion assumed using the Neuber equation and cyclic stress-strain data, the local strain-concentration factors are calculated as

$$K_\epsilon = 4.1 \text{ (cold start)}$$

$$K_\epsilon = 3.5 \text{ (hot start)}$$

The total strain ranges around the inclusion are

$$\Delta\epsilon_t = K_\epsilon \epsilon = 4.1 \times 0.00265$$

$$= 0.011 \text{ (cold start)}$$

$$\Delta\epsilon_t = K_\epsilon \epsilon = 3.5 \times 0.0021$$

$$= 0.0074 \text{ (hot start)}$$

At $\Delta\epsilon_t = 0.011$, N_f (cold) = 2100 cycles (from pure fatigue curve at 230 °F).

At $\Delta\epsilon_t = 0.0074$, N_f (hot) = 2200 cycles (from pure fatigue curve at 800 °F).

Total fatigue damage can be calculated as

$$D' = \frac{N(\text{cold})}{N_f(\text{cold})} + \frac{N(\text{hot})}{N_f(\text{hot})}$$

$$= \frac{105}{2100} + \frac{183}{2200} = 0.133$$

This shows that only 13% of the life to initiation has been expended. Therefore, initiation could not occur due to pure low-cycle fatigue.

2. *Hold-Time-Modified Fatigue*
In this case, pure fatigue damage occurring at 230 °F due to cold starts remains

the same, but the creep-fatigue damage occurring at 800 °F derives from both the cold starts and the hot starts because, for both types of starts, the cycles pass through 800 °F. Hence the total number of hot-start cycles is now redefined to include the cold-start cycles. The value of K_ϵ remains the same for cold starts, but for the hot starts K_ϵ is slightly modified because the cyclic stress-strain curve has now changed to include creep effects. The strain ranges are calculated as

$$\Delta\epsilon_t = K_\epsilon\epsilon = 4.1 \times 0.00265 = 0.011 \text{ (cold)}$$

$$\Delta\epsilon_t = K_\epsilon\epsilon = 4.2 \times 0.0021 = 0.0088 \text{ (hot)}$$

At $\Delta\epsilon_t = 0.011$, N_f (cold) = 2100 (from pure fatigue curve at 230 °F).

At $\Delta\epsilon_t = 0.0088$, N_f (hot) = 250 (from hold-time-modified fatigue curves at 800 °F).

$$D' = \frac{105}{2100} + \frac{288}{250} = 1.2$$

This shows that 120% of the crack-initiation life has been expended. Hence, crack initiation could easily have occurred due to low-cycle fatigue if creep effects are taken into account.

3. *Creep-Rupture Damage*
From a Larson-Miller rupture curve (e.g., mean curve in Fig. 1.8), the time to rupture at a stress of 49,600 psi at 800 °F can be determined to be about 432,000 h. Assuming steady operation at 800 °F and little or no relaxation of initial tangential stresses, the life fraction consumed in creep-rupture damage can be estimated as 106,000/432,000 = 0.24. The total damage due to low-cycle-fatigue and creep-rupture damage can be calculated as 1.44, or 144%.

The above example illustrates the general approach used in applying damage-summation methods to estimations of crack-initiation life. In calculating the over-all life of a heavy-section component, this approach may be combined with crack-growth-based approaches to predict total life.

Failure-Mechanism Maps

Based on a review of all the available low-cycle-fatigue data, Miller *et al* have concluded that most of the failures reported are either predominantly fatigue- or creep-controlled and that the region of interaction between the two mechanisms is rather small (Ref 20). The fatigue failures are characterized by transgranular fracture, whereas the creep failures are characterized by intergranular fracture. Mixed-mode fractures, indicative of creep-fatigue interaction, have been observed only over a narrow range of test conditions. In many laboratory tests, high strain ranges and short hold times are employed in order to get quick results. These failures have tended to be fatigue-dominated. Service conditions, on the other hand, involve long hold times (low frequencies) and small strain ranges, which are conducive to creep. Based on these considerations, the various failure regions can be delineated in the form of a fracture map, as shown in Fig. 4.38 (Ref 134). Small strain ranges and long hold times are found to promote creep damage, whereas large strain ranges and short hold times promote fatigue damage. Only in the regions near the boundary between the two phenomena is creep-fatigue interaction expected to occur. This explains why procedures assuming additive damage have worked at least as well as, if not better than, those assuming interactive damage mechanisms.

Thermal Fatigue

Most low-cycle-fatigue problems in high-temperature machinery involve thermal as well as mechanical loadings. In other words, the material is subjected to cyclic temperature simultaneously with cyclic stress. Analysis of these loadings and consideration of the attendant fatigue damage become very complex, and gross simplifications are often introduced. In the past, thermal fatigue traditionally has been treated as being synonymous with isothermal low-cycle fatigue at the maximum temperature of the thermal

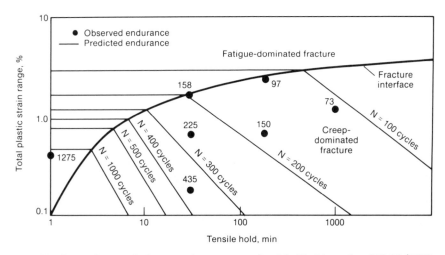

Fig. 4.38. Creep-fatigue failure-mechanism map for 1Cr-Mo-V steel at 565 °C (1050 °F) (Ref 134).

cycle. Life-prediction techniques also have evolved from the low-cycle-fatigue literature. More recently, advances in finite-element analysis and in servohydraulic test systems have made it possible to analyze complex thermal cycles and to conduct thermomechanical fatigue (TMF) tests under controlled conditions. The assumed equivalence of isothermal LCF tests and TMF tests has been brought into question as a result of a number of studies. It has been shown that for the same total strain range, the TMF test can be more damaging under certain conditions than the pure LCF test.

Spera has defined thermal fatigue as "the gradual deterioration and eventual cracking of a material by alternate heating and cooling during which free thermal expansion is partially or fully constrained" (Ref 135). Constraint of thermal expansion causes thermal stresses which may eventually initiate and propagate fatigue cracks. Because thermal cycles usually result in appreciable inelastic strains and cause failure in 10^4 to 10^5 cycles, thermal fatigue may be viewed as a form of low-cycle fatigue. However, in some cases, such as thermal striping, it can occur at frequencies near 1 Hz and give rise to high-cycle fatigue cracking. Constraint of free thermal expansion and contraction is an essential ingredient of thermal-fatigue damage. Constraints may be external or internal, with the former generally being

resorted to in laboratory tests to simulate the internally occurring constraints in actual components. In a typical laboratory test, a uniaxial specimen—usually a tubular or hourglass-shape specimen—is heated uniformly across its test section while constraining forces are applied through an end grip. This technique is referred to as thermomechanical fatigue (TMF) testing. In modern machines, the temperature and strain cycles can be applied independently according to predetermined programs. In components operated at high temperatures, internal constraints to expansion and contraction of an element arise from adjacent material elements which either are at a different temperature (e.g., surface vs interior of a rotor) or are made of a different material (e.g., dissimilar welds in boiler tubes, and austenitic stainless steel cladding on ferritic steel in pressure vessels). Internal constraints also can be present in laboratory tests such as those using tapered disk specimens. Such tests are known as thermal stress fatigue (TSF) tests.

In the TMF test, the temperature is changed from a maximum to a minimum value concurrently with an independent variation of the strain from a minimum to a maximum value. Two simple waveforms in TMF tests are shown and compared with the LCF test in Fig. 4.39. If maximum temperature corresponds to peak compression,

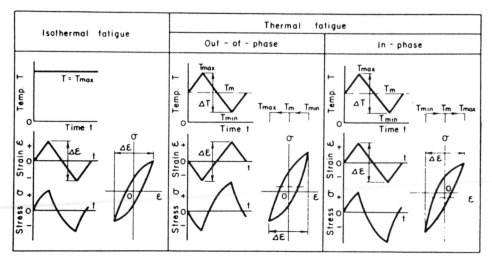

Fig. 4.39. Schematic diagrams showing waveforms of temperature, strain, and stress in thermal and isothermal fatigue tests.

as in the center diagram in Fig. 4.39, this is known as "out-of-phase" cycling. If the maximum tensile stress occurs at the peak temperature, it is known as "in-phase" cycling (diagram at right). Depending further on when the hold time is superimposed, various cycle shapes are possible, as described later.

Thermal-Stress Fatigue

In a common test approach for generating thermal-fatigue data under transient thermal conditions, tapered disk or wedge-shape specimens are used in combination with fluidized baths (Ref 136 and 137). The specimens are subjected to alternate heating and cooling shocks by immersing them in fluidized baths held at different temperatures. Unfortunately, direct measurement of stress or strain in the bath is not possible in these tests, so that the results obtained cannot be expressed in a quantitative manner. Quantification requires highly complex analysis when loading and temperature conditions induce nonlinear deformations. The analysis must take into account the nonuniform, three-dimensional geometry in the thermal and stress analyses, incorporate material-property variations with temperature, and account for the accumulation of time-independent and time-dependent non-

linear deformations. Several publications have dealt with the analytical procedures for calculating the timewise and spatial distributions of stresses and strains in tapered disk specimens (Ref 138 to 141). The accuracy of these procedures cannot, however, be evaluated, because no independent and alternative results are available for comparison.

The only validation for the stress-analysis procedure consists of coupling it with various life-prediction methods (damage rules), predicting the number of cycles to failure, and then comparing the predictions with actual observations. Any discrepancies between predictions and actual observations may be the results of inaccuracies in the damage rules assumed or in the analytical procedure, and the two effects cannot be examined independently. Despite these problems, reasonable success has been claimed in predicting failure lives of specimens of Nimonic 90, In 100, coated In 100, B 1900, MAR-M 200, and alloy T 111 (Ref 142 and 143). Spera and Cox have developed a computer program called TF LIFE which can be used to predict the thermal fatigue lives of metals and components (Ref 143). This program is used as a subroutine with a main program supplied by the user. The main program calculates input cycles of temperature and total strain for TF LIFE, which then calculates stress cycle, creep and plastic strain damage, and cyclic life. The life-prediction model uses the linear damage-summation method for estimating the damage caused by creep and fatigue. The creep damage is calculated using modified versions of the life-fraction rule and the universal slopes relationship. A unique feature of this program is that it incorporates several alternative failure criteria such as surface-crack initiation, interior-crack initiation, and complete fracture of both unnotched and notched specimens. In a qualitative sense, thermal-stress-fatigue tests have been used extensively to evaluate the relative susceptibilities to cracking of different materials, heat treatments, and micro-

structures. Bizon and Spera have performed extensive evaluations of the thermal-fatigue susceptibilities of a variety of nickel- and cobalt-base alloys and ranked them as shown in Fig. 4.40 (Ref 144). Beck and Santhanam have used the test to identify the optimum casting parameters for alloy MAR-M 509 (Ref 145).

Thermomechanical Fatigue

Thermomechanical fatigue (TMF) testing simulates many of the features of the general fatigue problem, yet retains the relative simplicity and ease of data gathering and interpretation associated with axial strain cycling of smooth laboratory specimens. In these tests, the temperature, strain, and stress can be measured directly at any point

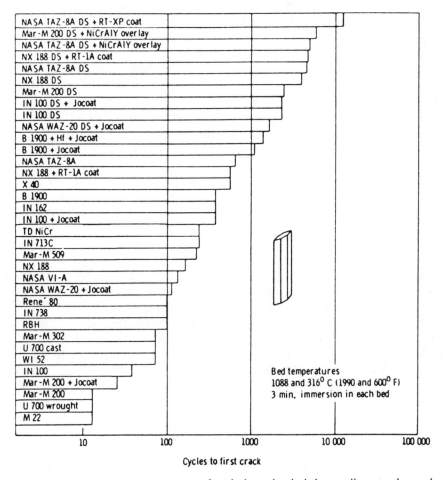

Fig. 4.40. Comparative resistances of nickel- and cobalt-base alloys to thermal-stress fatigue (Ref 144).

within the cycle. Furthermore, any two of these variables can be controlled independently of time.

TMF testing typically starts with a structural analysis of a component, resulting in a strain-temperature-time cycle at a critical location. This cycle is then applied to a uniaxial specimen in a servohydraulic machine. The temperature as well as the strain should have no spatial variation in the specimen test section, so that the local region can be represented in a quantitative fashion. A description of a typical test setup and procedure can be found in Ref 146.

There is now ample evidence to show that the fatigue-endurance curves obtained in TMF tests can be quite different from those obtained in isothermal LCF tests. Results reported by Thomas, Bressers, and Raynor (Ref 147) on Inconel 738 LC indicate that TMF testing at temperatures from 500 to 850 °C (930 to 1560 °F) was more damaging than isothermal tests at 850 °C, although not by a significant degree. In Udimet 710, out-of-phase (OP) TMF cycles resulted in much shorter lives than those obtained in isothermal tests, even at the peak temperature of 980 °C (1800 °F), as shown in Fig. 4.19 (Ref 63). In 2¼Cr-1Mo steels it has been found that the isothermal fatigue life at 540 °C (1000 °F) and a frequency of 0.5 cpm is nearly equal to the in-phase (IP) fatigue life at 300 to 540 °C (570 to 1000 °F) and a frequency of 0.5 cpm at the higher strain range, and equal to the OP fatigue life at the lower strain range. The IP cycles were more damaging than the OP cycles at the high strain range, but the behavior was reversed at low strain ranges (Ref 35). The situation with respect to type 316 stainless steel was found to be exactly the opposite of that for 2¼Cr-1Mo steel. It is evident from these facts that the thermal-fatigue lives of various materials cannot be reliably evaluated on the basis of isothermal low-cycle fatigue life at the peak temperature and at the same strain range.

Based on the results of an extensive study of many materials, Kuwabara, Nitta, and Kitamura (Ref 148) have classified them into four groups according to their relative lives in IP cycles (maximum strain at maximum temperature) and OP cycles (maximum strain at minimum temperature), as follows:

1. Type I — materials for which IP life is shorter than OP life at lower strain ranges
2. Type O — materials for which OP life is shorter than IP life at lower strain ranges
3. Type E — materials for which IP and OP lives are nearly equal
4. Type E' — materials for which IP life is shorter at higher strain ranges but nearly equal to OP life at lower strain ranges.

The representative examples of these four types are illustrated in Fig. 4.41. Classifications are the same whether the total strain range or the inelastic strain range is employed. High tensile strains at high temperature (IP) would favor creep, whereas high tensile strains at low temperature (OP) would favor cracking of oxide and hence accelerated environmentally induced damage during subsequent high-temperature exposure. Hence, Kuwabara *et al* rationalized that type I behavior was the result of creep damage whereas type O behavior was indicative of environmentally enhanced damage and type E behavior was exhibited when neither creep nor environment had a significant effect on damage.

The classification, however, was not rigid, because with increasing temperature (T_{max}) or hold time, the TMF life characteristics shifted from type O or type E to type I behavior, reflecting an increasing creep contribution. Accordingly, life-prediction methods based on fatigue damage alone, such as the Coffin-Manson equation or the method of universal slopes in combination with isothermal LCF data obtained at the peak temperature, were sufficient for safe prediction of the lives of type O and type E materials. Low-strength, high-ductility materials such as Cr-Mo, Cr-Mo-V, and Ni-Mo-V steels and austenitic stainless steels fall in this category. On the other hand, in

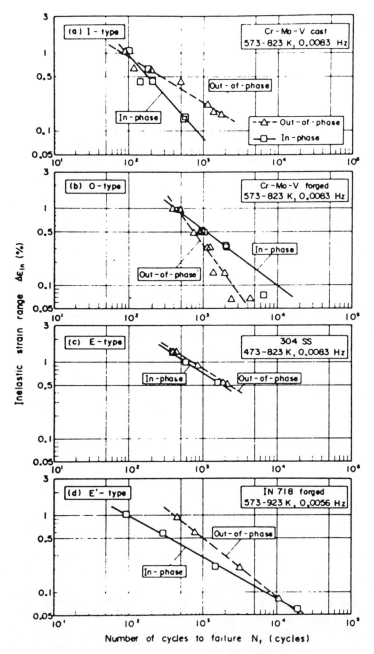

Fig. 4.41. Typical examples of the four types of thermal-fatigue-life characteristics in the inelastic-strain-range-vs-life relationship (Ref 148).

materials exhibiting type I behavior (e.g., high-strength, low-ductility materials), these methods dangerously underpredicted fatigue lives.

A comprehensive study of the TMF behavior of gas-turbine blade alloy Inconel 738 and vane alloy GTD 111 has been concluded recently (Ref 146 and 149). The data show that the type of cycle, peak temperature, frequency, and hold time affect the lives of the materials and that for a given strain range the life can vary by as much as two orders of magnitude. It also was noted that the maximum tensile stress during a given cycle was a better index than strain range for ranking the behavior under the various conditions. These results are described in more detail in Chapter 9.

TMF Life Prediction

Several reviews have surveyed the promising high-temperature methods of predicting fatigue life (Ref 150 to 154). None of these methods has received a strong collective endorsement in these reviews, although the Ostergren net-hysteresis-energy method and the damage-rate method may be identified as the leading ones. The most obvious ones to try for TMF situations are the ones which have shown success for isothermal LCF situations. However, attempts to correlate N_f with total or plastic strain range typically result in many problems, because isothermal results generally do not follow the same strain-life curve as TMF results and a number of different types of TMF cycles can result in as many separate curves. Halford and Manson have successfully applied the strain-range-partitioning (SRP) method to TMF testing of type 316 stainless steel (Ref 155). Nitta *et al* tried SRP with a variety of materials and found that it works well for high-ductility materials but not for nickel-base superalloys (Ref 156). This result is consistent with the experience of others in applying SRP to nickel-base alloys even under isothermal conditions (Ref 157 and 158). Bill *et al* tried several methods with nickel-base superalloy MAR-M 200 and concluded that none was applicable to TMF (Ref 159). Jaske (Ref 160) found that, even with a ductile carbon steel, simple strain-based techniques do not work as well as stress-strain-based ones such as the Smith-Watson-Topper parameter (Ref 161). Sehitoglu and Morrow used the maximum tensile stress to successfully correlate isothermal and TMF crack-initiation lives for carbon steels (Ref 162). The problems in trying to develop a universal method for life prediction for all materials and components have been reviewed by Russell (Ref 146), who points out the need for developing specific techniques focused on specific material and applications.

Fatigue-Crack Growth

Although the S-N curves have been used in the past as the basic tool for design against fatigue, their limitations have become increasingly obvious. One of the more serious limitations is the fact that they do not distinguish between crack initiation and crack propagation. Particularly in the low-stress regions, a large fraction of a component's life may be spent in crack propagation, thus allowing for crack tolerance over a large portion of the life. Engineering structures often contain flaws or cracklike defects which may altogether eliminate the crack-initiation step. A methodology that quantitatively describes crack growth as a function of the loading variables is, therefore, of great value in design and in assessing the remaining lives of components.

LEFM Approach

The growth of long cracks under low stresses is particularly amenable to evaluation in terms of the crack stress-intensity factor K. The cyclic equivalent of K in fatigue is its range, ΔK. As the crack grows— i.e., as its length, a, increases, the value of K increases and so does the crack-growth rate da/dN. A schematic plot of da/dN vs ΔK is shown in Fig. 4.42 (Ref 163). The curve is approximately sigmoidal, with the upper and lower asymptotes defined by, respectively, the static fracture condition K_{Ic} and a stress-intensity threshold ΔK_{Th} below which significant crack growth does not occur. In the intermediate region, a power-law relationship (the Paris law; Ref 164) is often obeyed, so that

$$\frac{da}{dN} = C'\Delta K^n \qquad \text{(Eq 4.44)}$$

where C' and n are material constants determined from experiments. For many metals and alloys, n is on the order of 3 to 4, with a lower-boundary value of 2. For high-strength or embrittled materials, n can rise above 4 in rare instances. C' as in Eq 4.44 is generally proportional to $(1/E)^n$.

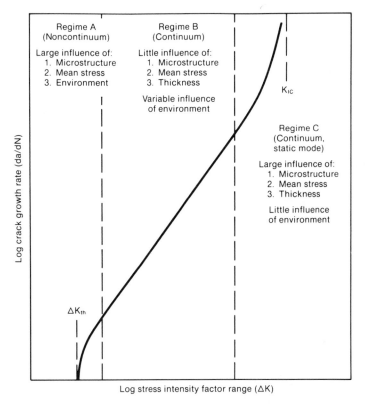

Fig. 4.42. Schematic fatigue-crack-growth curve (based on Ref 163; cited in Ref 14).

Structure sensitivity is exhibited at the two extremes of the curve, where either ΔK_{Th} or K_{Ic} is approached, but between these extremes, little effect of microstructure or mean stress is found. If the expression for K given in Eq 2.10 is substituted into Eq 4.44 and integration is performed, we get (Ref 165)

$$\int_{a_i}^{a_c} a^{-n/2} da = C' \sigma^n M^{n/2} dN \quad \text{(Eq 4.45)}$$

where M is a parameter approximately related to the flaw-shape parameter through an expression of the form $M = 1.21\pi/Q$. The resulting expressions for cyclic life are

$$N_f = \left[\frac{2}{(n-2)C'M^{n/2}\sigma^n} \right]$$

$$\times \left[\frac{1}{(a_i)^{(n-2)/2}} - \frac{1}{(a_c)^{(n-2)/2}} \right]$$

$$\text{[for } n \neq 2] \quad \text{(Eq 4.46)}$$

and

$$N_f = \frac{1}{C'M\sigma^2} \ln \frac{a_c}{a_i} \quad \text{[for } n = 2]$$

$$\text{(Eq 4.47)}$$

Example:

For a 12% Cr steel rotor, determine the critical crack size a_c for a radial axial bore crack growing under a tangential stress of 30% of the yield strength, σ_y. How many fatigue cycles can be endured if the inspection of the rotor reveals a 1-in.-deep crack with a length-to-depth ratio, c/a, of 4? The following information is given: $\sigma_y = 85.5$ ksi; the critical plane-strain stress-intensity $K_{Ic} = 105$ ksi$\sqrt{\text{in.}}$; for a surface flaw with c/a = 4, the flaw-shape parameter Q = 1.43.

Answer:

The tangential stress $\sigma = 0.3\sigma_y = 26.7$ ksi. Substituting for K_{Ic}, Q, and σ in the appropriate expression for 12% Cr rotor steel gives

$$a_c = \frac{K_{Ic}^2 Q}{1.21\pi\sigma^2}$$

$$= \frac{105^2 \times 1.43}{1.21 \times 3.17 \times (26.7)^2} = 2.51 \text{ in.}$$

$$\frac{da}{dN} = C'\Delta K^n = 3.69 \times 10^{-9} K^{2.4}$$

$$C' = 3.69 \times 10^{-9}$$

$$n = 2.4$$

$$Q = 1.43$$

$$M = 1.21\pi/Q = 2.6$$

$$a_i = 1 \text{ in.}$$

$$a_c = 2.51 \text{ in.}$$

Substituting the above into Eq 4.46 yields

$$N = 2.76 \times 10^4 \text{ cycles}$$

Appendix G of Section III and Appendix A of Section XI of the ASME Boiler and Pressure Vessel Code give procedures for summing the extent of crack growth resulting from loading cycles. The route is optional and applies to defects in thick-section pressure vessels. The calculations apply only where growth is described by LEFM. The crack must be progressively updated—i.e., for each transient, the calculated incremental growth Δa is added to the initial crack length a, and ΔK is reassessed for the new crack length. Other factors which must be included are the flaw shape, compliance changes over the entire crack length, frequency, mean load, and environmental effects.

J-Integral Correlations

For shorter cracks under stresses approaching and beyond the yield strength and for long cracks in ductile materials approaching instability, LEFM criteria are violated. Recently, attempts have been made to use the elastic-plastic crack-characterization parameter J to describe crack growth outside the LEFM region. The cyclic version of the J-integral, ΔJ, was introduced by Dowling and Begley, who used it to correlate the fatigue-crack-growth behavior of specimens of different geometries (Ref 166 and 167) under elastic-plastic and fully plastic conditions. They showed that the crack-growth rate could be related to J through the expression

$$\frac{da}{dN} = C''(\Delta J)^{n'} \qquad \text{(Eq 4.48)}$$

where C'' and n' are empirical constants. The total value of cyclic ΔJ_t was viewed as consisting of the elastic component, expressed as ΔJ_e, and the plastic component, ΔJ_p, so that

$$\Delta J_t = \frac{\Delta K^2}{E} + \Delta J_p \qquad \text{(Eq 4.49)}$$

Applying this idea to a center-cracked plate under reversed-stress fatigue, Dowling (Ref 168) described the expression for ΔJ as

$$\Delta J_t = \left(\frac{3.2\Delta\sigma^2}{2E} + \frac{5.0\Delta\sigma\Delta\epsilon_p}{1+\beta}\right)a \quad \text{(Eq 4.50)}$$

where the first term denotes the elastic contribution to ΔJ, the second term denotes the plastic contribution to ΔJ, and β is the cyclic strain-hardening coefficient. In the absence of plastic strain, Eq 4.50 reduces simply to $\Delta K^2/E$, and thus $n' = n/2$.

For uniformly stressed components with very small cracks, Skelton has defined a parameter called the equivalent stress-intensity range, ΔK_{eq}, such that it takes into account both the elastic and plastic contributions, as follows:

$$\Delta K_{eq} = (\sigma + E\Delta\epsilon_p)\sqrt{\pi a} \qquad \text{(Eq 4.51)}$$

With some minor simplifications, and by comparison with Eq 4.50, this can be written as

$$\Delta K_{eq}^2 = \left(\frac{\sigma^2}{2E} + \frac{E\Delta\epsilon_p^2}{2} \right) 2\pi aE \quad \text{(Eq 4.52)}$$

which reduces to

$$\Delta K_{eq} = \sqrt{E\Delta J_t} \quad \text{(Eq 4.53)}$$

The equivalence of the J-based approach and the equivalent stress-intensity-based approach has been discussed by Skelton (Ref 13).

Crack-growth rates measured under linear-elastic conditions and under elastic-plastic and fully plastic conditions using A533B steel could be plotted on a single trend curve over several orders of magnitude in growth rate (Ref 168). These results are shown in Fig. 4.43. The method thus provides a convenient means of using laboratory data obtained from specimens under linear-elastic conditions to predict the behavior of components under elastic-plastic and plastic conditions, and vice versa. The approach has now been extended to high temperature data for several other alloys (Ref 169 to 174), and the results have been compiled by Skelton, as shown in Fig. 4.44 (Ref 13). The exponent n' in Eq 4.48 has been found to have an approximate value of 1.3 (Ref 13). Other fatigue investigations using the J-integral as the crack-tip parameter include the work of Sadananda and Shahinian (Ref 175) on cold worked type 316 stainless steel at 595 °C (1100 °F), the studies of Taira *et al* (Ref 171) and of Jaske (Ref 172) on type 316 stainless steel at 600 and 650 °C (1110 and 1200 °F), and the investigation of Sadananda and Shahinian (Ref 176) on Udimet 700. Good correlation of crack growth with ΔJ was indicated in some of these studies, but the correlation was not unique in the others.

Crack Growth in Fully Plastic Cycling

Under fully plastic cycling conditions and for small a/w ratios, it has been found that da/dN is no longer proportional to $\sqrt{a}$, as in LEFM, but varies linearly with a, as given by

$$\frac{da}{dN} = Ha \quad \text{(Eq 4.54)}$$

where H is a constant whose value is dependent on the material, the test temperature, and the strain range. This is not surprising, because for ductile steels, the value of n in Eq 4.44 is approximately 2, which leads to a value of n' in Eq 4.48 of about 1. Hence, the proportionalities da/dN $\propto$ J $\propto$ $\Delta\sigma\Delta\epsilon_p a$ result. For a given material at a given temperature, H varies systematically with the strain range (see Fig. 4.45) in such a manner that Eq 4.54 can be rewritten as

$$\frac{da}{dN} = C\Delta\epsilon_p^p a \quad \text{(Eq 4.55)}$$

where C and p are constants. The value of p varies from 1 to 2, and its specific values

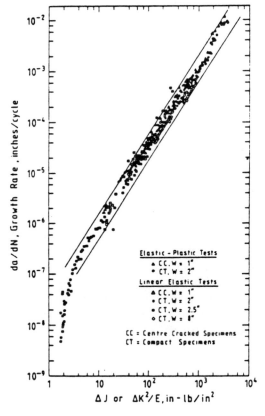

Fig. 4.43. Results of 20 experiments showing correlation of fatigue-crack-propagation rates in A533B steel in terms of cyclic J for a variety of specimen configurations (Ref 168).

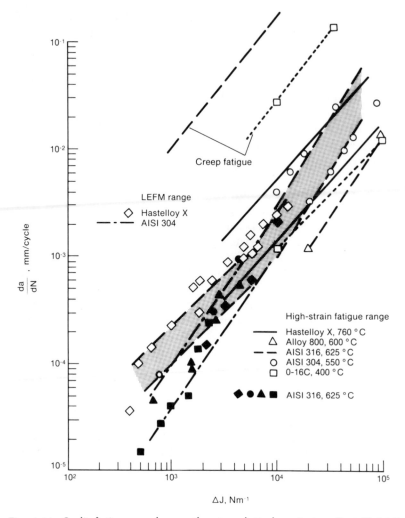

Fig. 4.44. Cyclic fatigue-crack-growth rates plotted against cyclic J (Ref 13).

for a variety of materials are summarized in Ref 177.

By taking into account the shape of the crack front and integrating Eq 4.55 between the limits of the initial crack size a_i and the final crack size a_c, a generalized expression for predicting crack-growth life is obtained:

$$\ln \frac{a_c}{a_i} = YC\Delta\epsilon_p^p(N_f - N_i) \quad \text{(Eq 4.56)}$$

where Y is a crack-front-shape factor whose values have been reported to be 1, 0.25, and 0.5 for straight-front, semicircular, and semielliptical cracks, respectively. This equation can be used to separate crack initiation and crack propagation in smooth-bar test specimens. If we assume a typical value of 10 μm (395 μin.) for a_i and about two-thirds of the specimen diameter for a_c, the time spent in crack propagation can be separated out. It is also interesting to observe that because a_c/a_i varies logarithmically with N, crack length will increase exponentially as N increases, implying that a major portion of the fatigue life is spent in propagation of small cracks. Experiments on austenitic steels have shown that the crack-growth life as given by Eq 4.56 is nearly identical to the total smooth-bar-specimen life at high strain ranges. On the other hand, at low strain ranges, the smooth-bar

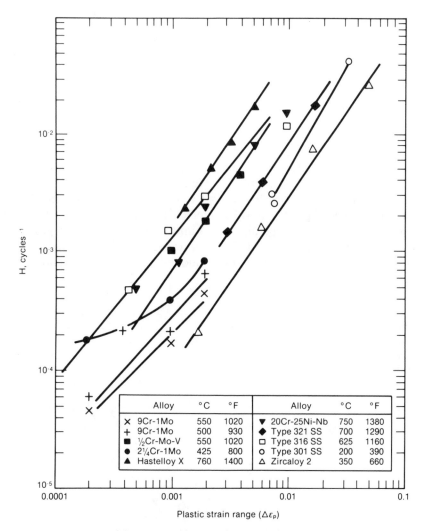

Alloy	°C	°F	Alloy	°C	°F
X 9Cr-1Mo	550	1020	▼ 20Cr-25Ni-Nb	750	1380
+ 9Cr-1Mo	500	930	◆ Type 321 SS	700	1290
■ ½Cr-Mo-V	550	1020	□ Type 316 SS	625	1160
● 2¼Cr-1Mo	425	800	○ Type 301 SS	200	390
▲ Hastelloy X	760	1400	△ Zircaloy 2	350	660

Fig. 4.45. Variation of fatigue-crack-growth rate with plastic strain (Ref 11).

life far exceeds the crack-growth life. These results support the statements made early in this chapter that at stresses and strains above the endurance, crack initiation occurs readily and that the total life is governed by the crack-propagation life. At stresses and strains near the endurance limit, most of the specimen life is expended in crack initiation.

Example:
A smooth-bar specimen of a Cr-Mo steel with a circular cross section is being subjected to push-pull testing at a nominal plastic strain range $\Delta\epsilon_p$ of 0.01. It has been determined from prior tests that for this material C = 2 and p = 1.2. Calculate the expected life assuming that crack initiation will occur at a = 0.001 in. and that the specimen will fail at a_c = 0.66 in. Calculate the life of a notched bar, given that the strain-concentration factor K_ϵ = 4.

Answer:
Rewriting Eq 4.56, we get

$$N_f = \frac{1}{YC\Delta\epsilon_p^p} \ln \frac{a_c}{a_i}$$

Substituting Y = 1 (assuming a straight crack front), C = 2, $\Delta\epsilon_p$ = 0.01, p = 1.2, a_c = 0.66, and a_i = 0.001, we get N_f = 815 cycles for the smooth bar. For the notched bar, effective $\Delta\epsilon = \Delta\epsilon_p \times K_\epsilon$ = 0.04. Substituting the revised value of $\Delta\epsilon$, we get N = 154 cycles for the notched bar.

Effects of Test Variables on LEFM Growth

The effects of test temperature, cyclic frequency, stress ratio (R), waveform (hold time), and environment on the crack-growth behavior in the Paris-law region have been investigated in several studies. Increasing temperature has been found to increase crack-growth rates. Illustrations of this behavior are provided for wrought 2¼Cr-1Mo steel and for several nickel-base superalloys in Fig. 4.46 to 4.48. For cast 2¼Cr-1Mo steel (Ref 178), wrought 2¼Cr-1Mo steel

(Ref 4), and Cr-Mo-V steel rotor forgings (Ref 179), crack-growth rates are reported to have increased approximately by a factor of 2 to 4 upon heating from room temperature to about 540 to 595 °C (1000 to 1100 °F). In type 316 stainless steel (Ref 180), crack-growth rate increased by a factor of 30 at low ΔK, but at higher ΔK values the difference was reduced to a factor of 6 for a temperature increase from room temperature to 625 °C (1155 °F). A number of studies have evaluated temperature effects in nickel- and cobalt-base alloys. A hundredfold increase in growth rate was observed in Hastelloy, Multimet (Ref 181), and cobalt-base Haynes alloy (Ref 182) due

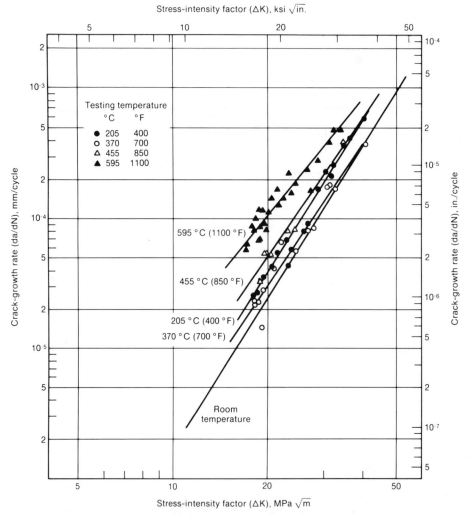

Fig. 4.46. Effect of temperature on fatigue-crack-growth behavior of 2¼Cr-1Mo steel (Ref 4).

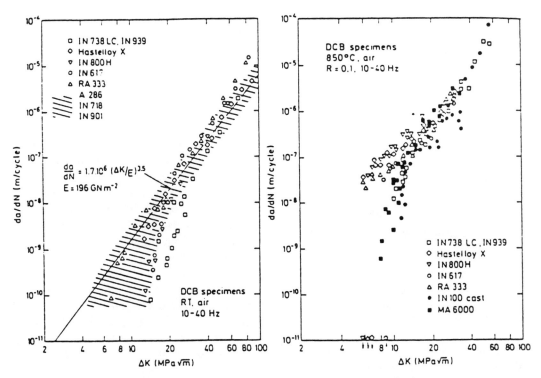

Fig. 4.47. Fatigue-crack-growth rates of long cracks for various high-temperature alloys in air at (left) room temperature and (right) 850 °C (1560 °F) (Ref 186).

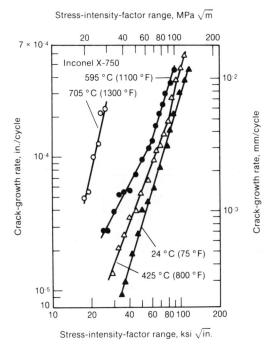

Fig. 4.48. Fatigue-crack-growth rates for Inconel X-750 as a function of stress-intensity-factor range at a cycling frequency of 0.17 Hz (Ref 185).

to an increase in temperature from room temperature to 870 °C (1600 °F). Variation of crack-growth rates at a fixed value of ΔK has been evaluated for Inconel alloys 600, 625, 718, and X-750 (Ref 183 and 184). The observed variation could be rationalized in terms of the temperature dependence of Young's modulus and the yield strength (Ref 184). Crack-growth rates in Inconel X-750 have been investigated at temperatures from 24 to 705 °C (75 to 1300 °F) by Shahinian (Ref 185). Bressers, Remy, and Hoffelner have compiled crack-growth-rate data for a number of nickel-base alloys (Ref 186). Their results showed that crack-growth rate increased by a factor of 25 on heating from room temperature to about 900 °C (1650 °F). Because the reported data have been obtained at various ΔK ranges and temperature ranges, it is difficult to compare the various types of materials directly. At a constant ΔK (arbitrarily chosen as 30 MPa$\sqrt{m}$, or 27 ksi$\sqrt{in}$.), a clear trend of crack-growth-rate increase with increasing temperature can be seen as

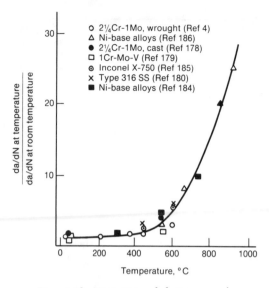

Fig. 4.49. Variation of fatigue-crack-growth rates as a function of temperature at $\Delta K = 30$ MPa $\sqrt{m}$ (27 ksi $\sqrt{in.}$).

shown in Fig. 4.49. In this figure it can be seen that at temperatures up to about 50% of the melting point (550 to 600 °C, or 1020 to 1110 °F), the growth rates are relatively insensitive to temperature, but the sensitivity increases rapidly at higher temperatures. The crack-growth rates for all the materials at temperatures up to 600 °C relative to the room-temperature rates can be estimated by a maximum correlation factor of 5 (2 for ferritic steels).

The effect of cycling frequency on the crack-growth rates of austenitic stainless steels has been investigated by James (Ref

187 to 189). The LMFBR Nuclear Systems Materials Handbook employs a frequency/temperature correlation factor to obtain the desired crack-growth relationship for specific conditions (Ref 190). The crack-growth data are normalized with respect to data at a standard frequency of 0.67 Hz. A typical plot of the relative crack-growth behavior versus cyclic frequency for types 304 and 316 stainless steels is shown in Fig. 4.50. The results indicate an increase in crack-growth rate of only a factor of 8, when the frequency is decreased by five orders of magnitude at 540 °C (1000 °F). The effect of frequency in a cobalt-base alloy has been investigated, and a saturation effect at high frequencies has been found, as shown in Fig. 4.51 (Ref 182). Similar effects have been reported for IN-738 (Ref 191).

Type 304 stainless steel exhibits an increase in crack-growth rate with increasing value of the R ratio over the range 0.15 to 0.75. Similar behavior occurs in ferritic steels. To account for the effect of R, the use of the Walker effective stress-intensity factor in place of ΔK in the Paris-law equation has been proposed such that

$$\frac{da}{dN} = C\Delta K_{eff}^{n} \qquad (Eq\ 4.57)$$

where

$$\Delta K_{eff} = \Delta K_{max}(1 - R)^{m} \qquad (Eq\ 4.58)$$

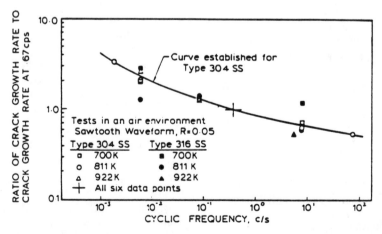

Fig. 4.50. Comparison of relative frequency effects on fatigue-crack growth in types 304 and 316 stainless steels over the temperature range 700 to 922 K (Ref 190).

with m being an empirical constant dependent on the material and the test temperature and ΔK_{max} being the maximum stress-intensity range. The applicability of this parameter over a range of R values from 0.063 to 0.807 has been demonstrated by James (Ref 188). The effect of R, however, may reach saturation values and the crack-growth rates may not increase exponentially as R approaches unity (Ref 192). The applicability of this approach to a ferritic steel has been demonstrated by Tomkins, as shown in Fig. 4.52 (Ref 14).

A combined parameter that empirically takes into account the effects of temperature, frequency, and R ratio has been proposed by Carden (Ref 193), as follows:

$$\frac{da}{dN} = XP\alpha \qquad (Eq\ 4.59)$$

$$P = \left\{ \left[A_1 \exp\left(\frac{-\Delta H_1}{R'T} \right) \right. \right.$$
$$\left. + A_2 \exp\left(\frac{-\Delta H_2}{R'T} \right) \right] \left(\frac{1}{\nu} \right)^{A_3}$$
$$\left. + C_1 \exp\left(\frac{\Delta H_3}{R'T} \right) \left(\frac{1}{\nu} \right) \right\}$$
$$\times [(\Delta K_{max}^2 - \Delta K_{Th}^2)(1 - R)^{2m'}]^m$$
$$(Eq\ 4.60)$$

where X, α, A_1, A_2, A_3, C_1, ΔH_1, ΔH_2, ΔH_3, m, and m' are constants independent of temperature; T is temperature; ν is frequency; R' is the universal gas constant; ΔK_{max} and ΔK_{Th} are the maximum stress intensity and the threshold intensity, respectively; and R denotes the stress ratio. Based on this parameter, Carden has correlated the crack-growth data for type 304 stainless steels from many sources over a wide range of conditions, as shown in Fig. 4.53. This method, however, is outdated, requires determination of too many empirical constants, and hence is not very useful.

Near-Threshold Crack Growth

Most engineering applications in which analyses of fatigue-crack-growth life have

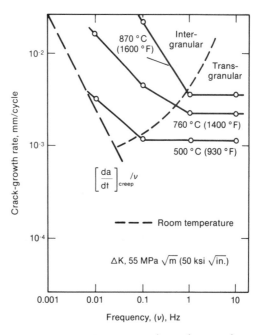

Fig. 4.51. Frequency dependence of fatigue-crack-growth rate for a cobalt-base superalloy (Ref 182).

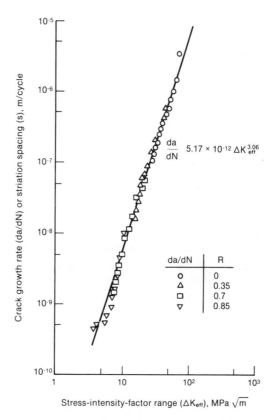

Fig. 4.52. Effect of R ratio on crack-growth rate and striation spacing for A533B steel at 25 °C (77 °F) (Ref 14).

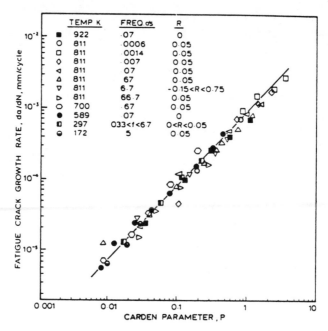

Fig. 4.53. Use of the Carden parameter P for correlation of crack-growth behavior of annealed type 304 stainless steel tested in air over a wide range of temperatures, cyclic frequencies, and stress ratios (Ref 193).

been performed have employed LEFM along with the assumption that, in the stress-intensity range of practical interest, the Paris-Erdogen law is applicable. This latter assumption has been partly necessitated by the detection limits imposed by current non-destructive inspection techniques. With the recent advances in NDT methods, the resolution of flaw detection has been greatly improved. Consequently, crack-growth rates at low ΔK levels, approaching the threshold value of ΔK, have received increasing attention. For components operating at low ΔK levels (low stress or small cracks), the near-threshold crack-growth life would be an important component of the total life. Concern about this problem has been raised by preliminary observations that small cracks grow more rapidly and would lead to failure in shorter times than would be suggested by the application of the Paris-Erdogen law, which deals primarily with the behavior of large cracks (Ref 194). Because of a lack of sufficient data, the validity of these claims has not yet been sufficiently ascertained. Apart from the crack-growth aspect, however, a number of studies have shown that the threshold ΔK itself can be shifted to

higher or lower values depending on the R ratio, frequency, temperature, environment, microstructure, and a host of other variables. Results from these studies are highlighted here.

The effects of test temperature and R ratio on near-threshold fatigue-crack growth in Cr-Mo-V rotor steels have been investigated by Liaw et al (Ref 195 and 196). The effect of the R ratio on ΔK_{Th} is illustrated in Fig. 4.54(b). With increasing R ratio, ΔK_{Th} decreased, the effect being more pronounced at higher temperatures. With increasing R, the effect of temperature became less important. As a function of increasing temperature, ΔK_{Th} decreased, reached a plateau, and then increased again at higher temperatures (see Fig. 4.54a). Increasing R also had the effect of increasing the near-threshold fatigue-crack growth rates.

The effects of temperature on ΔK_{Th} in numerous superalloys (Inconel 617, Inconel 718, IN-738, Incoloy 901, IN-939, and others) in the absence of environmental effects have been evaluated (Ref 186). Results from this study are shown in Fig. 4.55.

Environment also has been found to have an effect on ΔK_{Th} and on the near-

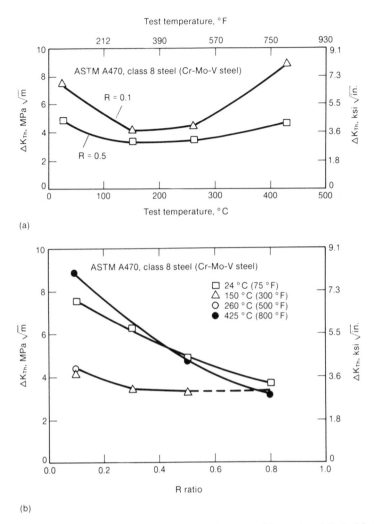

Fig. 4.54. Effects of test temperature (a) and R ratio (b) on the threshold ΔK for crack growth in 1Cr-Mo-V steel (Ref 195 and 196).

threshold crack-growth rate. Ritchie *et al* (Ref 197 and 198) have compared the effects of air and hydrogen on these parameters in 2¼Cr-1Mo steels. Moist air was found to increase ΔK_{Th} and reduce the crack-growth rates. These effects were attributed to formation of oxide corrosion products in the crack, which reduced the effective ΔK to a value below the applied ΔK. Similar effects, caused by oxide-induced crack closure, also have been found in 1Cr-Mo-V steels at 550 °C (1020 °F) tested over a range of frequencies (Ref 77), in stainless steels at elevated temperatures, and in 2¼Cr-1Mo steels tested in H_2S-containing environments at 425 °C (800 °F) (Ref 199). Decreasing frequency has the

effect of allowing more environmental interaction in the presence of air, and thus the beneficial effect of oxide or corrosion-product blocking becomes more pronounced as the frequency is lowered. The subject of environmental effects on fatigue has been reviewed by Marshall (Ref 200), and further coverage of the same is not needed here.

Thermal-Fatigue-Crack Growth

A limited amount of thermal-fatigue-crack-growth data are available, primarily for gas-turbine alloys based on thermal-stress-fatigue tests and thermomechanical strain-controlled tests. Woodford and Mowbray conducted thermal-stress-fatigue tests using

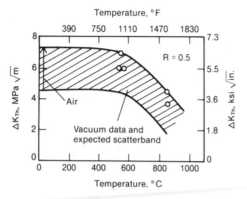

Fig. 4.55. Effect of test temperature on threshold ΔK for growth of long cracks in nickel-base superalloys Inconel 617, Inconel 718, IN-738, Incoloy 901, IN-939, and others (Ref 186).

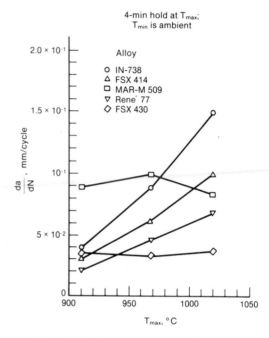

Fig. 4.56. Variation of steady-state crack growth (over 2.5 to 7.5 mm, or 0.1 to 0.3 in.) in several alloys with maximum temperature in thermal shock (Ref 201).

tapered disks in a fluidized bed on several alloys and showed that the crack-growth rates varied as a function of the peak temperature, as shown in Fig. 4.56 (Ref 201). An increase in the hold time caused further crossovers in the growth rate. The effect of dendrite-arm spacing on the crack-growth rates in MAR-M 509 was studied by Beck and Santhanam (Ref 145).

Thermomechanical strain cycling in ½ Cr-Mo-V steels in vacuum between 250 and 550 °C (480 and 1020 °F) showed that at a plastic-strain range of 0.001, crack-growth rates were only 1/3 those observed under isothermal strain cycling at 550 °C (Ref 202). Very similar growth rates were found in air and in steam. In type 316 stainless steels, crack-growth rates at a given strain range and initial crack size were higher under thermomechanical cycling in the range 400 to 625 °C (750 to 1160 °F) than under isothermal cycling at 625 °C (Ref 13).

Various researchers have monitored crack initiation and growth from artificial defects on the surfaces of smooth axial specimens tested in thermomechanical strain-controlled tests (Ref 203 to 206). The materials tested were two nickel-base superalloys, B1900 + Hf and directionally solidified MAR-M 200 + Hf, and cobalt-base superalloy MAR-M 509. They found that with all conditions being equal, crack-growth rate was a function of strain-intensity-factor range, regardless of the nominal strain range (Ref 203). The more ductile MAR-M

509 performed better in IP cycling than in OP cycling, whereas for the brittle B1900 + Hf alloy, the reverse was true (Ref 203 and 204). Some investigators have been successful in applying LEFM to thermomechanical fatigue-crack growth. Marchand and Pelloux, using stress-controlled tests, found that after allowance was made for crack-closure mechanisms, the crack-growth-rates-vs-ΔK behavior for Inconel X-750 was identical for IP, OP, and isothermal cycles (Ref 207). For ductile materials, elastic-plastic fracture-mechanics parameters have been successfully applied to thermomechanical fatigue. Specifically, Sehitoglu and Morrow have used crack-opening displacement for carbon steel (Ref 162), and Okazaki and Koizumi have used the J-integral approach for a low-alloy steel (Ref 208).

Crack Growth In Creep-Fatigue

The case of interaction between creep and fatigue is still in the rudimentary stages of formulation, but the work of James and Jones (Ref 209) provides a means of predicting crack growth in cyclically loaded structures with hold times at the maximum load. His work is, however, based on LEFM. A methodology is lacking for covering the entire range of behavior from short-time cyclic effects to long-term nonlinear creep responses. The efforts of Saxena, Williams, and Shih (Ref 210) come closest to the desired end, and the following discussion will be based on the phenomenology developed by these investigators.

Following Saxena *et al* (Ref 210), Wells *et al* have described the methodology for creep-fatigue-crack growth in Cr-Mo-V steels (Ref 211). Saxena *et al* assume that the crack growth Δa in a cycle that includes a hold time t_h is given by the sum of the fatigue contribution (with zero hold time) and the creep contribution, expressed as

$$\Delta a = \left(\frac{da}{dN}\right)_0 + \int_0^{t_h} C_5 C(t)^m dt \quad \text{(Eq 4.61)}$$

where $C(t)$ is the crack-tip driving force for creep, as described in Chapter 3, and C_5 and m are equation coefficients. Substituting for $C(t)$ results in

$$\Delta a = \left(\frac{da}{dN}\right)_0$$
$$+ C_5 \int_0^{t_h} \left[\frac{K^2(1-\nu^2)}{E(n+1)t} + C^*\right]^m dt$$
$$\text{(Eq 4.62)}$$

This integral is difficult to evaluate for arbitrary values of m. Hence it is broken up into two extreme cases of (1) small-scale creep, where C^* has little effect, and (2) large-scale creep, where C^* has a predominant effect.

For small-scale creep occurring at short times, C^* can be neglected and Eq 4.62 can be written as

$$\Delta a = \left(\frac{da}{dN}\right)_0 + C_5 \left[\frac{K^2(1-\nu^2)}{E(n+1)}\right]^m$$
$$\times \int_0^{t_h} t^{-m} dt \quad \text{(Eq 4.63)}$$

By integrating the second term between limits and substituting for

$$C_5 \frac{[(1-\nu^2)/E(n+1)]^m}{1-m} = C_2$$

we get the expression

$$\Delta a = \left(\frac{da}{dN}\right)_0 + C_2 K^{2m} t_h^{1-m} \quad \text{(Eq 4.64)}$$

Under steady-state creep, as t_h becomes large, crack growth is controlled by C^* and the crack growth during a cycle will vary linearly with t_h. Hence, the C^* term from Eq 4.62 is separated out and added as a linear component to get the final equation

$$\Delta a = \frac{da}{dN} = \left(\frac{da}{dN}\right)_0 + C_2 K^{2m} t_h^{1-m}$$
$$+ C_3 C^{*m} t_h \quad \text{(Eq 4.65)}$$

The first term in Eq 4.65 is a pure-fatigue contribution reflecting no effect of hold time and corresponding to crack-growth behavior at short hold times and high frequencies. The second term shows a nonlinear power-law dependence of crack-growth rate on hold time and pertains to intermediate hold times and frequencies where creep-fatigue interaction is present. The third term, containing C^*, shows a linear dependence of crack growth on hold time and corresponds to purely creep-dominated crack growth occurring at long hold times and low frequencies. Corresponding to these three terms, the plot of log da/dN vs log t_h should have three regions, as shown schematically in Fig. 4.57. This behavior has been experimentally confirmed in terms of frequency dependence of da/dN, as shown previously in Fig. 4.27 (Ref 69) for the iron-base alloy A-286. This figure shows that the crack-growth rate is a continuous function of frequency, even though the actual mechanism varies from transgranular fatigue at one end of the spectrum to intergranular creep at the other end of the spectrum. Hence, no discontinuous changes occur in the crack-growth rates, as confirmed by the work of James.

The first term (pure fatigue) in Eq 4.65 can be readily evaluated if the Paris-law coefficients are available. The last term

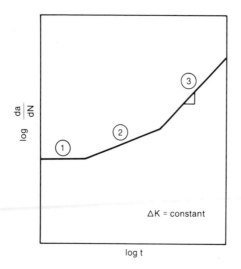

Fig. 4.57. Schematic illustration of the influence of hold time on crack growth per cycle (Ref 211).

ΔK = constant

vs-hold-time behavior for different a/w ratios and ΔK levels, for various assumed values of C_2. The best prediction is obtained when $C_2 = 5 \times 10^{-7}$. The predicted curves and the actual data for this value of C_2 agree closely, as shown in Fig. 4.58 (Ref 212). The applicable crack-growth law for the Cr-Mo-V steel at 540 °C (1000 °F) can, therefore, be written as

$$\frac{da}{dN} = 7.32 \times 10^{-9}\Delta K^{2.35}$$
$$+ 5 \times 10^{-7}\Delta K^{4/3}t_h^{1/3}$$
$$+ 0.013C^{*2/3}t_h \qquad \text{(Eq 4.67)}$$

Similar expressions have been derived by Wells et al (Ref 211) for types 316 and 304 stainless steels. For type 304 stainless steel, the da/dN behavior at 540 °C (1000 °F) is expressed as

$$\frac{da}{dN} = 1.1 \times 10^{-9}\Delta K^3$$
$$+ 10^{-6}\Delta K^{1.26}t_h^{0.37}$$
$$+ 0.04C^{*0.63}t_h \qquad \text{(Eq 4.68)}$$

Jaske (Ref 215) has developed a crack-tip-interaction model to account for creep-fatigue effects in type 316 stainless steel. He shows that hold times can be expected to accelerate da/dN or have no effect on da/dN, depending on the relative size of the zone of intense deformation at the crack tip during cyclic and monotonic loading. This model qualitatively accounts for all of the observed effects of hold time and frequency on crack growth in type 316 stainless steel. Jaske's approach is similar to those used by Willenborg et al (Ref 216) and by Wheeler (Ref 217) to account for the effects of crack retardation due to overload during fatigue-crack propagation at low temperatures. Further work is needed to achieve full quantification of such a model.

(pure creep) in Eq 4.65 also can be readily evaluated from the coefficients of the da/dt-vs-C* relationship. The only term whose value is not readily available is the coefficient C_2 in the middle term in Eq 4.65. The value of C_2 is therefore determined by curve-fitting of the da/dN-vs-t behavior for various assumed values of C_2 and fitting of the predictions of the actual behavior obtained from experiments (Ref 212). The curve-fitting procedure is briefly illustrated below with respect to a 1Cr-Mo-V rotor steel. Saxena et al and Swaminathan have published a comprehensive set of data on crack-growth rates with hold times ranging from 5 s to 24 h (Ref 210, 213, and 214). These data have been used for curve-fitting. Based on the Paris-law behavior and the creep-crack-growth behavior of Cr-Mo-V steels at 540 °C (1000 °F), Eq 4.65 can be written as

$$\frac{da}{dN} = 7.32 \times 10^{-9}\Delta K^{2.35} + C_2\Delta K^{4/3}t_h^{1/3}$$
$$+ 0.012C^{*2/3}t_h \qquad \text{(Eq 4.66)}$$

Using the handbook solutions for K and C* (and using the appropriate Norton-law coefficients to compute C*) for compact-type specimens, one can predict the da/dN-

Nomenclature

a — Crack length (or depth)

a_i — Initial crack length

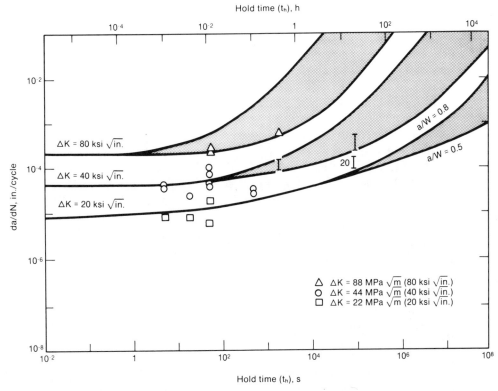

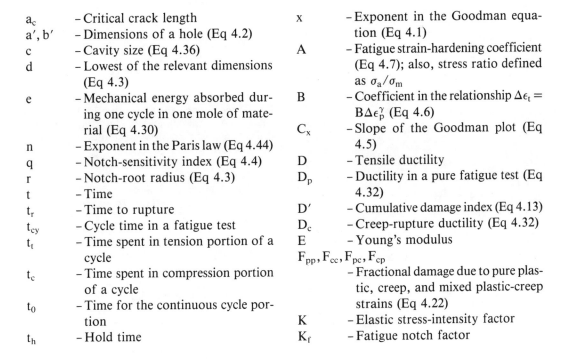

Solid lines are predictions from Eq 4.65.

Fig. 4.58. Comparison of observed and predicted influence of hold time on crack growth per cycle at selected ΔK levels for a Cr-Mo-V rotor steel at 540 °C (1000 °F) (Ref 212).

a_c	– Critical crack length
a', b'	– Dimensions of a hole (Eq 4.2)
c	– Cavity size (Eq 4.36)
d	– Lowest of the relevant dimensions (Eq 4.3)
e	– Mechanical energy absorbed during one cycle in one mole of material (Eq 4.30)
n	– Exponent in the Paris law (Eq 4.44)
q	– Notch-sensitivity index (Eq 4.4)
r	– Notch-root radius (Eq 4.3)
t	– Time
t_r	– Time to rupture
t_{cy}	– Cycle time in a fatigue test
t_t	– Time spent in tension portion of a cycle
t_c	– Time spent in compression portion of a cycle
t_0	– Time for the continuous cycle portion
t_h	– Hold time

x	– Exponent in the Goodman equation (Eq 4.1)
A	– Fatigue strain-hardening coefficient (Eq 4.7); also, stress ratio defined as σ_a/σ_m
B	– Coefficient in the relationship $\Delta\epsilon_t = B\Delta\epsilon_p^\gamma$ (Eq 4.6)
C_x	– Slope of the Goodman plot (Eq 4.5)
D	– Tensile ductility
D_p	– Ductility in a pure fatigue test (Eq 4.32)
D'	– Cumulative damage index (Eq 4.13)
D_c	– Creep-rupture ductility (Eq 4.32)
E	– Young's modulus
$F_{pp}, F_{cc}, F_{pc}, F_{cp}$	– Fractional damage due to pure plastic, creep, and mixed plastic-creep strains (Eq 4.22)
K	– Elastic stress-intensity factor
K_f	– Fatigue notch factor

K_t – Elastic stress-concentration factor

K_σ – Stress-concentration factor

K_ϵ – Total strain-concentration factor

$K_{\epsilon p}$ – Plastic strain-concentration factor

M   – Flaw-shape-related parameter (Eq 4.45)

N_0 – Number of cycles to failure in pure fatigue

N_h – Number of cycles to failure in a creep-fatigue test

N   – Number of actual cycles

N_f – Number of cycles to failure in a fatigue test

$N_{pp}, N_{cc}, N_{pc}, N_{cp}$

  – Numbers of cycles to failure under pure plastic, creep, plastic-creep, and creep-plastic strain components

Q   – Activation energy

R   – Stress ratio defined as $\sigma_{min}/\sigma_{max}$

R' – Universal gas constant

T   – Temperature

Y   – Crack-front-shape factor (Eq 4.56)

ΔJ_t – Total value of cyclic J-integral (Eq 4.49)

ΔJ_e – Elastic component of cyclic J-integral

ΔJ_p – Plastic component of cyclic J-integeral (Eq 4.49)

α_1, α_2 – Exponents in the Coffin-Manson law (Eq 4.8, 4.9)

β – Cyclic strain-hardening exponent (Eq 4.7)

$\Delta\epsilon_t$ – Total strain range

$\Delta\epsilon_e$ – Elastic strain range

$\Delta\epsilon_p$ – Plastic strain range

$\Delta\epsilon_{in}$ – Inelastic strain range

$\Delta\epsilon_p'$ – Equivalent plastic strain range (Eq 4.19)

$\Delta\epsilon_c$ – True tensile creep strain (Eq 4.32)

$\dot\epsilon$ – Strain rate

$\dot\epsilon_{eq}$ – Equivalent strain rate

$\Delta\epsilon_{pp}, \Delta\epsilon_{cc}, \Delta\epsilon_{pc}, \Delta\epsilon_{cp}$

  – Pure plastic, creep, and mixed plastic-creep strains

γ – Exponent in the relationship $\Delta\epsilon_t = B\Delta\epsilon_p^\gamma$ (Eq 4.6)

ν – Cycling frequency; also, Poisson's ratio

ν_t – Tension-going frequency (Eq 4.19)

ν_c – Compression-going frequency (Eq 4.19)

σ_a – Stress amplitude, or one-half of alternating stress range

σ_m – Mean stress

σ_{max} – Maximum stress

σ_{min} – Minimum stress

$\Delta\sigma$ – Total stress range

$\sigma_1, \sigma_2, \sigma_3$ – Alternating principal stresses

σ_e – Fatigue (endurance) limit

σ_y – Yield strength

σ_u – Ultimate tensile strength

ΔK – Stress-intensity-factor range

ΔK_{eq} – Equivalent stress-intensity-factor range (Eq 4.51)

ΔK_{Th} – Threshold stress-intensity-factor range

Other constants are as defined in text.

References

1. J. Goodman, *Mechanics Applied to Engineering*, 9th Ed., Longmans, Green and Co., New York, 1930

2. J. Gerber, cited in Ref 3, p 325 (no independent reference found)

3. G.E. Dieter, Jr., *Mechanical Metallurgy*, McGraw-Hill, New York, 1961, p 325

4. Various sources as cited in *Atlas of Fatigue Curves*, H.E. Boyer, Ed., American Society for Metals, Metals Park, OH, 1986

5. *Failure Analysis: The British Engine Technical Reports*, compiled by F.R. Hutchings and P.M. Unterweiser, American Society for Metals, Metals Park, OH, 1981, p 99-119

6. C.E. Inglis, Stresses in a Plate Due to the Presence of Cracks and Sharp Corners, *Trans. Inst. Naval Arch.*, Vol 55 (Pt 1), 1913

7. H. Neuber, *Kerbspannungalebre*, Springer, Berlin, 1937

8. R.E. Peterson, *Stress Concentration Design Factors*, John Wiley & Sons, New York, 1953

9. F.A. McClintock and A.S. Argon, *Mechanical Behavior of Materials*, Addison-Wesley, Reading, MA, 1966, p 411

10. G. Sines, *Proc. Soc. Exp. Stress Anal.*, Vol 14 (No. 1), 1956

11. R.P. Skelton, High Strain Fatigue Testing at Elevated Temperature, *Trans. Indian Inst. Metals*, Vol 35 (No. 6), Dec 1982, p 519-534

12. R.P. Skelton, The Prediction of Crack Growth Rates From Total Endurances in High Strain Fatigue, *Fatigue Engg. Mater. Struct.*, Vol 2, 1979, p 305-319

13. R.P. Skelton, Crack Initiation and Growth During Thermal Cycling in Fatigue at High Temperatures, in *Fatigue at High Temperatures*, R.P. Skelton, Ed., Applied Science Publishers, London, 1983, p 1-62

14. B. Tomkins, Fatigue: Introduction and Phe-

nomenology, in *Creep and Fatigue in High Temperature Alloys*, J. Bressers, Ed., Applied Science Publishers, London, 1981, p 73-110

15. R. Basquin, cited in Ref 11 (no independent reference found)
16. S.S. Manson, NASA TN2933, National Aeronautics and Space Administration, 1953
17. L.F. Coffin, Jr., *Trans. ASME*, Vol 76, 1954, p 931
18. S.S. Manson, *Exp. Mech.*, Vol 5 (No. 7), 1965, p 193
19. H. Terenishi and A.J. McEvily, in *Proceedings of International Conference on Low Cycle Fatigue Strength and Elastic-Plastic Behavior of Materials*, DVM, Stuttgart, 1979, p 25
20. D.A. Miller, R.H. Priest, and E.G. Ellison, A Review of Material Response and Life Prediction Techniques Under Fatigue-Creep Loading Conditions, *High Temp. Mater. Proc.*, Vol 6 (No. 3 and 4), 1984, p 115-194
21. C.E. Jaske and N.D. Frey, High Cycle Fatigue of 2¼Cr-1Mo Steel at Elevated Temperature, Paper C46/80, *Proceedings of the Fourth International Conference on Pressure Vessel Technology*, Institute of Mechanical Engineers, London, 1980
22. R.H. King and A. Smith, Thermal Fatigue Processes and Testing Techniques, in *IOM/ISI Conference on Thermal and High Strain Fatigue*, London, 1967, p 364-376
23. E.G. Ellison, A Review of the Interaction of Creep and Fatigue, *J. Mech. Engg. Sci.*, Vol 11, 1969, p 318-339
24. B. Tomkins and J. Wareing, Elevated Temperature Fatigue Interactions in Engineering Materials, *Met. Sci.*, Vol 11, 1977, p 416-424
25. E.M. Wundt and E. Krempl, *Hold Time Effects in High Temperature Low Cycle Fatigue — A Literature Survey and Interpretive Report*, STP 489, American Society for Testing and Materials, Philadelphia, 1971
26. L.F. Coffin, Fatigue at High Temperature — Prediction and Interpretation, *Proc. Inst. Mech. Eng.*, Vol 188, 1974, p 109-127
27. A.D. Batte, Creep-Fatigue Life Predictions, in *Fatigue at High Temperature*, R.P. Skelton, Ed., Applied Science Publishers, London, 1983
28. C.E. Jaske, Fatigue Curve Needs for High Strength, 2¼Cr-1Mo Steel for Petroleum Process Vessels, in *Fatigue Initiation, Propagation and Analysis for Code Construction*, MPC, Vol 29, American Society of Mechanical Engineers, New York, 1988, p 181-195
29. E.G. Ellison and A.J.F. Patterson, *Proc. Inst. Mech. Engg.*, Vol 190, 1976, p 321-350
30. W.J. Plumbridge, R.H. Priest, and E.G. Ellison, *Proceedings of the Third International Conference on the Mechanical Behavior of Materials*, ICM3, Vol 2, 1978, p 129
31. M.M. Leven, *Exp. Mech.*, Vol 353, Sept 1973
32. L.D. Kramer, D.D. Randolph, and D.A. Weisz, "Analysis of the Tennessee Valley Authority, Gallatin Unit 2 Turbine Rotor Burst," Winter Annual Meeting of ASME, New York, Dec 5-10, 1976
33. D.P. Timo, "Design Philosophy and Thermal Stress Considerations of Large Fossil Turbines," Report 84T16, General Electric Co., Schenectady, NY, 1984
34. G. Thomas and R.A.T. Dawson, The Effect of Dwell Period and Cycle Type on High Strain Fatigue Properties of 1CrMoV Rotor Forgings at 500-550 °C, in *Proceedings of the International Conference on Engineering Aspects of Creep*, Vol 1, Institute of Mechanical Engineers, Sheffield, Pub 1980-5, Mechanical Engineering Publishers, London, 1980, p 167-175
35. K. Kuwabara, A. Nitta, and T. Kitamura, "The Evaluation of Thermal Fatigue Strength of a 2¼Cr-1Mo Steel under Creep-Fatigue Conditions," Report E2797007, Central Research Institute for the Electric Power Industry (Japan), Tokyo, Apr 1980
36. C.R. Brinkman, J.P. Strizak, and M.K. Booker, AGARD Conf. Proc., No. 243, NATO, 1978
37. K.D. Challenger, A.K. Miller, and C.R. Brinkman, *Trans. ASME, J. Engg. Mater. Tech.*, Vol 103, 1981, p 7
38. H.G. Edmunds and D.J. White, *J. Mech. Engg. Sci.*, Vol 8 (No. 3), 1966, p 310-321
39. D.A. Miller and D. Gladwin, "Remaining Life of Boiler Pressure Parts Creep-Fatigue Effects," Report RP2253-1, Vol 5, Electric Power Research Institute, Palo Alto, CA, 1987
40. J. Wareing, *Met. Trans.*, Vol 8A, 1977, p 711-721
41. C.R. Brinkman, G.E. Korth, and R.R. Hobbins, *Nucl. Tech.*, Vol 16, 1972, p 299-307
42. Y. Asada and S. Mitsuhaski, in *Fourth International Conference on Pressure Vessel Technology*, Vol 1, 1980, p 321
43. C.R. Brinkman and G.E. Korth, *Met. Trans.*, Vol 5, 1974, p 792
44. J. Wareing, *Met. Trans.*, Vol 6A, 1975, p 1367
45. J. Wareing, H.G. Vaughan, and B. Tomkins, Report NDR-447S, United Kingdom Atomic Energy Agency, 1980
46. I.W. Goodall, R. Hales, and D.J. Walters, Proc. IUTAM 103, 1980
47. D.S. Wood, J. Wynn, A.B. Baldwin, and P. O'Riordan, *Fatigue Engg. Mater. Struct.*, Vol 3, 1980, p 89
48. J. Wareing, *Fatigue Engg. Mater. Struct.*, Vol 4, 1981, p 131
49. J.K. Lai and C.P. Horton, Report RD/L/R/200S, Central Electricity Generating Board, UK, 1979
50. C.E. Jaske, M. Mindlin, and J.S. Perrin, "Development of Elevated Temperature Fatigue Design Information for Type 316 Stainless Steel," International Conference on Creep and Fatigue, Conference Pub. #13, Institute of

Mechanical Engineers, London, 1973, p 163.1-163.7

51. S. Majumdar and P.S. Maiya, *Trans. ASME, J. Engg. Mater. Tech.*, Vol 102 (No. 1), 1980, p 159

52. V.B. Livesey and J. Wareing, *J. Met. Sci.*, Vol 17, 1983, p 297

53. D. Gladwin and D.A. Miller, *Fatigue Engg. Mater. Struct.*, Vol 5, 1982, p 275-286

54. D. Gladwin and D.A. Miller, unpublished work cited in Ref 20

55. W.J. Ostergren, *ASTM J. Test. Eval.*, Vol 4, 1976, p 327-339

56. J.M. Hyzak and H.L. Bernstein, AGARD Conf. Proc., No. 243, NATO, 1978

57. V.T.A. Antnunes and P. Hancock, AGARD Conf. Proc., No. 243, NATO, 1978

58. D.C. Lord and L.F. Coffin, *Met. Trans.*, Vol 4, 1973, p 1647-1653

59. C.S. Kortovich and A.H. Sheinker, AGARD Conf. Proc., No. 243, NATO, 1978

60. C.H. Wells and C.P. Sullivan, *ASM Trans. Qtrly.*, Vol 60, 1967, p 217-222

61. C.H. Wells and C.P. Sullivan, *Fatigue at High Temperatures*, STP 459, American Society for Testing and Materials, Philadelphia, 1968, p 59

62. D.C. Lord and L.F. Coffin, Jr., Low Cycle Fatigue Hold Time Behavior of Cast Rene 80, *Met. Trans.*, Vol 4, 1973, p 1647

63. R. Viswanathan, C.G. Beck, and R.L. Johnson, "Low Cycle Fatigue Behavior of Udimet 710 at Elevated Temperatures," Research Report 79-1D 4 STABL-R3, Westinghouse Research Laboratories, Pittsburgh, June 1979

64. G.A. Whitlow, R.L. Robinson, W.H. Pridemore, and J.M. Allen, Intermediate Temperature Low Cycle Fatigue Behavior of Coated and Uncoated Nickel Base Superalloys in Air and Corrosive Sulfate Environments, in *Advances in Life Prediction Methods*, D.A. Woodford and J.R. Whitehead, Ed., American Society of Mechanical Engineers, New York, 1983

65. M.Y. Nazmy, High Temperature Low Cycle Fatigue of In 738 and Application of Strain Range Partitioning, *Met. Trans. A*, Vol 14A, 1983, p 449-481

66. M.Y. Nazmy, The Applicability of Strain Range Partitioning to High Temperature Low Cycle Fatigue Life Prediction of IN 738 Alloy, *Fatigue Engg. Mater. Struct.*, Vol 4 (No. 3), 1981, p 253-256

67. L. Remy, F. Rezai Aria, R. Danzer, and W. Hoffelner, Comparison of Life Prediction Methods in Mar M509 Under High Temperature Fatigue, in *High Temperature Alloys for Gas Turbine and Other Applications*, Part II, W. Betz *et al*, Ed., D. Riedel Publishing Co., Boston, 1986, p 1617-1628

68. Y. Kadoya *et al*, Creep Fatigue Life Predic-

tion of Turbine Rotors, in *Life Assessment and Improvement of Turbogenerator Rotors for Fossil Plants*, R. Viswanathan, Ed., Pergamon Press, New York, 1985, p 3.101-3.114

69. H.D. Solomon and L.F. Coffin, *Fatigue at Elevated Temperatures*, STP 520, American Society for Testing and Materials, Philadelphia, 1973, p 112

70. L.F. Coffin, *Fatigue at Elevated Temperatures*, STP 520, American Society for Testing and Materials, Philadelphia, 1973, p 5

71. D.S. Wood *et al*, *Proceedings of the Conference on the Influence of Environment on Fatigue*, Institute of Mechanical Engineers, London, 1977, p 11-20

72. C. Levaillant, B. Rezqui, and A. Pineau, in *Proceedings of the Third International Conference on the Mechanical Behavior of Materials*, ICM3, No. 2, 1979, p 163-172

73. C.E. Jaske and R.C. Rice, in *ASME-MPC Symposium on Creep-Fatigue Interaction*, MPC-3, Metal Properties Council, New York, 1976, p 101

74. R.B. Scarlin, in *Proceedings of the Fourth International Conference on Fracture*, ICF4, Vol 2, 1977, p 849

75. D.L. Harrod and M.J. Manjoine, in *ASME-MPC Symposium on Creep-Fatigue Interaction*, MPC-3, Metal Properties Council, New York, 1976, p 87

76. M. Gell and G.R. Leverant, in *Fatigue at Elevated Temperatures*, STP 520, American Society for Testing and Materials, Philadelphia, 1973, p 37

77. J.R. Haigh, R.P. Skelton, and C.E. Richards, *J. Mater. Sci. Engg.*, Vol 26, 1976, p 167

78. ASME Boiler and Pressure Vessel Code, Section III, Code Case N-47, 1974

79. E.L. Robinson, Effect of Temperature Variation on the Creep Stength of Steel, *Trans. ASME*, Vol 160, 1938, p 253-259

80. M.A. Miner, *J. Appl. Mech.*, Vol 12 (No. 3), 1945, p A159-A167

81. S. Taira, *Creep in Structures*, Academic Press, 1962, p 96-124

82. R.M. Curran and B.M. Wundt, "Interpretive Report on Notched and Unnotched Creep-Fatigue Interspersion Tests in CrMo, 2¼Cr-1Mo and Type 304 Stainless Steel," Publication MPC-8, American Society of Mechanical Engineers, p 218-314

83. A.S. Warnock *et al*, Jersey Central Power and Light Co., Werner Station, Unit #4 HP-RH Rotor Life Assessment, in *Life Assessment and Extension of Fossil Power Plants*, R.B. Dooley and R. Viswanathan, Ed., Report CS 5208, Electric Power Research Institute, Palo Alto, CA, 1987, p 693-724

84. L.F. Coffin, Prediction Parameters and Their Application to High Temperature Low Cycle Fatigue, in *Proceedings of Second Interna-*

tional Conference on Fracture, Brighton, London, Chapmans Hall, 1969, p 643-654

85. L.F. Coffin, Fatigue at High Temperature — Prediction and Interpretation, *Proc. Inst. Mech. Eng.* (London), Vol 188, 1974, p 109-127

86. L.F. Coffin, The Prediction of Wave-Shape Effects in Time Dependent Fatigue, in *Proceedings of the Second International Conference on Mechanical Behavior of Materials*, Federation of Material Sciences, Boston, 1976, p 866-870

87. L.F. Coffin, The Concepts of Frequency Separation in Life Prediction for Time Dependent Fatigue, in *ASME-MPC Symposium on Creep-Fatigue Interaction*, MPC-3, Metal Properties Council, New York, 1976, p 349-363

88. L.F. Coffin, Instability Effect in Thermal Fatigue, in *Thermal Fatigue of Materials and Components*, STP 612, American Society for Testing and Materials, Philadelphia, 1976, p 227-230

89. R.V. Priest and E.G. Ellison, *Res. Mech.*, Vol 4, 1982, p 127-150

90. S.S. Manson, in *Fatigue at Elevated Temperatures*, STP 520, American Society for Testing and Materials, Philadelphia, 1973, p 744-782

91. S.S. Manson, G.R. Halford, and A.C. Nachtigall, Advances in Design for Elevated Temperature Environments, in *ASME National Congress on Pressure Vessels and Piping*, 1975, p 17

92. S.S. Manson, G.R. Halford, and M.H. Hirschberg, NASA Report TMX-67838, National Aeronautics and Space Administration, 1971

93. G.R. Halford, M.H. Hirschberg, and S.S. Manson, in *Fatigue at Elevated Temperatures*, STP 520, American Society for Testing and Materials, Philadelphia, 1973, p 659

94. G.R. Halford, J.F. Saltsman, and M.H. Hirschberg, NASA Technical Note 73737, National Aeronautics and Space Administration, 1977

95. G.R. Halford and S.S. Manson, in *Thermal Fatigue of Materials and Components*, STP 612, American Society for Testing and Materials, Philadelphia, 1976, p 239

96. S.S. Manson and G.R. Halford, Treatment of Multiaxial Creep Fatigue by Strain Range Partitioning, in *ASME-MPC Symposium on Creep-Fatigue Interaction*, MPC-3, Metal Properties Council, New York, 1976, p 299

97. J.F. Saltsman and G.R. Halford, Application of Strain Range Partitioning to the Prediction of MPC Creep-Fatigue Data for 2¼Cr-1Mo Steel, in *ASME-MPC Symposium on Creep-Fatigue Interaction*, MPC-3, Metal Properties Council, New York, 1976, p 283

98. W.J. Ostergren, A Damage Function and Associated Failure Equations for Predicting Hold Time and Frequency Effects in Elevated Temperature Low Cycle Fatigue, *ASTM J. Test. Eval.*, Vol 4 (No. 5), 1976, p 327-339

99. W.J. Ostergren, Correlation of Hold Time Effects in Elevated Temperature Fatigue Using a Frequency Modified Damage Function, in *ASME-MPC Symposium on Creep-Fatigue Interaction*, MPC-3, Metal Properties Council, New York, 1976, p 179

100. V. Bisego, C. Fossati, and S. Ragazzoni, An Energy Based Criterion for Low Cycle Fatigue Damage Evaluation, in *Material Behavior at Elevated Temperatures and Component Analysis*, Y. Yamada, R.L. Roche, and F.L. Cho, Ed., Book No. H 00217, PVP Vol 60, American Society of Mechanical Engineers, New York, 1982

101. G. Bartoloni and G. Ragazzoni, Low Cycle Fatigue of 1CrMoV Steam Turbine Rotor Steel, in *Creep and Fracture of Engineering Materials and Structures*, Proceedings of Second International Conference, Pineridge Press, London, p 1029-1042

102. R.H. Priest, D.J. Beauchamp, and E.G. Ellison, Damage During Creep-Fatigue, in *Advances in Life Prediction Methods* (ASME conference), Albany, American Society of Mechanical Engineers, 1983, p 115-122

103. R. Hales, in *International Conference on Creep and Fracture of Engineering Materials*, Swansea, England, 1984

104. S.M. Beech and A.D. Batte, in *International Conference on Creep and Fracture of Engineering Materials*, Swansea, England, 1984, p 1043-1054

105. R.H. Priest and E.G. Ellison, *Mater. Sci. Engg.*, Vol 49, 1981, p 7

106. D.A. Miller, Fatigue-Creep Failure Mechanism Maps for 20%Cr/25%Ni/Nb Stainless Steel at 593 and 640 °C, in *Advances in Life Prediction Methods* (ASME conference), Albany, American Society of Mechanical Engineers, 1983, p 157-163

107. R. Hales, *Fatigue Engg. Mater. Struct.*, Vol 6, 1983, p 121

108. G. Schlottner and R.E. Seeley, Estimation of Remaining Life of High Temperature Steam Turbine Components, in *Residual Life Assessment, Non Destructive Examination and Nuclear Heat Exchanger Materials*, C.E. Jaske, Ed., PVP Vol 98-1, American Society of Mechanical Engineers, 1985, p 35-45

109. K. Yamaguchi, S. Nishijima, and K. Kanazawa, Prediction and Evaluation of Long Term Creep Life, in *International Conference on Creep*, Tokyo, Apr 1986, p 47-52

110. S. Majumdar and P.S. Maiya, *Can. Met. Qtrly.*, Vol 18, 1979, p 57-64

111. S. Majumdar and P.S. Maiya, A Damage Equation for Creep-Fatigue Interaction, in *ASME-MPC Symposium on Creep-Fatigue*

Interaction, MPC-3, Metal Properties Council, New York, 1976, p 323

112. S. Majumdar and P.S. Maiya, in *Proceedings of the Second International Conference on Mechanical Behavior of Materials*, ICM2, 1976, p 924-928

113. S. Majumdar and P.S. Maiya, ASME-CSME Pressure Vessel and Piping Conference, PVP–PB 028, 1978

114. S. Majumdar and P.S. Maiya, in *Proceedings of the Third International Conference on Mechanical Behavior of Materials*, ICM3, Vol 12, 1979, p 101-109

115. K. Kuwabara and A. Nitta, "Estimation of Thermal Fatigue Damage in Steam Turbine Rotors," Report 277001, Central Research Institute for the Electric Power Industry (Japan), July 1977

116. K.N. Melton, Strain Wave Shape and Frequency Effects on the High Temperature Low Cycle Fatigue Behavior of 1CrMoV Ferritic Steel, *Mater. Sci. Engg.*, 1982, p 21-28

117. J.R. Ellis, M. Jakub, C.E. Jaske, and D.A. Utah, Elevated Temperature Fatigue and Creep Fatigue Properties of Annealed 2¼Cr-1Mo Steel, in *Structural Steels for Service at Elevated Temperature in Nuclear Power Generation*, A.O. Schaeffer, Ed., American Society of Mechanical Engineers, New York, 1975, p 213-246

118. J.F. Saltsman and G.R. Halford, NASA Technical Note TMX 73474, National Aeronautics and Space Administration, 1976

119. J.F. Saltsman and G.R. Halford, NASA Technical Note TMX 71898, National Aeronautics and Space Administration, 1976

120. M.Y. Nazmy and C. Wuthrich, The Predictive Capability of Three High Temperature Low Cycle Fatigue Models in the Alloy IN 738, in *Mechanical Behavior of Materials IV*, J. Carlsson and N.G. Ohlson, Ed., Pergamon Press, New York, 1984

121. M. Marchionni, D. Ranucci, and E. Picco, High Temperature Fatigue Life Prediction of a Nickel Base Superalloy by the Strain Range Partitioning Method, in *High Temperature Alloys for Gas Turbines and Other Applications*, proceedings of conference at Liege, Belgium, Oct 6-9, 1986, W. Betz *et al*, Ed., D. Riedel Publishing Co., Boston, 1986, p 1629-1638

122. K. Kuwabara and A. Nitta, in *Proceedings of the Third International Conference on Mechanical Behavior of Materials*, ICM3, Vol 2, 1978

123. M.F. Day and G.B. Thomas, AGARD Conf. Proc., NATO, 1978, p 243

124. P.O. Persson, C. Persson, G. Burman, and Y. Lindblom, The Behavior of Nimonic 105 and Inco 738 LC Under Creep and LCF Testing, in *High Temperature Alloys for Gas Turbines and Other Applications*, proceedings of conference at Liege, Belgium, Oct 6-9, 1986, W. Betz *et al*, Ed., D. Riedel Publishing Co., Boston, 1986, p 1501-1516

125. H. Fischmeister, R. Danzer, and BuchMayr, Life Time Prediction Models, in *High Temperature Alloys for Gas Turbines and Other Applications*, proceedings of conference at Liege, Belgium, Oct 6-9, 1986, W. Betz *et al*, Ed., D. Riedel Publishing Co., Boston, 1986, p 495-550

126. T. Sato, K. Takeishi, and T. Sakon, Thermal Fatigue Life Prediction of Air Cooled Gas Turbine Vanes, in *ASME Joint Power Generation Conference 85 GT-17*, American Society of Mechanical Engineers, New York, 1985

127. F. Masuyama, K. Setoguchi, H. Haneda, and F. Nanjo, Findings on Creep-Fatigue Damage in Pressure Parts of Long Term Service Exposed Thermal Power Plants, *Trans. ASME, J. Pressure Vessel Tech.*, Vol 107, Aug 1985, p 260-270

128. D.S. Wood, Materials Data for Severe Variable Loading, in *Recent Developments in High Temperature Design Methods* (conference proceedings), Institute of Mechanical Engineers, London, 1977, p 65-70

129. L.K. Severud, Application of American Design Codes for Elevated Temperature Environments, Paper C 327/80, in *Engineering Aspects of Creep*, Institute of Mechanical Engineers, Sheffield, 1980

130. S. Fawcet, Development of Fast Reactor Codes in the U.K., Paper 218/80, in *Engineering Aspects of Creep*, Institute of Mechanical Engineers, Sheffield, 1980

131. H. Neuber, Theory of Stress Concentration for Shear Strained Prismatical Bodies with Arbitrary Non-Linear Stress-Strain Law, *Trans. ASME*, Series E, Vol 28, 1961, p 544-550

132. D.C. Gonyea, Method for Low Cycle Fatigue Design Including Biaxial Stress and Notch Effects, in *Fatigue at Elevated Temperatures*, STP 520, American Society for Testing and Materials, Philadelphia, 1973, p 678-687

133. E.E. Zwicky, Jr., ASME Paper 67 WA/PVP-6, ASME, New York

134. D.A. Miller, A Creep-Fatigue Failure Mechanism Map for 1Cr-Mo-V at 565°C, *Mater. Sci. Engg.*, Vol 54, 1982, p 273-278

135. D.A. Spera, What Is Thermal Fatigue?, in *Thermal Fatigue of Materials and Components*, D.A. Spera and D. Mowbray, Ed., STP 612, American Society for Testing and Materials, Philadelphia, 1976, p 3-9

136. E. Glenny *et al*, A Technique for Thermal

Shock and Thermal Fatigue Testing Based on the Use of Fluidized Beds, *J. Inst. Metals*, Vol 57, 1958, p 294-302

137. M.A.H. Howes, in *Fatigue at Elevated Temperatures*, STP 520, American Society for Testing and Materials, Philadelphia, 1973, p 242-254

138. P.W.H. Howe, in *Thermal and High Strain Fatigue*, The Metals and Metallurgy Trust, 1967, p 122-141

139. D.A. Spera, "The Calculation of Thermal Fatigue Life Based on Accumulated Creep Damage," NASA TMX 52558, National Aeronautics and Space Administration, 1969

140. D.F. Mowbray and D.A. Woodford, Observation and Interpretation of Crack Propagation Under Conditions of Thermal Strain, in *International Conference on Creep and Fatigue*, Institute of Mechanical Engineers, London, 1973, p 179.1-179.11

141. D.F. Mowbray and J.E. Meconelee, Non-Linear Analysis of a Tapered Disc Thermal Fatigue Specimen, in *Thermal Fatigue of Materials and Components*, STP 612, American Society for Testing and Materials, Philadelphia, 1976, p 10-29

142. D.A. Spera, Comparison of Experimental and Theoretical Thermal Fatigue Lives for Five Nickel Base Alloys, in *Fatigue at Elevated Temperatures*, STP 520, American Society for Testing and Materials, Philadelphia, 1973, p 648-656

143. D.A. Spera and E.C. Cox, Description of a Computerized Method for Predicting Thermal Fatigue Life of Metals, in *Thermal Fatigue of Materials and Components*, STP 612, American Society for Testing and Materials, Philadelphia, 1976, p 69-85

144. P.T. Bizon and D.A. Spera, Thermal Stress Behavior of Twenty Six Superalloys, in *Thermal Fatigue of Materials and Components*, D.A. Spera and D.F. Mowbray, Ed., STP 612, American Society for Testing and Materials, Philadelphia, 1976, p 106-122

145. C.G. Beck and A.T. Santhanam, Effect of Microstructure on the Thermal Fatigue Resistance of Cast Cobalt Base Alloy, Mar M 509, in *Thermal Fatigue of Materials and Components*, D.A. Spera and D.F. Mowbray, Ed., STP 612, American Society for Testing and Materials, Philadelphia, 1976, p 123-140

146. E.S. Russell, Practical Life Prediction Methods for Thermo-Mechanical Fatigue of Gas Turbine Buckets, in *Life Prediction for High Temperature Gas Turbine Materials* (conference proceedings), V. Weiss and W.T. Bakker, Ed., Report AP 4477, Electric Power Research Institute, Palo Alto, CA, 1985, p 3.1-3.17

147. G.B. Thomas, J. Bressers, and D. Raynor, Low Cycle Fatigue and Life Prediction Methods, citation of unpublished data of A. Samuelsson, L.E. Larsson, and L. Lundberg, in *High Temperature Alloys for Gas Turbines*, R. Brunetaud, Ed., D. Riedel Publishing Co., Dardrecht, Netherlands, 1982, p 291-317

148. K. Kuwabara, A. Nitta, and T. Kitamura, Thermal Mechanical Fatigue Life Prediction in High Temperature Component Materials for Power Plants, in *Advances in Life Prediction*, D.A. Woodford and R. Whitehead, Ed., American Society of Mechanical Engineers, New York, 1985, p 131-141

149. G.T. Embley and E.S Russell, Thermal Mechanical Fatigue of Gas Turbine Bucket Alloys, in *First Parsons International Turbine Conference*, Dublin, Ireland, June 1984, p 157-164

150. H.L. Bernstein, "An Evaluation of Four Current Models to Predict the Creep Fatigue Interaction in Rene 95," Report AFML-TR 79.4075, U.S. Air Force Materials Laboratory, June 1979

151. National Materials Advisory Board, "Analysis of Life Prediction Methods for Time Dependent Fatigue Crack Initiation in Nickel Base Superalloys," NMAB 347, National Research Council, Washington, 1980

152. G.J. Lloyd and J. Wareing, Life Prediction Methods for Combined Creep Fatigue Endurance, *Met. Tech.*, Aug 1981, p 297-305

153. J.M. Hyzak and D.A. Hughes, "An Evaluation of Creep Fatigue Life Prediction Models for Solar Central Receiver," Report SAND 81-8220, Sandia Laboratories, Albuquerque, Sept 1981

154. R.H. Priest and E.G. Ellison, An Assessment of Life Analysis Techniques for Fatigue Creep Situations, *Res. Mechanica*, Vol 4, 1982, p 127-150

155. G.R. Halford and S.S. Manson, Life Prediction of Thermal-Mechanical Fatigue Using Strain Range Partitioning, in *Thermal Fatigue of Materials and Components*, D.A. Spera and D.F. Mowbray, Ed., STP 612, American Society for Testing and Materials, Philadelphia, 1976, p 239-254

156. A. Nitta, K. Kuwabara, and T. Kitamura, The Characteristics of Thermal Mechanical Fatigue Strength in Superalloys for Gas Turbines, in *International Gas Turbine Congress*, 83-Tokyo-IGTC-99, Tokyo, 1983, p 765-772

157. M.F. Day and G.B. Thomas, Creep-Fatigue Interactions in Alloy In 738LC, in *Characterization of Low Cycle High Temperature Fatigue by the Strain Range Partitioning Method*, AGARD Conf. Proc., No. 243, NATO, Aug 1978

158. J.M. Hyzak and H.L. Bernstein, An Analysis of the Low Cycle Fatigue Behavior of the Superalloy Rene 95 by the Strain Range Partitioning Method, in *Characterization of Low Cycle High Temperature Fatigue by the Strain Range Partitioning Method*, AGARD Conf. Proc., No. 243, NATO, Aug 1978

159. R.C. Bill, M.J. Verrilli, M.A. McGraw, and G.R. Halford, "Preliminary Study of Thermo Mechanical Fatigue of Polycrystalline Mar M 200," NASA Technical Paper 2280, National Aeronautics and Space Administration, 1984

160. C.E. Jaske, Thermal Mechanical Low Cycle Fatigue of AISI 1010 Steel, in *Thermal Fatigue of Materials and Components*, D.A. Spera and D.F. Mowbray, Ed., STP 612, American Society for Testing and Materials, Philadelphia, 1976, p 170-198

161. K.N. Smith, P. Watson, and T.H. Topper, *J. Materials*, Vol 5 (No. 4), Dec 1970, p 767-778

162. H. Sehitoglu and J. Morrow, Characterization of Thermomechanical Fatigue, in *Thermal and Environmental Effects in Fatigue: Research-Design Interface*, C.E. Jaske et al, Ed., American Society of Mechanical Engineers, 1983

163. T.C. Lindley, C.E. Richards, and R.O. Ritchie, in *Proceedings of Conference on Mechanics and Physics of Fractures*, Cambridge, 1975

164. P. Paris and F. Erdogen, A Critical Analysis of Crack Propagation Laws, *Trans. ASME*, Dec 1963, p 528-534

165. E.T. Wessel, W.G. Clark, Jr., and W.K. Wilson, "Engineering Methods for the Design and Selection of Materials Against Fracture," Report DDC-AD801001, Westinghouse Research Laboratories, Pittsburgh, 1966

166. N.E. Dowling, Geometry Effects and the J Integral Approach to Elastic Plastic Fatigue Crack Growth, in *Cracks and Fracture*, STP 601, American Society for Testing and Materials, Philadelphia, 1976, p 19-32

167. N.E. Dowling and J.A. Begley, Fatigue Crack Growth During Gross Plasticity and the J Integral, in *The Mechanics of Crack Growth*, STP 590, American Society for Testing and Materials, Philadelphia, 1976, p 82-103

168. N.E. Dowling, Crack Growth During Low Cycle Fatigue of Smooth Axial Specimens, in *Cyclic Stress-Strain and Plastic Deformation Aspects of Fatigue Crack Growth*, STP 637, American Society for Testing and Materials, Philadelphia, 1977, p 97-121

169. J.S. Huang and R.M. Pelloux, Low Cycle Fatigue Crack Propagation in Hastelloy X at 25 and 760 C, *Met. Trans. A*, Vol 11, 1980, p 899-904

170. M.J. Douglass and A. Plumtree, Accumulation of Fracture Mechanics to Damage Accumulation in High Temperature Fatigue, in *Fracture Mechanics*, STP 677, American Society for Testing and Materials, Philadelphia, 1979, p 68-84

171. S. Taira, R. Ohtani, and T. Komatsu, Application of J Integral to High Temperature Crack Propagation, *Trans. ASME, J. Engg. Mater. Tech.*, Series H, Vol 101, 1979, p 162-167

172. C.E. Jaske, Creep-Fatigue Crack Growth in Type 316 Stainless Steel, in *Advances in Life Prediction Methods*, international conference held in Albany, NY, Apr 1983, D.A. Woodford and J.R. Whitehead, Ed., American Society of Mechanical Engineers, New York, 1983, p 93-103

173. Y. Asada et al, Fatigue Crack Propagation Under Elastic-Plastic Medium and Elevated Temperature, in *Fourth International Conference on Pressure Vessel Technology*, Vol 1, Institute of Mechanical Engineers, London, 1981, p 347-352

174. R. Ohtani, Crack Propagation Under Creep Fatigue Interaction, in *Engineering Aspects of Creep*, Vol 2, Institute of Mechanical Engineers, London, 1980, p 17-22

175. K. Sadananda and P. Shahinian, Application of J Integral Parameter to High Temperature Fatigue Crack Growth in Cold Worked Type 316 Stainless Steel, *Int. J. Fracture*, Vol 15, 1979, p 81-84

176. K. Sadananda and O. Shahinian, Hold Time Effects on High Temperature Fatigue Crack Growth in Udimet 700, *J. Materials*, Vol 13, 1978, p 2347-2357

177. J. Wareing, Mechanisms of High Temperature Fatigue and Creep Fatigue Failure in Engineering Materials, in *Fatigue at High Temperature*, R.P. Skelton, Ed., Applied Science Publishers, London, 1983, p 135-186

178. W.A. Logsdon, P.K. Liaw, A. Saxena, and V.E. Hulina, Residual Life Prediction and Retirement for Cause Criteria for Upper Casings, *Engg. Fract. Mech.*, Vol 25, 1986, p 259-288

179. R.C. Schwant and D.P. Timo, Life Assessment of General Electric Large Steam Turbine Rotors, in *Life Assessment and Improvement of Turbo Generator Rotors for Fossil Plants*, R. Viswanathan, Ed., Pergamon Press, New York, 1985, p 3.25-3.40

180. G.J. Lloyd and J.D. Walls, "Section Thickness Effects and the Temperature Dependence of Fatigue Crack Growth in 316 Stainless Steel," Report R 335 (R), United Kingdom Atomic Energy Agency, 1979

181. D.A. Jablonski, J.V. Carisella, and R.M. Pelloux, Fatigue Crack Propagation at Elevated Temperature in Solid Solution Strengthened Superalloys, *Met. Trans. A*, Vol 8, 1977, p 893-900

182. T. Ohmura, R.M. Pelloux, and N.J. Grant,

High Temperature Fatigue Crack Growth in a Cobalt Base Superalloy, *Engg. Fract. Mech.*, Vol 5, 1973, p 909-922

183. P. Shahinian, Fatigue Crack Growth Characteristics of High Temperature Alloys, *Met. Tech.*, Vol 5, 1978, p 372-380

184. G.J. Lloyd, High Temperature Fatigue and Creep Fatigue Crack Propagation: Mechanics, Mechanisms and Observed Behaviour of Structural Materials, in *Fatigue at High Temperature*, R.P. Skelton, Ed., Applied Science Publishers, London, 1983, p 187-258

185. P. Shahinian, *Met. Sci.*, Vol 5, 1978, p 372

186. J. Bressers, L. Remy, and W. Hoffelner, Fatigue Dominated Damage Processes, in *High Temperature Alloys for Gas Turbines and Other Applications*, proceedings of conference held in Liege, Belgium, Oct 6-9, 1986, W. Betz *et al*, Ed., D. Riedel Publishing Co., Boston, 1986, p 441-468

187. L.A. James, Hold Time Effects on the Elevated Temperature Fatigue Crack Propagation of Type 304 Stainless Steels, *Nucl. Tech.*, Vol 16, 1972, p 521-530

188. L.A. James, Frequency Effects on the Elevated Temperature Crack Growth Behavior of Austenitic Stainless Steel — A Design Approach, *J. Pressure Vessel Tech.*, Vol 101, 1979, p 171-176

189. L.A. James, The Effect of Stress Ratio on the Elevated Temperature Fatigue Crack Propagation of Type 304 Stainless Steels, *Nucl. Tech.*, Vol 14, 1972, p 163-170

190. *Nuclear Systems Materials Handbook*, Vol 1 and 2, Hanford Engineering Development Laboratory (HEDL), TID 2666, Hanford, WA

191. R.D. Scarlin, in *Proceedings of the Fourth International Conference on Fracture*, ICF4, Vol 2, 1977, p 849

192. W.H. Bamford, The Effect of Pressurised Water Environment on Fatigue Crack Propagation of Pressure Vessel Steels, in *The Influence of Environment on Fatigue*, Institute of Mechanical Engineers, London, 1977

193. A.E. Carden, Parametric Analysis of Fatigue Crack Growth, Paper C 324/73, in *International Conference on Creep and Fatigue in Elevated Temperature Applications*, American Society of Mechanical Engineers, 1973

194. J. Lankford, Relevance of the Small Crack Problems to Life Time Prediction in Gas Turbines, in *Proceedings of Conference on Life Prediction for High Temperature Gas Turbine Materials*, V. Weiss and W.T. Bakker, Ed., EPRI AP 4477, Electric Power Research Institute, Palo Alto, CA, Apr 1986

195. P.K. Liaw, Mechanisms of Near Threshold Fatigue Crack Growth in a Low Alloy Steel, *Acta Met.*, Vol 33 (No. 8), 1985, p 1489-1502

196. P.K. Liaw, A. Saxena, V.P. Swaminathan, and T.T. Shih, Effects of Load Ratio and Temperature on the Near Threshold Fatigue Crack Propagation Behavior in a CrMoV Steel, *Met. Trans. A*, Vol 14A, 1983, p 1631-1640

197. R.O. Ritchie, S. Suresh, and C.M. Moss, Near Threshold Fatigue Crack Growth in 2¼ Cr-1Mo Pressure Vessel Steel in Air and Hydrogen, *Trans. ASME, J. Engg. Mater. Tech.*, Vol 102, 1980, p 293

198. S. Suresh, G.F. Zamiski, and R.O. Ritchie, Oxide Induced Crack Closure, A Mechanism for Near Threshold Corrosion Fatigue Crack Growth, *Met. Trans.*, 1982

199. R. Viswanathan, J.D. Landes, D.E. McCabe, and S.J. Hudak, The Effect of Hydrogenous Environments on the Fracture Toughness of Power Plant Steels, in *Advances in Fracture Research*, Cannes, France, Vol 4, Theme 7, Pergamon Press, 1981, p 1945

200. P. Marshall, The Influence of Environment on Fatigue, in *Fatigue at High Temperature*, P. Skelton, Ed., Applied Science Publishers, London, 1983, p 259-304

201. D.A. Woodford and D.F. Mowbray, Effect of Material Characteristics and Test Variables on Thermal Fatigue of Cast Superalloys, *Mater. Sci. Engg.*, Vol 16, 1974, p 5-43

202. R.P. Skelton, Environmental Crack Growth in 0.5CrMoV Steel During Isothermal High Strain Fatigue and Temperature Cycling, *Mater. Sci. Engg.*, Vol 35, 1978, p 287-298

203. C.A. Rau, A.E. Gemma, and G.R. Leverant, Thermal Mechanical Fatigue Crack Propagation in Nickel and Cobalt Base Superalloys Under Various Strain, Temperature Cycles, in *Fatigue at Elevated Temperature*, STP 520, 1973, p 166-178

204. G.R. Leverant, T.C. Strangman, and B.S. Langer, Parameters Controlling the Thermal Fatigue Properties of Conventionally Cast and Directionally Solidified Turbine Alloys, in *Superalloys: Metallurgy and Manufacture*, Proceedings of the Third International Symposium, B. Kear *et al*, Ed., 1976, p 285-295

205. T.E. Strangman and S.W. Hopkins, Thermal Fatigue of Coated Superalloys, *Ceram. Bull.*, Vol 55 (No. 3), Mar 1976, p 304-307

206. A.E. Gemma, F.X. Ashland, and R.M. Masci, The Effects of Stress Dwells and Varying Mean Strain on Crack Growth During Thermal Mechanical Fatigue, *ASTM J. Test. Eval.*, Vol 9 (No. 4), July 1981, p 209-215

207. N. Marchand and R.M. Pelloux, "Thermal Mechanical Fatigue Crack Growth in Inconel X-750," NASA Contractor Report 174740, Oct 1984

208. M. Okazaki and T. Koizumi, Effect of Strain Wave Shape on Thermal Mechanical Fatigue

Crack Propagation in Cast Low Alloy Steel, *Trans. ASME, J. Engg. Mater. Tech.*, Vol 105 (No. 2), Apr 1983, p 81-87

209. L.A. James and D.P. Jones, Fatigue Crack Growth Correlations for Austenitic Stainless Steels in Air, in *Predictive Capabilities in Environmentally Assisted Cracking*, PVP Vol 99, American Society of Mechanical Engineers, 1985, p 363-414

210. A. Saxena, R.S. Williams, and T.T. Shih, A Model for Representing and Predicting the Influence of Hold Times on Fatigue Crack Growth Behavior at Elevated Temperature, in *Fracture Mechanics: Thirteenth Conference*, STP 743, American Society for Testing and Materials, Philadelphia, 1981, p 86-99

211. C.H. Wells *et al*, unpublished work, Failure Analysis Associates, Palo Alto, CA, 1987

212. F. Ammirato *et al*, "Life Assessment Methodology for Turbogenerator Rotors," Vol 1: "Improvements to the SAFER Code Rotor Life Time Prediction Software," EPRI CS/EL 5593, Electric Power Research Institute, Palo Alto, CA, Mar 1988

213. A. Saxena and J.L. Bassani, Time Dependent Fatigue Crack Growth Behavior at Elevated Temperature, in *Fracture: Interaction of Microstructure, Mechanisms and Mechanics*, AIME/ASME conference, Los Angeles, 1984, p 357-383

214. V.P. Swaminathan, T.T. Shih, and A. Saxena, Low Frequency Fatigue Crack Growth Behavior of A470 Class 8 Rotor Steel at 538°C (1000°F), *Engg. Fract. Mech.*, Vol 16 (No. 6), 1982, p 827-836

215. C.E. Jaske, A Crack-Tip-Zone Interaction Model for Creep-Fatigue Crack Growth, *Fatigue Engg. Mater. Struct.*, Vol 6 (No. 2), 1983, p 159-166

216. J. Willenborg, R.M. Engle, and H.A. Wood, "Crack Growth Retardation Model Using an Effective Stress Concept," Report # AFFDL-TM-71-1-FBR, Wright-Patterson Air Force Base, Ohio, 1971

217. O.E. Wheeler, "Spectrum Loading and Crack Growth," ASME Paper # 71-Met-X, American Society of Mechanical Engineers, New York

5

Life Prediction for Boiler Components

General Description

The function of a boiler is to convert water into superheated steam, which is then delivered to a steam turbine. A schematic illustration of a boiler is shown in Fig. 5.1. Coal, oil, or natural gas with preheated air is burned in the furnace. The combustion gases flow up through the furnace and evaporate the water into steam inside the furnace waterwall tubes. At the roof of the furnace, the gas flow is made horizontal across banks of secondary superheater and reheater tubes. The gases are then turned downward, where they encounter the primary superheater and the economizer. Before exiting through the stack, the combustion gases are fed through an air preheater and then through various cleaning devices.

The gas flow past the fire side of the boiler tubes originates in the waterwall section and ends with the economizer. The flow of the working medium (water or steam) within the water side of the boiler tubes involves a cyclic regenerative Rankine cycle. Low-pressure steam from the turbine exit is processed into feedwater via the condenser, feedwater heaters, and deaerators. The feedwater is then pumped into the

economizer, where it is heated and sent to the furnace waterwall tubes. From here, the high-pressure steam goes through the superheater into the turbine inlet, completing the cycle. For the superheater, reheater, and economizer sections, the working fluid in the tubes is discharged at the inlet and outlet ends into large-diameter piping to ensure mixing. These pipes are known as headers.

Boiler designs vary depending on the flow conditions, operating temperatures and pressures, type of duty, and other requirements. In the once-through boiler, feedwater is completely converted to steam in each boiler tube, en route to the turbine. In the drum boiler, only partial conversion of water to steam occurs in the tubes. The steam/water mixture is discharged by the waterwall tubes into a drum, where saturated steam is separated before being superheated and taken to the turbine, while the water is recirculated through the waterwall tubes for steam generation.

The once-through design is less tolerant of feedwater contaminants because many of the nonvolatile solutes are deposited inside the boiler tubes, contributing to corrosion and overheating. Contamination is therefore minimized by condensate polishing and maintenance of tight condensers. The drum-

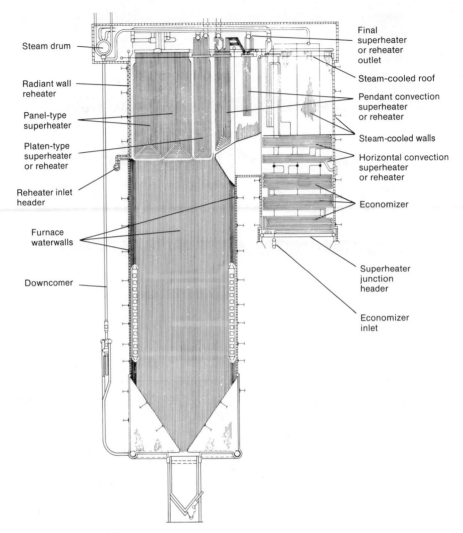

Fig. 5.1. Typical cross section of large drum-type utility boiler, showing major water-
and steam-cooled tube circuits.

type boiler can handle contaminants more readily because nonvolatile solutes remain in the water for periodic removal by blowdown.

Boilers also may be classified by their operating pressures relative to the critical steam pressure of 22.12 MPa (3208 psi). Thus, subcritical boilers employ a nominal 16.6-MPa (2400-psi) steam cycle, whereas most supercritical boilers employ a 24-MPa (3500-psi) cycle. Subcritical boilers use drums for steam separation, whereas supercritical operation dictates the use of the once-through type of design. In addition to the above, boiler designs also vary depend-

ing on the fuel used (coal, oil, or gas) and on the cyclic duty employed.

The major problem in all boilers with respect to availability is the failure of boiler tubes. From a life-extension and safety point of view, the critical components are the large-diameter, thick-wall piping known as headers. In addition, pipes that carry superheated steam to the turbines, known as main steam pipes and hot reheat pipes, are also subject to high-temperature problems. This chapter will describe the high-temperature material problems and damage-assessment techniques employed for boiler tubes, headers, and steam pipes. Because the distinction among these components is essentially one of size and not one of function, and because the steels employed are similar in composition, much of the discussion will be on a common basis.

Materials and Damage Mechanisms

The material grades that are used widely in fossil boiler construction are listed in Table 5.1. Material selection can vary within the boiler, depending on duty requirements, economics, and the availability of a component in the sizes required. Boiler tubes are seamless, extruded tubes and can vary all the way from carbon steel in the low-temperature waterwall sections to austenitic stainless steels in the finishing stages of the superheater. Most of the header pipes are made of $1\frac{1}{4}$Cr-$\frac{1}{2}$Mo or $2\frac{1}{4}$Cr-1Mo steels. For steam pipes, austenitic stainless steels occasionally have also been employed for supercritical conditions. Pipes generally are extruded, although for large-diameter piping, plate-formed and seam-welded piping have been employed in some designs. Forgings are used as reinforcing rings around nozzles and openings, fittings, valves, and flanges. Castings are used only occasionally as valve or pump bodies, where the ease of fabrication may be important. Because extensive welded tube-to-tube and tube-to-pipe joints are present in the boiler, the weldability of the materials involved is a major consideration in their selection.

Although tubes and pipes have some unique problems of their own, two fundamental and common factors that dictate material selection are creep strength and resistance to oxidation at high temperatures. The ASME Boiler and Pressure Vessel Code, Paragraph A-150 of Section I, clearly spells out the criteria for determining allowable stresses. These criteria were described in Chapter 3. A comparison of the allowable stresses at various temperatures for commonly used steels is shown in Fig. 5.2. Although the ASME code pro-

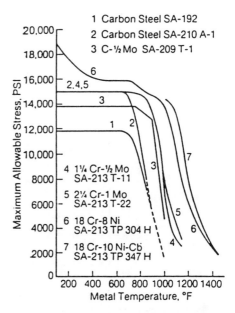

Fig. 5.2. Effect of temperature on ASME Boiler and Pressure Vessel Code allowable stress for several grades of steel tubing.

vides allowable stress values and allowable temperature values, boiler manufacturers have adopted maximum allowable temperature values that are lower than the highest ASME temperature values, to provide for oxidation resistance. These values, shown in Table 5.2 (Ref 1), correspond to the metal temperature on the outside surface. They represent design limits which take into consideration the heat-transfer design analysis and actual material properties. The data in Table 5.2 should be considered in conjunction with several footnotes included in Ref 1. Specific problems with respect to boiler tubes, headers, and steam pipes will be discussed in the following sections.

Boiler-Tube Failure Mechanisms

Failure of boiler tubes is the foremost cause of forced boiler outage in the United States and in most other countries in the world (Ref 1 and 2). In the United States alone, the cost penalty due to these failures is estimated to be in excess of 5 billion dollars per year in replacement power and maintenance costs. The failure mechanisms in

Table 5.1. Materials used in boiler construction

Alloy	Product form	ASME or ASTM specification	Grade	Minimum tensile strength, ksi	Minimum yield strength, ksi	Composition(a), % C	Mn	P	S	Si	Ni	Cr	Mo
Carbon steels													
Low strength	Tubes	SA-192	...	(47)	(26)	0.06-0.18	0.27-0.63	0.048	0.058	0.25	...	...	...
	Tubes (ERW)	SA-178	A	...	...	0.06-0.18	0.27-0.63	0.050	0.060	...	...	...	...
	Tubes (ERW)	SA-226	...	(47)	(26)	0.06-0.18	0.27-0.68	0.050	0.060	...	...	...	...
Intermediate strength	Tubes	SA-210	A-1	60	37	0.27	0.93	0.048	0.058	0.10 Min	...	...	...
	Tubes (ERW)	SA-178	C	60	37	0.35	0.30	0.050	0.060	...	...	...	...
	Pipe	SA-106	B	60	35	0.30	0.29-1.06	0.048	0.058	0.10 Min	...	...	...
	Castings(b)	SA-216	WCA	60	30	0.25	0.70	0.040	0.045	0.60	...	...	...
	Structural shapes	A36	...	58	36	0.26	...	0.040	0.05	...	...	...	...
High strength	Pipe	SA-106	C	70	40	0.35	0.29-1.06	0.048	0.058	0.10 Min	...	...	...
	Plate	SA-299	...	75	40	0.30	0.86-1.55	0.035	0.040	0.13-0.33	...	...	...
	Plate	SA-515	70	70	38	0.35	0.90	0.035	0.04	0.13-0.33	...	...	...
	Forgings	SA-105	...	70	36	0.35	0.60-1.05	0.040	0.050	0.35	...	...	...
	Castings(b)	SA-216	WCB	70	36	0.30	1.00	0.040	0.045	0.60	...	...	...
Ferritic alloys													
C-0.5Mo	Tubes	SA-209	T1	55	30	0.10-0.20	0.30-0.80	0.045	0.045	0.10-0.50	...	...	0.44-0.65
1Cr-½Mo	Forgings	SA-336	F12	70	40	0.10-0.20	0.30-0.80	0.040	0.040	0.10-0.60	...	0.80-1.10	0.45-0.65
	Tubes	SA-213	T12	60	30	0.15	0.30-0.61	0.045	0.045	0.50	...	0.80-1.25	0.44-0.65
	Pipe	SA-335	P12	60	30	0.15	0.30-0.61	0.045	0.045	0.50	...	0.80-1.25	0.44-0.65
	Plate	SA-387	12Cl2	65	40	0.17	0.36-0.69	0.035	0.040	0.13-0.32	...	0.74-1.21	0.40-0.65
	Forgings	SA-182	F12	70	40	0.10-0.20	0.30-0.80	0.040	0.040	0.10-0.60	...	0.80-1.25	0.44-0.65
1.25Cr-0.5Mo	Tubes	SA-213	T11	60	30	0.15	0.30-0.60	0.030	0.030	0.50-1.00	...	1.00-1.50	0.44-0.65
	Pipe	SA-335	P11	60	30	0.15	0.30-0.60	0.030	0.030	0.50-1.00	...	1.00-1.50	0.44-0.65
	Plate	SA-387	11Cl2	75	45	0.17	0.36-0.69	0.035	0.040	0.44-0.86	...	0.94-1.56	0.40-0.70
	Forgings	SA-182	F11	70	40	0.10-0.20	0.30-0.80	0.040	0.040	0.50-1.00	...	1.00-1.50	0.44-0.65
	Castings(b)	SA-217	WC6	70	40	0.20	0.50-0.80	0.040	0.045	0.60	...	1.00-1.50	0.45-0.65

Alloy	Product form	Spec.	Grade	Tensile	Yield	C	Mn	P	S	Si	Ni	Cr	Mo
2.25Cr-1Mo	Tubes	SA-213	T22	60	30	0.15	0.30-0.60	0.030	0.030	0.50	...	1.90-2.60	0.87-1.13
	Pipe	SA-335	P22	60	30	0.15	0.30-0.60	0.030	0.030	0.50	...	1.90-2.60	0.87-1.13
	Plate	SA-387	22Cl1	60(c)	30(c)	0.17	0.27-0.63	0.035	0.035	0.50	...	1.88-2.62	0.85-1.15
		SA-387	Cl2	75(d)	45(d)						...		
	Forgings	SA-182	F22	75	45	0.15	0.30-0.60	0.040	0.040	0.50	...	2.00-2.50	0.87-1.13
	Castings(b)	SA-217	WC9	70	40	0.18	0.40-0.70	0.040	0.045	0.60	...	2.00-2.75	0.90-1.20
5Cr-0.5Mo	Tubes	SA-213	T5	60	30	0.15	0.30-0.60	0.030	0.030	0.50	...	4.00-6.00	0.45-0.65
9Cr-1Mo	Tubes	SA-213	T9	60	30	0.15	0.30-0.60	0.030	0.030	0.25-1.00	...	8.00-10.00	0.90-1.10
Austenitic stainless alloys													
18Cr-8Ni	Tubes	SA-213	TP304H	75	30	0.04-0.10	2.00	0.040	0.030	0.75	8.00-11.00	18.00-20.00	...
	Pipe	SA-376	TP304H	75	30	0.04-0.10	2.00	0.040	0.030	0.75	8.00-11.00	18.00-20.00	...
	Plate	SA-240	304	75	30	0.08	2.00	0.045	0.035	1.00	8.00-10.50	18.00-20.00	...
		SA-240	304H	75	30	0.04-0.10	2.00	0.045	0.030	1.00	8.00-12.00	18.00-20.00	...
	Forgings	SA-182	F304H	75	30	0.04-0.10	2.00	0.040	0.030	1.00	8.00-11.00	18.00-20.00	...
18Cr-10Ni-Ti	Tubes(e)	SA-213	TP321H	75	30	0.04-0.10	2.00	0.040	0.030	0.75	9.00-13.00	17.00-20.00	...
18Cr-10Ni-Cb	Tubes(f)	SA-213	TP347H	75	30	0.04-0.10	2.00	0.040	0.030	0.75	9.00-13.00	17.00-20.00	...
16Cr-12Ni-2Mo	Tubes	SA-213	TP316H	75	30	0.04-0.10	2.00	0.040	0.030	0.75	11.00-14.00	16.00-18.00	2.00-3.00
	Pipe	SA-376	TP316H	75	30	0.04-0.10	2.00	0.040	0.030	0.75	11.00-14.00	16.00-18.00	2.00-3.00
	Forgings	SA-182	F316H	75	30	0.04-0.10	2.00	0.040	0.030	1.00	10.00-14.00	16.00-18.00	2.00-3.00
	Plate	SA-240	316H	75	30	0.04-0.10	2.00	0.045	0.030	1.00	10.00-14.00	16.00-18.00	2.00-3.00
	Structural sheet	A167	316L	70	25	0.03	2.00	0.045	0.03	1.00	10.00-14.00	16.00-18.00	2.00-3.00
25Cr-12Ni	Castings	SA-351	CH20	70	30	0.20	1.50	0.040	0.040	2.00	12.00-15.00	22.00-26.00	...

(a) Single values shown are maximums. (b) Residual elements not to exceed 1.00%. (c) Annealed. (d) Normalized. (e) Titanium content not less than four times carbon content and not more than 0.60%. (f) Cb + Ta not less than eight times carbon content and not more than 1.00%.

Table 5.2. Maximum tube-metal temperatures permitted by ASME code and boiler manufacturers (Ref 1)

Tube steel type	ASME specification No.	ASME °F (°C)	Babcock and Wilcox °F (°C)	Combustion Engineering °F (°C)	Riley Stoker °F (°C)
Carbon steel	SA-178 C	1000 (538)	950 (510)	850 (454)	850 (454)
Carbon steel	SA-192	1000 (538)	950 (510)	850 (454)	850 (454)
Carbon steel	SA-210 A1	1000 (538)	950 (510)	850 (454)	850 (454)
C-Mo	SA-209 T1	1000 (538)	. . .	900 (482)	900 (482)
C-Mo	SA-209 T1a	1000 (538)	975 (524)	. . .	. . .
Cr-Mo	SA-213 T11	1200 (649)	1050 (566)	1025 (552)	1025 (552)
	SA-213 T22	1200 (649)	1115 (602)	1075 (580)	1075 (580)
Stainless	SA-213 321H	1500 (816)	1400 (760)	. . .	1500 (816)
Stainless	SA-213 347H	1500 (816)	. . .	1300 (704)	. . .
Stainless	SA-213 304H	1500 (816)	1400 (760)	1300 (704)	. . .

boiler tubes, and their characteristics and remedies, have been surveyed and documented in a manual (Ref 1). The major failure categories noted in this survey are listed in Table 5.3. Of these mechanisms, only stress-rupture, fire-side corrosion, and thermal fatigue are specifically related to the high-temperature environment. Because the aforementioned manual deals extensively with all the failure mechanisms, only a few mechanisms which relate to high-temperature operation will be briefly described here.

Stress-Rupture Failures. The strength of a boiler tube depends on the level of stress as well as on temperature when the tube-metal temperatures are in the creep range. Because an increase in either stress or temperature can reduce the time to rupture, attention must be given to both factors during investigation of a failure by a stress-rupture mechanism. Failures by stress-rupture mechanisms are encountered predominantly in steam-cooled superheater and reheater sections where tube operating temperatures are in the creep range. However, stress-rupture also can occur in water-cooled tubing if abnormal heat-transfer conditions result in an increase in tube operating temperature. The circumferential hoop stress in a tube is determined by the diameter and thickness of the tube. Should the thickness of the tube be decreased by corrosion or

erosion, the hoop stress, and hence the likelihood of failure, will increase.

The significant subcategories of stress-rupture failure mechanisms are:

1. Short-term overheating
2. High-temperature creep.

The term "overheating failure" is often misused but generally means a failure resulting from operation of a tube at a temperature higher than expected in design selection of the tube steel for a period of time sufficient to cause a stress-rupture failure. Time at temperature is an important factor, and these types of failures are often called "short-term" and "long-term" overheating failures. However, because no time-duration criteria have been established for distinguishing short-term from long-term failures, considerable confusion can result when an attempt is made to categorize a particular failure—especially if it is neither a very short-term nor a very long-term failure.

Short-Term Overheating. A short-term-overheating failure is one in which a single incident or a small number of incidents exposes the tube steel to an excessively high temperature (hundreds of degrees above normal) to the point where deformation or yielding occurs. Overheating results from abnormal conditions such as loss of coolant flow and excessive boiler-gas temperature.

Table 5.3. Failure mechanisms for boiler tubing (Ref 1)

Stress-rupture
- Short-term overheating
- High-temperature creep
- Dissimilar-metal welds

Water-side corrosion
- Caustic corrosion
- Hydrogen damage
- Pitting (localized corrosion)
- Stress-corrosion cracking

Fire-side corrosion
- Low temperature
- Waterwall
- Coal ash
- Oil ash

Erosion
- Fly ash
- Falling slag
- Sootblower
- Coal particle

Fatigue
- Vibration
- Thermal
- Corrosion

Lack of quality control
- Maintenance cleaning damage
- Chemical excursion damage
- Material defects
- Welding defects

These abnormal conditions are created by the following circumstances:

1. Internal blockage of the tube
2. Loss of coolant circulation or low water level
3. Loss of coolant due to an upstream tube failure
4. Overfiring or uneven firing of boiler fuel burners.

The first three circumstances produce so-called starvation or low-coolant-flow failures. A tube can be blocked by erection and repair debris, tools, steel shot, preboiler oxides, deposits from carryover or spray water, or loose pieces of internal nonpressure-part hardware such as bolts, nuts, and steel plates. In pendant superheater tubes, blockage also can occur as a result of condensate not being completely boiled out, especially during rapid boiler start-ups. In tubing containing water, blockage will reduce coolant flow, which will result in film boiling and produce local metal temperatures approaching the furnace-gas temperature.

Loss of coolant circulation can have several causes, such as low drum-water level or a failure in the same tube at a different location. Inadequate coolant turbulence or circulation in a region of high heat flux can result in a deviation from the normal nucleate type of boiling condition that is desired inside a water-cooled tube. A departure-from-nucleate-boiling (DNB) condition results when steam bubbles formed on the hot tube surface begin to interfere with the flow of water coolant to the tube surface. The bubbles can eventually cover the inside surface and produce a film of steam which restricts the flow of heat away from the tube. When film-type boiling exists in a water-cooled tube, the metal temperature can exceed 540 °C (1000 °F). Film boiling is more likely to occur in horizontal tubes which are heated from above or in inclined tubes which have low coolant turbulence/flow characteristics. Film boiling also can be produced when overfiring or uneven firing of fuel burners results in regions of high heat flux.

In general, short-term-overheating failures involve considerable tube deformation in the form of metal elongation and reduction in wall area or cross section. Such failures often are characterized as having knife-edge fracture surfaces. Figure 5.3(a) illustrates the elongation and deformation normally encountered with short-term overheating. Wall thinning and local bulging precede the actual fracture, because the strength of the material is reduced at the higher temperature. A fishmouth appearance with thin-edge fracture surfaces and considerable swelling is typical for a ferritic steel tube that has failed before its temperature has exceeded the upper critical temperature (Ac_3). If, however, the tube temperature was high enough to transform the iron in the steel from ferrite to austen-

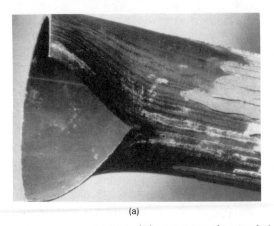
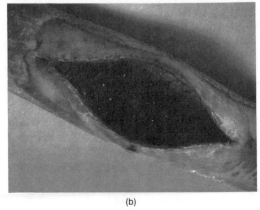

(a)   (b)

(a) Typical short-term overheating failure. (b) Typical long-term creep failure.

Fig. 5.3. Creep-rupture failures in boiler tubes (Ref 1).

ite, there will be no noticeable "necking down," or reduction in wall thickness, of the fracture edges. There will still be metal elongation and tube swelling so that measurement of the tube diameter will show an increase. A metallurgical analysis of the microstructure of the steel should be performed to confirm that the tube temperature prior to failure was high enough to transform the ferrite to austenite.

Changes in tube ID and OD measurements can be indicators of overheating. Increases of 5% or more are indicative of short-term overheating. Also, significant microstructural changes in carbon steel will occur when the steel is overheated, and these changes can be used to estimate the metal temperature at failure. A normalized microstructure for carbon steel boiler tubing consists of ferrite and pearlite phases. Above the lower critical temperature (Ac_1), the pearlite will begin to transform to austenite. At the upper critical temperature (Ac_3), the conversion to austenite is complete. Upon the rapid cooling that occurs when the tube bursts, the austenite will transform to martensite. If the relative amounts of ferrite and martensite can be determined by microstructural analysis, and if the alloy composition is known, an iron-carbon equilibrium phase diagram can be used to estimate the metal temperature at the time the tube burst.

Because short-term overheating failures can result from several root causes, the corrective actions necessary to prevent recurrence can vary. In general, quality control measures should be enforced to prevent tube blockages, low coolant-flow rates, low drum-water levels, and excessive firing rates. Maintenance procedures should be followed during welding of tube joints to prevent tools, cutting debris, and weld spatter from entering the tube circuit. Operating instructions should be followed to avoid low drum-water levels, excessive firing rates, improper fuel-burner operation, excessive desuper-heating sprays, or low heat transfer through waterwalls.

Boiler regions with high heat flux may require redesign to prevent film boiling. Tube design with internal ribbing or rifling should be considered to promote coolant turbulence. Rifled tubes produce a swirling flow which forces water droplets toward the inner tube surface and prevents formation of a steam film. Relocation of an inclined or horizontal tube away from a high-heat-flux area may be necessary to prevent film boiling within the tube.

Furnace slagging, fuel-burner position, or sliding-pressure operation should be controlled to limit the amount of desuperheating spray water required to control the final main-steam-outlet temperature and prevent overheating in the tube circuit prior to the attemperation injection point. Changes in furnace operating parameters should be carefully considered for their possible effects on tube-metal temperature.

High-Temperature Creep. Boiler-tube failures can result from high-temperature creep of the superheater and reheater tube steel. Metal degradation and permanent deformation will occur with time depending on the actual temperature and stress levels. If temperatures and stresses exceed design-selection values, the tube steel will exhibit a higher creep rate and will fail earlier than expected. Predominant locations for creep failure are: (1) just prior to a change to a higher grade of steel; (2) just prior to the final outlet header, where the tubing and steam temperatures are the highest; and (3) where the radiant-heat effect can result in high tube temperatures.

High-temperature-creep failures sometimes are called "long-term" or "extended" overheating failures. Such a failure results from a relatively continuous extended period of slight overheating (differential between design and actual operating temperatures), a slowly increasing level of temperature or stress, or accumulation from several periods of excessive overheating. The creep damage occurs along the grain boundaries of the steel and is aligned 90° from the direction of applied tensile stress. Creep deformation results in little or no reduction in wall thickness but produces measurable creep elongation or increases in diameter in ferritic steel tubes. Stainless steel does not exhibit very much deformation in long-term failures.

High-temperature creep develops from insufficient boiler-coolant circulation, elevated boiler-gas temperature, or material properties that are inadequate for the actual operating conditions. These abnormal conditions are created by the following circumstances:

1. Internal restriction of tube-coolant flow by scale, debris, or condensate
2. Reduction of heat-transfer capability due to internal (steam-side) surface oxide scales or chemical deposits
3. Periodic overfiring or uneven firing of fuel burners
4. Blockage or laning of boiler gas passages
5. Operation of a tube material at higher-than-allowable temperatures
6. Increases in stress due to wall thinning.

High-temperature creep usually results in a longitudinal fracture on the heated side of the tube. The extent of the fracture may vary and have different physical appearances. A small fracture will form a blister-type opening, whereas a large fracture exhibits a wide, gaping, fishmouth-type appearance. The fracture surface has thick edges or thick lips because the creep damage creates link-ups of individual voids and black oxide-filled cracks. Secondary cracking adjacent to the main fracture is extensive and is a positive indication of creep, although the absence of longitudinal cracks in the brittle iron oxide scale does not mean that creep swelling has not occurred. Failures of intermediate duration at moderately high temperatures will exhibit some deformation and wall-thickness reduction. Figure 5.3(b) shows a high-temperature-creep failure due to overheating over a substantial length of the tube. The fishmouth opening exposes the thick-edge fracture surfaces. The thickness of the tube at the fracture edge is an indicator of a very long-term failure. Metallurgical analysis should be performed to confirm suspected high-temperature-creep failures and to provide clues to their actual root causes.

Prevention of high-temperature-creep failure involves keeping tube-metal stress and temperature within the capabilities of the tube material. Overheating and/or overstressing of the tube material beyond its design limits as established by ASME or the boiler manufacturer accelerates creep deformation and results in premature tube failure. Corrective action for control of high-temperature-creep failure depends on the specific cause for overheating or overstressing. Failures from overheating caused by internal-flow restrictions or tube heat-transfer reductions can be eliminated by removal of the scale, debris, or deposits that have accumulated inside the tube. High-pressure fluid flushing or chemical cleaning may be necessary to restore the design coolant flow or tube heat-transfer characteristics. Failures from overstressing caused by wall thinning can be controlled by applying ultrasonic wall-thickness measurements with residual-life estimates. Such residual-life estimation techniques have been successfully used in many countries to determine which tubes are beyond a critical wall thickness and should be immediately replaced to prevent in-service failure within the next operating period. This technique is described later in this chapter.

Failures from creep damage caused by periods of operation at metal temperatures above the design limit can be controlled by restoring boiler design conditions or by upgrading tube material. Measurements of actual tube-metal temperature can show where design limits are being exceeded. When actual temperatures cannot be reduced, the tube material should be replaced with a higher-chromium-content ferritic steel or an austenitic stainless steel. Residual-life estimates can be performed to determine when tube failures can be expected so that corrective actions can be taken prior to their occurrence.

Boiler-tube failures also can result from high-temperature creep at or adjacent to supports or attachments welded to superheater tubes. A welded attachment can increase the circumferential stress in the tube due to (1) differences in thermal-expansion properties, (2) residual stresses from weld shrinkage, (3) stress concentrations from rough weld contours, and (4) restraints to uniform tube expansion. The welded attachment also can transfer more heat to the tube and result in a higher metal temperature at the weld region. These increases in stress and temperature conditions at the weld attachment can accelerate creep damage to the tube steel and significantly decrease its stress-rupture life. A more complete coverage of this subject can be found in Ref 1.

Failures in Dissimilar-Metal Welds. Dissimilar-metal welds are widely used in power-station boilers for joining ferritic steel tubes to austenitic steel tubes in the superheater/ reheater (SH/RH) sections. The austenitic steel tubing is used in the final stages of the SH/RH, where increased resistance to creep and oxidation is needed. Economic considerations dictate the use of low-alloy ferritic steel tubing in the earlier stages, where temperatures are lower. A dissimilar-metal weld is formed between the ferritic steel tube and the austenitic steel tube (see Fig. 5.4) using ferritic, austenitic, or Inconel-type filler metal. Due to differential thermal expansion between the ferritic and austenitic tubes, failures occur in the ferritic steel heat-affected zone. The following facts have been established regarding dissimilar-metal weld failures: (1) use of Inconel filler metals results in weld lives three to four times longer than those obtained with other filler metals; (2) failures are brittle, with little evidence of wall thinning, necking, or other deformation; (3) the fracture front occurs at a location one to two grains away from the fusion line in the HAZ of the ferritic steel tube; and (4) the fracture front is intergranular with austenitic and ferritic filler metals but follows a continuous interface of carbides in the case of Inconel filler metals (Ref 3 and 4).

One of the significant results of the extensive field inspections and sample examinations that have been performed is the development of a procedure for prediction of damage in service, or PODIS (Ref 4). This procedure allows calculation of the total damage to dissimilar-metal welds arising from steady-state loads and cyclic loads. The loading history is defined in terms of time; weld-metal temperature; changes in weld-metal temperature; number and types of cycles, and axial stresses at the weld due to pressure, dead weight, and restrained thermal-expansion loads within the assembly. The information needed is obtained from plant records, design drawings, sacrificial sample examinations, and over-all inspection of the boiler. It is assumed that a theoretical estimation of damage can be obtained by linear additon of three components, as follows:

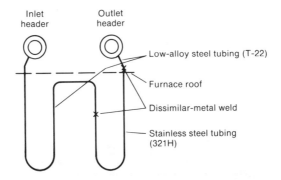

Fig. 5.4. Typical dissimilar-metal weld locations and failures.

1. D_I – Intrinsic damage due to differential thermal expansion when temperature changes are applied to the DMW.
2. D_P – Damage caused by the primary (load-controlled) steady-state components of system load. System loads in this context are those arising from the location of the weld within a tube assembly, plus the axial effect of pressure. Loads in this category are dead weights and axial pressure loads.
3. D_S – Damage caused by the secondary (strain-controlled) cyclic components of system load. Loads in this category are those due to restrained thermal expansion within the tube assembly and thermally induced system movements external to the assembly.

Essentially, D_I, D_P, and D_S correspond to damage components due to stress relaxation, steady-state creep, and fatigue, respectively. Their actual magnitudes are calculated using equations and charts derived empirically from data on service samples and laboratory samples of known history. The three components can be added linearly to calculate the total damage, D. A value of D = 1 corresponds to through-the-wall fracture. Predictions from the PODIS procedure have been verified on a number of service-returned samples. The PODIS procedure not only gives an estimate of life expenditure but also, and more important, allows identification of the root cause of failure.

Remedies that have been suggested include: (1) relocation of welds to lower-temperature or lower-stress areas; (2) modification of welding procedures so as to promote an innocuous carbide distribution in the HAZ, in the case of Inconel-filler welds; (3) avoidance of decarburization near the fusion line by control of the postweld heat treatment; (4) avoidance of temperature excursions and cycling; (5) use of improved filler metals; and (6) improvements in weld geometry designed to reduce stresses at the fusion-line interface. These remedies and their effective-

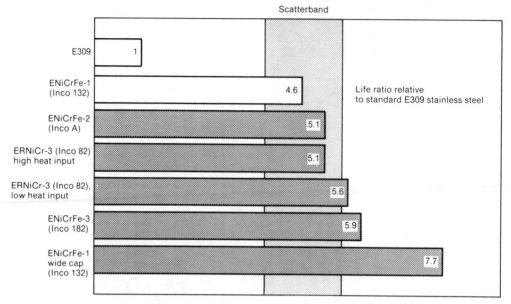

For details of this test, see Vol 3 of Ref 4.

Fig. 5.5. Relative performance of dissimilar-metal welds made with different commercial filler metals and having different geometries under accelerated discriminatory testing (Ref 5).

ness have been discussed elsewhere (Ref 4 and 5). The relative performance of various filler metals is illustrated in Fig. 5.5 (Ref 5). Clearly, the nickel-base filler metals are superior to the austenitic stainless steel filler metal. Even further improvements can be achieved by implementing a wide cap geometry.

Fire-Side Corrosion of Waterwall Tubes. Boiler-tube failures can result from fire-side waterwall corrosion that causes wastage of the external metal surface. One possible mode for the development of such corrosion can become operative when a reducing atmosphere is present in the burner zone as a result of incomplete combustion. The atmosphere contains high levels of carbon monoxide and also delivers unburned particles of coal to the tube surface. The incompletely burned coal particles release volatile sulfur compounds and chloride compounds (if present), which cause sulfidation and accelerated metal corrosion. Sodium and potassium pyrosulfate compounds, which have low melting points (below 425 °C, or 800 °F), are generally

held responsible for the attack. The attack is also characterized by formation of abnormally thick iron oxide and iron sulfide scales. Tube thinning eventually leads to bursting of the tubes.

Operational remedies for this problem include better coal grinding, proper adjustment of fuel distribution to the burners, increasing the flow and redistribution of the secondary air, and bleeding of air into the sidewall areas. Other preventive activities include thermal-sprayed coatings, installation of more corrosion-resistant ferritic alloys, and use of coextruded or clad tubing. Spray coatings generally contain aluminum or an iron-chromium-aluminum-molybdenum alloy. Coextruded tubes with mild steel on the inside and type 304 stainless steel on the outside have shown considerable promise.

Fire-Side Corrosion of Superheater/ Reheater Tubes. Corrosion of SH/RH tubes by liquid ash deposits at high temperature can cause wall thinning and contribute to stress-rupture failures by increasing the hoop stress. This type of "liquid phase" corrosion occurs in the range of metal temperatures of from 595 to 705 °C (1100 to 1300 °F). Fire-side corrosion problems often occur when a change in fuel supply or type is made which results in the formation of a more aggressive ash than was anticipated in boiler design. Because corrodent species and mechanisms are different in the cases of coal-ash corrosion and oil-ash corrosion, these two phenomena are discussed separately in the following sections.

Coal-Ash Corrosion. In coal-fired boilers, corrosive conditions are created by the following circumstances: (1) operation with coals that produce corrosive ash products; and (2) operation that produces tube-surface temperatures between 595 and 705 °C (1100 and 1300 °F) (maximum corrosion rates occur at 650 °C, or 1200 °F).

Deposit-related molten salt attack concerns the development of conditions beneath a surface deposit which are conducive to the formation of a low-melting salt of the type $(Na,K)_3Fe(SO_4)_3$. These alkali-iron trisulfates form by reaction of alkali sulfates with iron oxide in the presence of SO_3 and have a minimum melting temperature (1:1 mixture of sodium and potassium salts) of 552 °C (1026 °F), whereas the melting points of the simple sulfates are 884 °C (1623 °F) for Na_2SO_4 and 1069 °C (1956 °F) for K_2SO_4. The level of SO_3 in the ambient gas necessary for the stability of the alkali-iron trisulfates ranges from about 250 ppm at 480 °C (900 °F) to 1000 to 1500 ppm at 635 °C (1175 °F). Although the SO_3 levels in the flue gas are expected to be much lower, catalytic oxidation of SO_2 in the stagnant zones beneath a layer of deposit is believed to generate SO_3 levels sufficiently high to favor the formation of the liquid trisulfates at temperatures up to 705 °C (1300 °F). Above this temperature, the required SO_3 concentrations cannot be sustained, and the trisulfates become unstable, decomposing to the alkali sulfates, which are solid.

The temperature dependence of coal-ash corrosion is shown in Fig. 5.6 (Ref 6). The shape of the curve is much the same for low-alloy ferritic steels and for austenitic

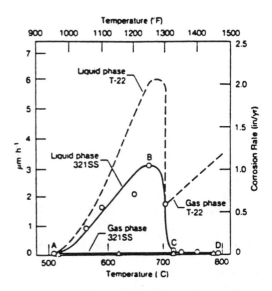

Fig. 5.6. Temperature dependence of fire-side corrosion for 2¼Cr-1Mo ferritic steel and type 321 austenitic stainless steel (Ref 6).

stainless steels such as types 304 and 347, although the attack is minimized with increasing chromium content of the alloy (Ref 7). The onset of accelerated attack is associated with the appearance of molten sulfate phase. The attack reaches a maximum around 650 °C (1200 °F) and then decreases as a result of dissociation of the sulfate phase into solid deposits due to insufficient pressure of SO_3.

Because the tube-metal temperatures slowly increase with time due to steam-side oxide buildup, innocuous deposits may suddenly become aggressive upon reaching the melting point. The corrosion rates may thus be suddenly accelerated in service. More important, the corrosion problem is the result of high temperatures due to overheating. Overheating of the superheater can occur as a result of poor initial boiler design, when slagging problems are experienced, and when adjustments are made to the fireball to increase the heat transferred in the pendant tubes to allow the steam temperature to be attained. Operational difficulties of this type may arise when a change is made in the feed coal. Overheating of reheaters can occur in rapid start-up situations, when the combustion gas temperature at the reheater reaches its maximum value before full steam flow through the reheater is achieved.

Experience has shown that not all coals are necessarily expected to form aggressively corrosive molten sulfates under similar circumstances. Studies by Borio *et al* (Ref 8) in particular have shown that the participation of species such as CaO and MgO in the formation of the complex sulfates leads to elevation of the melting points of the sulfates, and so reduces the possibility of molten salt attack. Borio's work led to the concept of blending coals to adjust the levels of Ca + Mg, depending on the sodium, potassium, and sulfur contents of the coal, to reduce the "corrosion index." A correlation was found between coal composition and corrosion in terms of: (1) acid-soluble sodium and potassium, (2) calcium and magnesium content, and (3) iron content, serving as an index for sulfur in the coal. A nomograph was developed in which the amounts of these components could be used to read a "corrosion index" from 0 (no corrosion) to 20 as a basis for comparing coals.

A strong correlation has also been found between the corrosion rate of superheater/reheater tubes and the chlorine content of the coal (Ref 9). Current understanding is that the chlorine promotes the release of both sodium and potassium into the flame. Hence, chlorine is thought to act as a sort of catalyst for the molten trisulfate attack, rather than participating directly. There is also evidence that HCl formed in the flame can destroy the Fe_2O_3 layer that normally exists on a steel surface, thereby exposing it to additional oxidation attack (Ref 10). The potential for corrosion from chlorine is recognized by boiler manufacturers, and a limit of 0.3% chlorine in the coal is usually set.

Based on the severity of the corrosion problem as determined by the rate of the corrosion and the extent over which it occurs, several options for corrective action are available. These include:

1. Using thicker tubes of the same material
2. Shielding the tubes with clamp-on tube protectors
3. Coating the tubes with thermal-sprayed corrosion-resistant material
4. Blending coals to reduce corrosive constituents in the ash
5. Replacing the tubes with tubes made of a higher-grade alloy or a coextruded tube steel
6. Lowering the metal temperature by lowering the final steam outlet temperature
7. Redesigning the superheater or reheater to modify the heat-transfer rates and lower the metal temperature.

Using thicker tubes of the same material and shielding tubes with clamp-on tube protectors have limited appeal because of the continued and/or additional maintenance costs. Lowering the final steam outlet temperature or blending coals may not be practical due to lower unit output, higher unit heat rate, or limitations of fuel-supply contracts. Coating tubes with thermal-sprayed corrosion-resistant materials provides only short-term protection, and recoating is necessary for continued operation. Replacing tubes with tubes made of more corrosion-resistant steels is sometimes worthwhile to eliminate the problem or significantly extend the time until replacement.

Increasing the chromium content of the steel at its outside surface improves its resistance to coal-ash corrosion, but alternative materials must also possess adequate creep strength because the tubes operate at metal temperatures in the creep range. Coextruded tubing with a creep-resistant inner layer and a corrosion-resistant outer layer has been successfully used in the United Kingdom and in the United States. A substrate of type 310H austenitic stainless steel over an outer layer of 25Cr-20Ni steel has been used in Europe. In the United States a clad tubing has been used which has a 50Cr-50Ni exterior surface. A welding method for joining coextruded tubing to stainless steel using

ENiCrFe-3 filler metals has been developed and has performed well in service. The initial cost of the coextruded tube material is estimated to be four to five times greater than that of single-layer austenitic stainless steel, so that application of coextruded tubing may depend on life-cycle costs including additional outage costs.

Oil-ash corrosion results when molten slag containing vanadium compounds forms on the tube surface. Accelerated corrosion occurs by the fluxing action of molten sodium-vanadium compounds on the protective oxide scale on the tube steel. The corrosion process is believed to be a catalytic oxidation of the metal by reaction with vanadium pentoxide (V_2O_5) or complex vanadates or vanadylvanadates. The corrosive slag forms under the following circumstances: (1) operation with oil that contains high levels of vanadium, sodium, and sulfur; and (2) operation that produces tube-surface temperatures above 595 °C (1100 °F).

The severity of oil-ash corrosion is affected by a number of factors such as the temperature, the chloride content of the fuel oil, the amount of excess air available for the formation of V_2O_5, and the amounts of vanadium, sodium, and sulfur in the fuel oil. The greatest wastage occurs when the ratio of sodium oxide (Na_2O) to vanadium oxide (V_2O_5) is about 1:5. Increases in tube-metal temperature due to steam-side oxidation can lead to a sudden rash of oil-ash corrosion failures after eight to ten years of boiler operation.

The use of additives as a corrective action has had successful effects and has proven to be economically feasible. Addition of magnesium compounds results in formation of a magnesium vanadate complex ($3MgO \cdot V_2O_5$) that has a higher melting point so that liquid phases do not exist at superheater and reheater temperatures. The magnesium is generally added as MgO dispersed in oil, but oil-soluble forms of magnesium, magnesium-base additives, combinations of other metals in oil-soluble form, and dry powders have been found to be effective in reducing oil-ash corrosion.

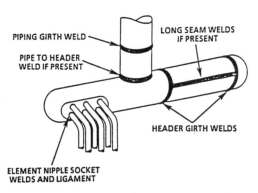

Fig. 5.7. Schematic illustration of an elevated-temperature header (courtesy of B.W. Roberts, Combustion Engineering, Inc.).

Higher grades of alloy and coextruded tubing have also been used to resist or reduce the corrosion rate.

Header-Damage Mechanisms

A schematic illustration of a header is shown in Fig. 5.7. A header is essentially a pipe to which tubes are welded, spaced either axially or circumferentially. The spacing between the tubes is known as the ligament. In addition to the tubes, other pipe-to-pipe connections are also present, either integral with the header pipe or welded to it. These branch connections can be T-shape connections, as shown in Fig. 5.7, or of a Y-shape configuration. Numerous pipe-to-pipe and pipe-to-tube weldments are normally present, as shown in the figure.

A detailed survey of 62 U.S. utilities conducted by Roberts, Ellis, and Viswanathan (Ref 11) has delineated the key damage mechanisms and locations, as summarized in Table 5.4 (Ref 12). In this table, the "survey percentage" represents the percentage of the responding utilities reporting specific problems. Creep at weldments and thermal fatigue in header-body ligaments have emerged as the principal damage mechanisms. The ligament-cracking problem has been found to be much more widespread since the time of the above survey.

Initial signs of creep-related distress in headers often appear at welds — welds at stub-tube inlets, long seams, header branch

Table 5.4. Long-term damage in elevated-temperature headers (Ref 11 and 12)

Location	Survey percentage	Damage mechanism
Stub tube/header weld, tube side	40	Creep-cavitation in the HAZ
Stub tube/header weld, header side	34	Creep-cavitation in the HAZ
Cracking of ligaments between tubes	21	Thermal fatigue
Longitudinal seam welds	3	Creep-cavitation in HAZ and weld metal
Girth butt welds	3	Creep-cavitation in HAZ and weld metal
All other	10	
Branch connections, saddle and crotch positions	Unknown	Creep-cavitation in the HAZ
Header body swelling	Unknown	Thermal softening
Other locations	Unknown	Unknown

connections, or girth butt joints. Creep damage in the base metal generally occurs by softening and header-body swelling preceding visible creep cavitation. Creep cavitation without prior swelling in the base metal is observed only in regions of high local stress concentration or stress multiaxiality. With the exception of some cases of long seam welds, creep damage in welds is invariably manifested on the outside surface as cavities, cracks, or, in extreme cases, steam leaks. Except in regard to long seam welds, concern about catastrophic bursts has been minimal. Although weld-related cracking is generally detectable and repairable, and although it does not have as great an impact on the over-all component life as does header-body base-metal deterioration, it is important from a life-assessment point of view for the following reasons:

1. Because weld failures are often the forerunners of damage in the body, they can provide an index of creep damage and remaining life in the base metal.
2. Failure of welds at crucial and multiple locations may constitute the end of the life of the header, regardless of the condition of the base metal.
3. The need for frequent weld repair may prove uneconomical and justify retirement of a header.

Due to the above reasons, creep-damage assessment of welds has received considerable attention. There are several unique problems associated with the over-all life assessment of welded components, which will be described in a later section. Damage characteristics at various locations are briefly reviewed here on the basis of the extensive information contained in Ref 13.

Stub-Tube Welds. Cracking at both the tube and header sides of stub-tube welds is the most common type of creep damage in high-temperature headers. Although such cracking may lead to steam leakage and forced outages, it is easily detected and repaired. The cracking may be attributed to any one of several causes, including improper seating of the stub tube, inadequate tube flexibility, improper support of the header, bowing, weld-fabrication defects, and locally excessive temperatures. Metallography in several instances has shown the creep damage to consist of cavitation and microcracking at the prior austenite grain boundaries in the heat-affected zone.

Longitudinal Seam Welds. Plate-formed and seam-welded headers have been used in some designs. Detailed failure reports are available for only one incidence (Ref 13 and 14), in which a crack, 864 mm (34 in.) long, in the weld seam had led to a major leak in a secondary superheater outlet header made of $2\frac{1}{4}$ Cr-1Mo steel after 187,000 hours of service. This failure occurred as a result of creep-rupture. The damage was confined to the weld metal, with no evidence of damage in the heat-affected zone or base metal. The cracking had apparently initiated just below the outer surface, broke through to the outer surface at an early stage, and then propagated to the inner surface. The reason for the subsurface crack initiation was believed to be the inferior material properties at the location due to a lower carbon content and tempering of the weld head by subsequent passes. The most severe cracking occurred at the centerline of the weld beads, presumably as a result of impurity segregation during the last stages of solidification. Boat-shape samples removed at locations away from the cracked area showed several degrees of cavitation. For each location, a life fraction consumed could be estimated on the basis of the model and the plots described later in this chapter. These values were borne out by subsequent isostress-rupture tests of samples from the various locations (Ref 13).

Girth Welds. Both axial and circumferential cracks have been observed in damaged girth butt welds, with cracking being found in the weld metal and/or the HAZ. The axial cracking has been attributed to internal pressure loading and pipe swelling, whereas the circumferential cracking has been associated with combined pressure and piping system loads. Several instances of girth weld cracking have been reviewed (Ref 13). In one instance, circumferential cracking along the coarse-grain HAZ was attributable to stress-relief cracking prior to service. Axial creep cracking across the weld metal has been attributed to a combination of pipe swelling and poor weld ductility. Circumferential cracking in the intercritical regions of the HAZ has also been observed in both Cr-Mo-V and Cr-Mo steels. This type of cracking, known as type IV cracking, occurs at the edge of the HAZ adjacent to the unaffected parent metal. Type IV cracking is generally attributed to localized creep deformation in a "soft" zone in the intercritical region under the action of bending stresses. Field experience suggests that Cr-Mo-V steels may be more susceptible to cracking than Cr-Mo steels and that operation at 565 °C (1050 °F) rather than at 540 °C (1000 °F) might further exacerbate the problem. Because most of the headers in the United States are made of Cr-Mo steels and operate at 540 °C (1000 °F), the problem has not been encountered to any significant degree.

Branch-Connection Welds. Several instances of cracking in branch-connection welds have been observed. Such cracking has occurred on both the header side and the branch side (Ref 13), and in both the HAZ and the weld metal.

Summary of Creep Cracking in Header Welds. Numerous instances of cracking at various locations in header welds, as described above, have been reviewed by Ellis *et al* (Ref 13). The salient facts brought out in this review are as follows. (1) Most weld failures are creep failures and are clearly evidenced by creep cavitation. (2) Cracks can occur in the weld metal, in the coarse-grain HAZ, or in the intercritical zone of the HAZ (type IV cracking). (3) Cracks in the weld metal are generally attributable to lower strength or lower ductility of the weld metal. (4) Cracks in the HAZ can arise as a result of hoop stresses, system bending stresses, or residual stresses due to stress relief. (5) Frequently, the direction of alignment of creep cavities, which is normal to the tensile loading direction, gives a clue to the nature of the system stresses involved.

Ligament Cracking. The problem of ligament cracking in high-temperature boiler headers was first recognized in late 1983 when investigation of a header showing large creep deformation also revealed cracks in and around tube bore holes. In this particular header, it was determined that all inspected areas had ligament cracks extending from tube hole to tube hole. The cracks originated from inside the header, extending axially in the tube-hole penetrations and radially from these holes into the ligaments, as shown in Fig. 5.8. All available information concerning the ligament-cracking problem has been reviewed by Ryder (Ref 15). The following summary is based almost entirely on his review, because there is little other published information on the subject.

Since the initial findings in 1983, 376 high-temperature headers have been exam-

Fig. 5.8. Ligament cracking in a header (courtesy of G. Harth, Babcock and Wilcox).

ined. Of the 157 secondary superheater outlet headers inspected, 28% were found to contain ligament cracking. Of the remaining 219 headers, only 3% were found to have been damaged by this phenomenon, indicating that secondary superheater outlet headers are more susceptible to this form of cracking. Headers made of 2¼Cr-1Mo and 1¼Cr-½Mo steels have been found to be equally prone to cracking. No unique correlation could be found between the age of the header and the susceptibility to cracking. The cycling history, on the other hand, seems to be a major contributing factor to cracking.

The occurrence of ligament cracking has often been found in conjunction with stub-tube weld cracking, which is a form of damage known to be related to creep deformation of the header, implying that creep deformation may be significant for ligament cracking. However, there is also a similarity between the appearances of ligament cracking in high-temperature headers and that in economizer inlet headers (low-temperature headers). The latter type of cracking is known to be caused by fatigue. Furthermore, metallographic evidence ob-

tained from the cracks in high-temperature headers indicates a fatiguelike propagation mode. Thus a controversy exists in regard to the mode of propagation of the cracking.

Ligament cracks generally initiate in the tube bore holes and are oriented parallel with the axis of the tube hole. The initial cracks usually are numerous and may extend to the inside surface of the header, exhibiting a characteristic "starburst" pattern when the hole is viewed from inside the header. In most cases these initial cracks are not visible unless the oxide layer is removed. The oxide layer usually is thickest at the location where ligament cracking is observed, implying that this type of cracking is associated with high temperature. Some of the initial cracks subsequently grow deeper into the ligaments between the holes, both inside the bore and on the inside surface of the header. Link-up of cracks between holes on the inside surface then leads to crack propagation from the inside of the header to the outside. Steam leakage can occur when these cracks link up with those associated with the outside surface stub-tube welds.

Detection of ligament cracks requires inspection of the tube-hole penetrations after some tubes have been cut off. The oxide on the inside surface is carefully removed to avoid smearing over of any shallow cracks. A high-sensitivity dye penetrant is then applied and the penetration is examined with a fiberscope. If any cracks are detected, knowledge of the length-to-width ratio of these defects provides a qualitative measure of the degree of damage. For large cracks, volumetric ultrasonic examination of the header using pulse-echo techniques can be utilized to determine the crack depth. A crack-growth analysis can then be performed to determine the remaining life or to determine the effects of changes in operating conditions to extend header life. Findings from several recent investigations are helpful in identifying the "suspect" locations that need evaluation. It has been observed that the susceptibility to cracking could be related to locally high metal temperatures on the inside walls of the headers at the tube intersections and to the type of tube penetration design. Ultrasonic examination of tubes to characterize the oxide-scale thickness (and hence temperature) and design drawings can be used to pinpoint locations in the header that need further evaluation.

Several circumstances associated with propagation of ligament cracks suggest that it may be attributable to a fatigue mechanism. The crack propagation is invariably transgranular. No evidence of creep cavitation ahead of the cracks has been reported. Three-dimensional elastic finite-element stress analyses carried out by some investigators have shown that the cyclic stresses resulting from start-stop cycles might be sufficiently high to propagate cracks with initial depths approaching 2.5 mm (0.1 in.). Beach markings also have been observed on fracture surfaces. Based on these markings and on oxide dating techniques, it has been suggested that the cracks could be initiated after as little as 25% of the total component life, followed by an extended period of crack propagation.

The cause of ligament-crack initiation is clouded in much controversy. Thermal fatigue (creep-fatigue) is considered to be a major contributing cause, although the specific nature of the thermal transients responsible for the thermal fatigue has not been identified. Start-stop transients, temperature fluctuations during operation (minor in magnitude but large in number), and abnormal thermal shocks (few but severe) are being investigated as the possible root causes. Thermal shock loads have been known to cause similar cracking patterns in economizer inlet headers. Another mechanism being investigated is based on periodic rupture of oxide scales due to steam temperature cycles, which results in "oxide notches"; these notches are then believed to promote crack initiation.

Investigations into the ligament-cracking problem are proceeding in a number of organizations. Definition of the key parameters causing initiation and propagation is being attempted by installation of thermocouples in previously analyzed headers, metallographic examination of cracked material, measurement of the residual mechanical properties of ligament material, study of the formation and fracture properties of the steam-side oxide layer, and stress analysis of various header configurations under transient conditions. Once these parameters have been quantified they will have to be incorporated into life-prediction codes.

Current practices for disposition of cracked headers include immediate retirement, weld repair, and continued operation. At the present time, in view of the many uncertainties described above, there is no clear-cut basis for preferential justification of any of these practices. Use of replacement headers made of a modified 9% Cr steel (P91) appears to be a promising method of alleviating the problem. This aspect is discussed in more detail in Chapter 8.

Damage Mechanisms in Steam Pipes

Steam pipes carry steam from the boiler to the turbines. They are straight pipes with some elbows and bends, but do not have any tube connections. The principal problem areas therefore are girth welds, bends and elbows, and long seam welds (if present). The incidence of creep damage at girth welds is quite similar to that previously described for headers. Bends and elbows, being more highly stressed areas, also have shown a tendency for creep damage. In addition, if they contain long seam welds, they are usually the first locations where leakage occurs. The most significant problem has been with respect to the long seam-welded steam pipes. Two recent catastrophic failures of seam-welded steam pipes have generated great concern about the integrity of steam pipes. Based on limited details, the early criteria for identifying pipes at risk were set forth (Ref 16) as follows:

1. Piping made of SA387 (1¼Cr-½Mo) steel, which was designed between 1952 and 1967 according to ANSI B31.1, Pressure Piping Cole

2. The presence of long seam welds in piping carrying steam at temperatures above 510 °C (950 °F) for 1¼Cr-½Mo steel and above 540 °C (1000 °F) for 2¼Cr-1Mo steel

3. Piping welded by the submerged-arc process.

These criteria were, however, considered inadequate for screening and evaluation of pipes because: a large fraction of the utility piping met the above criteria, and complete inspection of such piping was deemed to be prohibitively expensive; optimized inspection techniques had not been designed; and the procedures and data needed for evaluation and disposition of flaws in pipes had not been adequately delineated.

Extensive pipe inspections and metallographic evaluations (Ref 16) have shown the following salient points:

1. Seam-welded pipes may contain a variety of fabrication flaws such as slag inclusions, lack of fusion, incomplete weld penetration, welding toe cracks, solidification cracks, and aligned inclusions which can serve as preferred sites for creep-crack initiation and propagation.

2. Flaws have been found to be present in the weld metal, in the base metal, and at the fusion line. The fusion-line flaws have shown a greater tendency to grow by creep in service compared with flaws at other locations.

3. There is no clear-cut association between fabrication flaws and creep cavities.

4. Cavities, if present, may be found at the outside surface, at the inside surface, or at midwall. Evidence of distress is not always apparent from surface examination.

5. There appears to be an incubation period associated with the growth of creep damage emanating from pre-existing flaws. The incubation period appears to be shorter for fusion-line flaws.

6. Local weld repairs and aligned inclusions appear to be contributing causes for creep-crack growth. This observation, coupled with other literature observations, that impurities, post-weld heat treatments, and over-all material degradation affect crack growth, suggests that the integrity of the welds may be affected by all of these factors.

A step-by-step road map for the evaluation of steam pipes (Ref 16) has been put together on the basis of all the above observations and of the various life-assessment methodologies described later in this chapter.

Physical Metallurgy of Boiler Steels

In the present context, it would be impractical to attempt to discuss all of the various steels used in boiler applications. Therefore, because the steels most commonly employed in the creep regime are of the approximate compositions 2¼Cr-1Mo and 1¼Cr-½Mo, the basic metallurgical aspects of only these two steels will be described here.

Continuous cooling transformation diagrams for these steels are presented in Fig. 5.9 (Ref 17). The 2¼Cr-1Mo steel has higher hardenability than the 1¼Cr-½Mo steel. In actual applications, boiler tubes are

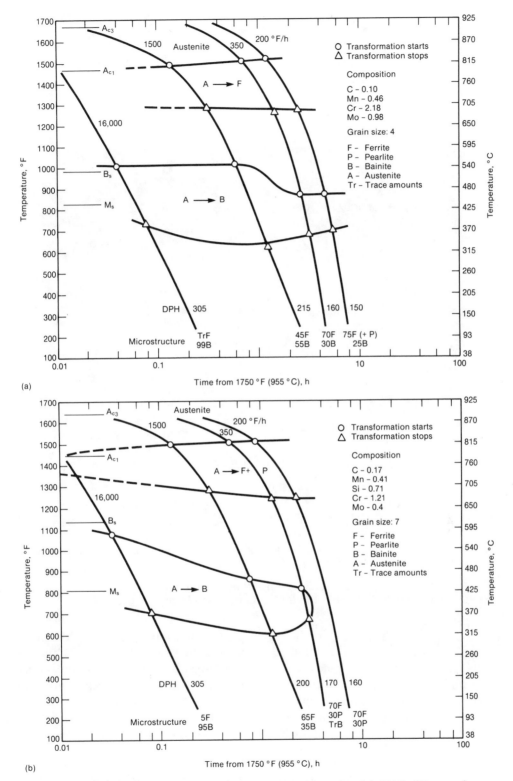

Fig. 5.9. Continuous cooling transformation diagrams for (a) 2¼Cr-1Mo steel austenitized at 955 °C (1750 °F) and (b) 1¼Cr-½Mo steel austenitized at 910 °C (1675 °F) (Ref 17).

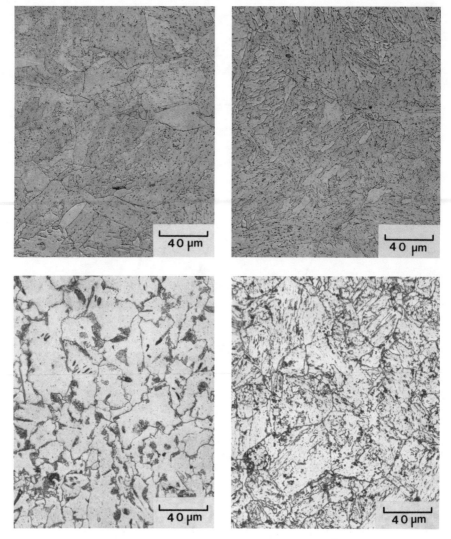

At top: 2¼Cr-1Mo steel plate; base metal at left, heat-affected zone at right. At bottom: 1¼Cr-½Mo steel tube; base metal at left, weld-heat-affected zone at right.

Fig. 5.10. Typical base-metal and HAZ microstructures in 1¼Cr-½Mo and 2¼Cr-1Mo steels (courtesy of F.V. Ellis, Combustion Engineering, Inc.).

used mostly in the annealed condition, whereas piping is used mostly in the normalized-and-tempered condition. Bend sections used in piping, however, are closer to an annealed condition than to a normalized condition. As a result of the cooling rates employed in these treatments, the microstructures in 2¼Cr-1Mo steel consist essentially of bainite/ferrite mixtures, with the bainite content being small (<15%) in the annealed condition but increasing to much higher levels in the normalized-and-tem-

pered condition. In 1¼Cr-½Mo steel, annealed microstructures consist mainly of ferrite-pearlite aggregates, whereas normalized-and-tempered microstructures consist of ferrite-bainite aggregates. In the case of weldments, however, high local heat inputs and high cooling rates during welding generate a coarse-grain region adjacent to the fusion line which may be fully bainitic. Typical base-metal and HAZ microstructures for 1¼Cr-½Mo and 2¼Cr-1Mo steels are shown in Fig. 5.10. Bainitic microstructures have better creep resistance under high-stress, short-time conditions but degrade more rapidly at high temperatures than pearlitic structures. As a result, ferrite-pearlite material has better intermediate-term, low-stress creep resistance. Because both microstructures will eventually spheroidize, it is expected that over long service lives the two microstructures will converge to similar creep strengths. Based on the limited data presented in Fig. 3.5, this convergence can be estimated to occur in about 50,000 h at 540 °C (1000 °F).

The creep strength of Cr-Mo steels derives mainly from two sources: solid-solution strengthening of the matrix ferrite by carbon, molybdenum, and chromium; and precipitation hardening by carbides. Although a number of different types of carbides may be present, the principal carbide phase responsible for strengthening is a fine dispersion of M_2C carbides, where M is essentially molybdenum. The initial microstructure consists of bainite and ferrite containing Fe_3C carbides, epsilon carbides, and fine M_2C carbides. With increasing aging in service, or tempering in the laboratory, a series of transformations of the carbide phases takes place (Ref 18), as described by the sequence

$$\epsilon \text{ carbide} \rightarrow Fe_3C \rightarrow \begin{array}{c} M_7C_3 \rightarrow M_6C \\ \uparrow \\ Fe_3C \\ + \\ |MO_2C| \rightarrow M_{23}C_6 \end{array}$$

where M is mostly chromium. Such an evolution of the carbide structure results in coarsening of the carbides, changes in the matrix composition, and an over-all decrease in creep strength. The time-temperature kinetics of carbide evolution in 2¼ Cr-1Mo steels in both bainitic and martensitic conditions have been delineated by Baker and Nutting, as shown in Fig. 5.11 (Ref 18). They also observed that the higher creep resistance of bainite compared with that of martensite could be related to a greater persistence of the MO_2C carbides in bainite. The Baker-Nutting diagram is a useful tool for estimating the service conditions of a steel component, based on x-ray analysis of the carbides.

Damage and Life Assessment of Boiler Components

Various techniques for assessment and monitoring of creep and creep-fatigue damage in boiler components have become available in recent years (Ref 19). In the

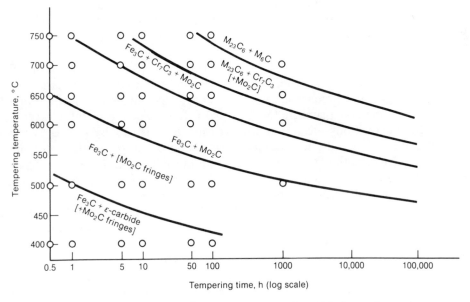

Fig. 5.11. Isothermal diagram showing sequence of carbide formation on tempering of normalized 2¼ Cr-1Mo steel (Ref 18).

application of these techniques, there are basic differences in objectives and approaches with respect to heavy-section components such as headers and steam pipes as opposed to superheater and reheater tubes, which have much thinner walls. Pipe failures can be catastrophic, can endanger human safety, and can lead to extended forced outages. Tube failures, on the other hand, do not pose a threat to human safety, but, because of their high frequency, their economic impact in terms of forced-outage costs is enormous. Repair of a leaking tube may involve two to three days of forced outage, at a cost exceeding $1 million. Tubes are less expensive than piping, are available with shorter lead times, and also can be stocked as spare parts. Tube repairs are carried out more easily than pipe repairs. Tube-life assessment can easily include destructive tests, because sacrificial tubes are readily available. Nevertheless, the need to pinpoint the critical locations for sampling, amidst miles of tubing, makes it highly desirable to use destructive tests in conjunction with nondestructive evaluations. Removal of samples for destructive tests is much more difficult in the case of pipes; hence, emphasis has to be placed more on nondestructive procedures for a different reason than for tubes. The predominant damage mechanism for tubes is creep. On the other hand, pipes are subject to low-cycle-fatigue damage (creep-fatigue) because of their larger section sizes and the associated thermal and stress gradients. Most of the available life-prediction techniques — analytical methods, rupture tests, and metallographic methods — lead only to estimates of crack-initiation life for headers and steam pipes. Estimation of total life requires crack-growth assessment.

For tubes, crack-initiation-based approaches are adequate for remaining-life prediction because, for thin sections, creep-rupture life can be considered to be identical to crack-initiation life. For pipes, however, both crack initiation and crack propagation must be considered. The techniques discussed in the subsequent subsections on calculational methods, extrapolation of

statistics of past failures, dimensional measurements, metallographic methods, methods based on temperature estimation, postservice creep and rupture tests, and removal of samples from components deal mainly with crack initiation and the incipient damage events preceding it. Assessments made using these techniques must be supplemented by crack-growth analysis in the case of heavy-section components. Furthermore, if nondestructive evaluation of the component has already revealed the presence of cracks or flaws, crack-initiation-based assessment procedures may be irrelevant.

A complete matrix of equipment areas to be evaluated in a boiler, together with inspection techniques and anticipated primary failure mechanisms, are summarized in Table 5.5 (Ref 20). Because the procedures for conventional NDE techniques are well documented in the literature and are beyond the scope of this book, they will not be discussed further.

Unique Problems in Assessment of Welds

Because most of the problems encountered in heavy-section piping occur at welded joints, damage-assessment techniques need to focus on these regions rather than on the base metal. There are several problems unique to weldments that make both material characterization and stress analysis extremely challenging.

The microstructure of a fusion weld in a pipe can be very complex and may comprise seven distinct zones, as shown in Fig. 5.12. In such a weld, the metal adjacent to the fusion zone is rapidly heated to temperatures approaching the melting point of the weld metal and then is cooled at rates determined by the conductivity (and hence the mass) of the surrounding metal. The microstructure that develops in the heat-affected zone is determined by the thermal cycle, the kinetics of austenite formation, the grain-growth kinetics, and the relevant continuous cooling transformation reaction. The welding thermal cycle produces peak temperatures and cooling rates that are highest at the fusion boundary. In a single-pass

Table 5.5. Equipment areas to be evaluated in a boiler, inspection techniques, and primary failure mechanisms expected (Ref 20)

Boiler area	Tube samples required	Boat samples for testing	Replicas	Inspection methods(a)	Primary failure mechanism(b)
Secondary superheater:					
Outlet header		X	X	ABCHEFK	123
Tube bank(s)	X			G	1237
Inlet header			X	EHK	123
Attemperator				CF	125
Connection piping		X	X	GM	123
Reheater/superheater:					
Outlet header(s)		X	X	ABCEHFK	123
Tube bank(s)	X			G	12347
Inlet header			X	EHK	123
Attemperator				CF	125
Connection piping		X	X	GM	123
Primary superheater:					
Outlet header(s)			X	ABCEFHK	123
Tube bank(s)	X			G	1234
Inlet header(s)			X	EHK	235
Connection piping			X	GM	235
Economizer:					
Inlet header(s)				AFK	2356
Tube bank(s)	X			GL	2346
Outlet header(s)				AF	2346
Furnace enclosure tubes	X			DGL	2345
Convection pass enclosure				DGL	234
Drum				EHJ	2356
Waterwall headers or collection headers				AC	23
Waterwall tubes at attachments	X			DL	23

(a) A – Header stub and hard hole cap removal and internal inspection. B – Header dimensional measurements. C – Ultrasonic flaw detection (angle/beam). D – Radiography. E – Dye penetrant. F – Fiber optic probe. G – Ultrasonic thickness testing (scope-type). H – Magnetic particle. I – Field alloy detector. J – Wet-fluorescent magnetic particle. K – Stub tube magnetic particle. L – Tube removal at attachments. M – Strain monitoring (dimensional).
(b) 1 – Creep. 2 – Fatigue. 3 – Corrosion. 4 – Erosion. 5 – Thermal shock. 6 – Deposition. 7 – DMW.

weld in ferritic steel, four distinct regions can be identified in the HAZ alone (see Fig. 5.12):

- Coarse-grain HAZ—a zone of material several grains in width adjoining the fusion boundary where grain growth has occurred as a result of high-temperature austenitization
- Grain-refined region—material just beyond the coarse-grain region that has recrystallized at a lower austenitization temperature
- Intercritical region—an area adjoining

the base metal where, due to even lower temperatures, the austenite phase has only partly transformed
- Tempered heat-affected zone.

The HAZ of a multiple-pass weld is even more complex, and, as a result of multiple heat treatments, may contain more than four regions. The weldment is thus a composite material consisting of the base metal, three or more HAZ regions, and the weld metal. Mechanical properties can vary from zone to zone in an unpredictable manner, and modeling of the mechanical behavior

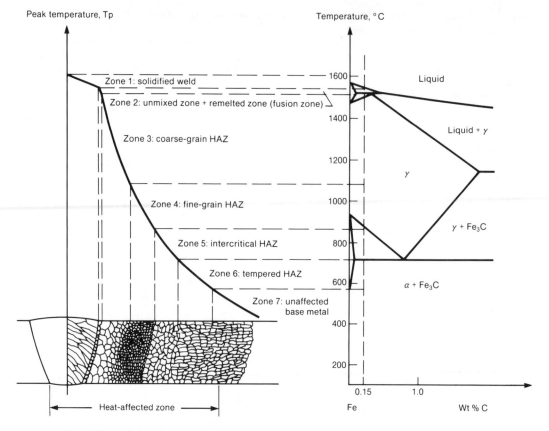

Fig. 5.12. Schematic diagram of the unique heat-affected zones expected in a low-alloy steel (courtesy of S.R. Paterson, APTECH Engineering).

of the composite configuration can be a nightmare for the stress analyst. Cracking in header welds has been observed in all of the zones described above. In addition, the fusion line itself, being the last area to solidify, provides an ideal location for collection of impurities, inclusions, and other debris.

The stress state at a pipe weld is multiaxial because of the complex geometry and the influence of the internal pressure and piping-system loads. Because stress state is known to influence rupture life, application of uniaxial rupture data to complex weldments detracts from the accuracy of the analysis.

As a result of the variable creep resistance of the base metal, weld metal, and HAZ, redistribution of the initial elastic stresses can take place in complex ways. These redistributions can be different for girth welds and for seam welds. It has been

shown (Ref 21) that in girth welds the pressure stresses in the axial and hoop directions peak at the most creep-resistant location (i.e., the HAZ) and are lowest at the least creep-resistant location (i.e., the weld metal). In contrast, the work of Wells (Ref 22) suggests that in seam welds the peak stresses after stress redistribution occur at the weakest location (i.e., the weld metal) whereas the lowest stresses are encountered at the strongest location (i.e., the HAZ). Furthermore, the near-midwall cusp regions (at weld-bead intersections) at the fusion line of a double-V seam weld undergo a peak hydrostatic stress as a result of stress distribution and weld-metal creep resistance that is lower than that of the base metal.

Material-property data pertaining to welds are very scanty. The limited data that are available relate only to pure base metal, only to weld metal, or only to simulated HAZ material. It is not even clear that the use of these data for prediction of the behavior of composite welds is relevant. What is needed is generation of data in the laboratory under conditions that simulate those encountered in service and yet are sufficiently accelerated to produce results in short periods of time. This is far from easy to do. Attempts to grow cracks in laboratory samples at locations similar to those of field failures have often been unsuccessful. Thus, neither the kind of data that needs to be generated nor the manner of obtaining it is readily apparent. In the absence of relevant data, the most convenient practice has been simply to assume that a weld is always weaker than the base metal and then model the weld behavior simply by modifying the base-metal properties by arbitrary safety factors.

Welds also serve as locations for fabrication defects. These defects can act as stress concentrations and preferred sites for damage. They are often undetectable by conventional NDE techniques, and yet there is no effective way of taking this into account during stress analysis. Recent work by Henry, Ellis, and Lundin (Ref 22a) has shown that weld metal in submerged-arc test welds in 2¼Cr-1Mo steel made with an "acid-type" flux exhibited creep-rupture strengths even lower than the lowest values obtained in the base metal. The degradation in rupture strength was accompanied by a decrease in rupture ductility. Use of a "neutral" flux resulted in rupture properties approaching the lower-bound values for the base metal; use of a "basic-type" flux resulted in rupture properties approaching the mean values for the base metal. The adverse effect of the acid flux was attributed to the presence of large amounts of manganese sulfide inclusions, which promoted creep-cavitation.

Calculational Methods

The analytical methods are aimed at estimating life expenditure based on operating history, component geometry, and material properties. Although uncertainties in these parameters, as well as in the calculational methods used, lead to inaccuracies in the results, analytical procedures are a necessary first step in identifying potential damage sites and damaging transients. Without such analysis, isolated problems could be found using nondestructive examination, but it would be difficult to identify the root cause of the problem.

For headers, the principal loadings that cause stresses are due to thermal transients and pressure. Ideally, it would be desirable to continuously monitor and record these loadings (steam pressure and temperature, flow, and outside metal temperature). Boiler stress and condition analyzers (BSCA) which perform such monitoring and calculate the life expenditure on a continuous basis have been developed recently (Ref 23 and 24). Most power plants, however, have not readily accepted this concept because of the cost involved. Furthermore, for plants which have already operated for many years, a procedure for ascertaining the current condition is needed, although the BSCA can monitor further damage in the future. A common approach therefore is to monitor the plant operation over a brief period considered to be "typical" of the plant. Then, from the record, the number of each type of cycle (hot starts, warm starts, cold starts, steady state, load fluctuations, etc.) can be determined and included in the analysis. With these loadings, heat-transfer and stress analysis can be performed on the component and the stresses then can be utilized for computing damage due to fatigue, ratcheting, and creep-rupture.

Once the loading, geometry, and material properties of the component are known, the question arises as to what type of analysis should be performed. A three-dimensional finite-element analysis including plasticity and creep, even for simple loadings, can be time-consuming and expensive. If plasticity can be ignored, the computing procedure is much simpler. ASME Code Case N-47, for instance, allows an elastic analysis and provides simplified rules for accommodating creep and fatigue. This procedure, however, overestimates the steady-state stresses and leads to unduly pessimistic life predictions (Ref 25). It is possible to take plasticity and creep into account, yet simplify the analysis procedure by simplification of the component geometry. One such procedure and a computer code named CREPLACYL have been described in the literature (Ref 25 and 26). The stress-analysis procedure is based on a one-dimensional, generalized plane-strain solution for a hollow cylinder. The program performs transient stress analysis with inelastic material behavior (including plasticity and creep). For analysis of highly stressed areas such as tees and bore holes, stress-concentration factors are utilized with respect to the hoop stress, and the axial stresses are suitably adjusted to ensure axial equilibrium. Results from this simplified procedure are purported to agree reasonably well with those from more elaborate three-dimensional analyses (Ref 25).

Typically, the various types of start-ups in a boiler are categorized into hot starts, warm starts, and cold starts. Definitions vary, but essentially they are related to the temperature difference between the start-up and the final operating conditions. An illustrative set of definitions might be as follows:

- Start-up after 8 h at 425 °C (800 °F)–hot start (daily)
- Start-up after 8 to 72 h at 260 °C (500 °F)–warm start (after a weekend)
- Start-up after 72 h or more at 21 °C (70 °F)–cold start (weekly)

For instance, a complete weekend shutdown followed by a start-up from room temperature would constitute a warm start, whereas an overnight lay-up at 425 °C (800 °F) followed by a start-up would be defined as a hot start. The number of each type of start-up can be calculated from plant records.

During a start-stop or other cycle of a heavy-wall component such as a header, the damage mechanisms due to creep and fatigue can be broken up into three components: (1) pure fatigue damage due to repeated cycles of imposed strain (stress) range; (2) creep damage by stress relaxation from the peak stress to the steady-state stress during the holding period at the operating temperature; and (3) stress-rupture (creep) damage at the steady-state stress during steady operation. If the transients are not severe and the thermally induced strains do not exceed the yield strain, then the stress-relaxation component (item 2) becomes negligible. If the peak temperature is below the creep regime, then both the stress-relaxation and creep-rupture components (items 2 and 3) will be negligible and only pure fatigue will need to be considered. If the plant operates under steady base-load conditions, only the creep component is of importance.

For assessing creep-rupture life consumption alone, the time-temperature history of the component is reviewed. The creep-life fraction consumed for each time-temperature segment of the component history is then calculated, and all fractions are summed using the lower-bound ISO or other standard rupture data in conjunction with the life-fraction rule. The stress value used may be the mean-diameter elastic hoop stress, the hoop stress taking time-dependent redistribution into account, or a reference stress. These procedures were described adequately in Chapter 3.

The steps involved in calculating the applicable strain range for fatigue-life calculations include: (1) determining the history of the changes in the hoop strain ϵ_h, the radial strain ϵ_r, and the axial strain ϵ_θ at the critical location of the component; (2) selecting the point at which conditions are extreme for the cycle, either maximum or minimum, referred to by the subscript i; and (3) determining the history of the changes in strain differences by subtracting the value at time i from the corresponding values at each point in time during the cycle. These strain-difference changes may be designated as

$$\Delta(\epsilon_r - \epsilon_\theta) = (\epsilon_r - \epsilon_\theta) - (\epsilon_r - \epsilon_\theta)_i \quad \text{(Eq 5.1)}$$

$$\Delta(\epsilon_\theta - \epsilon_h) = (\epsilon_\theta - \epsilon_h) - (\epsilon_\theta - \epsilon_h)_i \quad \text{(Eq 5.2)}$$

$$\Delta(\epsilon_h - \epsilon_r) = (\epsilon_h - \epsilon_r) - (\epsilon_h - \epsilon_r)_i \quad \text{(Eq 5.3)}$$

The maximum value among the three quantities on the left-hand sides of the above equations during a given cycle gives the applicable $\Delta\epsilon_{max}$ in accordance with the Tresca criterion. Alternatively, the equivalent strain range can be computed using the Von Mises criterion and Eq. 5.1 to 5.3, as follows:

$$\Delta\epsilon_{eq} = \frac{\sqrt{2}}{3} \{ [\Delta(\epsilon_r - \epsilon_\theta)]^2 + [\Delta(\epsilon_\theta - \epsilon_h)]^2$$
$$+ [\Delta(\epsilon_h - \epsilon_r)]^2 \}^{1/2} \qquad \text{(Eq 5.4)}$$

Both the Tresca criterion (Ref 25) and the Von Mises criterion (Ref 26) have been used in the literature.

For calculation of life under pure fatigue, the number of allowable cycles to failure, N_f, is estimated by entering the appropriate $\Delta\epsilon$-vs-N_f curve at the maximum value of $\Delta\epsilon$ or $\Delta\epsilon_{eq}$ for each type of transient. The fractional damages due to the various types of transients can be calculated and summed to arrive at the total fatigue damage. The fatigue curves used for this may be the $\Delta\epsilon$-vs-N_f curves for pure fatigue at the appropriate temperature, corrected by a factor of 2 in strain or a factor of 20 in cycles. The correction factors are intended to account for the effects of environment, specimen size, surface finish, and data scatter (Ref 25).

The specific fatigue curves used vary from user to user and are based on unpublished experience. Possible bases for these curves are the design curves contained in the ASME Boiler and Pressure Vessel Code, Section VIII, Division 2, and the curves in ASME Code Case N-47. The ASME code contains design fatigue curves for a variety of heat treated steels, including 2¼Cr-1Mo steel, for temperatures up to 370 °C (700 °F). ASME Code Case N-47 contains design fatigue curves for annealed 2¼Cr-1Mo steel for temperatures up to 595 °C (1100 °F). Figure 5.13 shows both the design fatigue curve applicable to heat treated 2¼Cr-1Mo steel from Section VIII, Division 2, and the design fatigue curves for annealed 2¼Cr-1Mo steel from Code Case N-47 (Ref 27).

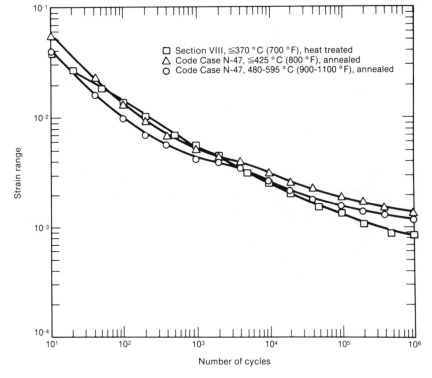

□ Section VIII, ≦370 °C (700 °F), heat treated
△ Code Case N-47, ≦425 °C (800 °F), annealed
○ Code Case N-47, 480-595 °C (900-1100 °F), annealed

Fig. 5.13. Design fatigue curves for 2¼Cr-1Mo steel, from Section VIII (Division 2) and Code Case N-47 of the ASME Boiler and Pressure Vessel Code (Ref 27).

These curves incorporate the safety factors (2 in $\Delta\epsilon$, 20 in N_f) mentioned previously.

For creep-fatigue situations, two alternative routes are possible. In the first route, the following steps can be taken: (1) the pure fatigue component is calculated as discussed above; (2) the stress-relaxation damage component is calculated by linear summation of the time life fractions expended at the various stress levels during the stress relaxation; and (3) the rupture damage is calculated by summing the time life fractions expended at the steady, relaxed stress. Linear addition of the three components gives the cumulative life fraction expended. When this value approaches unity, failure is deemed to have occurred, from a crack-initiation point of view.

In the second route, procedures similar to (but not necessarily the same as) either of the options (elastic or inelastic) described in ASME Code Case N-47 can be employed (see the subsection "Design Rules for Creep-Fatigue" under "Creep-Fatigue Interaction" in Chapter 4). If the elastic option is used, fatigue curves incorporating hold-time effects (see Fig. 4.16) and additional factors of safety are used to calculate the fatigue damage. Creep damage is calculated using time life-fraction rules. Linear addition of the damage fractions is used with total damage, $D = 1$, as the failure criterion. In the inelastic option, pure fatigue curves similar to those shown in Fig. 5.13 are employed to compute fatigue damage, and failure is governed by the experimentally determined values of D at failure. This failure envelope has not been defined experimentally for many steels, particularly in the heat treated condition. An example of the types of results obtained can be illustrated by use of the data on annealed material according to ASME Code Case N-47, as shown in Fig. 5.14. In the absence of such data for heat treated low-alloy steels, the elastic route has been found to be more attractive by some investigators (Ref 26).

Figure 5.15, from the work of Masuyama *et al* (Ref 28), is intended to show that the actual failures of various components at the

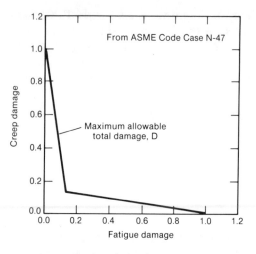

Fig. 5.14. Creep-fatigue damage envelope for 2¼ Cr-1Mo steel, from Code Case N-47 of the ASME Boiler and Pressure Vessel Code.

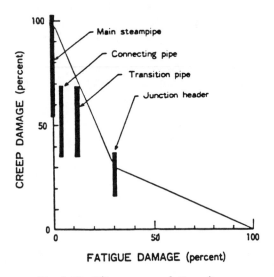

Fig. 5.15. Bilinear creep-fatigue linear damage curve and validation of actual failures for a type 316 stainless steel component (Ref 28).

Eddystone Power Plant could be predicted by the above type of analysis using a bilinear creep-fatigue damage curve. Results of their creep-fatigue analysis agreed with the actually observed degrees of damage in the various components.

Examples of fatigue data from tests of 2¼ Cr-1Mo and 1Cr-½Mo steels incorporating hold times were presented in Fig. 4.16

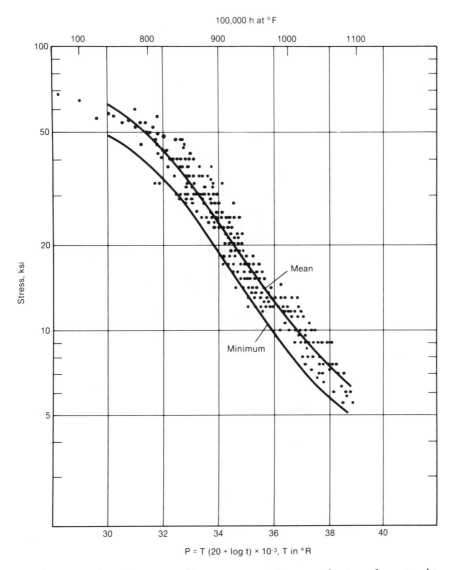

100,000 h at °F

Fig. 5.16. Variation of Larson-Miller rupture parameter with stress for wrought 1¼ Cr-½ Mo-Si steel (Ref 29).

and 4.17 in Chapter 4. Stress-rupture data needed for calculating creep-rupture damage fractions are shown in Fig. 5.16 and 5.17 (Ref 29 and 30). Low-cycle fatigue data from tests with hold times, as shown in Fig. 4.16, generally are converted into working curves of life consumption that define constant damage levels for various ramp rates, such as those shown in Fig. 5.18 for 2¼ Cr-1Mo steel header T sections (Ref 26). These cyclic-life-expenditure curves are included for illustration purposes only, and are not to be

used in a generic sense. They are specific to certain geometric, operational, and material variables that have been assumed for purposes of their derivation.

Extrapolation of Statistics of Past Failures

For components, such as boiler tubes, which are numerous in a given plant, which perform identical functions, and whose failure is not critical to plant safety, a history of past failure rates can provide a useful index

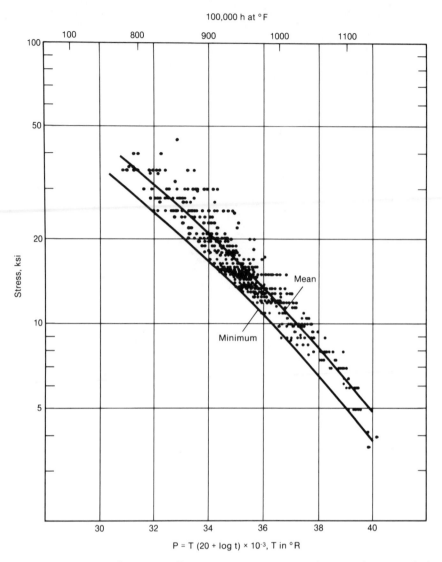

Fig. 5.17. Variation of Larson-Miller rupture parameter with stress for annealed 2¼Cr-1Mo steel (Ref 30).

of potential future failure rates. An example of a plot of cumulative percent failures vs time for dissimilar-metal weld failures in boiler tubing is shown in Fig. 5.19 (Ref 31). This plot is based on a sample population of 9474 welds made with austenitic filler metals and 103,554 welds made with Inconel filler metals in U.K. plants as reported by Nicholson and Price (Ref 32). It can be seen from the figure that a significant percentage of failures in austenitic welds do not occur until about 2×10^5 h of service and that Inconel welds would have much longer lifetimes at 565 °C (1050 °F). If an acceptable

limit for cumulative failures could be established, such data could be used to determine when corrective actions are necessary. In applying such methods, one should be careful to include only components of similar composition, heat treatment, operating history, and failure mechanism in the population. Unfortunately, detailed information of this type is rarely collected in plants and hence this method has only limited value.

Dimensional Measurements

Dimensional measurements sometimes can provide estimates of the remaining lives of

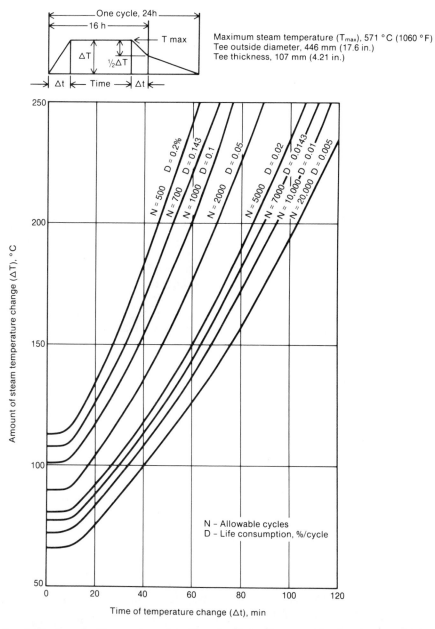

Fig. 5.18. Sample life-consumption chart for superheater outlet header tees with four steam leads from a 350-MW, 16.5-MPa (2.4-ksi), 565/540 °C (1050/1000 °F) middle load unit (Ref 26).

boiler components. At present, some utilities determine component strain from diametral measurements on pipes made under ambient conditions with bow micrometers. An accuracy of 0.1% can be achieved with this method. The basis of such analysis is determination of the accumulated strain from the measured distortions. The strain is then compared with the expected failure strain for the material under service conditions. If measurements are made at frequent intervals, the creep rate for the material can be established and the remaining life can be estimated on the basis of an assumed Monkman-Grant constant. A sudden increase in the creep rate may be indicative of entry into

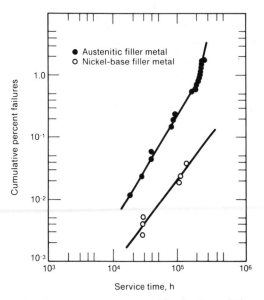

Based on experience reported by the Central Electricity Generating Board (U.K.) with dissimilar-metal fusion welds in 2¼Cr-1Mo steel tubes (Ref 32). Data are for 9474 welds made with austenitic filler metals and 103,554 welds made with nickel-base (Inconel) filler metals.

Fig. 5.19. Use of statistics of past failures to predict future failure rates (Ref 31).

the tertiary creep stage and hence provide a forewarning of failure.

Based on a generalized damage model proposed by Rabotnov (Ref 33) and Kachanov (Ref 34), Cane has proposed a model for life prediction based on component strain or strain-rate measurements (Ref 35), as follows:

$$\text{LFR} = \left(1 - \frac{t}{t_r}\right) = \left(1 - \frac{\epsilon}{\epsilon_r}\right)^{\epsilon_r/\epsilon_s}$$

$$\text{(Eq 5.5)}$$

and

$$\dot{\epsilon}_i t = \epsilon_s \left(\frac{t}{t_r}\right)\left(1 - \frac{t}{t_r}\right)^{(\epsilon_s/\epsilon_r)-1} \quad \text{(Eq 5.6)}$$

where LFR is life fraction remaining, t is service time, t_r is expected rupture life, t/t_r is life expended, ϵ is measured strain at time t, ϵ_r is rupture strain, $\epsilon_s = \dot{\epsilon} t_r$ (the Monkman-Grant constant, where $\dot{\epsilon}$ is the minimum

creep rate), and $\dot{\epsilon}_i$ is the strain rate measured at time t. For relatively ductile materials, such as normalized-and-tempered boiler steels, the LFR was found to be relatively insensitive to the chosen value of ϵ_r. ϵ_s has been shown to be a constant, particularly at low stresses, and can be assumed to be about 3% (or 0.03) for low-alloy steels. From Eq. 5.5 and 5.6, the LFR can be calculated on the basis of a one-time measurement of strain or strain rate.

Example:
Dimensional measurements on a uniaxially loaded component after 100,000 h of service show a creep strain of 1%. Assume a Monkman-Grant constant of 3% and calculate the remaining life. Show that the value is insensitive to the choice of rupture strain, ϵ_r. The following information is known:

$$t = 100,000 \text{ h}$$

$$\epsilon = 1\%$$

$$\epsilon_s = 3\%$$

$$\epsilon_r = 10\%$$

Substituting into Eq. 5.5, we get

$$\text{LFR} = \left(1 - \frac{1}{10}\right)^{10/3} = 0.7$$

Expended life fraction (0.3) = 100,000 h

$$\text{Remaining life} = \frac{0.7}{0.3} \times 100,000$$

$$= 233,000 \text{ h}$$

If we assume that $\epsilon_r = 20\%$, then

$$\text{LFR} = \left(1 - \frac{1}{20}\right)^{20/3} = 0.71$$

It can be seen that the LFR is relatively insensitive to the chosen value of rupture strain, ϵ_r, in the Cane model.

In applying the Cane model to actual components, the multiaxial state of stress has to be taken into account and the appropriate values of ϵ and ϵ_s must be substituted into the equation. An example of this procedure is given in Ref. 35.

For practical application of strain-based techniques, there are several limitations. The initial dimensions of the component usually are not known precisely except within a range of tolerances. Although calculation of strains from dimensional changes may be straightforward for simple geometries, such calculations for complex shapes can be time-consuming and expensive. The failure strain (i.e., the creep ductility) is a function of several metallurgical, geometric, and operational variables and hence can be estimated only within large degrees of scatter. Finally, the dimensional measurements are not capable of resolving localized damage such as damage in weld heat-affected zones.

The applications and limitations of both off-line and on-line strain-monitoring techniques have been reviewed by Cane and Williams (Ref 36). They estimate that, to obtain an accuracy of 10% in predicted life, it is necessary to achieve an accuracy of better than 0.01% in strain. Such accuracy requires measurement of creep rates from 10^{-7} to 10^{-9} per hour.

For accurate measurement of strains when the plant is "off load," Cane and Williams have described a convenient system using "creep pips." This system involves fiducial markers attached to the component surface that can be measured during inspection periods. The markers should be unaffected by oxidation and should not interfere with conventional NDE procedures. At present, these markers are projections welded to or peened into the surface and are tipped with Stellite or austenitic materials to minimize oxidation. The pips are positioned across the section of interest and their spacing can be measured at each outage, using micrometers or bow gages, to an accuracy of ± 25 μm (± 1.0 mil). Developments in the areas of stereophotography and laser holography that can permit more accurate and detailed measurements in the laboratory based on photographic or holographic rec-

ords taken on site have also been reviewed (Ref 36).

On-line monitoring of strain has been limited by a lack of availability of strain gages that can withstand the high-temperature environment and temperature fluctuations. Conventional resistance-type strain gages are usable at temperatures up to only 300 °C (570 °F). Capacitance-type strain gages have been successfully used in the laboratory at temperatures up to 600 °C (1110 °F) and for times up to 20,000 h. Their in-plant use has been more limited, but is promising (Ref 36).

In the case of superheater/reheater tubes, fire-side corrosion can lead to a decrease in wall thickness and a consequent increase in stress and decrease in rupture life. Assuming a linear corrosion rate and the applicability of the linear damage rule, Moles and Westwood (Ref 37) have derived a relationship for estimation of remaining life under wall-thinning conditions, as follows:

$$t_{nr} = \frac{1}{K'} \{ 1 - [1 + K'(n-1)t_r]^{1/(1-n)} \}$$

(Eq 5.7)

where K' is wall-thinning rate (h^{-1}), n is stress sensitivity (Norton law exponent), t_r is time to rupture for a tube with no wall thinning, and t_{nr} is service rupture life under wall-thinning conditions. The wall-thinning rate K' is defined by the expression

$$K' = \frac{w_i - w_f}{w_i t_{op}}$$

where w_i is initial tube wall thickness, w_f is final tube wall thickness, and t_{op} is operating time in service (h). K' is assumed to be constant, and n (the stress sensitivity) ranges from 4 to 8 for ferritic steel tubes. An n value of 4 is generally used because this parameter has little effect on service-life predictions where high wall-thinning rates are encountered.

The above model is very useful for estimating the remaining life of tubing in a SH/RH section when a tube in that section has failed as a result of high-temperature fire-side corrosion. The service life of the tube that has failed, t_{nr}, is its operating time in service, t_{op}. The wall-thinning rate K' can be calculated by measuring the final wall thickness of the failed tube. Assuming an n value of 4, the time to rupture, t_r, for a tube with no wall thinning can be calculated. Using this value of t_r, a curve is constructed on a graph of K'-vs-t_{nr} values which then can be used to estimate the remaining life of a tube that has not failed in that SH/RH section. The present ultrasonic thickness measurement of an unfailed tube is obtained and used as the final tube wall-thickness value in calculating the value of K' for that tube and obtaining its t_{nr} value. This technique has been applied widely in the Ontario Hydro System (Ref 37).

Example:
A reheater tube failed after 57,000 h during which the wall had thinned from 3.81 to 1.02 mm as a result of hot corrosion. Calculate the expected rupture life for a similar tube without wall thinning. Assume that n = 4. Calculate the remaining life for an adjoining tube that has thinned to a wall thickness of 2.49 mm.

Answer:
For the failed tube,

$$K' = \frac{w_i - w_f}{w_i t_{op}} = \frac{3.81 - 1.02}{3.81 \times 57,000}$$
$$= 128 \times 10^{-7} \text{ h}^{-1}$$

$$n = 4$$

$$t_{nr} = 57,000 \text{ h}$$

Substituting K', n, and t_{nr} into Eq 5.7, we get, if no wall thinning has occurred, a time to rupture of

$$t_r = 1.3 \times 10^6 \text{ h}$$

For the unfailed tube,

$$K' = \frac{w_i - w_f}{w_i t_{op}} = \frac{3.81 - 2.49}{3.81 \times 57,000}$$
$$= 61 \times 10^{-7} \text{ h}^{-1}$$

Substituting $K' = 61 \times 10^{-7}$, n = 4, and $t_r = 1.3 \times 10^6$, we can calculate service rupture life as

$$t_{nr} = 106,600 \text{ h}$$

Because the tube has already operated for 57,000 h, the remaining life is 49,600 h.

Metallographic Methods

Metallographic methods have been developed that can correlate either cavitation evolution or changes in carbide spacings with creep-life expenditure. It has been observed that, in boiler piping, cavitation is the principal damage mechanism at brittle zones, weld heat-affected zones, and high-stress regions in the base metal (Ref 13). In the other cases, carbide coarsening was found to be a better indicator of life consumption. Results pertaining to both of these changes are described below.

Creep-Cavitation Model for Heat-Affected Zone. The evolution of creep cavitation in 2¼Cr-1Mo steels in the simulated-HAZ condition has been investigated by Lonsdale and Flewitt (Ref 38). Uniaxial creep tests were conducted at temperatures ranging from 838 to 923 K and at stresses from 55 to 76 MPa. The number of cavities was measured by scanning electron microscopy (SEM) examination of samples at various stages of the creep curve. It was observed that the number of cavities per square millimetre, N, was related to the minimum creep rate and time t (in seconds), independent of stress, by the equation

$$N = (3.3 \times 10^5 \, \dot{\epsilon}t) - (3.3 \times 10^3) \quad \text{(Eq 5.8)}$$

In a plant component, if N can be determined at the critical location, the instantaneous value of $\dot{\epsilon}$ can be calculated, because t is known. This value can then be used with a strain-based failure criterion to define the remaining life.

The first published attempt to relate creep-life consumption of plant components to cavitation was that of Neubauer and Wedel (Ref 39). They characterized cavity evolution in steels at four stages—i.e., isolated cavities, oriented cavities, linked cavities (microcracks), and microcracks—as shown in Fig. 5.20. Based on extensive observations on steam pipes in German power plants, they estimated the approximate time intervals required for the damage to evolve from one stage to the next under typical plant conditions. Using this experience, they formulated recommendations corresponding to the four stages of cavitation. For class A damage, no remedial action would be required. For class B damage, consisting of oriented cavities, reinspection within 1½ to 3 years would be required. For class C damage, repair or replacement would be needed within six months. For class D damage, immediate repair would be required. Wedel and Neubauer built considerable conservatism into their recommendations and viewed their technique more as a monitoring technique than as a life-prediction technique. Their procedure, because of its simplicity, has found worldwide support with power-plant operators.

To provide a theoretical and quantitative basis for cavity evolution, Cane *et al* used a constrained-cavity-growth model and proposed the following relationship (Ref 40 and 41):

$$A = 1 - \left(1 - \frac{t}{t_r}\right)^{(\lambda'-1)/n\lambda'} \qquad \text{(Eq 5.9)}$$

where A is the number fraction of cavitated boundaries, t/t_r is the expended creep-life fraction, n is the stress exponent for creep, and $\lambda' = \epsilon_r/\epsilon_s$ (where ϵ_r is the strain at rupture and ϵ_s is the secondary creep strain— i.e., $\dot{\epsilon} \times t_r$). For assumed values of $\lambda' = 2.5$ and n = 3, they found good correlation between the theoretically predicted shape of the A-vs-t/t_r curve and the curve based on experimental data (Ref 40). A consolidation

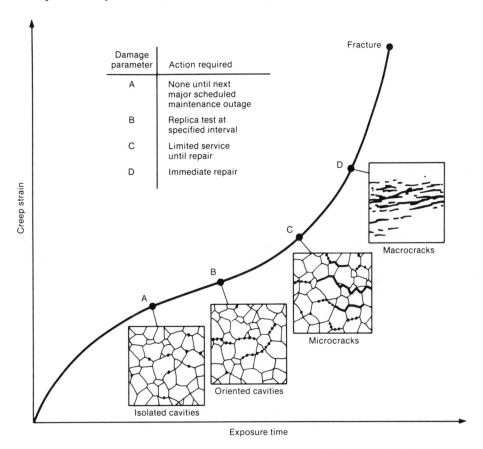

Fig. 5.20. Creep-life assessment based on cavity classification (Ref 39).

of their preliminary results is presented in Fig. 5.21 (Ref 12). Unfortunately, values of n and λ' that are specific to a given heat of steel used in a plant cannot be determined nondestructively. Hence, plant remaining life will have to be estimated with reference to a master plot based on laboratory data, as shown in Fig. 5.21. Alternatively, if A could be monitored on a plant component at several intervals, specific values of λ' and n could be ferreted out for future extrapolations.

Extensive laboratory tests have been carried out by Shammas *et al* (Ref 41) on 1Cr-0.5Mo steels to verify the model proposed by Cane *et al*. Simulated HAZ materials containing different impurity levels were creep tested in the range 550 to 605 °C (1020 to 1120 °F). The tests were interrupted at various creep-life fractions and the cavity distributions were measured on metallographic sections. The volume fraction of cavities, f_v, as well as the number fraction of cavitated boundaries, A, were both measured because either of them could be used as the damage parameter in the model. In the final analysis, A was chosen as the damage parameter for the following reasons: (1) f_v is very sensitive to specimen preparation and etching conditions whereas A is not; (2) cavity link-up can lead to erroneous estimation of f_v, but has no effect on A; (3) it is more tedious and time-consuming to measure f_v than to measure A; and (4) measurement of A requires only an optical microscope, which can be transported to the site, whereas measurement of f_v may require a scanning electron microscope (SEM), which is more expensive to buy, maintain, and transport.

Initial examination of the results of Shammas showed the data to have too much scatter to verify the life-prediction model. The data could nevertheless be used empirically, because distinct correlations between A and t/t_r could be established at different temperatures, as shown in Fig. 5.22 (Ref 42). The lower limit lines for the data at 550 and 575 °C (1020 and 1065 °F) could be described by the equations

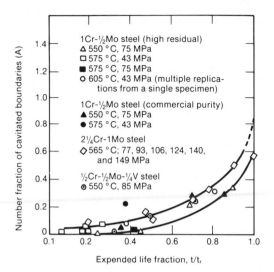

Fig. 5.21. Evolution of creep-cavitation damage with expended life fraction for ferritic steels (Ref 12).

$$A = 0.51 \left(\frac{t}{t_r} \right) - 0.095 \text{ [at 550 °C]}$$

$$\text{(Eq 5.10a)}$$

and

$$A = 0.78 \left(\frac{t}{t_r} \right) - 0.134 \text{ [at 575 °C]}$$

$$\text{(Eq 5.10b)}$$

Shammas *et al*, however, believe that grouping of the data into two sets on the basis of temperature is not justified in view of the data scatter. They have plotted all the data together in the form of a scatterband whose lower limits are defined by the equation

$$A = 0.517 \left(\frac{t}{t_r} \right) - 0.186 \quad \text{(Eq 5.10c)}$$

A method of assessment that is an alternative to the A parameter has also been proposed by Shammas. The damage classifications have been correlated with life fractions (see Fig. 5.23), and thus a life-fraction range has been established for each class. By the use of Fig. 5.23, it can be established that the Wedel-Neubauer clas-

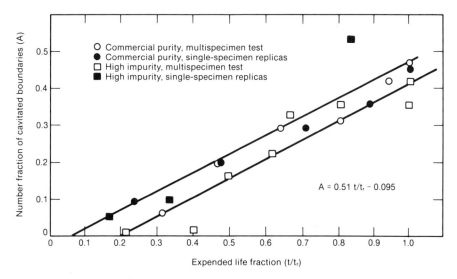

Fig. 5.22. Evolution of creep-cavitation damage with expended life fraction for 1Cr-½Mo steels tested at 550 °C (1020 °F) (Ref 42).

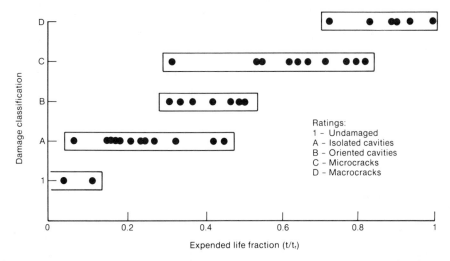

Fig. 5.23. Correlation between damage classification and expended creep-life fraction for 1¼Cr-½Mo steels (Ref 41).

sifications of material condition (undamaged, class A, class B, class C, and class D) correspond roughly to expended-life-fraction (i.e., t/t_r) values of 0.12, 0.46, 0.5, 0.84, and 1, respectively.

Three methods have been suggested by which the results of Shammas *et al* can be applied to a plant component to determine its remaining life. As a first step it is convenient to express the life fraction expended, t_{exp}/t_r, in terms of the remaining life as follows:

$$t_{rem} = t_{exp}\left(\frac{t_r}{t_{exp}} - 1\right) \qquad \text{(Eq 5.11)}$$

The first method of estimating the remaining life is to measure A and then apply the model by combining Eq. 5.11 with Eq. 5.9:

$$t_{rem} = t_{exp}\left[\frac{1}{1 - (1 - A)^{n\lambda'/(\lambda' - 1)}} - 1\right]$$

$$\text{(Eq 5.12)}$$

Unfortunately, values of n and λ' specific to the steel used in the component cannot be obtained without conducting creep tests. Hence, conservative values of n = 3 and $\lambda' = 1.5$ have to be assumed. Equation 5.11 now becomes

$$t_{rem} = t_{exp} \left[\frac{1}{1 - (1 - A)^9} - 1 \right]$$

$$\text{(Eq 5.13)}$$

By substituting the observed value of A and the known value of t_{exp}, the remaining life can be estimated.

The second method is to use the correlation between A and t_{exp}/t_r given by Eq 5.10. Substituting Eq 5.10 into Eq 5.11, we get

$$t_{rem} = t_{exp} \left(\frac{0.51}{A + 0.095} - 1 \right) \quad \text{(Eq 5.14)}$$

By substituting the observed value of A and the known value of t_{exp}, the remaining life can be estimated.

The third method is to use the correlation between the Neubauer and Wedel classification and t_{exp}/t_r provided in Fig. 5.23. The t_{exp}/t_r value corresponding to the class of cavitation observed can be plugged into Eq 5.10 to determine directly the remaining life.

Example:
Replica examination of a superheater outlet header which had been in service for 100,000 h showed the presence of class A (isolated) cavities. The number fraction of boundaries containing creep cavities was estimated to be 0.109. Estimate the remaining life using the three methods discussed above.

Answer:
Method 1: A = 0.109; t_{exp} = 100,000 h. Substituting these into Eq 5.13, we get t_{rem} = 54,770 h.
Method 2: A = 0.109; t_{exp} = 100,000 h. Substituting these into Eq 5.14, we get t_{rem} = 150,000 h.

Method 3: From Fig. 5.20, t_{exp}/t_r = 0.46, corresponding to cavity classification A. Hence, t_{rem} = 117,000 h.

Method 1 gives a very conservative value, whereas methods 2 and 3 agree closely. Comparison of predicted values with actual values of remaining life during validation tests performed by Shammas *et al* also showed that the predictions from the theoretical model were always overly conservative.

In using methods 1 and 2, the specific procedure used for measuring A is crucial, as described in Ref 41. The A parameter is defined as the number fraction of cavitating grain boundaries encountered in a line parallel with the direction of maximum principal stress. A magnification of 400× is used to maintain a balance between the need to resolve and identify cavities and the need to maintain a large field of view at one time. The need to ultimately make measurements at the site using optical microscopes also dictates limiting the magnification to about 400×. Classification of boundaries as damaged or undamaged also requires considerable judgment, and specific rules have to be enunciated to make the measurement procedure consistent. Counting boundaries other than prior austenite grain boundaries can lead to different results. A minimum of 400 boundaries must be counted to achieve an accurate result, and a sufficient number of parallel, nonoverlapping traverses must be made to achieve this target for each measurement. Use of other magnifications and procedures is likely to lead to different correlations between A and t/t_r. The procedure for measurement of A thus needs to be standardized. A round-robin exercise aimed at evolving a standard procedure is now underway. Method 3, using the Neubauer and Wedel classification, is far simpler to use. At the present time, the models described here are in their infancy and have not been sufficiently validated. Additional data are also required for other steels used in piping to verify the general applicability of the correlations described. Numerous potential limitations of the model and sources of scatter in the data have been discussed by Shammas *et al* (Ref 41). However, because the data presented here include two steels of widely different impurity concentrations, it is anticipated that the scatterband encompasses most 1Cr-½Mo steels in service. Use of the lower limit line together with methods 2 and 3 described above will result in reasonable estimates of remaining life. Limited data on HAZ samples from 2¼Cr-1Mo steels also have shown that the cavitation behavior can be described by Eq 5.10c. Extension of this correlation to weld metal, to steels with fine grain size, and to other steels, however, would lead to errors.

Damage Model for Base Metal. Thermal softening of the base metal can result in a variety of microstructural changes in the steel, such as changes in composition, structure, size, and spacing of carbides; in ferrite composition; in solid-solution strength; and in lattice parameter. All of these changes have been monitored by Askins *et al* in 1Cr-½Mo steel samples subjected to interrupted creep testing at a variety of stresses and temperatures (Ref 43). The most meaningful and usable parameter was determined to be the mean interparticle spacing between carbides. The highlights of the model relating interparticle spacing to creep life are as follows (Ref 43 and 44).

The presence of precipitates was postulated to result in a "threshold" stress which must be exceeded to allow dislocations to climb over the particles so that

$$\sigma_0 = \frac{\alpha'\mu b}{\lambda} \qquad \text{(Eq 5.15)}$$

where α' is a constant, μ is the shear modulus, b is the Burgers vector, and λ is the mean interparticle spacing. The creep-rate equation under the effective stress can be written as

$$\dot{\epsilon} = B\left(\sigma - \frac{\alpha'\mu b}{\lambda}\right)^n \qquad \text{(Eq 5.16)}$$

where σ is the applied stress and B is the constant containing the temperature dependence, defined as

$$B = B_0 \exp(k_A T) \qquad \text{(Eq 5.17)}$$

Askins *et al* found that the kinetics of carbide coarsening (see Fig. 5.24) could be described as

$$\lambda_t^3 = \lambda_0^3 + C_0 \exp(\beta T)t \qquad \text{(Eq 5.18)}$$

where λ_t is the instantaneous interparticle spacing at time t, λ_0 is the spacing at t = 0, T is temperature in K, and C_0 and β are constants. Substituting Eq 5.18 and 5.17 into Eq 5.16, we get

$$\dot{\epsilon} = B_0 \exp(k_A T)$$
$$\times \left(\sigma - \frac{\alpha'\mu b}{[\lambda_0^3 + C_0 \exp(\beta T)t]^{1/3}}\right)^n$$
$$\text{(Eq 5.19)}$$

For the 1Cr-½Mo steel investigated, it was found that

$$B_0 = 4.50 \times 10^{-37} \text{ MPa}^{-5.6} \cdot \text{h}^{-1}$$
$$\text{(for true strain)}$$

$$k_A = 0.0594 \text{ K}^{-1}$$

$$n = 5.6$$

$$\lambda_0 = 0.1375 \ \mu\text{m}$$

$$C_0 = 1.33 \times 10^{-26} \ \mu\text{m}^3 \cdot \text{h}^{-1}$$

$$\beta = 0.0534 \text{ K}^{-1}$$

$$\alpha' = 0.15$$

By substituting the values of B_0, k_A, σ, α', λ_0, C_0, and β, Eq 5.19 can be integrated between limits of t = 0 and t = t, and the strain accumulated up to that time can be determined. Because the creep rate $\dot{\epsilon}$ is known, the failure time t_r at any arbitrarily selected value of failure strain, say 10%,

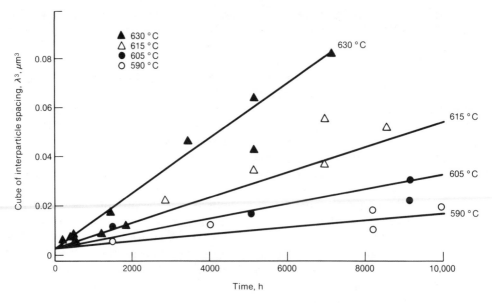

Fig. 5.24. Carbide-coarsening kinetics for a 1Cr-½Mo steel (Ref 43).

can be calculated. Using the above model, reasonable agreement was demonstrated between rupture-life predictions from precipitate size and actual rupture lives determined by experiment.

The Askins model is based on the premise that once the kinetics of carbide growth are known, the creep rate and hence rupture life can be calculated. The initial carbide spacing λ_0 is usually unknown. Therefore, monitoring of the carbide spacing λ_t at a given location as a function of time or at different locations of known temperature will be necessary in order to determine the carbide-growth kinetics.

For application of the model to a field component, samples or replicas from three locations of different temperatures will have to be removed and the carbide spacing λ_t measured. From these values, the constants λ_0, C_0, and β in the carbide-coarsening-kinetics equation (Eq 5.18) can be determined. The service applied stress and the local temperature where remaining life estimates are to be made are expected to be known. Values for A_0, k_A, and n will have to be assumed. All these values are substituted into Eq 5.19 to compute a creep curve for the material. From the creep curve, the time to reach a given critical strain or the time to rupture can be estimated.

The carbide-coarsening model is at present ridden with numerous limitations. Because carbide distributions in steels are nonhomogeneous and the starting microstructures for different components are never the same, it is inevitable that the carbide-coarsening kinetics specific to the component must be determined by taking samples or replicas from locations of known temperature. This can hardly be achieved in practice, because local temperature measurements in components are rarely made. Besides, if the temperatures and stresses are known, the expended life fraction can be calculated directly instead of using the carbide-coarsening model, and the answers are expected to be at least as accurate, if not more so. Even after the carbide-coarsening kinetics have been determined, the other constants needed to exercise Eq 5.19, such as B_0, k_A, n, and α', still have to be assumed using bounding values of data obtained on other heats. These assumptions may also lead to errors. The failure criterion assumed in terms of a critical strain is arbitrary. The carbide-coarsening model thus contains too many constants which are difficult to obtain and evaluate. The best way to use this method is to use it in conjunction with other life-estimation methods. Alternatively, the carbide-coarsening kinetics can be used simply as a temperature monitor, as described in a later section.

One of the earliest qualitative approaches to estimation of remaining life was that suggested by the work of Toft and Mardsen (Ref 45). They procured samples from service-exposed 1Cr-½Mo boiler tubes containing a range of microstructures and tested them to determine the remaining rupture life. The results showed a clear correlation (Fig. 5.25) between the rupture strength and the microstructural category (based on the degree of carbide spheroidization) (Ref 45). These graphs could be extrapolated to service conditions and a prediction of remaining creep life could be made for each microstructural category. In practice, it is only necessary to assign a component's microstructure to a category by comparison with a set of standard micrographs and to read off the predicted life under the applicable operating conditions. This procedure is analogous to that developed later by Neubauer and Wedel for cavitation damage.

Field Implementation of Metallographic Methods. Implementation of the metallographic methods can be done by removal of samples or nondestructively by replication. Although samples can be removed from most components, there are situations in which replication may provide the only possible approach to microstructural evaluation—e.g., when the removal of a sample is geometrically difficult or is liable to affect the integrity of the component, or when repeated observations are required. The two major applications of replication techniques are (1) the study of microstructure (creep cavitation, precipitate spacing, grain size, etc.) using both optical and electron microscopy, and (2) the examination and identification of small second-phase particles by extraction techniques.

The principle of the surface replication technique is illustrated in Fig. 5.26. The objective is to reproduce as faithfully as possible the surface topography of a specimen on a film which can subsequently be examined in the microscope. Hence, a very high surface quality is crucial. Surface oxides as well as decarburized zones must be removed prior to replication of the component surface. The basic steps in surface replication typically include: (1) grinding of the selected surface on 400- or 600-grit emery paper and polishing with 6- and 1-μm diamond paste; (2) repeated polishing (three to four times) with a suspension of γ-alumina and etching with 2 to 5% nital (for steels); (3) replication of the prepared surface by firmly pressing onto it a cellulose acetate replicating film (25 to 50 μm, or 1 to 2 mils, thick) softened by immersion in acetone; (4) peeling off the replica when dry and mounting it on a slide; (5) vapor deposition of the primary replica in vacuum using chromium, aluminum, or gold/palladium on the impression side to improve contrast; and (6) coating with carbon to improve surface reflectivity. After step 6, the replica is suitable for examination in an optical microscope. If examination in a transmission electron microscope (TEM) is required, the gold-coated or gold-coated-and-carbon-shadowed replica can be partially immersed in acetone. This dissolves away the acetate film, leaving behind a positive image of the replica for TEM examination.

Instead of the two-stage procedure of metal deposition followed by carbon coating, single-stage metal or carbon coatings on the impression side also produce acceptable micrographs (Ref 46). Less-elaborate equipment and shorter times are required for the single-stage procedure. Acceptable micrographs also can be produced by painting the back side of the replica with flat black spray paint. Further SEM examination, however, would require recoating of the impression side using an evaporated coating in the usual way.

Studies of precipitate morphology, composition, and structure can be performed using a carbide-extraction replica procedure. Following normal surface preparation, the area of interest is heavily etched in 5% nital (low-alloy ferritic steels). The area is then washed with alcohol and dried in air, and a dilute solution of 1% polyvinyl formal (Formvar resin) in chloroform is

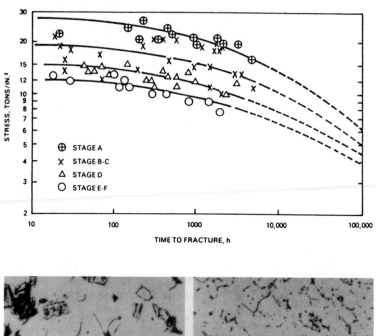

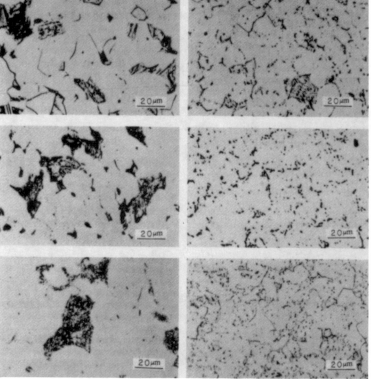

Stage A (top left): Ferrite and very fine pearlite; microstructure of new tube. Stage B (middle left): First signs of carbide spheroidization, usually accompanied by precipitation at the grain boundaries. Stage C (bottom left): Intermediate stage; appreciable spheroidization of pearlite, but some carbide plates still evident. Stage D (top right): Spheroidization complete, but the carbides are still grouped in a pearlitic pattern. Stage E (middle right): Carbides are dispersed, leaving little trace of original pearlitic areas. Stage F (bottom right): Size of some of the carbide particles has increased markedly due to coalescence.

Fig. 5.25. Above: Graph showing results of creep-rupture tests on 1Cr-½ Mo steel after removal from boiler tubes (tests at 510 °C, or 950 °F). Below: Micrographs showing stages of spheroidization present in samples at beginning of test (Ref 45).

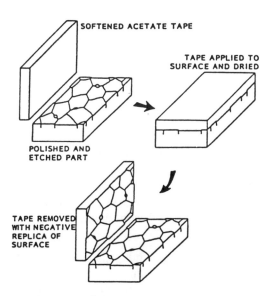

SOFTENED ACETATE TAPE

TAPE APPLIED TO SURFACE AND DRIED

POLISHED AND ETCHED PART

TAPE REMOVED WITH NEGATIVE REPLICA OF SURFACE

Fig. 5.26. Illustration of the principle of acetate replication.

applied to the surface. This deposits a fine film, which, when hardened, is backed with softened cellulose acetate. Once dry, the composite film can be stripped off and returned to the laboratory. It is then coated with carbon on the impression side, cut into small pieces, mounted on grids, and washed in acetone. The acetate foil is dissolved, leaving behind the extracted particles embedded in the carbon film. After drying, both microstructural and chemical investigations of the precipitate can be carried out using scanning transmission electron microscopy (STEM) combined with energy-dispersive x-ray analysis (EDAX). The carbide-extraction technique suffers from two limitations. First, despite extreme care, very fine carbides such as VC and Mo_2C may sometimes escape the extraction process; and second, due to the more complex nature of the process compared with surface replication, it has not yet found widespread application in the field.

Several alternative surface-preparation techniques have been explored, mainly with a view to reducing the preparation time (Ref 46). Grinding wheels used in dentistry and in die profiling can be employed. The normal sequence involves grinding on 120-, 240-, 400-, and 600-grit papers followed by polishing with 9-, 3-, and 1-μm diamond paste carried on disks of felt polishing cloth. Use of die profilers with coarse diamond paste involves grinding on 120- and 240-grit papers followed by polishing with silicon carbide and then with 60-, 9-, and 1-μm diamond paste on cloth disks. Use of electropolishing can further reduce the number of steps in the sequence to grinding on 120-, 240-, and 400-grit papers, polishing with silicon carbide, and electropolishing (10 to 50 s). The best results from electropolishing have been obtained with a solution of reagent-grade perchloric acid (78 ml), distilled water (120 ml), ethanol (700 ml), and butylcellulosolve (100 ml) and a current of 0.5 A for 15 s (Ref 46). Field electropolishing units have been found to be capable of producing excellent-quality metallurgical finishes on ferrous and nonferrous materials in very short times (Ref 46 and 47). Many electropolishing solutions, however, require careful handling, transporting, and disposal, which might be disadvantageous in some field applications.

Replication procedures have been described in detail by a number of authors (Ref 13, 46, 48, and 49). A detailed evaluation of several alternative procedures, types of equipment, and lists of vendors has been provided by Clark and Cervoni (Ref 46). Procedures generally vary from one laboratory to another, and there is a real need to evolve a standardized procedure for industry-wide use. ASTM recently issued an emergency interim standard for plant replication (Ref 50).

The plastic replication technique offers several advantages. It is capable of producing the same high-quality micrographs achieved in conventional metallography, as illustrated in Fig. 5.27. It is applicable to many types of components, materials, and material conditions. It is totally nondestructive and hence can be used for periodic monitoring. The replicas provide a permanent record of observations and can be stored. Because replicas can be examined at high magnification in scanning electron microscopes, a very high degree of resolu-

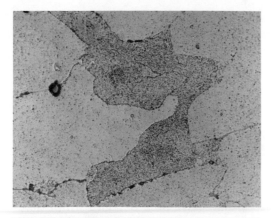

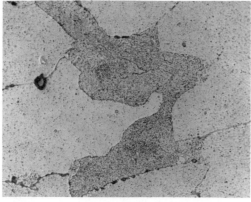

Fig. 5.27. Creep-cavitation damage in a cracked tee section of a desuperheater inlet header shown by conventional metallography (above) and plastic replication (below).

has also been performed by *in situ* examination of components using a portable microscope (Ref 51). This procedure is less expensive and speedier but is more limited in terms of accessibility and resolution of damage.

The first requirement for the application of metallographic methods to plant components is identification of the locations to be sampled or replicated. Although preinspection stress analysis is helpful, location selection still requires judgment that comes only with experience. Recommended locations include:

1. Welds at known hot spots—e.g., stub-tube penetrations for tubes carrying steam at temperatures considerably higher than the mixed-steam temperature
2. Welds associated with known stress concentrations
3. Expected regions of high stress—e.g., bends, section transitions, and locations near support failures
4. Locations with prior history of failure
5. High-risk locations—e.g., longitudinal welds in seamed piping
6. Locations with high exposure to personnel traffic
7. Locations with flaw indications based on NDE techniques.

tion can be achieved. This technique is capable of detecting damage in highly localized regions. It also has a few limitations. Replicas can be taken only at accessible locations. They give no information regarding volumetric damage. Indiscriminate replication can be expensive and counterproductive. Careful analysis and judgment are needed in selecting locations for replication. Because there is some variability in interpretation of replicas, evaluations can sometimes differ. At present the plastic replication technique is used principally for reproducing surface features such as creep cavities, cracks, and gross microstructural features. Application of carbide extraction replicas to plant situations will require further developmental work.

In lieu of replication, field metallography

Once the locations requiring inspection have been identified, the exact sampling positions must be selected. Positions must be chosen which are likely to represent the most damaged parts of the component. In the case of a branch weld, for instance, such positions are most often the saddle flank position and the crotch corners. Stub welds are found to crack most readily in the ligament between stub tubes in the same circumferential row, where the hoop stress in the header acts normally to the zones of stress concentration in usual component geometries.

Methods Based on Temperature Estimation

One of the crucial parameters in estimation of creep life is the operating temperature.

Although steam temperatures are occasionally measured in a boiler, local metal temperatures are rarely measured. Due to load fluctuations and steam-side oxide-scale growth during operation, it is also unlikely that a constant metal temperature is maintained during service. It is therefore more convenient to estimate an "equivalent" or "mean" metal temperature in service after the fact, by examination of such parameters as hardness, microstructure, and thickness of the steam-side oxide scale (for tubes). Because the changes in these parameters are functions of time and temperature, their current values may be used to estimate an "equivalent" thermal history for a given operating time. The estimated temperature can then be used in conjunction with standard stress-rupture data to estimate the remaining life. Several methods for estima-

tion of metal temperature have been reviewed elsewhere (Ref 52) and are briefly described in this section.

Steam-Side Oxide-Scale Thickness. Extensive data indicate that in relatively pure steam, the growth of oxide scales is a function of temperature and time of exposure alone and is presumed to obey specific rate laws. Several expressions have been proposed in the literature to describe oxide-scale-growth kinetics (Ref 53 to 56). A compendium of the proposed expressions is given in Table 5.6. The general forms of the expressions and the standard rate laws to which they approximately reduce in the temperature range 510 to 620 °C (950 to 1150 °F) are shown in Table 5.7. Plots of oxide-scale thickness vs time based on these rate laws are shown in Fig. 5.28 to 5.30 (Ref 53 to 56).

Table 5.6. Expressions for oxide-growth kinetics in Cr-Mo steels

No.	Expression(a)	Steel	Temperature range, °C (°F)	Units	Reference
1	$\log x = -7.1438 + 2.1761 \times 10^{-4}\, T(20 + \log t)$	1-3%Cr	Below FeO formation	x in mils, T in °R	53
2	$y^2 = kt$			T in K	54
	For 1Cr-½Mo:				
	$\log k = (-7380/T) + 2.23$ [T ≤ 585 °C (1085 °F)]	1Cr-½Mo	585 (1085)		
	$\log k = (-48{,}333/T) + 49.28$ [T > 585 °C (1085 °F)]	1Cr-½Mo	585 (1085)		
	For 2¼Cr-1Mo:				
	$\log k = (-7380/T) + 1.98$ [T ≤ 595 °C (1103 °F)]	2¼Cr-1Mo	595 (1103)		
	$\log k = (-48{,}333/T) + 49.2$ [T > 595 °C (1103 °F)]	2¼Cr-1Mo	595 (1103)		
3	$\log x = -6.8398 + 2.83 \times 10^{-4}\, T(13.62 + \log t)$	2¼Cr-1Mo	429-649 (800-1200)	x in mils, T in °R	55
4	$x = \dfrac{cpt}{1 + pt} + Et$	2¼Cr-1Mo	428-593 (800-1100)	x in μm, T in °F	56

where $\log$ (coefficient) $= b_0 + b_1 T + b_2 T^2$, with b_0, b_1, and b_2 values as follows:

	b_0	b_1	b_2
c	13.2413	-2.5800×10^{-2}	1.4319×10^{-5}
p	-5.7267	4.7931×10^{-3}	-2.0905×10^{-6}
E	6.6488	-2.4771×10^{-2}	1.5425×10^{-5}

(a) x is scale thickness; y is metal loss (penetration); T is temperature; t is time, in hours; all logarithms to base 10. °R = °F + 460; K = °C + 273; 1 mm = 10^3 μm = 40 mils; y = 0.42x.

Table 5.7. General forms of the oxide correlations for Cr-Mo steels (Ref 52)

General form of expression(a)	Approximate oxide-growth law
Rehn and Apblett (Ref 53):	
$\log x = A + B$ (LMP)	$x^3 = kt$
Paterson and Rettig (Ref 55):	
$\log x = A + B$ (LMP)	$x^{2.1 \text{ to } 2.6} = kt$
Dewitte and Stubbe (Ref 54):	
$x^2 = kt$	$x^2 = kt$
D.I. Roberts (Ref 56):	
$x = \dfrac{At}{B + Ct} + Dt$	$x = kt$

(a) x is oxide-scale thickness; t is time; k is oxide-scale-growth-law rate constant; A, B, C, and D are coefficients; LMP is Larson-Miller parameter.

The various oxide-growth laws have been reviewed and compared by Viswanathan, Foulds, and Roberts (Ref 52). The variability in estimates of mean operating temperature using the different correlations is illustrated in Fig. 5.31 for oxide thicknesses of 300, 600, and 1200 μm (11.8, 23.6, and 47.2 mils) on the interior surfaces of 2¼ Cr-1Mo steel superheater/reheater tubes with an actual operating time of 11.4 years (100,000 h). The variability in mean operating temperature is about 24 °C (43 °F) at a scale thickness of 300 μm and increases to as much as 40 °C (72 °F) as the scale thickness increases to 1200 μm.

It is important to note that the steam-side external oxide layer can be used to estimate only an "equivalent" exterior metal temperature and provides qualitative information on relative thermal gradients across the tube wall. Disruption of oxide scales by spalling or by chemical cleaning can cause errors in temperature estimates. The oxide-thickness correlations have all been developed on the basis of exposure times representing a small fraction of the total service life (<15%). The databases used in deriving the various correlations are different and hence lead to large variations in estimated temperatures. Preference for any

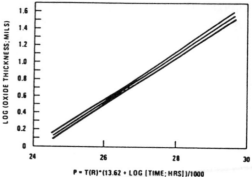

Fig. 5.28. Empirical relationship between oxide thickness and a time-temperature parameter for 2¼ Cr-1Mo steel in isothermal steam environments (Ref 55).

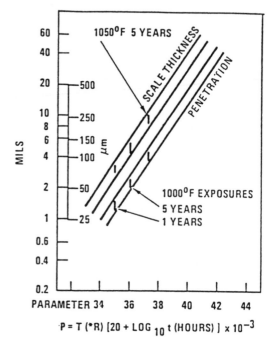

Fig. 5.29. Correlation between thickness of magnetite oxide scale and Larson-Miller parameter for 1 to 3% Cr steels (Ref 53).

correlation is subjective. In spite of these limitations, the oxide-scale measurement has become a standard tool in life prediction of SH/RH tubes. In particular, with the recent developments in nondestructive measurement of interior-surface scale by ultrasonic techniques, this method has become even more popular. Values of oxide-

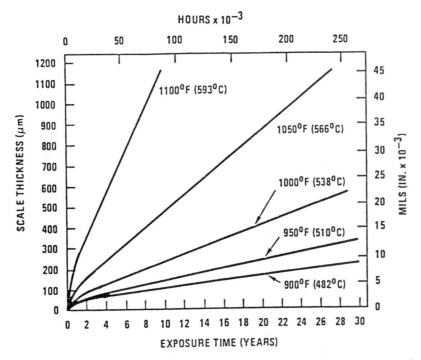

Fig. 5.30. Correlation (from Ref 56) between oxide growth and exposure time for 2¼ Cr-1Mo steel in steam at various temperatures (Ref 52).

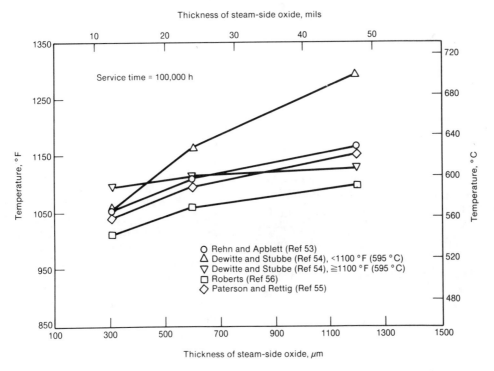

Fig. 5.31. Estimates of mean operating temperature made by various correlations for 2¼ Cr-1Mo steel tubes having inside-surface oxides with thicknesses of 300, 600, and 1200 μm (11.8, 23.6, and 47.2 mils) after operation for 100,000 h (Ref 52).

scale thickness estimated from ultrasonic measurements agree closely with those measured by destructive examination.

Several proprietary and nonproprietary codes have been developed to predict the lives of SH/RH tubes based on oxide-scale measurements. To illustrate the general methodology, the principal steps underlying a code known as TUBELIFE (Ref 55) will be discussed. After careful review of plant records, review of drawings, and inspection of tube banks, tubes are selected for laboratory examination. Metallographic cross sections of the samples are prepared and the tube dimensions and scale thicknesses are measured. Alternatively, the scale and wall thicknesses can be measured ultrasonically. The tube dimensions are used to calculate the effective current stress using a maximum-elastic-hoop-stress formula or a Tresca reference-stress formula. The scale thickness is used in conjunction with the rate law proposed in expression 3 of Table 5.6 to determine the current effective temperature of the tube. The temperature and stress values are then extrapolated back to the initial conditions assuming linear growth of steam scale and fire-side wastage and the known heat-transfer properties of the steel and the oxide scale. The service life up to the present time is divided into three-month intervals. For each interval, the oxide-scale thickness and wall thinning under the nonisothermal conditions are estimated using linear kinetics of scale growth and hot corrosion. The corresponding equivalent temperature is calculated using the isothermal kinetics given by expression 3 in Table 5.6. By iterating this process for every three-month interval, the average metal temperature and stress for every three-month period are estimated, and a history of temperatures and stresses as functions of time are established. The life fraction expended at each three-month interval is computed from:

1. The temperature increase caused by the steam-side scale
2. The stress increase caused by fire-side corrosion
3. The stress-rupture curve at each calculated temperature based on standard data.

These life fractions are summed to evaluate the expended life up to the present time. The remaining life is estimated by extrapolating into the future by the same technique. When the total expended life fraction reaches a value of unity, the end of life is deemed to have been reached. A simplified example considering only the oxide-scale growth is presented below to illustrate the methodology.

Example:
A boiler tube made of 2¼Cr-1Mo steel, operating at a nominal temperature of 1000 °F and a hoop stress of 5 ksi, was examined after ten years of service. The oxide-scale thickness, x, on the steam side was found to be 20 mils (0.02 in.). Calculate the life expended based on five-year intervals.

Answer:
From Fig. 5.28, the time-temperature parameter P corresponding to the current oxide thickness of 20 mils is 28,800:

$$T(13.62 + \log t) = 28{,}800$$

t = 87,600 h; substituting for log t, we get T = 1551 °R (1091 °F).

The equivalent temperature at the end of five years of operation can be determined by repeating the same procedure, using a different value of oxide thickness. Assuming a linear oxide-scale-growth law, the oxide-scale thickness x at the end of the first five years (43,800 h) is estimated as 10 mils (0.01 in.). The corresponding value of P can be estimated as 27,800, using Fig. 5.28. Because the values of x, t, and P are known, the value of T can be calculated as 1522 °R (1062 °F). The tube is thus assumed to have operated at equivalent temperatures of 1062 °F for the first five years and 1091 °F for the next five years.

From the Larson-Miller rupture curve for 2¼Cr-1Mo steel (see Fig. 5.17), the value of t_r at 1062 °F and 5 ksi is 420,099 h, and the value of t_r at 1091 °F and 5 ksi is 139,636.

Life fraction expended after first five
  years = 43,800/420,900 = 0.1
Life fraction expended during next five
  years = 43,800/139,636 = 0.31
Total life fraction expended at the end
  of ten years = 0.41
Remaining life = 0.6/0.4 × 10 = 15 years
In actuality, because of continued scale buildup and wall thinning due to wastage, the remaining life will be less than 15 years.

Actual tube-metal temperatures in boilers depend on scale thickness, as well as on tube geometry, heat flux, and the thermal conductivities of the scale and the metal. Under known heat-flux conditions, tube-metal temperatures may be estimated from the steam temperature–oxide correlation. This can be done by assuming a "thermal" film between oxide and metal resulting in a temperature gradient between steam and metal quantitatively estimated [for example, after French (Ref 57)] as

$$\Delta T = K''x \qquad \text{(Eq 5.20)}$$

where ΔT is the temperature gradient across the oxide-metal "thermal" film, x is the oxide thickness, and K'' is a constant dependent on tube geometry and heat flux. Of course, Eq 5.20 is unnecessary under low-heat-flux conditions such as in the penthouse of a fossil boiler furnace, where $\Delta T \simeq 0$. Another point to note is that these "oxide" methods are applicable only to low-alloy steels (those containing less than about 3% Cr). The scale-growth kinetics of 9Cr-1Mo steel are expected to be significantly different. The oxide-scale method has the advantages of simplicity and reproducibility. Also, oxide scale can be measured nondestructively.

In situations where a chemical cleaning procedure has been carried out and the scale formed up to that point has been removed, two alternatives are available. If the thickness of the oxide scale has been measured prior to cleaning, life consumption up to that point can be computed and added to the value computed subsequent to the cleaning. Alternatively, the scale thickness prior

to cleaning can be estimated on the basis of the observed scale growth since the cleaning, and the expended life fractions before and after cleaning can be computed separately and then summed. Knowledge of the heat flux is also needed for accurate estimation of metal temperatures.

A life-assessment procedure based on similar principles, but taking into account the observation that liquid-phase corrosion may suddenly accelerate upon reaching a critical temperature due to oxide-scale growth, has been described by Alice, Janiszewski, and French (Ref 58).

Hardness-Based Techniques. The strength of low-alloy steel changes with service exposure in a time- and temperature-dependent manner. Thus, any measure of change in strength during service (e.g., change in hardness) may be used to estimate a "mean" operating temperature for the component. This approach is particularly suitable when strength changes in service occur primarily as a result of carbide precipitation and growth (microstructural coarsening) and strain-induced softening can be neglected.

The tempering responses of steels at typical service temperatures, as evidenced by hardness changes influenced by time (t) and temperature (T) of exposure, often are described by the Larson-Miller parameter P — e.g., $P = T(20 + \log t)$, where T is in °R. A correlation between hardness H and the Larson-Miller parameter P can be obtained by aging a given material with initial hardness H_0 (at t = 0) at temperature T and measuring the change in hardness as a function of time t. The resulting relationship is $H = f(P)$. This relationship, however, is unique to the starting material condition represented by the initial hardness H_0.

Figure 5.32 is a schematic illustration showing a typical experimentally derived $H = f(P)$ correlation obtained on material having an initial hardness of H_0. P_0 is defined with no physical significance as being the value of P at which $H = H_0$. Described below are three cases illustrating the usefulness and limitations of such a correlation.

Case 1. Assume a tube with a known initial preservice hardness (equal to the initial

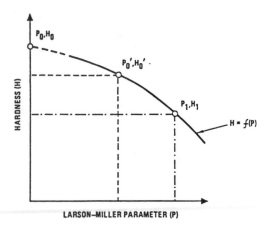

Fig. 5.32. Schematic diagram showing a correlation between hardness, H, and the Larson-Miller parameter, P (see text for interpretation) (Ref 52).

hardness of the "correlation" material, H_0) and a measured hardness H_1 after service time t_1. The mean operating temperature T is estimated from

$$T = P_1/(20 + \log t_1) \qquad \text{(Eq 5.21)}$$

where P_1 is the value of P at $H = H_1$ obtained from the curve $H = f(P)$.

Example:
Consider a low-alloy steel tube with an initial hardness of $H_0 = 330$ DPH and a characteristic tempering (softening) curve defined by

$$H = 960 - 0.02P \qquad \text{(Eq 5.21a)}$$

where H is Vickers hardness (DPH) and P is the Larson-Miller parameter, $P = T(20 + \log t)$, where T is in °R and t is in hours. For a tube of a similar alloy with the same initial hardness (330 DPH) and a hardness of $H_1 = 165$ DPH after a service time of 80,000 h, the "mean" operating temperature T is calculated from

$$\frac{960 - 165(= H_1)}{0.02} = P_1$$

$$= T(°R)(20 + \log 80,000)$$

and thus $T = 1596$ °R $= 1136$ °F.

Case 2. Assume a tube with a known initial preservice hardness (lower than the initial hardness of the "correlation" material, H_0) and a measured hardness H_1 after service time t_1. If the hardness of the tube was H_0' before service, P_1 is estimated as the value of P at $H = H_1$, from $H = f(P)$ as for case 1 above, and P_0' is calculated as the value of P at $H = H_0'$. This P_0' value is used to estimate an equivalent or fictitious time t_1' required to realize a decrease in hardness from H_0 to H_0' at the mean service temperature T. The fictitious time t_1' may be obtained from

$$P_1 = T[20 + \log (t_1 + t_1')] \qquad \text{(Eq 5.22)}$$

and

$$P_0' = T(20 + \log t_1') \qquad \text{(Eq 5.23)}$$

i.e., from

$$\frac{P_1}{P_0'} = \frac{20 + \log (t_1 + t_1')}{20 + \log t_1'} \qquad \text{(Eq 5.24)}$$

The fictitious time t_1' is then used to determine T from Eq 5.22.

It should be noted that if the initial preservice hardness exceeds the initial hardness of the material used to derive the correlation ($H_0' > H_0$), the correction for a different initial hardness is impossible to make because the form of the parameter P cannot accommodate a negative fictitious time t_1'. In general, however, for evaluating ex-service or in-service tubes where $t_1' \ll t_1$, the correlation $H = f(P)$ may be used as is with fair predictive accuracy. Nevertheless, for applicability to short-term aging predictions as well, it is obviously advantageous for correlations between hardness and tempering parameter to be derived on material with high initial hardness H_0 (t = 0), such as untempered, as-quenched, or as-normalized material. This may be done by generating aging data on as-tempered (softened) material but including a tempering-treatment equivalent or fictitious time for the specific data point in the calculation of P for the correlation.

Example:
Consider an alloy steel tube with an initial hardness of $H_0' = 300$ DPH and a hardness of 165 DPH after 80,000 h. Using the same calibration (i.e., Eq 5.21a) obtained for a steel with an initial hardness of $H_0 = 330$ DPH, the "equivalent" temperature can be calculated as follows:

1. The fictitious time t_1' required for the hardness to decrease from 330 DPH (correlation "initial" material) to 300 DPH (actual "initial" material) at "mean" temperature T is given by

$$\frac{960 - 300(= H_0')}{0.02} = P_0'$$

$$= T(°R)(20 + \log t_1')$$

and thus

$$33 \times 10^3 = T(20 + \log t_1')$$

$$(Eq\ 5.22a)$$

2. The decrease in hardness over the 80,000-h service period can be written as (Eq 5.22)

$$\frac{960 - 165(= H_1)}{0.02} = P_1$$

$$= T(°R)[20 + \log (t_1' + 80,000)]$$

and thus

$$39.75 \times 10^3$$

$$= T[20 + \log (t_1' + 80,000)]$$

$$(Eq\ 5.23a)$$

3. Equations 5.22a and 5.23a are combined to solve for t_1':

$$\frac{33 \times 10^3}{39.75 \times 10^3} = \frac{20 + \log t_1'}{20 + \log (t_1' + 80,000)]}$$

$$(Eq\ 5.24a)$$

and thus $t_1' \simeq 4.5$ h.

4. Therefore, from Eq. 5.23a, we get

$$39.75 \times 10^3 = T[20 + \log (80,004.5)]$$

and thus T = 1596 °R = 1136 °F.

Case 3. Assume a tube with an unknown preservice hardness and a hardness H_1 after a service time t_1. This is a commonly encountered circumstance where tubes have been in service for a long period of time and where the original material description does not include hardness data. The simplest approach is to assume a preservice hardness that is typical of the starting material conditions (e.g., annealed, normalized and tempered, etc.). This reasonably assumed value of hardness, H_0', is then used as described in case 2 above and the "mean" operating temperature is estimated from Eq 5.21, 5.22, and 5.23. Again, if a rough estimate of t_1' indicates that $t_1' \ll t_1$, the correlation may be used as is—i.e., assuming that $H_0' = H_0$. This assumption causes a negligible error in the temperature estimation, as illustrated in the following example.

Example:
Suppose that the starting material hardness was 240 DPH and that we use Eq 5.21a (for a starting hardness of 330 DPH) directly, with no correction t_1' as described in case 2. In this situation, for a service period of 80,000 h and an ex-service hardness of 165 DPH, we would obtain

$$\frac{960 - 165}{0.02} = P = T(°R)(20 + \log 80,000)$$

and thus T = 1596 °R = 1136 °F. If we correctly account for the start-up (t = 0) hardness of the service material (240 DPH), as was done in case 2 (this can be done only if we know the start-up hardness of the service material), we can estimate t_1' to be 360 h. This is still a very small fraction of the service life of 80,000 h, and its consideration still gives T = 1596 °R = 1136 °F (i.e., no difference in estimated "mean" temperature).

Another available option when the correlation H = f(P) is linear (i.e., dH/dP =

constant) is to make two hardness measurements H_1 and H_2 at different times t_1 and t_2 during service, making sure that the time interval $t_2 - t_1$ is large enough to cause a measurable change in hardness and is also representative of the long-term operating characteristic with regard to "mean" service temperature T. T may then be estimated from

$$T = \frac{H_2 - H_1}{m \log t_2/t_1} \qquad \text{(Eq 5.25)}$$

where $m = dH/dP = $ constant.

A specific location for which the initial material condition and hardness are known and relatively constant is the as-welded heat-affected zone (HAZ) of a weldment. A correlation obtained for this zone is useful for estimation of the mean tube service temperature in the vicinity of a weld. The application may thus be as in case 1 above (no postweld heat treatment) or, more likely, as in case 2 where, although H_0' (preservice hardness after PWHT) may not be available, the generally known PWHT can be used to estimate P_0' for use in Eq 5.22 to 5.24. If the PWHT is unknown, a reasonable assumption of it may be made, as explained earlier.

Roberts *et al* have developed a correlation between hardness and service exposure in the form of the Larson-Miller parameter applicable to HAZ material of nominal composition 2¼Cr-1Mo and 9Cr-1Mo material (Ref 59). The database used consists of laboratory-aged and ex-service dissimilar-metal weldments (DMWs), and the hardness technique employs a 100-g load and a 136° diamond pyramid indenter. In addition to these correlations, laboratory data generated in the United Kingdom (Ref 43 and 60) on normalized-and-tempered 1Cr-½Mo steel using a 20- or 30-kg load and a similar indenter were analyzed. The data used were obtained for aging temperatures ranging from 575 to 650 °C (1065 to 1200 °F). The latter correlation is presented here (Fig. 5.33) for as-normalized material with the aging-exposure Larson-Miller parameter adjusted to include the tempering treatment of 3 h at 640 °C (1185 °F) for each data point. Figure 5.33 is a graphical representation of the H = f(P) correlations and the data used to derive them. The data

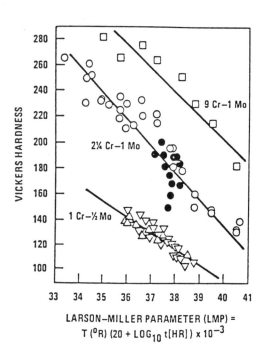

☐ 9 Cr–1 Mo HAZ LABORATORY
   DATA, 100 g LOAD

○ 2¼ Cr–1 Mo HAZ LABORATORY
   DATA, 100 g LOAD

● 2¼ Cr–1 Mo HAZ EX–SERVICE
   WELDMENT DATA, 100 g LOAD

△ 1 Cr–½ Mo NORMALIZED
   MATERIAL LABORATORY DATA,
   20 kg LOAD

▽ 1 Cr–½ Mo NORMALIZED
   MATERIAL LABORATORY DATA,
   30 kg LOAD

} LMP INCLUDES
  TEMPER
  TREATMENT
  –3 h, 640°C

Fig. 5.33. Correlations between hardness and the Larson-Miller parameter for 1Cr-½Mo, 2¼Cr-1Mo, and 9Cr-1Mo steels (Ref 52).

for $2\frac{1}{4}$Cr-1Mo and 9Cr-1Mo steels are for "as-quenched" HAZ starting materials with approximate initial hardnesses (H_0) of 330 DPH for $2\frac{1}{4}$Cr-1Mo and 380 DPH for 9Cr-1Mo and were obtained using a 100-g load and a 136° diamond pyramid indenter. The data for 1Cr-$\frac{1}{2}$Mo steel (expected to be the same as for $1\frac{1}{4}$Cr-$\frac{1}{2}$Mo) were generated for normalized-and-tempered material (1 h at 960 °C or 1760 °F, air cool, plus 3 h at 640 °C or 1185 °F, air cool) with a hardness of 160 DPH (HV). The "as-normalized" hardness (H_0) for the representation of the 1Cr-$\frac{1}{2}$Mo material in Fig. 5.33 is estimated to be approximately 210 HV.

For 1Cr-$\frac{1}{2}$Mo (or $1\frac{1}{4}$Cr-$\frac{1}{2}$Mo) steel in the "as-normalized" condition (H_0 = 210 HV) where t = 0 and hardness is measured with a 20-kg load and a Vickers indenter.

Hardness (HV)

$$= 595.453 - 1.2603 \times 10^{-2}\ P$$

$$\text{(Eq 5.26)}$$

where $P = T(20 + \log t)$ (the Larson-Miller parameter as defined above, where T is in °R and t is service time in hours). For $2\frac{1}{4}$Cr-1Mo in the "as-quenched" HAZ condition (H_0 = 330 DPH) where t = 0 and hardness is measured with a 100-g load and a Vickers indenter,

Hardness (DPH)

$$= 961.713 - 2.0669 \times 10^{-2}\ P$$

$$\text{(Eq 5.27)}$$

where P is the Larson-Miller parameter as defined above. For 9Cr-1Mo steel in the "as-quenched" HAZ condition (H_0 = 380 DPH) where t = 0 and hardness is measured with a 100-g load and a Vickers indenter,

Hardness (DPH)

$$= 933.0 - 1.825 \times 10^{-2}\ P \quad \text{(Eq 5.28)}$$

where P is the Larson-Miller parameter as defined above.

In addition to the correlations represented by Eq. 5.26 to 5.28, the Central Electricity Generating Board of the United Kingdom has recently produced a correlation for a commercial cast of 1Cr-$\frac{1}{2}$Mo steel in the simulated-HAZ-plus-tempered condition (5 min at 1300 °C or 2370 °F, oil quench, plus 2 h at 650 °C or 1200 °F, air cool) with an initial hardness (H_0) of 240 HV (Ref 61). This correlation, which was obtained from laboratory creep-tested specimens with hardness measurements made in the unstressed shoulder area using a 20-kg load, is given by

Hardness (HV)

$$= 160.8 + 0.02793P - 0.0000019P^2$$

$$\text{(Eq 5.29)}$$

where P is a Larson-Miller parameter different from that defined above and given by

$$P = T(°C)(11 + \log t) \quad \text{(Eq 5.30)}$$

where t is time, in hours, measured after tempering. Equation 5.29 can be applied to the HAZ of a 1Cr-$\frac{1}{2}$Mo (or $1\frac{1}{4}$Cr-$\frac{1}{2}$Mo) weldment in service.

For application of these tempering curves to estimation of mean tube service temperature, it is important to emphasize again that the fictitious time correction t_1' (Eq 5.22 and 5.23) for any as-fabricated starting material condition different from the one for which curves were obtained may be a small fraction of the service time; for example, $t_1'/t_1 < 1\%$ when a correlation obtained on "as-normalized" material is applied to "normalized-and-tempered" material (3 h at 650 °C, or 1200 °F) that has been in service for 100,000 h at a mean temperature of 550 °C (1020 °F). This implies that tempering curves for a given material in such a case may be used, with no consideration of initial material fabrication condition, to give reasonable estimates of mean service temperature (an example of this is described in case 3, above). This also indicates that correlations obtained for heat-affected zones may be applied to base metal after long-term service.

The hardness method may be used nondestructively for estimation of exterior-surface operating temperature. When performed destructively, this method can provide tube-wall operating thermal gradients. This method is relatively inexpensive, requiring no sophisticated equipment, but has several limitations. At low loads (e.g., 100 g), the sensitivity of hardness to local microstructural variation can result in significant scatter and poor reproducibility. This is less so at high loads (e.g., 20 kg). Strain-softening effects, if present, can cause an erroneous increase in the estimated mean service temperature. It is recommended that, where correlations are obtained from hardness measurements using low loads (<5 kg), the load used for determining temperature should be equal to that used for deriving the correlation. For nondestructive use of this method, tube access may be limited.

Microstructural Catalog. Toft and Mardsen demonstrated that there are basically six stages of spheroidization of carbides in ferritic steels. Using a Sherby-Dorn-type parametric equation, they showed that it was possible to get a reasonable correlation of microstructure with a "weighted average" service temperature (Ref 45). Similar semiquantitative and qualitative approaches involving microstructural "cataloging" as a function of service history have been utilized by others (Ref 62). An example of the approach used by the State Electricity Commission of Victoria in Australia is shown in Fig. 5.34. The microstructures corresponding to various time-temperature combinations—i.e., various values of the parameter where P is defined as

$$P = \log t - \frac{C}{T} \qquad \text{(Eq 5.31)}$$

where T is temperature in K, C = 12,370, and t is time in hours—are cataloged on the basis of laboratory samples. The microstructure found in the service component is then compared with the catalog to deter-mine the appropriate value of P. Because the service exposure time t is known, the "equivalent" service temperature T can be estimated. This procedure again suffers from the limitation that the current state of the microstructure depends on the starting microstructure prior to service, which is often unknown and may not necessarily correspond to the reference starting microstructure in the laboratory samples. One way to circumvent this problem is to remove samples from the relatively colder regions of the component and assume that they represent the typical starting microstructure. If the typical starting structure is reasonably close to the starting structure in the laboratory samples, then the temperature estimate is expected to be good. In the absence of such information, a catalog of microstructural changes for different starting structures, and a scatterband of behavior, must be established.

Example:
SEM examination of a sample from a failed superheater tube shows the microstructure to correspond to P = −10.71 following 30,000 h of service. Estimate the service temperature.

Answer:
Substitution of P = −10.71 and t = 30,000 into Eq. 5.31 gives T = 813 K = 540 °C.

Interparticle-Spacing Measurements. A method for estimating creep rate based on mean interparticle spacing between carbides, λ, was described earlier. The cubic growth law described earlier, with the appropriate constants reported by Askins for 1Cr-½Mo steels (Ref 43), can be expressed as

$$\lambda^3 = (0.1375)^3 + 1.33 \times 10^{-26}$$
$$\times \exp(0.0534T)t \qquad \text{(Eq 5.32)}$$

where λ is in μm, T is in K, and t is in hours. Estimation of a mean service tube temperature with an exposure time t can be made by measurement of λ and use of Eq 5.32. Because the constants in the above equation may be different for different ini-

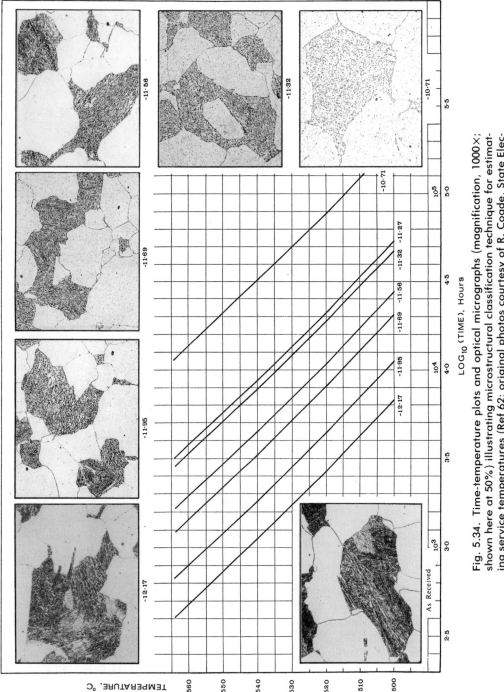

Fig. 5.34. Time-temperature plots and optical micrographs (magnification, 1000×; shown here at 50%) illustrating microstructural classification technique for estimating service temperatures (Ref 62; original photos courtesy of R. Coade, State Electricity Commission of Victoria, Australia).

tial microstructures, the equation may be used in the form

$$\frac{d\lambda^3}{dt} = C_0 \exp(\beta T) \qquad \text{(Eq 5.33)}$$

By measuring λ at different time intervals, the constants specific to the steel can be evaluated. Alternatively, if small slivers can be removed from the component and subjected to accelerated aging, the particle-coarsening kinetics specific to the steel can be established and then applied to temperature estimation.

Example:
Replica observations on a component after 1400 h of service show the average carbide spacing to be 0.2715 μm. Subsequent observations after another 5600 h show the spacing to have increased to 0.4309 μm. Estimate the average service temperature.

Answer:

$$\text{At 1400 h, } \lambda_1^3 = 0.02 \ \mu m^3$$

$$\text{At 7000 h, } \lambda_2^3 = 0.08 \ \mu m^3$$

$$d\lambda^3 = \lambda_2^3 - \lambda_1^3 = 0.06 \ \mu m^3$$

From Eq 5.32,

$$\frac{d\lambda^3}{dt} = 1.33 \times 10^{-26} \exp(0.0534T)$$

Substituting for $d\lambda^3/dt$, and solving the above equation for T, we get T = 902 K = 629 °C.

Measurement of Interfacial Carbide Size in Dissimilar-Metal Welds. This is a special case of a zone in which a structure parameter can be measured more easily than in normal base metal. The interface region between a nickel-base weld and 2¼Cr-1Mo steel base metal has been shown to develop a planar array of carbides parallel with and very close to (1 HAZ grain diameter from) the fusion line (Ref 59). These carbides,

mainly $M_{23}C_6$ and M_6C carbides, are commonly referred to as type I carbides. Foulds *et al* have developed correlations describing the growth of type I carbides as a function of temperature (Ref 63). A schematic illustration of type I interfacial carbides and the types of measurements made on them is presented in Fig. 5.35. The correlations obtained are given by

$$M_j^3 = 2.1056 \times 10^{12} \exp\left(\frac{-32,830}{T}\right)t$$

$$\text{(Eq 5.34)}$$

and

$$m_i^3 = 9.252 \times 10^{11} \exp\left(\frac{-32,760}{T}\right)t$$

$$\text{(Eq 5.35)}$$

where M_j is the length of the major axis, in μm; m_i is the length of the minor axis, in μm; T is the aging (service) temperature, in K; and t is the aging (exposure) time, in hours. The type I carbide growth kinetics are shown in Fig. 5.36. If M_j or m_i can be measured, T can be estimated using the equations above. Alternatively, if M_j (m_i) is measured at two different values of t, T can be estimated from the derivative

$$\frac{dM_j^3}{dt} = 2.1056 \times 10^{12} \exp\left(\frac{-32,830}{T}\right)$$

$$\text{(Eq 5.36)}$$

Extraction replicas can be used for nondestructive application of this method to tube outside surfaces. Destructive measurements across the thickness can be useful for estimating thermal gradients across the tube wall. These procedures, however, are difficult to standardize. For example, selection of a carbide to be considered as a type I carbide, choice of a micrograph to be representative of a given tube location/structure, and selection of the actual measuring technique are all very subjective decisions.

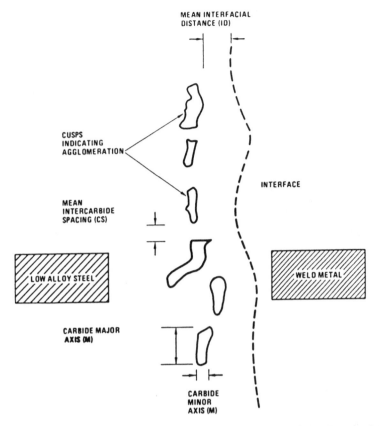

Fig. 5.35. Schematic illustration of an array of type I carbides near the interface in a dissimilar-metal weld (nickel-base weld metal, low-alloy steel base metal) and the relevant structure parameters (Ref 52 and 63).

The procedure for measurement is tedious and requires expensive equipment such as a SEM or TEM.

Analysis of Carbides. As discussed previously, the carbides in steel continuously evolve into higher-alloy carbides. The evolution of these carbides in time-temperature space is described by the Baker-Nutting-type diagrams (Ref 18). The carbides present at a given instant in a service component can be determined either by x-ray diffraction analysis of residues extracted from samples or by electron diffraction analysis performed on carbide extraction replicas. This information can be utilized to make estimates of service temperature. Attempts have been made by Askins *et al* to utilize parameters such as matrix-solute compositions, ratios of carbide phases, and matrix-lattice parameters as quantitative indices of temperature during aging of 1Cr-½Mo steels (Ref 43). Their data indicate a rapid initial decline in matrix-solute content and an increase in the $M_{23}C_6/M_3C$ ratio with time and temperature of aging. Changes in the matrix-solute content were also found to be reflected in changes in lattice-parameter values. Although these general trends could be confirmed, excessive scatter in the results did not permit quantitative correlation of any of these parameters with aging history. Improved procedures for carbide extraction as well as a larger database on samples with various initial compositions and heat treatments are needed before the microstructural techniques can be used for plant life assessment. It has been shown that even trace amounts of tramp elements such as phosphorus can affect the kinetics of evolution of the carbide phases (Ref 64).

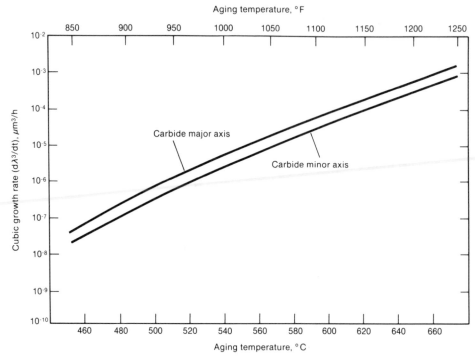

Fig. 5.36. Structure-coarsening kinetics for type I carbides in a dissimilar-metal weld (nickel-base filler metal, 2¼Cr-1Mo steel base metal) (Ref 52 and 63).

Postservice Creep and Rupture Tests

A very common way of estimating the remaining life under creep conditions is the use of stress-rupture tests on the service-exposed material. In the case of tubes and thin components, this method provides a direct estimate of the expended creep life. In the case of thick-section components, estimates of the expended and remaining lives pertain only to crack initiation. These estimates must then be coupled with crack-growth analysis to determine the over-all remaining life.

An approach commonly employed in the past is to test a single specimen from the service-exposed component to rupture under accelerated conditions. The life fraction expended is calculated as the ratio of the time to rupture of the service-exposed specimen to the expected time to rupture for virgin material under the same conditions. Using the life-fraction rule, the remaining life fraction is calculated by subtracting the expended life fraction from unity. An alter-

native approach is to draw a line parallel with the virgin-material rupture data and passing through the single data point. By extrapolating to the service conditions, the remaining life is estimated. These two methods are equivalent because assumption of the life-fraction rule implies that the postservice test data would be parallel with the virgin-material data. Because virgin-material data specific to a heat of steel are never available, the scatterband of ISO data is used as the reference curve, with the minimum line being used for the calculations. Results of postservice tests on samples from a number of headers and steam pipes have been reported (Ref 13, Vol 1). Other results too numerous to mention are also available in the literature. The main problems with this approach are as follows. Very often the data from the service-exposed material fall within the scatterband of the ISO standard material data. Most of the data reported in the literature have been generated from tests in which stress alone or both temperature and stress were varied. When these data are plotted as log t_r vs log stress or in the form of the Larson-Miller plots, the curve for the service-exposed material is below that of the ISO band, but very rarely parallel with it. The curves for the service-exposed and ISO data converge at large values of the Larson-Miller parameter or at low stresses. This behavior is only to be expected in view of the invalidity of the life-fraction rule under stress-varied conditions, as discussed in detail in Chapter 3. Due to these limitations, use of a single specimen in conjunction with the life-fraction rule is subject to considerable inaccuracy.

Example:
A boiler penthouse tube has been in service for 100,000 h at 1000 °F. An accelerated test on a sample from the tube tested at 1200 °F causes specimen rupture in 500 h. Given that, from ISO data, time to rupture for virgin material at 1200 °F is 1000 h, calculate the remaining life under service conditions.

Answer:
Fractional life expended $= 500/1000 = 0.5$

Remaining life at 1000 °F $= 100,000$ h

A more accurate procedure involves testing several samples using the isostress method. The principle of the isostress-rupture test has been described in Chapter 3. Essentially, this method involves conducting accelerated stress-rupture tests at higher temperatures and at stress levels close to the service stress and extrapolating the plot of log t_r vs T to the service temperature to estimate the remaining life. Such plots have been observed to be linear for many of the ferritic steels (see Fig. 3.33, Chapter 3) (Ref 13, 25, 65, and 66). Because extraction of large samples from critical components is not practical, the use of miniature specimens has been demonstrated (see Fig. 5.37). Miniature-specimen tests conducted in inert environments yield results identical with those obtained from conventional samples tested in air (Ref 13). Care must be exercised to ensure that the sample is representative of the bulk microstructure of the area being

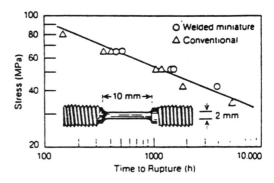

Fig. 5.37. Comparison of results of stress-rupture test on conventional and miniature specimens of 1Cr-½Mo steel at 630 °C (1165 °F) (Ref 13).

evaluated. Any inhomogeneities, such as those in grain size and inclusion content, must be typical of the bulk material. The specimen diameter should be well in excess of the microstructural parameters such as grain size and inclusion size.

The isostress-rupture method offers the advantage that a database on virgin-material properties is not required. Furthermore, the life-fraction rule is not utilized because remaining-life estimates are obtained directly from the data. If isostress tests are conducted in air using standard ASTM specimens, oxidation during testing may lead to conservative predictions of remaining life with respect to heavy-section pipes and headers. The data on the effect of specimen size presented in Fig. 3.21 (Chapter 3) suggest that a correction factor of 2 to 3 to remaining life may be appropriate. These correction factors are in the process of being developed and finalized on the basis of research and development studies of oxidation and specimen-size effects (Ref 67).

In the case of tubes, longitudinal as well as chordal samples have been utilized. Standard full-size chordal samples can be obtained only from large-diameter tubes. Hence, longitudinal samples are often employed. If damage has progressed to the point of forming cavities and cracks, the damage will be anisotropic, and longitudinal samples may give optimistic estimates of remaining life. On the other hand, if damage consists merely of softening, either type of sample can be used. Burst tests also can be conducted on tube samples by pressurizing end-capped samples with argon or steam. Such tests are more realistic but also more expensive, requiring specialized test facilities. Reasonable correlations between the results of uniaxial tests and burst tests have been established and can be readily used in conjunction with uniaxial tests to predict the remaining lives of tubes (Ref 68). Uniaxial tests and burst tests on samples removed from tubes which previously have been internally pressurized to produce creep damage have been shown to result in similar predictions of remaining life. Similar comparisons (Ref 69) using ex-service tube material have shown that the rupture lives of uniaxial specimens and of tubes are comparable if the stress applied to the uniaxial specimens is the same as the equivalent stress in the tube computed on the basis of the mean-diameter hoop-stress formula, using internal diameter instead of mean diameter. If the mean diameter is used to calculate the equivalent stress, the uniaxial specimens fail in shorter times compared with the tubes, which indicates that this procedure may be conservative.

Walser and Rosselet have observed a systematic variation of the Norton law stress exponent, n, for creep rupture with the duration of service exposure of a boiler steel, as shown in Fig. 5.38 (Ref 70). If a larger database relating n to service time-temperature history could be established, measurement of n alone based on rupture tests of small samples taken from a component could be used to estimate the effective operating temperature and hence the creep life expended. In order to determine the value of n, however, several specimens must be tested at different stresses. These same specimens could be better utilized to get direct estimates of remaining life by conducting temperature-varied isostress-rupture tests, as discussed previously.

In the course of conducting rupture tests, it is also desirable to acquire information relating to the minimum creep rate, the rupture strain, and the Monkman-Grant

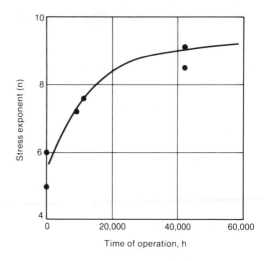

Fig. 5.38. Variation of Norton law stress exponent for rupture as a function of prior operating time at 535 °C (995 °F) (Ref 70).

strain. These parameters may provide additional input for life estimation using the strain-based techniques previously discussed. Such information is also useful in calculating the crack-tip driving force C^* or C_t and thus estimating crack-growth rates in components.

Removal of Samples From Components

Many of the evaluation procedures described earlier require removal of samples from components. In the case of superheater and reheater tubes, a section of the tube can be cut and removed for evaluation. An equal length of new tubing can be welded on to complete the circuit. In the case of headers and steam pipes, either small samples gouged from the surface or through-wall core-plug samples can be trepanned, depending on the type of evaluation being performed. The tools and techniques needed for sample removal have been described by Der *et al* (Ref 48). The core-plug samples are helpful in characterizing through-wall damage, whereas the smaller surface samples are adequate for characterizing only surface damage.

Small surface samples, termed "cone" samples by Der *et al*, have been found to be useful in optical metallography, transmission microscopy, and fractography. It has been determined that a conical sample approximately 5 mm (0.2 in.) in both base diameter and height would be adequate for these purposes. The types of samples removed are illustrated in Fig. 5.39. Sample cutting is achieved by a compressed-air-motor-driven sapphire cutting nozzle using a stream of 53-μm (2.1-mil) silicon carbide abrasive powder as the cutting medium, operated at a pressure of 500 kPa (5 bar). Following sample removal, the defect left in the component may be machined to a hemispherical hole, using fine-grade tungsten/steel burrs, to minimize any stress concentration.

To allow both an assessment of the through-thickness microstructure and post-exposure creep testing of samples from a component such as a header body, core-plug samples up to 60 mm (2.35 in.) in diameter may be removed using a portable trepanning tool such as the one depicted at top in Fig. 5.40 (Ref 48). This tool consists essentially of a hollow cylindrical saw cutter mounted on a drilling stand, which can be strapped to the component by adjustable chains. The cutting tool is lowered by hand and rotated at up to 100 rpm by a compressed air supply through a drive mechanism located above the cutter. To prevent the sample from falling back into the component, it is secured by a 3-mm- (0.12-in.-) diam rod which passes through the center of the cutter and is screwed into a previously drilled and tapped blind hole in the center of the plug surface. Selection of the location for sampling must be done with the utmost care so that the integrity of the component is not compromised. The location of sample removal must be clearly identified in order to keep track of the orientation of the specimen with respect to the component. After the core plug has been removed the component can be weld re-

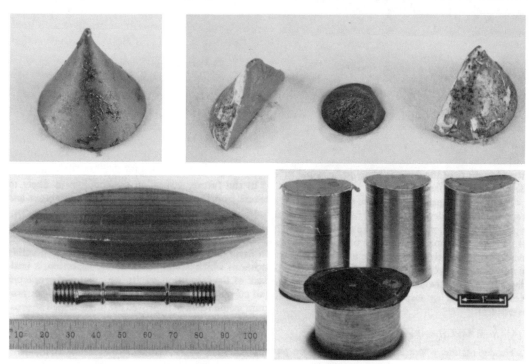

Top left: Cone sample. Top right: Slices from cone samples for foils. Bottom left: Boat sample. Bottom right: Core-plug samples.

Fig. 5.39. Various types of samples removed from headers (Ref 48) (courtesy of P.E.J. Flewitt and R.T. Townsend, CEGB, UK, and B.W. Roberts, Combustion Engineering, Inc.).

Fig. 5.40. Above: Portable trepanning tool. Below: Sample removed using tool shown above (Ref 48) (courtesy of P.E.J. Flewitt and R.D. Townsend, CEGB, UK).

paired by welding onto it a dummy oversize stub. An example of a core-plug sample is shown at bottom in Fig. 5.40.

Fracture-Mechanics Approach

Components which are subject to stress and temperature gradients (typical of headers and pipes) generally do not fail by creep rupture. It is more likely that, at the end of the predicted creep-rupture life, a crack will initiate at the critical location, propagate, and eventually cause failure. The difference between the actual life and the life predicted from creep-rupture data would thus correspond to the crack-propagation life. A similar situation exists when failures originate at pre-existing defects. In such cases, the life-assessment procedure should involve a fracture-mechanics approach. A value of critical crack size is established, and then crack-growth analysis is performed to determine the remaining life. The remaining life is the time required for an initial crack detected during inspection to grow to the critical crack size.

Determination of Critical Crack Size. Several alternative criteria can be used to establish a value for the critical crack size. These criteria include: (1) a flow-stress-governed failure criterion; (2) a J_{Ic}-controlled failure criterion; and (3) a limiting creep-crack-growth-rate criterion. The first two criteria are employed in a scenario in which rupture occurs during or immediately following a start-up transient, in the absence of creep. The third criterion is used in a scenario where failure occurs by creep-crack growth under operating conditions. The lowest value of critical crack size determined by use of these criteria is then used for remaining-life analysis. The use of this combined failure criterion to define the safe operating pressure for a pipe is illustrated in Fig. 5.41.

Rupture in the Absence of Creep. For rupture in the absence of creep, two different methods have been proposed. In the first method, proposed by Battelle investigators (Ref 71 and 72), a flow-stress-dependent failure criterion is employed. In the second method, a fracture-mechanics approach is used.

In the Battelle method, the following equations have been found to apply:

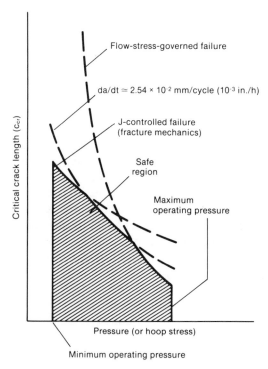

Fig. 5.41. Combined failure criterion for a pipe under internal pressure.

$$\frac{K_C^2 \pi}{8C\bar{\sigma}^2} = \ln \sec\left(\frac{\pi M_T \sigma_T}{2\bar{\sigma}}\right) \quad \text{(Eq 5.37)}$$

for a through-wall flaw, and

$$\frac{K_C^2}{8c_{eq}\bar{\sigma}^2} = \ln \sec\left(\frac{\pi M_p \sigma_P}{2\bar{\sigma}}\right) \quad \text{(Eq 5.38)}$$

for a surface flaw.

K_C is a toughness parameter estimated as

$$K_C^2 = \frac{12EC_v}{A_c} \quad \text{(Eq 5.39)}$$

Definitions of the other terms in these equations (and of the terms within these definitions) are as follows:

c is the half-length of the flaw.
$\bar{\sigma}$ is the flow stress of the material (yield strength + 10 ksi).
M_T is the Folias correction for a through-wall flaw.
σ_T is the hoop stress for failure for a through-wall flaw (i.e., PR/t).

P is the internal pressure.
R is the pipe radius.
t is the wall thickness.
E is the elastic modulus.
C_v is the Charpy V-notch shelf energy.
c_{eq} = A/2a, the equivalent crack half-length.
A is the actual area of the surface flaw.
a is the depth of the surface flaw.
A_c is the area of the fracture surface of a Charpy V-notch specimen.

$$M_p = \frac{1 - (a/M_T t)}{1 - (a/t)}, \text{ the Folias correction}$$

for a surface flaw.
σ_P is the hoop stress for failure for a surface flaw (i.e., PR/T).

Hence, from Eq. 5.37 to 5.39, given the Charpy V-notch shelf energy, the elastic modulus, the pipe dimensions, and the internal pressure, one can estimate the critical flaw size (i.e., depth and length of flaw) that will lead to failure. Alternatively, one can use the measured flaw size to estimate the internal pressure that will cause failure. This approach applies to low yield strengths, small defects, and high-toughness materials.

In the fracture-mechanics approach, it is assumed that rupture occurs when the applied J exceeds the J_c of the material. J_c for the material in question should be known from independent tests. The applied J is calculated on the basis of pressure and the flaw/pipe geometry. Because the J expressions contain crack dimensions, the critical crack size can be defined. A second criterion for instability is the tearing modulus described in Chapter 2. If the applied value of T as defined in Eq. 2.30 exceeds the tearing modulus of the material, then unstable crack propagation will result.

Rupture During Operation. When fracture occurs during normal operation, one cannot be sure that the crack-tip stress fields are dominated by J, because of the presence of time-dependent creep strains. Therefore, for practical purposes, it is assumed that fracture occurs when da/dt > 10^{-3} in./h. At this rate, there is plenty of experimental evidence that creep-crack-growth rates are controlled by C_t. Also, the crack-growth rate is sufficiently rapid so that not much crack-growth life remains once this rate has been attained. Thus, the crack size at which the crack-growth rate exceeds 10^{-3} in./h is taken as the critical crack size (Ref 73).

Creep-Crack-Growth Analysis. Figure 5.42 shows schematically the various steps involved in prediction of creep-crack-growth life. Step 1 consists of either generating or identifying from available data the creep-crack-growth and creep-deformation behavior of the material. The relevant mate-

rial parameters are A, n, b, and m, as defined in the figure. The elastic modulus, E, and the yield strength, σ_y, are also required. Step 2 consists of assembling the expressions needed for estimating C_t values. This procedure was discussed in detail in Chapter 3. Step 3 includes development of the residual-life curve. This curve may be in the form of remaining life as a function of initial crack size. Alternatively, given an initial crack size, the predicted crack size as a function of time may be estimated.

The basic expression that relates the creep-crack-growth rate da/dt to the crack-tip driving force C_t or C^* can be written as

Step 1: Generation of creep-crack-growth and creep-deformation data.

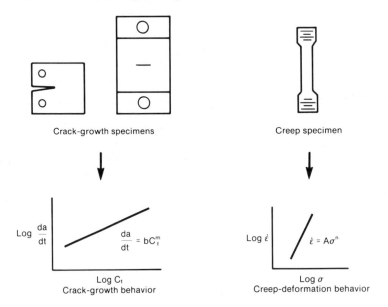

Step 2: Assemble the C^* or C_t expressions appropriate to the component.

Step 3: Combine data from Step 1 and C^* or C_t expressions from Step 2 to develop curves of residual life vs initial crack size.

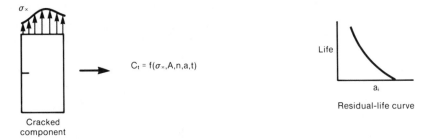

Fig. 5.42. Methodology for predicting crack-propagation life using time-dependent fracture-mechanics (TDFM) concepts (Ref 73).

$$\frac{da}{dt} = bC_t^m \text{ [same as Eq 3.80]} \quad \text{(Eq 5.40)}$$

where b and m are material constants and C_t is a function of crack depth a and time t. The remaining life, t_r, is given as follows:

$$t_r = \int_{a_i}^{a_c} \frac{da}{bC_t^m} \quad \text{(Eq 5.41)}$$

where a_i is the initial crack depth and a_c is the critical crack depth. Equation 5.38 can now be solved numerically to determine the remaining life.

In order to demonstrate the various steps of such an analysis, an example problem for a thick-wall cylinder with a longitudinal crack, taken from Ref 73, is presented in the next major section. Other examples of creep-crack analysis in headers are also described in Ref 73. A crack-growth methodology for crack propagation between bore holes (ligament cracking) has been illustrated by Saxena, Sherlock, and Viswanathan (Ref 74).

Leak-Before-Break. When a crack in a radial-axial plane breaks through the wall of a pressure vessel but continues to grow at a stable rate in the axial direction, a condition known as leak-before-break occurs — i.e., a slow leak of the fluid occurs but a large rupture does not. If the depth of the flaw is designated by a and the length of the flaw by 2c, the following conditions must be met in order to ensure a leak-before-break:

$$t_r = \int_{a_i}^{a_c} \frac{da}{bC_t^m} = \int_{c_i}^{c_{cr}} \frac{dc}{bC_t^m} \quad \text{(Eq 5.42)}$$

where a_i and a_c are the initial and critical crack depths, c_i is the maximum allowable initial crack half-length at inspection, and c_{cr} is the critical crack half-length. Usually, a_i is determined by inspection, and a_c and c_{cr} are determined by the methods discussed in the previous section. Hence, c_i is the only unknown factor in Eq. 5.42, and its value must be determined by solving the

equation. The value of c_i then becomes the critical crack half-length that must not be exceeded at the time of inspection to ensure a leak-before-break condition.

The above creep-crack-growth analysis, including leak-before-break, is conveniently performed with the help of computers. Application of this methodology to evaluation of hot reheat piping has been illustrated by Gaitonde (Ref 75).

Integrated Methodology for Life Assessment

The various techniques described previously can be integrated into a generic methodology for life assessment that consists of a phased three-level approach, as described in Chapter 1 (Ref 76). A brief outline of this procedure (based primarily on the description given in Ref 76), as applicable to superheater outlet headers, superheater/reheater tubes, and steam pipes, is presented below.

Superheater Outlet Headers. Specific flow charts for preliminary issues and for level I and level II assessments of superheater outlet headers are shown in Fig. 5.43. Level III assessment steps are presented in Fig. 5.44. The scheme for header-life assessment is based on the practice of the Central Electricity Generating Board of the United Kingdom (Ref 77) and on derivatives of that practice (Ref 78). It begins with the acquisition of service information and then answers the questions outlined in Fig. 5.43. This procedure addresses strictly the life expenditure with respect to creep. If creep-fatigue is of concern, the calculational procedures described earlier should be incorporated. Plant records of historical and current operation and maintenance are compiled and the appropriate data for superheater outlet headers examined. For headers, the most important service information includes boiler running hours, history of tube failures, details of past header repairs or replacement, compositional and dimensional checks, design parameters, and steam-temperature records. The key questions regarding these preliminary data are:

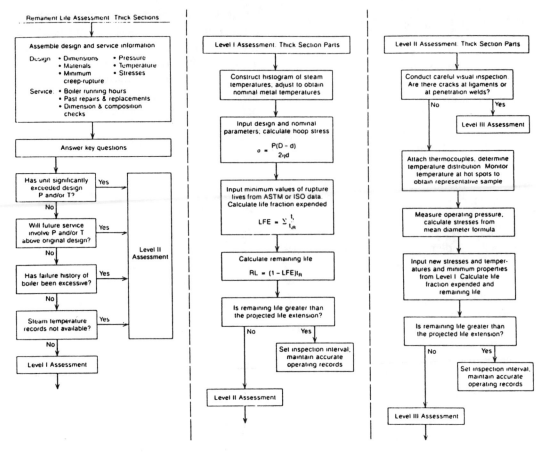

Fig. 5.43. Preliminary issues and level I and level II assessments of superheater outlet headers (Ref 76).

1. Has the unit significantly exceeded the design temperature and/or pressure conditions?
2. Will the future use (extended life) involve temperatures, pressures, or cycles outside the design envelope?
3. Has the failure history of the boiler been excessive?
4. Are steam-temperature records inadequate or unavailable?

A "yes" answer to any of these questions should indicate that level I assessment is inappropriate.

Assuming that the answers to all of the above questions are "no," the assessment procedure may start at level I. The first activity in level I assessment is the construction of a steam-temperature history—preferably in 10 °C (or 20 °F) increments,

because creep damage is very sensitive to temperature. The creep-life expenditure in each temperature and pressure regime is calculated from the design parameters, including minimum creep-rupture properties. The mean-diameter formula described in Chapter 3 may be used to calculate the nominal stress. The cumulative creep damage can be calculated by linear addition of expended life fractions at the various temperature intervals. Cumulative creep-fatigue-life expenditure can be computed from a knowledge of the various transients, the strain levels corresponding to each type of transient, and the low-cycle fatigue behavior of the material, using the procedures discussed earlier. If the remaining life fraction at the end of the analysis is equal to or less than the expected period of future service, then a level II analysis is performed.

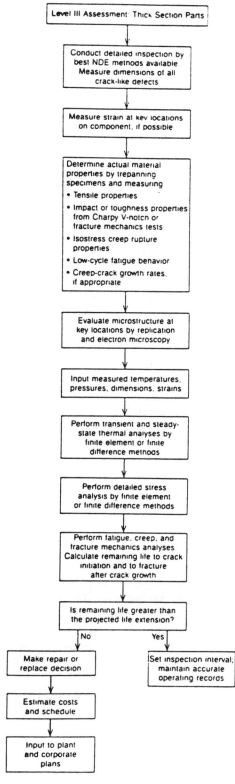

Fig. 5.44. Level III assessment of super-heater outlet headers (Ref 76).

The main difference between level I and level II assessments is that the level II assessment uses the temperature distribution along the header as measured with thermocouples. The thermocouples do not necessarily have to be installed if *in situ* devices are already in place; that is, their use will change from a primarily design-oriented function to an operational function. The stresses, minimum creep-rupture properties, and low-cycle-fatigue properties used in a level II assessment are the same as for level I. If the remaining life fraction at the end of this analysis is less than or equal to the period of anticipated future use, then a level III analysis is required.

A level III assessment (in the absence of obvious cracks) typically consists of the following elements:

1. A refined stress analysis by finite-element methods to identify and quantify the highly stressed regions.
2. Detailed inspection of the header by dye-penetrant, magnetic-particle, ultrasonic, and/or radiographic methods, especially at welds and penetrations.
3. Sampling of the header material for accelerated creep-rupture testing to establish the actual properties of the material.
4. Replication of the surfaces in high-stress, high-temperature regions so that high-resolution microscopy can be used to ensure that creep damage has not accumulated locally to the degree that microscopic cavities are visible or that the microstructure has not otherwise degenerated. Replication results can also supplement other remaining-life methods.

The closed-form stress analyses that were an integral part of level I and II evaluations can be used to establish initial inspection locations required early in the level III evaluation. Refined stress-analysis methods and more detailed inspections then follow, as required.

During the evaluation for creep damage, an integrated evaluation of other damage mechanisms will also be taking place. As a minimum, an early visual inspection of the header should be conducted to examine ligaments and penetration welds for cracking. Other NDE methods should be used as appropriate to examine attachments, nipples, and other components to ensure that there is no fatigue damage. Internal inspection may also be appropriate. The presence of cracks will necessitate a level III assessment that includes core-sample removal and a crack-propagation analysis. NDE techniques such as dye-penetrant and magnetic-particle inspection should be scheduled to coincide with a maintenance outage to confirm the conclusions from the visual inspection. Re-examination or recalculation should be scheduled well before the calculated end-of-life (margin to be chosen by the individual utility).

Steam Pipes. Although steam pipes are geometrically simpler than headers, degradation and cracking of pipes have traditionally occurred in weld heat-affected zones. Thus, special attention must be paid to careful inspection and evaluation of weldments and to the secondary system stresses that are associated with pipework-support structures. Due to the recent catastrophic ruptures of seam-welded hot reheat pipes at three power plants, prudence dictates that remaining-life assessment of steam piping should, at the very outset, be bifurcated into activities associated with seamless pipe and parallel activities pertinent to seam-welded pipe.

Level I assessment would thus start with a review of fabrication, construction, and operating history to determine whether the piping is seamless or seam welded. If it is seamless, the assessment proceeds along the path already described—i.e., in stages commencing with an evaluation based on design conditions, and thereafter becoming more accurate and detailed, through level III if necessary. Considerations of the effects of hangers, girth welds, and attachments are appropriate in Level I, and replication typically is an essential part of the inspection protocol for level II assessment. If level I and/or II assessment indicates no concern about creep damage in seamless pipe, then

"strain monitoring" is a suitable means for checking pipe condition on a continual basis. Outside-surface measurements on pipes fitted with creep "pips" is routinely used by the CEGB (Ref 78). Plots of distance between pips (strain) versus time will signal concern by upward changes of slope between inspections. A baseline condition of the pipe is thereby established, which will be particularly useful for future deliberations concerning life extension. If the piping is seam welded, the next step will be to inspect the weldments by quantitative methods. Activities and procedures should be consistent with the guidelines for inspecting seam-welded pipe (Ref 16). Inspection will almost certainly include a microstructural evaluation (by replication or excision of samples) as well as a search for defects or cracks by magnetic-particle, ultrasonic, and radiographic testing. Because seam-welded pipe is an important safety issue, the highest level of inspection capability and coverage should be invoked; probably 100% inspection of each line will be mandated. Analysis of the information will involve fracture-mechanics evaluation and creep-damage estimates. If actual material properties are required for a level III assessment, special care must be accorded the procedures for evaluating "clamshell" elbows (i.e., those fabricated by axial welding of two half-sections). Taking plug samples along the weld on the outer bend (the line of maximum stress concentration) will almost certainly make the component unsafe for future operation, even if it was not significantly degraded by the service exposure. Finally, a run/repair/replace decision will be made, along with a prescription for future inspection intervals. A road map for steam-pipe assessment is shown in Fig. 5.45 (Ref 16).

Superheater/Reheater Tubes. In boilers fired with gas and/or oil, creep can be the main damage mechanism because fire-side corrosion is not always important. In coal-fired units, fire-side corrosion may be more important, although the main problem is overheating. Creep damage, being dependent on both time and temperature, is well

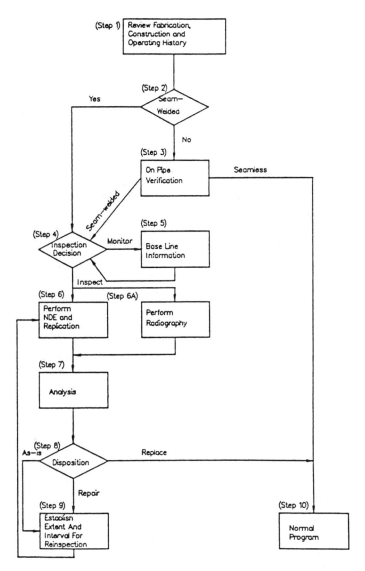

Fig. 5.45. Road map for steam-pipe assessment (Ref 16).

suited for assessment by the three-level approach. The basic parameters for level I are as follows:

1. Nominal tube dimensions (from drawings)
2. Design pressure and temperature
3. Nominally specified materials
4. Minimum creep-rupture properties.

Tube-metal temperatures are taken from past measurements or calculated from the design steam temperature (Ref 57). Operative stress is estimated from the mean-diameter formula (Chapter 3). The life fraction expended is then calculated. If the remaining life is less than the anticipated service extension, level II assessment is undertaken.

Level II differs from level I primarily in the use of temperatures from operating records. A temperature histogram is constructed in 10 °C intervals from steam-temperature records, and each temperature is corrected for the difference between steam temperature and metal temperature. Stresses are similarly adjusted from the records of operating (rather than design) pressures. Expended life is calculated for each temperature segment, and the resulting fractions are summed to arrive at the cumulative life fraction expended.

In coal-fired boilers, overheating is typically the main threat to superheater/reheater tube integrity, although fire-side corrosion is also a problem where high-chlorine coals are used. Failures associated with fire-side corrosion can still be thought of as long-term creep failures, except now the process is accelerated by the higher stresses brought about by wall thinning. For a level I assessment, records of the last inspection(s) will be required. The current status of tubes is estimated by assuming a linear rate of wall thinning, extrapolated from the original wall thickness and the thickness last measured. Long-term creep damage is estimated, as previously, from adjusted design temperature, but with the stresses calculated from the mean-diameter formula applied to the thinner part of the tube wall. In the absence of such information, a level II analysis must be undertaken.

Level II assessment for superheater and reheater tubes subject to fire-side corrosion or erosion begins with the results of ultrasonic inspection. The expended life fraction is calculated from the temperature histogram by summing the contributions of damage caused by time, temperature, and stress in each segment. In order to estimate the remaining life, an expected wall thickness must be obtained by extrapolation to the end of the anticipated service period. Alternatively, remaining life can be estimated directly from the current wall thickness if a previous time to failure in the boiler is known (Ref 37, 79, and 80). Should the superheater or reheater fail to pass the criterion that remaining life be equal to or greater than the life-extension period, level III analysis is performed. A refined stress analysis and more accurate measurement of temperature differentiate this stage from the preceding stage. Temperature measurements should pertain to the critical regions of the superheater or reheater; a knowledge of previous tube failures and of the flue-gas and steam temperature profiles across the subsystems will be valuable in identifying

these regions. The temperature measurements can be made with chordal thermocouples, or by cutting short tube sections from critical locations, measuring the thickness of the steam-side oxide scale, and converting that thickness to temperature at the oxide-metal interface (Ref 55 and 81). Wall thickness is measured directly on the excised sections, and by ultrasonic testing at other locations. Initial stresses (pressure and thermal), and stress redistribution after creep, are calculated (Ref 55 and 81) after the temperatures have been determined. Wastage rates also can be calculated from known temperatures (Ref 82). Expended life and remaining life are computed on the basis of linear damage summation, as before, with multiheat creep-rupture data. Further refinement of the evaluation, if needed, can then be accomplished by conducting accelerated creep-rupture tests using pressurized tube samples or uniaxially stressed chordal samples machined from thick-wall tubes.

Example Problem

This problem concerns an internally pressurized thick-wall cylinder containing a crack in the radial-axial plane (see Fig. 5.46). The

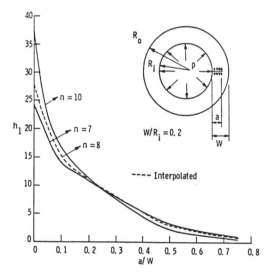

Fig. 5.46. h_1 functions for various n values for an axially cracked cylinder under internal pressure (Ref 73).

inside radius of the cylinder, R_i, is 381 mm (15 in.), and the outside radius, R_o, is 457 mm (18 in.). Thus, the wall thickness is 76 mm (3 in.). The pressure is assumed to be 13.79 MPa (2000 psi), and the wall temperature is 538 °C (1000 °F). The configuration of this cylinder somewhat resembles that of an actual header. However, the dimensions are imaginary and have been chosen such that known solutions for estimating the stress-intensity parameter, K, and the C^*-integral can be used. The format for describing the methodology will be a three-step approach, as outlined earlier. This example is reproduced from Ref 73.

Step 1: Identification of Material Properties

Table 5.8 lists all the material-property data to be used in solving the problem. These data pertain to samples taken from a location that was severely degraded (hot end) and a location that was less severely degraded (cold end) in an actual header pipe retired from utility service.

Step 2: Expression for Estimating C_t

The expression for the crack-tip parameter C_t is

$$C_t = \frac{4\alpha(1 - \nu^2)}{E(n - 1)} \frac{K^4}{W} (EA)^{2/(n-1)}$$

$$\times t^{-(n-3)/(n-1)} \frac{F'}{F} + C^* \quad \text{(Eq 5.43)}$$

where $\alpha = 0.2569$ for $n = 10.1$ and 0.281 for $n = 8$ (as shown in Ref 73); $\nu = 0.3$; and E, A, and n are given in Table 5.8. The width W in this case is equal to the wall thickness of 76 mm (3 in.).

For a W/R_i value of 0.2 and for $0 \leq a/W \leq 0.5$, the K expression is given by

$$K = p\sqrt{WF} \quad \text{(Eq 5.44)}$$

where p is internal pressure and F is calculated as follows:

$$F = 11.6\sqrt{\frac{a}{W}} \left[0.95 + 1.301\left(\frac{a}{W}\right) \right.$$

$$+ 10.66\left(\frac{a}{W}\right)^2$$

$$- 55.02\left(\frac{a}{W}\right)^3$$

$$\left. + 75.56\left(\frac{a}{W}\right)^4 \right] \quad \text{(Eq 5.45)}$$

$$F' = \frac{dF}{d(a/W)} \quad \text{(Eq 5.46)}$$

Equation 5.45 has been derived by conducting a polynomial fit to the numerical results of Kumar *et al* (Ref 83). The expression for estimating C^*, taken from the same source as for the K expression (Ref 83), is given by

$$C^* = A\left(1 - \frac{a}{W}\right) ah_1$$

$$\times \left(\frac{\sqrt{3}}{2} p \frac{R_i/W + a/W}{1 - a/W} \right)^{n+1}$$

$$\text{(Eq 5.47)}$$

where h_1 is a function of a/W, n, and W/R_i, and is plotted in Fig. 5.46 for a W/R_i value of 0.2. The h_1 values in Fig. 5.46 have been derived from two sources (Ref 83 and 84). The h_1 values for $0.125 \leq a/W \leq 0.75$ were obtained from the EPRI handbook (Ref 83). The limiting value of h_1 as $a/W \rightarrow 0$ is derived from the work of He and Hutchinson (Ref 84), who obtained an expression for calculating J for edge cracks in a semi-infinite plate loaded under uniform tension. Their expression leads to the following value of h_1:

$$h_1\left(\frac{a}{W} \rightarrow 0\right) = 1.2\pi\sqrt{n} \left[\frac{R_i^2 + R_o^2}{R_i(R_i + R_o)} \right]^{n+1}$$

$$\text{(Eq 5.48)}$$

Table 5.8. Summary of material constants at 540 °C (1000 °F) needed for prediction of

Material condition	σ_y		$E \times 10^3$			A	
	MPa	ksi	MPa	ksi	n	$MPa^{-n}\,h^{-1}$	$ksi^{-n}\,h^{-1}$
South end (cold)	131.0	19.0	140.6	20.4	10.1	1.462×10^{-24}	4.3×10^{-16}
North end (hot)(b)	131.0	19.0	140.6	20.4	8.0	4.49×10^{-20}	2.29×10^{-13}

(a) The numbers provided for b are those which correspond to C_t values in $J/m^2 \cdot h$ and da/dt in mm/h. The estimated to be the same for hot- and cold-end materials.

In deriving Eq 5.48 it was assumed that, as $a/W \to 0$, the C^* value for a crack in an internally pressurized cylinder equals the C^* value for an edge-cracked member loaded with a uniform stress equal in magnitude to the hoop stress generated by internal pressure.

By making the appropriate substitutions in Eq 5.43 for all constants, and combining Eq 5.43 with Eq 5.45 and 5.47 using a pressure value of 13.79 MPa (2000 psi), we get the following equations for determining C_t:

$$C_t = 0.157(t^{-0.78})F'F^3$$
$$+ 1.0 \times 10^{-7}h_1\left(1 - \frac{a}{W}\right)$$
$$\times \left(\frac{a}{W}\right)\left(\frac{5 + a/W}{1 - a/W}\right)^{11.1} \quad \text{(Eq 5.49)}$$

for the cold-end material, and

$$C_t = 0.25(t^{-0.714})F'F^3$$
$$+ 1.686 \times 10^{-5}h_1\left(1 - \frac{a}{W}\right)$$
$$\times \left(\frac{a}{W}\right)\left(\frac{5 + a/W}{1 - a/W}\right)^{9} \quad \text{(Eq 5.50)}$$

for the hot-end material. Equations 5.49 and 5.50 yield values of C_t in $J/m^2 \cdot h$. To obtain C_t in $in. \cdot lb/in.^2 \cdot h$, divide the entire right-hand side of the equation by approx-

imately 175. Table 5.9 lists the values of $F'F^3$ and h_1 as functions of a/W. We may also define $H(a/W,n)$ as follows:

$$H\left(\frac{a}{W},n\right) = h_1\left(1 - \frac{a}{W}\right)\left(\frac{a}{W}\right)$$
$$\times \left(\frac{5 + a/W}{1 - a/W}\right)^{n+1} \quad \text{(Eq 5.51)}$$

The values of H as functions of a/W are also listed in Table 5.9 for n = 8 and n = 10.1.

Step 3: Life Estimation

For purposes of demonstrating the method being used, the details of the calculation using cold-end material properties will be provided. Only the final results will be provided for the hot-end material.

By substituting Eq 5.51 into Eq 5.49, and substituting the result into the experimentally obtained relationship

$$\frac{da}{dt} = 5.94 \times 10^{-5}C_t^{0.751} \quad \text{(Eq 5.52)}$$

we get the following equation:

$$\frac{da}{dt} = 5.94 \times 10^{-5}$$
$$\times [0.157(t^{-0.78})F'F^3 + 10^{-7}\,H]^{0.751}$$
$$\text{(Eq 5.53)}$$

and hence

$$\int_{t_0}^{t_r} dt = \int_{a_i}^{a_c} \frac{da}{5.94 \times 10^{-5}[0.157(t^{-0.78})F'F^3 + 10^{-7}\,H]^{0.751}} \quad \text{(Eq 5.54)}$$

remaining creep-crack-growth life (Ref 73)

q	b(a)	m	D MPa^{-m}	D ksi^{-m}
0.751	5.94×10^{-5} (1.131×10^{-4})	5.4	8.36×10^{-15}	2.82×10^{-10}
0.546	1.89×10^{-4} (1.25×10^{-4})	5.4	8.36×10^{-15}	2.82×10^{-10}

numbers in parentheses are those corresponding to da/dt in in./h and C_t in in.·lb/in.^{2}·h. (b) Tensile properties are

Table 5.9. Values of h_1, $F'F^3$, and H as functions of a/W (Ref 73)

a/W	$F'F^3$	h_1 n = 8	h_1 n = 10.1	$H\left(\dfrac{a}{W}, n\right) = h_1\left(1 - \dfrac{a}{W}\right)\left(\dfrac{a}{W}\right)\left(\dfrac{5 + a/W}{1 - a/W}\right)^{n+1}$ n = 8	$H\left(\dfrac{a}{W}, n\right)$ n = 10.1
0.01	77.03	26.25	32.5	5.65×10^5	2.109×10^7
0.02	170.7	24.5	29.0	1.165×10^6	4.25×10^7
0.03	285.7	23.0	26.2	1.81×10^6	6.55×10^7
0.04	426.7	21.75	24.5	2.52×10^6	9.25×10^7
0.05	597.9	20.25	22.2	3.24×10^6	1.19×10^8
0.06	803.2	19.25	20.75	4.09×10^6	1.52×10^8
0.07	1.046×10^3	18.25	19.75	5.00×10^6	1.92×10^8
0.08	1.329×10^3	17.25	18.5	5.99×10^6	2.35×10^8
0.09	1.655×10^3	16.25	17.5	7.03×10^6	2.85×10^8
0.10	2.02×10^3	15.5	16.75	8.26×10^6	3.46×10^8
0.11	2.43×10^3	14.75	16.0	9.58×10^6	4.17×10^8
0.12	2.89×10^3	14.0	15.25	1.1×10^7	4.96×10^8
0.13	3.38×10^3	13.5	14.75	1.28×10^7	5.97×10^8
0.14	3.906×10^3	13.25	14.25	1.5×10^7	7.12×10^8
0.15	4.46×10^3	12.75	13.85	1.72×10^7	8.54×10^8
0.16	5.048×10^3	12.50	13.625	2.0×10^7	1.03×10^9

where t_r is remaining life, a_i is initial crack size, a_c is final (critical) crack size, and t_0 is the incubation time required for the crack to start growing upon application of the load. During the incubation time, crack extension either does not occur or occurs very slowly. At present, no approaches are available for predicting incubation time. In a conservative analysis, the incubation time may be essentially neglected by designating it to be 1 h. The choice of 1 h is convenient because it also gets rid of the singularity in the value of C_t at time t = 0. This singularity arises because C_t cannot be defined for t = 0, because the instantaneous response of a cracked body does not involve any creep, and is either elastic or elastic-plastic. Because time terms are involved on both sides of the integral, and cannot be separated, the following approximate method is used to solve for time.

We assume that the total crack extension $a_c - a_i$ occurs in m equal increments of Δa. The Δa increment is chosen to be sufficiently small such that C_t can be assumed to be constant during the time interval Δt_n required for the crack to grow a distance of Δa. If Δt_n represents the time interval for the nth increment of crack length, then Δt_n is given approximately as

$$\Delta t_n = \frac{\Delta a}{5.94 \times 10^{-5}(0.157 t_{n-1}^{-0.78} F'_{n-1} F^3_{n-1} + 10^{-7} H_{n-1})^{0.751}} \qquad \text{(Eq 5.55)}$$

The crack size at the end of the nth increment of crack growth, a_n, is related to the initial crack size, a_i, through the expression

$$a_n = a_i + n\Delta a \qquad \text{(Eq 5.56)}$$

and the total time elapsed at the end of the nth increment of crack growth, t_n, can be expressed as

$$t_n = t_0 + \sum_{n=1}^{n} \Delta t_n \qquad \text{(Eq 5.57)}$$

where t_0 is the incubation time required for crack growth to start. The terms F'_{n-1}, F_{n-1}, and H_{n-1} represent the values of F', F, and H corresponding to a crack depth of a_{n-1}. Equations 5.55 to 5.57 can be solved progressively by varying n from 0 to m. Thus, the a-vs-time curve can be generated. Table 5.10 summarizes the results of the calculations using the cold-end properties. The value selected for Δa was 0.76 mm (0.03 in.), and the value of a_i was also 0.76 mm.

Figure 5.47 shows the predicted crack-size-vs-time behavior for the axially cracked cylinder. These predictions are made by using both the cold-region and hot-region material properties. As expected, the predicted crack size increases much more quickly with time for the hot-region properties. To illustrate the effect of pressure,

Table 5.10. Predicted crack-depth-vs-time behavior for a cracked cylinder

i	Time (t_{n-1}), h	a_{n-1} mm	a_{n-1} in.
1	1	0.762	0.03
2	1.748×10^3	1.524	0.06
3	6.0×10^3	2.286	0.09
4	9.1×10^3	3.048	0.12
5	11.49×10^3	3.81	0.15
6	13.47×10^3	4.572	0.18
7	15.12×10^3	5.334	0.21
8	16.5×10^3	6.096	0.24
9	17.69×10^3	6.858	0.27
10	18.72×10^3	7.62	0.30
11	19.61×10^3	8.382	0.33
12	20.39×10^3	9.144	0.36
13	21.07×10^3	9.906	0.39
14	21.66×10^3	10.668	0.42
15	22.18×10^3	11.43	0.45
16	22.63×10^3	12.192	0.48
17	23.02×10^3	12.954	0.51

these calculations can be repeated for another assumed value of internal pressure, 8.96 MPa (1300 psi). Next, the remaining lives for four conditions (two levels of pressure with cold- and hot-region material properties) can be calculated as functions of various initial crack sizes. A final crack size of 13 mm (0.51 in.) was chosen for the present calculations. At this crack size, the remaining life is expected to be almost totally exhausted because of the high crack-growth rates (see Fig. 5.47). The plot of remaining life as a function of initial crack

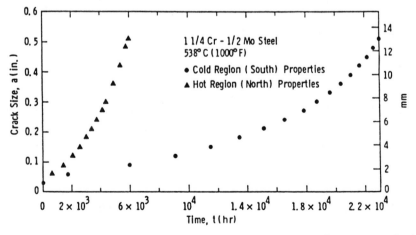

Fig. 5.47. Predicted crack size as a function of time for an internally pressurized cylinder (p = 13.79 MPa, or 2000 psi) (Ref 73).

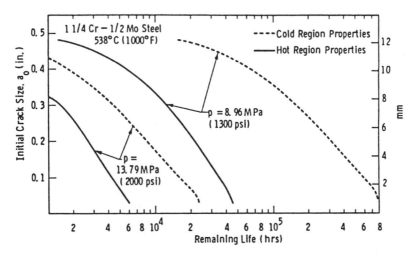

Fig. 5.48. Remaining life as a function of initial crack size for an internally pressurized cylinder (Ref 73).

size is shown in Fig. 5.48. Both pressure and the level of degradation significantly alter the predicted remaining life.

Nomenclature

a	– Crack depth
a_i	– Initial crack depth
a_f	– Final crack depth
a_c	– Critical crack depth
b	– Burgers vector (Eq 5.15); also, constant in Eq 5.40
c	– Crack half-length (Eq 5.37)
c_i	– Initial crack half-length
c_{cr}	– Critical crack half-length at failure
c_{eq}	– A/2a, equivalent crack half-length (Eq 5.38)
f_v	– Volume fraction of cavities
h_1	– f(a/w) (Eq 5.48)
k_A	– Constant (Eq 5.17)
m	– dH/dP, slope of plot of hardness vs time-temperature parameter (Eq 5.25); also, exponent in creep-crack-growth law (Eq 5.40)
m_i	– Carbide size along minor axis (Eq 5.35)
n	– Norton-law stress exponent (Eq 5.43)
p	– Internal pressure in a tube/pipe
t	– Time; also, wall thickness (in definition pertaining to σ_T in Eq 5.37)
t_r	– Time to rupture
t_{rem}	– Remaining life
t_{op}	– Time of operation; same as expended time, t_{exp}
t/t_r	– Life fraction expended
t_{nr}	– Rupture life under wall-thinning conditions (Eq 5.7)
w_i	– Initial tube wall thickness (in definition of K′ in Eq 5.7)
w_f	– Tube wall thickness after thinning (in definition of K′ in Eq 5.7)
x	– Oxide-scale thickness (Eq 5.20)
A	– Number fraction of cavitated boundaries (Eq 5.9); also, Norton-law coefficient (Eq 5.43 and Table 5.8)
A_c	– Area of fracture surface of a Charpy V-notch specimen (Eq 5.39)
C_v	– Charpy V-notch shelf energy (Eq 5.38)
C^*	– Integral defining crack-tip driving force for creep (Eq 5.47)
C_t	– Integral defining crack-tip driving force for creep (Eq 5.43)
E	– Young's modulus
F	– f(a/W) (Eq 5.45)
F'	– dF/d(a/W) (Eq 5.46)
H	– f(a/W,n) (Eq 5.51)
J	– J-integral as defined in Eq 2.26, Chapter 2
J_C	– Critical value of J for stable crack propagation
K	– Crack-tip stress-intensity parameter
K'	– Wall-thinning-rate constant (Eq 5.7)
K_C	– Fracture toughness

M_j — Carbide size along major axis (Eq 5.34)

M_T, M_p — Folias correction factors (Eq 5.37 and 5.38)

N — Number of cavities per mm^2 (Eq 5.8)

R — Radius of pipe

R_i — Inside radius of pipe

R_o — Outside radius of pipe

α — $f(n)$ (Eq 5.43)

α' — Constant (Eq 5.15)

β — Constant in carbide-coarsening kinetics equation (Eq 5.18)

ϵ_h — Hoop strain

ϵ_θ — Axial strain

ϵ_r — Radial strain; rupture strain

ϵ_{eq} — Equivalent strain range (Eq 5.4)

$\dot{\epsilon}$ — Minimum creep rate

ϵ_s — Monkman-Grant creep strain $= t_r$

λ — Carbide size in a dissimilar-metal weld (Eq 5.32); also, mean interparticle spacing (Eq 5.18)

λ' — ϵ_r/ϵ_s (Eq 5.9)

λ_0 — Initial interparticle spacing (Eq 5.18)

λ_t — Interparticle spacing at time t (Eq 5.18)

μ — Shear modulus

ν — Poisson's ratio

$\bar{\sigma}$ — Flow stress (yield strength + 10 ksi) (Eq 5.37)

σ_0 — Threshold stress (Eq 5.15)

σ_T — Hoop stress for failure for a through-wall flaw (Eq 5.37)

σ_P — Hoop stress for failure for a surface flaw (Eq 5.38)

References

1. G.A. Lamping and R.M. Arrowwood, Jr., "Manual for Investigation and Correction of Boiler Tube Failures," Report CS 3945, Electric Power Research Institute, Palo Alto, CA, Apr 1985

2. R.B. Dooley and D. Broske, Ed., "Boiler Tube Failures in Fossil Power Plants," Report CS 5500SR, Electric Power Research Institute, Palo Alto, CA, 1988

3. "Dissimilar Welds in Fossil-Fired Boilers," Report CS 3623, Electric Power Research Institute, Palo Alto, CA, July 1985

4. D.I. Roberts *et al*, "Dissimilar Weld Failure Analysis and Development Program," Report CS 4252, Vol 1-9, Electric Power Research Institute, Palo Alto, CA, Nov 1985

5. D.I. Roberts, R.H. Ryder, and R. Viswanathan, Dissimilar Metal Weld Failure Reduction, in "Boiler Tube Failures in Fossil Power Plants," R.B. Dooley and D. Broske, Ed., Report CS 5500SR, Electric Power Research Institute, Palo Alto, CA, 1988

6. W. Nelson and C. Cain, Jr., *Trans. ASME, J. of Engg. for Power*, Vol 82, Series A, 1960, p 194

7. J. Stringer, High Temperature Corrosion Problems in the Electric Power Industry and Their Solutions, in *High Temperature Corrosion*, NACE-6, 1983, p 389-397

8. R. Borio *et al*, "The Control of High-Temperature Fireside Corrosion in Utility Coal Fired Boilers," R & D Report 41, U.S. Office of Coal Research, Apr 1969

9. *The Control of High Temperature Fireside Corrosion*, 2nd Ed., Central Electricity Generating Board, UK, 1977

10. P. Mayer and A.V. Monolescu, Influence of Hydrogen Chloride on Corrosion of Boilers Steels in Synthetic Flue Gas, *Corrosion*, Vol 36 (No. 7), 1980, p 369-373

11. B.W. Roberts, F.V. Ellis, and R. Viswanathan, "Utility Survey and Inspection for Life Assessment of Elevated Temperature Headers," American Power Conference TIS 7795, Chicago, Apr 1985

12. R. Viswanathan and R.B. Dooley, "Creep Life Assessment Techniques for Fossil Power Plant Boiler Pressure Parts," International Conference on Creep, JSME, I. Mech. E., ASME, and ASTM, Tokyo, Apr 14-18, 1986, p 349-359

13. F.V. Ellis *et al*, "Remaining Life Assessment of Boiler Pressure Parts," Final Report RP 2253-1, Vol 1-5, Electric Power Research Institute, Palo Alto, CA, 1988

14. J. Alice, personal communication, Pennsylvania Electric Co., June 1985

15. R.H. Ryder, "Draft Status Report on the Ligament Cracking Problem," unpublished report, Electric Power Research Institute, Palo Alto, CA, 1988

16. L. Becker, S.M. Walker, and R. Viswanathan, "Guidelines for the Evaluation of Seam Welded Steam Pipes," Report CS 4774, Electric Power Research Institute, Palo Alto, CA, Mar 1987

17. F.E. Werner, "Austenite Transformations and Tempering of Some Low Alloy Steels Used in Steam Turbines," Research Report 11-0103-1-R1, Westinghouse Research Laboratories, Pittsburgh, June 1960

18. R.G. Baker and J. Nutting, The Tempering of 2¼Cr-1Mo Steel After Quenching and Normalizing, *J. Iron Steel Inst.*, July 1959, p 257-269

19. Numerous papers in *Life Extension and Assessment of Fossil Power Plants*, R.B. Dooley and R. Viswanathan, Ed., proceedings of EPRI-ASM-ASME conference, Washington, Report CS 5208, Electric Power Research Institute, Palo Alto, CA, Jan 1987

20. R.B. Dooley, personal communication, Electric Power Research Institute, Palo Alto, CA, 1987

21. B.J. Cane, in *Proceedings of International Conference on Welding Technology for Energy Applications*, ORNL Conference 820544, Gatlinburg, TN, May 1982, p 623-639

22. C.H. Wells, On the Life Prediction of Longitudinal Seam Welds, in *Life Extension and Assessment of Fossil Power Plants*, R.B. Dooley and R. Viswanathan, Ed., Report CS 5208, Electric Power Research Institute, Palo Alto, CA, Jan 1987

22a. J.F. Henry, F.V. Ellis, and C.D. Lundin, "The Effect of Inclusions, as Controlled by Flux Composition, on the Elevated Temperature Properties of Submerged-Arc Weldments," Weld Tech 88, International Conference on Weld Failures, British Welding Institute, London, Nov 1988

23. K.M. Dunn, J.R. Scheibel, and F. Schwartz, "Monitoring for Life Extension," TIS-PSG-85-001, Combustion Engineering, Windsor, CT, 1985

24. M.J. Davidson, T.J. Jones, Jr., D.D. Rosard, and J.R. Scheibel, Monitoring for Life Extension, *Trans. ASME, J. Pressure Vessel Tech.*, Aug 1985

25. F.V. Ellis, R.W. Loomis, and S. Tordonato, Life Extension: The CE Approach to the Analysis of Thick Walled Components, in *Life Extension and Assessment of Fossil Power Plants*, R.B. Dooley and R. Viswanathan, Ed., Report CS 5208, Electric Power Research Institute, Palo Alto, CA, 1987, p 335-350

26. J.D. Fox *et al*, "Prediction of Creep Fatigue Damage in Major Steam Generator Components," 38th American Power Conference, Chicago, Apr 1976

27. C.E. Jaske, "Fatigue Curve Needs for Higher Strength 2¼Cr-1Mo Steel for Petroleum Process Vessels," MPC Symposium on Fatigue Initiation, Propagation and Analysis for Code Construction, ASME Winter Annual Meeting, Chicago, Nov 1988

28. F. Masuyama, K. Setoguchi, H. Haneda, and F. Nanjo, Findings on Creep Fatigue Damage in Pressure Parts of Long Term Service Exposed Thermal Power Plants, in *Residual Life Assessment, Non Destructive Examination and Nuclear Heat Exchanger Materials*, C.E. Jaske, Ed., Proceedings of the 1985 Pressure Vessels and Piping Conference, PVP Vol 98-1, ASME, 1985, p 79-91

29. G.V. Smith, Ed., *Evaluation of the Elevated Temperature Tensile and Creep Rupture Properties of ½Cr-½Mo, 1Cr-½Mo, and 1¼Cr-½Mo-Si DS*, Metal Properties Council, ASTM Data Series DS 50, American Society for Testing and Materials, 1971

30. G.V. Smith, Ed., *Supplemental Report on the Elevated Temperature Properties of Chromium Molybdenum Steels*, Metal Properties Council, ASTM Data Series, DS 6S1, American Society for Testing and Materials, 1971

31. D.I. Roberts and R. Viswanathan, Dissimilar Weldments in Fossil Fired Power Plants, in *Proceedings of Seminar on Dissimilar Welds in Fossil Fired Boilers*, Report CS 3623, Electric Power Research Institute, Palo Alto, CA, July 1985

32. R.D. Nicholson and A.T. Price, "Service Experience of Nickel Based Transition Joints," Welding Institute Seminar, Leicester, England, June 1981

33. Y.N. Rabotnov, *Creep Problems in Structural Members*, North Holland Publishers, Amsterdam, 1968

34. L.M. Kachanov, *Isv. Acad. Nauk. USSR, Old Tekd. Nauk*, No. 8, 1958, p 26-31

35. B.J. Cane, "Remaining Creep Life Estimation by Strain Assessment of Plant," Report RL/L/2040 R81, Central Electricity Generating Board Research Laboratories, U.K., 1981

36. B.J. Cane and J.A. Williams, Remaining Life Prediction of High Temperature Materials, *Int. Mater. Rev.*, Vol 32 (No. 5), 1987, p 241

37. M.D.C. Moles and H.J. Westwood, "Residual Life Estimation of High Temperature Superheater and Reheater Tubing," CEA RP 78-66, Final Report of Ontario Hydro Research Div., Toronto, for the Canadian Electrical Assn., Montreal, Mar 1982, p 67-82

38. D. Lonsdale and P.E.J. Flewitt, Damage Accumulation and Microstructural Changes Occurring During the Creep of a 2¼Cr-½Mo Steel, *Mater. Sci. Engg.*, Vol 39, 1979, p 217-229

39. B. Neubauer and U. Wedel, Restlife Estimation of Creeping Components by Means of Replicas, in *Advances in Life Prediction Methods*, D.A. Woodford and J.R. Whitehead, Ed., American Society of Mechanical Engineers, New York, 1983, p 307-314

40. B.J. Cane and M.S. Shammas, "A Method for Remanent Life Estimation by Quantitative Assessment of Creep Cavitation on Plant," Report TPRD/L/2645/N84, Central Electricity Generating Board of U.K., Leatherhead Laboratories, June 1984

41. M. Shammas *et al*, "Remaining Life of Boiler Pressure Parts, HAZ Models," Final Report RP 2253-1, Vol 2, Electric Power Research Institute, Palo Alto, CA, 1988

42. R. Viswanathan, R.B. Dooley, and A. Sax-

ena, A Methodology for Evaluating the Integrity of Longitudinally Seam Welded Steam Pipes in Fossil Plants, *ASME J. Pressure Vessel Tech.*, Vol 110, 1988, p 283-290

43. M.C. Askins *et al*, "Remaining Life of Boiler Pressure Parts–Base Material Model," Report RP 2253-1, Vol 3, Electric Power Research Institute, Palo Alto, CA, 1988

44. J.M. Brear, S. Bruce, J.M. Silcox, and B.J. Cane, "Models for Determination of Creep Curves in Low Alloy Ferritic Steels Based on Carbide Coarsening," EPRI RP 2253-1, Report P2, Electric Power Research Institute, Palo Alto, CA, March 1984

45. L.H. Toft and R.A. Mardsen, The Structure and Properties of 1% Cr–0.5% Mo Steel After Service in CEGB Power Stations, in *Conference on Structural Processes in Creep*, JISI/JIM, London, 1963, p 275

46. M. Clark and A. Cervoni, "In-Situ Metallographic Examination of Ferrous and Non Ferrous Components," Ontario Hydro Research Div., Toronto, Report to the Canadian Electrical Assn., Montreal, 314 G429, Nov 1985

47. J.A. Janiszewski, W.J. Jeitner, and J.A. Alice, A Production Approach to On-Site Metallographic Analysis, in *Life Extension and Assessment of Fossil Power Plants*, R.B. Dooley and R. Viswanathan, Ed., Report CS 5208, Electric Power Research Institute, Palo Alto, CA, 1987, p 547-560

48. T.J. Der *et al*, "Methods of Sampling Service Components for Metallurgical Evaluation of Remanent Creep Life," SE/SSD/RN/81/067, Central Electricity Generating Board of U.K., Oct 1981

49. J.F. Henry, F.V. Ellis, and R. Viswanathan, "Field Metallography Techniques for Plant Life Extension," Nineteenth Annual Technical Meeting of the International Metallographic Society, Boston, 1986

50. R.T. Delong, "Standard Practice for Production and Evaluation of Field Metallographic Replicas," E12-87, American Society for Testing and Materials, Philadelphia, 1987

51. J.F. Delong, personal communication, Philadelphia Electric Co., Philadelphia, 1988

52. R. Viswanathan, J.R. Foulds, and D.A. Roberts, Methods for Estimating the Temperature of Reheater and Superheater Tubes in Fossil Boilers, in *Proceedings of the International Conference on Life Extension and Assessment*, The Hague, June 1988

53. I.M. Rehn and W. Apblett, "Corrosion Problems in Coal Fired Fossil Boiler Superheater and Reheater Tubes," Report CS 1811, Electric Power Research Institute, Palo Alto, CA, 1981

54. M. Dewitte and J. Stubbe, personal communication to D.A. Roberts of G.A. Technologies, San Diego, Laborolec Co., Belgium, 1986

55. S.R. Paterson and T.W. Rettig, "Remaining Life Estimation of Boiler Pressure Parts–2¼Cr-1Mo Superheater and Reheater Tubes," Project RP 2253-5, Final Report, Electric Power Research Institute, Palo Alto, CA, 1987

56. D.I. Roberts, "Magnetic Oxide Thickness Time-Temperature Models for 2¼Cr-1Mo Operating in High Pressure Steam," letter communication to the author, from G.A. Technologies, San Diego, 1986

57. D.N. French, *Metallurgical Failures in Fossil Fired Boilers*, 1st Ed., John Wiley & Sons, New York, 1983, p 249

58. J.A. Alice, J.A. Janiszewski, and D.N. French, "Liquid Ash Corrosion, Remaining Life Estimation and Superheater/Reheater Tube Replacement Strategy in Coal Fired Boilers," ASME Paper 85JPGC-PWR-3, ASME/IEEE Joint Power Generation Conference, Milwaukee, Oct 20-24, 1985

59. D.I. Roberts *et al*, "Dissimilar-Weld Failure Analysis and Development Program," Vol 2, "Metallurgical Characteristics," Report CS 4252, Electric Power Research Institute, Palo Alto, CA, Nov 1985

60. B.J. Cane, P.F. Aplin, and J.M. Brear, *J. Pressure Vessel Tech.*, Vol 107 (No. 3), Aug 1985, p 295

61. M.S. Shammas *et al*, "Estimating the Remaining Life of Boiler Pressure Parts," Project RP 2253-1, Final Report, Vol 4, Electric Power Research Institute, Palo Alto, CA, 1987

62. R. Coade, "Temperature Determination Based on Microstructural Changes Occurring in 1% Cr–0.5% Mo Steel," Report No. SO/85/87, State Electricity Commission of Victoria, Australia, Feb 1985

63. J.R. Foulds *et al*, "Dissimilar Weld Failure Analysis and Development," Vol 6, "Weld Condition and Remaining Life Assessment," Report CS 4252, Electric Power Research Institute, Palo Alto, CA, Aug 1988

64. R.A. Stevens and P.E.J. Flewitt, "The Effect of Phosphorus on the Microstructure and Creep Strength of 2¼Cr-1Mo Steels," SER/SSD/85/0020/R, Central Electricity Generating Board of U.K., Scientific Services Dept., Graves End, Kent, Mar 1985

65. T.P. Sherlock, personal communication, Babcock and Wilcox, Alliance, OH, 1986

66. K.N. Melton, Isostress Extrapolation of Creep Rupture Data, *Mater. Sci. Engg.*, Vol 59, 1983, p 143-149

67. J.M. Brear and B.J. Cane, unpublished results from a multiclient sponsored project at ERA Technology, Leatherhead, U.K., EPRI project reference No. RP 2253-4

68. J.M. Brear and B.J. Cane, same source as Ref 67, EPRI reference No. RP 2253-2

69. H. Grunloh, R.H. Ryder, and R. Viswanathan, unpublished work in progress, EPRI Project RP 2253-10

70. B. Walser and A. Rosselet, "Determining the Remaining Life of Superheated Steam Tubes by Creep Tests and Structural Examinations," Sulzer Technical Review, Sulzer Bros., Winterhur, Switzerland, Research No. 1978, 1978

71. C.E. Jaske, S.H. Smith, J.F. Kiefner, and A.R. Rosenfield, Life Assessment of High Temperature Fossil Plant Components, in *Progress in Flaw Growth and Fracture Toughness Testing*, STP 536, American Society for Testing and Materials, Philadelphia, 1973, p 439-453

72. J.F. Kiefner, C.R. Barnes, P.M. Scott, and G. Andres, Failure Stress Levels of Flaws in Pressurized Cylinders, in *Progress in Flaw Growth and Fracture Toughness Testing*, STP 536, American Society for Testing and Materials, Philadelphia, 1973, p 461-481

73. P.K. Liaw and A. Saxena, "Remaining Life Estimation of Boiler Pressure Parts–Crack Growth Studies," Report CS 4688, Electric Power Research Institute, Palo Alto, CA, July 1986

74. A. Saxena, T.P. Sherlock, and R. Viswanathan, Evaluation of Remaining Life of High Temperature Headers: A Case History, in *Life Extension and Assessment of Fossil Power Plants*, R.B. Dooley and R. Viswanathan, Ed., Report CS 5208, Electric Power Research Institute, Palo Alto, CA, 1987, p 575-605

75. R. Gaitonde, "Commonwealth Edison Program for Hot Reheat Pipe Integrity Evaluation," presented to the Edison Electric Institute, Metallurgy and Piping Task Force Meeting, Chicago, June 1985

76. W.P. McNaughton, R.H. Richman, C.S. Pillar, and L.W. Perry, "Generic Guidelines for the Life Extension of Fossil Fuel Power Plants," Report CS 4778, Electric Power Research Institute, Palo Alto, CA, Nov 1986

77. "Procedure for Boiler Creep Life Assessment," Generation Operation Memorandum 101, Issue 2, Central Electricity Generating Board of U.K., Aug 1982

78. D.J. Gooch and R.D. Townsend, CEGB Remanent Life Assessment Procedures, in *Life Extension and Assessment of Fossil Power Plants*, R.B. Dooley and R. Viswanathan, Ed., Report CS 5208, Electric Power Research Institute, Palo Alto, CA, 1987

79. M.D.C. Moles, D.V. Leemans, and H.J. Westwood, Residual Life Estimation of Boiler Tubing in Thermal Power Plants, in *Proceedings of the International Conference on Engineering Aspects of Creep*, Vol II, Paper C 236/80, Institute of Mechanical Engineers, London, 1980, p 241-247

80. D. Sidey, R.B. Dooley, and J.D. Parker, "Remaining Life Estimation of High Temperature Boiler and Turbine Components," Report CS 4207, Electric Power Research Institute, Palo Alto, CA, 1985, p 7.1-7.39

81. S.R. Patterson, T.W. Rettig, and K.J. Clarke, Creep Damage and Remaining Life Assessment of Superheater and Reheater Tubes, in *Life Extension and Assessment of Fossil Power Plants*, Report CS 5208, Electric Power Research Institute, Palo Alto, CA, 1987

82. J.A. Alice, J.A. Janiszewski, and D.N. French, "Liquid Ash Corrosion, Remaining Life Estimation and Superheater/Reheater Tube Replacement Strategy in Coal Fired Boilers," ASME Paper 85JPGC-PWR-3, ASME-IEEE Joint Power Generation Conference, Milwaukee, Oct 20-24, 1985

83. V. Kumar, M.D. German, and C.F. Shih, "An Engineering Approach to Elastic Plastic Fracture Analysis," Report NP 1931, Electric Power Research Institute, Palo Alto, CA, July 1981

84. M.Y. He and J.W. Hutchinson, in *Elastic Plastic Fracture: Second Symposium*, Vol 1, *Inelastic Crack Analysis*, STP 803, American Society for Testing and Materials, Philadelphia, 1983, p 1277-1305

6

Life Assessment of Steam-Turbine Components

The steam turbine is the device that converts the heat energy of the steam coming from the boiler into the mechanical energy of shaft rotation. It basically consists of a rotor from which project several rows of closely spaced blades (buckets). Between each two rows of moving blades there is a row of fixed vanes (nozzles) that projects inward from a circumferential housing. The vanes are carefully shaped to direct the flow of steam against the moving blades at an angle and at a velocity that will maximize the energy conversion. Because the steam's temperature and pressure decrease and the volume increases continuously by expansion through the turbine, the length of the blades increases progressively from the inlet to the outlet end of the turbine. A typical plant may have a high-pressure (HP) turbine, an intermediate-pressure (IP) turbine, and one or more low-pressure (LP) turbines connected in tandem, as illustrated in Fig. 6.1 (Ref 1). The LP section thus has the largest cross-sectional diameter, whereas the HP section has the smallest cross-sectional diameter. The various turbine sections are enclosed in casings (or shells) which are usually made in two half-sections bolted together. The key components of a steam turbine therefore are the rotor, blades, vanes, casing, and bolting.

In achieving increased unit capacity, efficiency, reliability, and cycling capability, developments in materials technology have been the focal point of industry's efforts. Larger unit sizes require the ability to produce the various components in large sizes without compromising quality. Increased efficiency, which usually results from higher steam pressures and temperatures, requires materials with greater creep resistance. Reliability, of course, involves a whole spectrum of material properties required to resist failure. The economically driven need for unit cycling in recent times has also required materials and designs with improved thermal fatigue resistance. This chapter will review the material-property requirements, damage mechanisms, and damage-assessment procedures for the key components of the steam turbine. Damage mechanisms and the general principles of remaining-life

Fig. 6.1. Modern 600-MW steam turbine with reheat (courtesy of C. Verpoort, ASEA–Brown Boveri Corp., Baden, Switzerland) (Ref 1).

assessment for steam turbines are discussed in several articles in the literature (Ref 2 to 5).

Materials and Damage Mechanisms

Most of the components of a steam turbine are made of steels containing various amounts of the principal alloying elements nickel, chromium, molybdenum, and vanadium. With the exception of some of the high-temperature rotors, bolting, blading, and valve stems, which are made of 12% Cr steels, all components are made of low-alloy steels.

Table 6.1 illustrates the potential damage mechanisms, failure criteria, and typical remedial actions pertaining to steam-turbine components (Ref 6). With the exception of LP rotors, fatigue, creep, and brittle fracture are the main concerns in turbine components. LP rotors operate below the creep regime and hence the possibility of brittle fracture at lower temperatures is of greater concern than creep. In the LP section of the turbine, particularly near the last stages, problems relating to low-temperature corrosion phenomena, such as corrosion fatigue and stress corrosion, are of great concern; these problems, however, are beyond the scope of this book and will not be treated here. It can also be seen from Table 6.1 that a number of failure criteria, such as crack initiation, crack propagation, economics of repairability, and feasibility of repair have been employed depending on the component, in line with the discussion in Chapter 1. The course of action generally involves a calculation of the expended life under the service conditions, followed by detailed inspections and corrective actions as necessary.

Rotors

Steam-turbine rotors are among the most critical and highly stressed components of the steam power plant. Failures of rotors have resulted in a wide spectrum of damage, ranging from lengthy and costly forced

outages to catastrophic bursts. Histories of known rotor failures and discussions of failure causes have been presented elsewhere (Ref 7 and 8). Because LP rotors operate at relatively low temperatures, only peripheral references to LP rotor behavior will be made here. The major focus will be on HP and IP rotors, which operate at temperatures close to 540 °C (1000 °F) at the steam inlet end.

Rotor Designs

Three rotor configurations typically employed are shown in Fig. 6.2 (Ref 9). These configurations are monoblock rotors in which the disks are integral parts of the rotor shaft (Fig. 6.2a), rotors in which disks are shrunk onto the rotor shaft (Fig. 6.2b), and rotors in which sections of the rotor are welded together (Fig. 6.2c). Until recently, monoblock rotors made by many manufacturers had central bores. A discussion of the design philosophy behind bored rotors has been presented by Timo (Ref 10). The reasons cited for having a bore

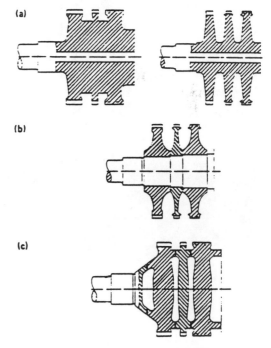

(a) Integral forging construction. (b) Shrunk-on-disk construction. (c) Welded-disk construction.

Fig. 6.2. Different types of rotor construction (Ref 9).

Table 6.1. Damage locations, causes, and remedies for steam turbine components (Ref 6)

Component	Position	Cause of life exhaustion	Time of remedial action	Remedial action	Action in regular inspection	Definition of life (limit of usage)
HP-IP rotor	Outer groove	Fatigue	① Calculation	Skin peeling	Nondestructive test	When skin peeling is no longer practicable
	Center bore	Fatigue, creep-crack propagation, and brittle fracture	③ Center bore inspection and calculation	Overboring	Nondestructive test	When overboring is no longer practicable
	Blade-groove shoulder	Creep	① Calculation	Detailed inspection and investigation	Nondestructive test	When crack initiation is confirmed
LP rotor	Center bore	Fatigue, crack propagation, and brittle fracture	③ Center bore inspection and calculation	Overboring	Nondestructive test	When overboring is no longer practicable
HP inner casing	Inner surface	Fatigue and creep	② Nondestructive test	Weld repairing	Nondestructive test	When repair is considered no longer realistic
	Female thread	Creep	① Calculation	Oversizing	Nondestructive test	When oversizing is no longer practicable
Nozzle block	Root of vane	Fatigue and creep	① Calculation	Detailed inspection and investigation	Nondestructive test, measurement of nozzle chamber deformation	When cracks are no longer repairable and/or deformation of nozzle chamber is significant

Component	Location	Failure mode				
HP-IP blades	Tenon and root of blade	Creep	① Calculation	Detailed inspection and investigation	Nondestructive test, hardness test	When cracks have initiated and/or any abnormality appears in detailed inspection
Main valves	Body	Fatigue and creep	② Nondestructive test	Weld repairing	Nondestructive test, hardness test	When weld repairing is no longer practicable and/or material has significant deterioration
	Female thread	Creep	① Calculation	Oversizing	Nondestructive test	When oversizing is no longer practicable
High-temperature bolts	Thread	Fatigue and creep	① Calculation	Replacement (destructive test for sampled bolts)	Nondestructive test	When crack initiation is anticipated

Timeline:

① Start of service
② Prediction of crack initiation (based on lower limits of life-prediction data)
③ Actual initiation of crack
④ Allowance for fracture
Fracture

were: (1) the bore hole ensures the removal of defective material from the center of the rotor and permits identification of the causes of defects; (2) the bore hole facilitates magnetic-particle and ultrasonic inspection, thus supplementing NDE data obtainable by ultrasonic inspection from the periphery; (3) samples of the bore material permit characterization of the current mechanical properties of the rotor as well as potential degradation occurring with long-term service; and (4) during an overhaul, the bore hole permits nondestructive inspection of the shaft—a procedure difficult to perform from the outside surface of a bladed shaft. Arguments against having a bore hole have been presented by Mayer (Ref 11), as follows: (1) the bore hole results in a substantial increase in the total tangential stress (steady centrifugal plus transient thermal) and a corresponding decrease in the tolerable crack size; (2) solid (unbored) shafts present an almost ideal condition at the center for ultrasonic inspection from the exterior; (3) improved steel-making practices in recent years have led to clean and homogeneous material of high fracture toughness at the centers of rotors, and it is no longer necessary to remove the core material; (4) specimens for mechanical tests can be obtained from other locations in the rotor; and (5) because of a combination of reduced stresses and cleaner material at the bore, the reliability and cycling ability of solid shafts are better than those of bored rotors. The controversy over rotor bores has abated substantially during the last several years, and now there is a general trend toward the use of solid (unbored) rotors for all current designs.

A recent catastrophic failure of a solid (unbored) LP rotor made of 2Ni-Cr-Mo-V steel in Germany has shown that solid rotors are not immune to failure. This failure occurred during a cold start and was attributed to a large flaw in a segregated area near the center of the rotor, which grew to critical size by fatigue in a radial-axial direction. Fracture occurred in a brittle manner as a result of a combination of high stress and low fracture toughness (Ref 12). This fail-

ure underscores the need to ensure high levels of cleanness and toughness so as to avoid failures even in unbored rotors.

Service Failure Experience

Typical configurations and failure locations for HP and IP rotors are illustrated in Fig. 6.3. The failure of a rotor can be defined as the inability of the rotor to perform its intended function reliably, safely, and economically. On the basis of this definition, it would be impossible to gather information regarding all rotor failures. Most retirements of rotors, some of which are done prematurely for various reasons, go unreported. A list of a few well-publicized failures is presented in Table 6.2 (Ref 7). Details regarding many of these failures have been described by Bush (Ref 8). Potential locations, causes, and current remedies for rotor failures are summarized in Table 6.3 (Ref 7).

Bore Cracking. Rotors of the 1950 vintage have poor quality of the near-bore material and are characterized by clusters of inclusions, segregation streaks, and high local concentrations of impurity elements. The inclusions serve as sites for nucleation of cracks. Crack propagation by low-cycle fatigue, creep, or a combination of both is also facilitated by inclusion clusters and severe embrittlement of the grain boundaries. Another reported cause of cracking in Cr-Mo-V steel rotors is poor creep-rupture ductility (notch sensitivity) resulting from a 1010 °C (1850 °F) austenitizing treatment used in the early 1950's (class C rotors). Most rotors made after the 1950's were austenitized at 955 °C (1750 °F), because this lower-temperature austenitizing treatment was found to reduce rupture strength and improve rupture ductility. With respect to the older rotors, however, utilities continue to face the dilemma of whether to run or retire these rotors. Utilities have retired numerous rotors of the old class C type. Other options have been exercised, including derating of the machine and removal of damaged bore material by local grinding, "bottle" boring, and overboring. No evi-

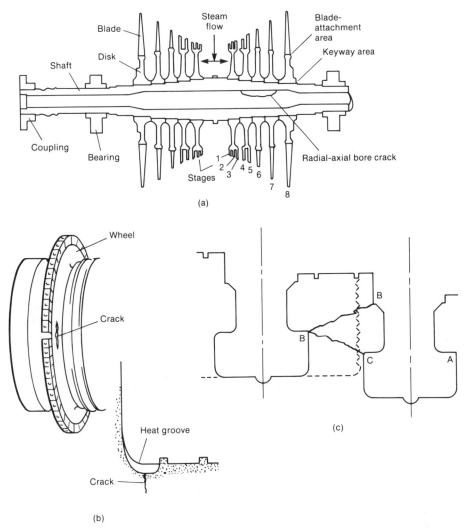

(a) Radial-axial bore crack. (b) Transverse crack in heat groove. (c) Creep cracking in blade-attachment area.

Fig. 6.3. HP-IP rotor configurations and cracking locations.

dence of creep-cavitation damage has been found, however, in the bore of any class of rotor.

Rim Cracking. The first reported rim failure occurred in the first reheat disk in an IP turbine. This failure occurred by creep-rupture due to the low ductility of the steel. This form of cracking was also identified in IP rotors at two stations. The cracks were located at the outer corners of the T grooves for the blades and usually started at the blade-entrance slots. Low creep strength in one case and poor creep ductility in the other case were identified as the cause of failure. The remedial action consisted of changing over to Cr-Mo-V steels, lowering the austenitizing temperature for Cr-Mo-V steel from 1010 to 955 °C (1850 to 1750 °F), and making design changes to lower the stresses at the T grooves. Creep-cavitation damage has recently been observed at the blade-attachment areas of some retired rotors even though they did not belong to "class C" (Ref 13). Current remedies for cracked rotors consist of grinding away the cracks and retrofitting with lighter-weight blades. Steam cooling of critical regions is also employed to reduce the local metal temperature.

Rotor Surface Cracking. Surface crack-

Table 6.2. Failures in steam-turbine rotors (based on Ref 7)

Rotor component	Name of unit	Year	Description of failure	Failure mechanism
LP shaft	Ridgeland #4	1954	Burst in service(a)	Brittle failure
	ENESA (Spain)	1950	Burst in test pit(a)	Brittle failure
	Nijmijen (Netherlands)	1950	Fracture in service	Brittle failure
	Name not available (Siemens)	1951	Burst in test pit	Brittle failure
	Conner Creek #16	1977	Crack 5 by 16 in.(b)	
	Didcot (England)	1972	Water induction	
	Unnamed (England)	1972	Water induction	
	Aberthaw (England)	1973	Water induction	
	Wurgassen (West Germany)	1974	Two shafts, transverse cracking(c)	Stress corrosion/corrosion fatigue; evidence of corrosion not always obvious
	Ferrybridge C (England)	1977	Three shafts, transverse cracking(c)	Same as above
	R.S. Wallace	1974	Transverse cracking(b)	Same as above
	Fort Martin #1	1976	Two shafts, transverse cracking(c)	Same as above
	Ravenswood #3	1978	Transverse cracking(c)	Same as above
	Astoria #5	1978	Transverse cracking(c)	Same as above
	Oak Creek #7	1980	Transverse cracking(c)	Same as above
	Waukegan #8	1981	Transverse cracking(c)	Same as above
	St. Clair #6	1981	Transverse cracking(b)	Same as above
	Pennelec	1981	Transverse cracking	Same as above
	State #4	1983	Transverse cracking(d)	Same as above
	Campbell	1984	Transverse cracking	Same as above
	Vohberg	1987	Catastrophic burst	Fatigue/brittle fracture
IP shaft	Shawnee #1	1954	Steeple fracture(c)	Creep-rupture
	Weadock #1	1955	Steeple fracture(c)	Creep-rupture
	Tanners Creek #1	1953	Wheel fracture	Creep-rupture
	Wagner #1	1974	Piece out of IP section	Creep-rupture
	Gallatin #2	1974	Burst in service(a)	Creep-fatigue/brittle fracture
	Muskingum River #2	1968	Transverse cracking(c)	
	Cumberland #2	1976	Coupling cracking(c)	Corrosion fatigue
	Mitchell #2	1980	Coupling cracking(c)	
	St. Clair #3	1983		
HP shaft	Philo #5	1962	Shaft transverse fracture(a)	Brittle fracture
	El Segundo #3 and #4	1978	Shaft wheel fracture(a)	
	Alamitos #3 and #4	1978	Shaft transverse cracking(a)	

(a) Crack plane in axial-radial direction. (b) Cracking discovered by nondestructive testing. (c) Rotor was shut down due to high vibration prior to complete fracture. (d) Cracking discovered after change out.

ing is a problem generally encountered in HP rotors and is attributed to thermal fatigue resulting from cycling. The cracks generally occur in the heat grooves and at relatively small radii at labyrinth seal areas along the rotor. These cracks are generally shallow and in most instances can be removed by machining (skin peeling). The skin-peeling process is repeated as cracking reappears in service until it is no longer practical or economical. To prevent recurrence of cracking, the radii are enlarged to reduce the stress concentration.

The Physical Metallurgy of Cr-Mo-V Rotor Steel

The evolution of steel compositions for high-temperature rotor applications has

Table 6.3. Types of cracking found in rotors (Ref 7)

Rotor component	Type of cracking	Cause of cracking	Remedial actions	
			Current rotors	New rotors
LP rotor/shaft	Radial-axial bore cracks	Poor toughness and transient thermal stresses	Retire; grind, over-bore, or bottle bore cracked areas	Improve toughness by control of cleanness, H_2 and temper embrittlement
	Transverse cracks	High-cycle fatigue with or without corrosion assistance	Retire; weld repair	Need to develop materials with improved pitting resistance; improve design to minimize stress concentration; reduce da/dn; coatings
HP/IP rotor	Radial-axial bore cracks	Creep with or without low-cycle fatigue; poor creep ductility due to faulty heat treatment (class c) coupled with poor center quality facilitating initiation and poor toughness	Retire; grind, over-bore, or bottle bore; derate machine; steam cool; control start-stops	Improve center quality; heat treatment has been modified
	Blade-groove-wall cracking	Poor creep ductility	Retire; machine cracks and use lighter blades; steam cool	Heat treatment has been modified
	Rotor surface cracking	Thermal fatigue	Machine cracks; enlarge radius	Improved materials with resistance to thermal fatigue

been reviewed by Timo, Curran, and Placek (Ref 14). The early use of carbon steels was superseded by Ni-Mo-V steels in the mid-1940's. With increasing demands on creep strength, a 1Cr-1Mo-0.25V steel was introduced in the early 1950's and has remained the industry standard ever since, although a few higher-alloy steel rotors (12% Cr) have been placed in service.

Achievement of the desired properties in Cr-Mo-V steel rotors is made possible by careful control of heat treatment and composition. Examination of the continuous cooling transformation diagram for Cr-Mo-V steel (Fig. 6.4) shows that for the normal range of air cooling rates employed for rotors, the predominant transformation product would be upper bainite (Ref 15 and 16). Oil quenching of rotors may shift the transformation product increasingly toward lower bainite, but it is unlikely that the cooling rates needed for formation of martensite (i.e., 20,000 °F/h) are ever encountered. In the United States, the usual practice has been to air cool the rotors from the austenitizing temperature in order to achieve a highly creep-resistant, but somewhat less tough, upper bainitic microstructure. On the other hand, European manufacturers have resorted to oil quenching of rotors from the austenitizing temperature, to achieve a better compromise between creep strength and toughness.

The final heat treatment of the forging usually consists of austenitizing at about 955 °C (1750 °F) followed by tempering in the range 675 to 705 °C (1250 to 1300 °F). The austenitizing and tempering treatments are chosen so as to achieve the desired strength without sacrificing ductility. It is well known that austenitizing at temperatures above 955 °C (1750 °F) can lead to reduced rupture ductility and notch sensitivity. Similarly, tempering at too low a temperature can lead to reduced ductility. A typical tempering curve for Cr-Mo-V

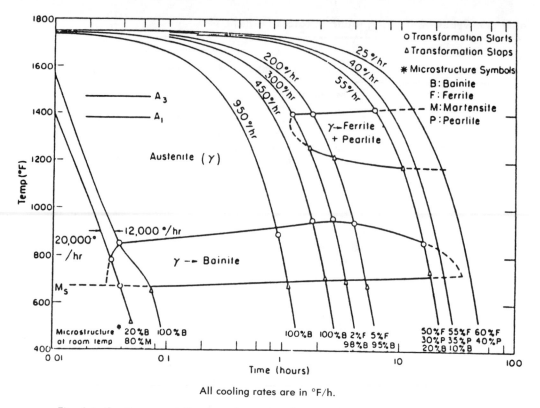

Fig. 6.4. Continuous cooling transformation diagram for Cr-Mo-V rotor steel (Ref 15).

rotor steel is shown in Fig. 6.5 (Ref 17). The microstructure resulting from the standard heat treatment is shown in Fig. 6.6. In the as-heat-treated condition, examination of this microstructure at higher magnifications would essentially reveal rounded Fe_3C particles as the major carbide phase, with small amounts of needlelike Mo_2C and VC carbides. Prolonged service exposure or tempering results in the appearance of M_7C_3 carbides and eventually the M_6C- and $M_{23}C_6$-type massive carbides.

Comparative evaluation of creep properties of Cr-Mo-V steels with martensite, bainite, and ferrite-pearlite as the principal microstructure have been conducted by numerous investigators, and the results have been reviewed elsewhere (Ref 18). There is consensus that upper bainitic structures provide the best creep resistance coupled with adequate ductility.

The element that contributes most significantly to the strength of Cr-Mo-V steels is vanadium. The effect of vanadium arises primarily from the formation of a stable and fine dispersion of V_4C_3-type carbides, and the V:C ratio in the alloy is therefore critical. Too low a V:C ratio results in the formation of excess M_3C-type carbides; too high a V:C ratio leads to excessive V_4C_3 precipitation, accompanied by high rupture strength and very low rupture ductility. Molybdenum suppresses ferrite formation and promotes bainite formation. It also contributes to strengthening via solid-solution effects and by precipitation as Mo_2C. Chromium contributes to solution strengthening as well as to the necessary oxidation resistance in steam. In-depth discussions of the roles of alloying elements in Cr-Mo-V steels can be found in other articles cited in Ref 18.

Among the deleterious elements, the most noteworthy are manganese, silicon, antimony, phosphorus, tin, sulfur, aluminum, and copper. Manganese, silicon, antimony,

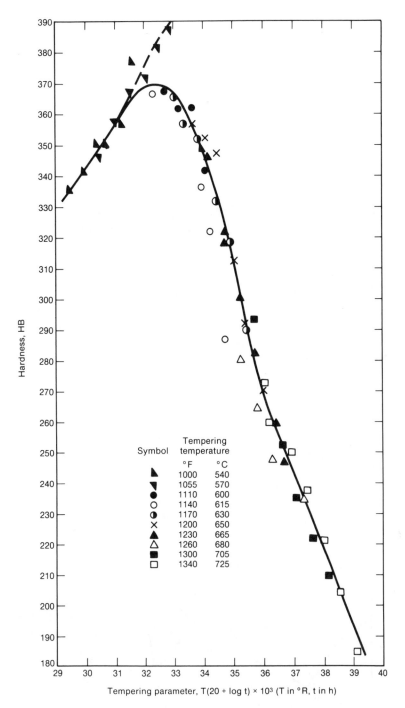

Fig. 6.5. Tempering behavior of Cr-Mo-V rotor steel (Ref 17).

phosphorus, and tin increase susceptibility to temper embrittlement. The effect of sulfur results from formation of MnS inclusions, which promotes crack nucleation. Many of these elements also lead to reduced rupture ductility. The effects of trace elements on rupture ductility have been reviewed in the literature (Ref 19) and are also discussed with respect to specific components in subsequent sections.

Fig. 6.6. Typical microstructure of Cr-Mo-V rotor steel. (500×; shown here at 67%)

Material Properties

The critical material properties for rotor integrity are toughness, resistance to crack initiation under creep and thermal-fatigue conditions, and resistance to subcritical crack propagation in creep and fatigue.

Toughness. A qualitative guarantee of adequate toughness in Cr-Mo-V steel rotors is provided by ASTM specification A470, class 8, which limits the FATT to 120 °C (250 °F) max. FATT values for the bore material normally range from 85 to 125 °C (185 to 260 °F). With the advancement of

fracture-mechanics technology it has now become possible to characterize toughness in terms of a critical crack size a_c. A typical loading sequence, illustrating the variations in temperature and stress (see Fig. 6.7; Ref 20) shows that the smallest value of a_c—i.e., the highest risk of brittle failure—occurs during transient conditions. Figure 6.7 is based on analysis of a cold-start sequence in the Gallatin rotor, which failed catastrophically. Region A consists of a warm-up period after which the roll-off commenced (region B). During roll-off, the rotor was gradually brought up to speed. Once the synchronous speed was reached, loading began in region C, approximately 3 h after the beginning of the warm-up period. Analysis of the transient conditions at the failure location (seventh row) showed that the stresses reached a peak value of about 520 MPa (74 ksi) 1½ h after the synchronous speed was attained. The temperature at the time of the peak stress was 130 °C (270 °F). It was estimated that the critical crack size reached its lowest value of 0.7 cm (0.27 in.) under these conditions.

Variations in temperature, stress, and material inhomogeneity along and across the rotor dictate that the a_c value for the rotor be computed for the worst combination of these variables. This is done by using the lower scatterband values of K_{Ic} shown

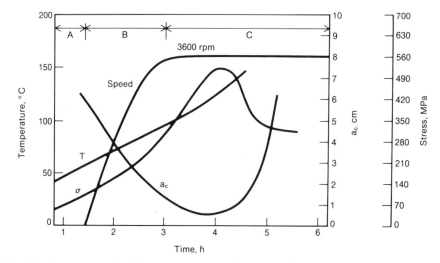

Fig. 6.7. Illustration of cold-start sequence and associated variations in stress (σ), temperature (T), and critical flaw size (a_c) as functions of time from start (Ref 20).

in Fig. 6.8 (Ref 21). An alternative method of estimating the lower-bound values of K_{Ic} for Cr-Mo-V steels (Ref 21a) is by use of the expression

$$K_{Ic} = \frac{6600}{60 - (T - FATT)}$$

where K_{Ic} is expressed in MPa$\sqrt{m}$ and T is expressed in °C. If the location where the worst combination of variables occurs is known, it is also possible to perform rotor-specific evaluations of toughness to reduce the conservatism.

Equipment manufacturers generally have records of the FATT value of the rotor material prior to service. Unfortunately, temper embrittlement during service increases the FATT value and decreases the K_{Ic} value, as shown in Fig. 6.9 (Ref 20). The limited data available from retired rotor evaluations have shown that the maximum temper embrittlement occurs at locations exposed to temperatures from 370 to 425 °C (700 to 800 °F). Estimation of FATT (or K_{Ic}) in the service-exposed condition at the location of concern is therefore critical for damage assessment. Currently available methods for doing this are discussed in a separate section on remaining-life-assessment methods.

Creep-Rupture and Stress-Rupture. Creep-rupture failures and evidence of creep damage at the blade-attachment areas of rotors have been observed in many instances (Ref 8 and 13). No clear evidence of creep damage in the bore has been documented, although concern remains regarding the possibility of crack initiation by this mechanism at the bore. In the high-temperature regions of HP/IP rotors, the relaxed long-term bore stresses and rim stresses are assessed against the creep-rupture data for the steel. The design stresses generally are based on the 10^5-h smooth-bar creep-rupture stress divided by some appropriate safety factor. The traditional approach is to use a Larson-Miller plot of the type shown in Fig. 6.10. The degree of conservatism implied in the process is unknown to the user. In-service

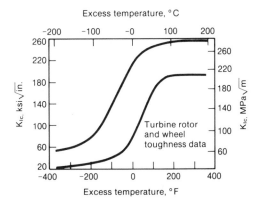

Fig. 6.8. Correlation of K_{Ic} values for rotor and disk steels with excess temperature, defined as the temperature of interest minus the FATT at that temperature (Ref 21).

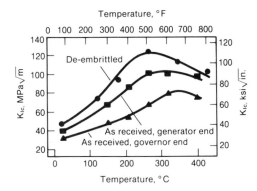

Fig. 6.9. Effect of temper embrittlement on fracture toughness of a Cr-Mo-V rotor steel (Ref 20).

degradation of the creep-rupture strength further clouds the issue. To avoid low-ductility notch-sensitive failures, manufacturers specify lower limits for rupture ductility. In addition, short-term notched-bar tests under specified conditions are also performed prior to acceptance of the rotor forging from the forging vendor. Such tests, however, may fail to predict the onset of notch sensitivity. Notch sensitivity is not an inherent property but depends on the temperature, stress, stress state, and strain rate. Methods for predicting rupture ductility, and the associated difficulties, were discussed in Chapter 3.

Low-Cycle Fatigue. The problem of low-cycle fatigue arises in a rotor primarily during transient conditions where, due to the

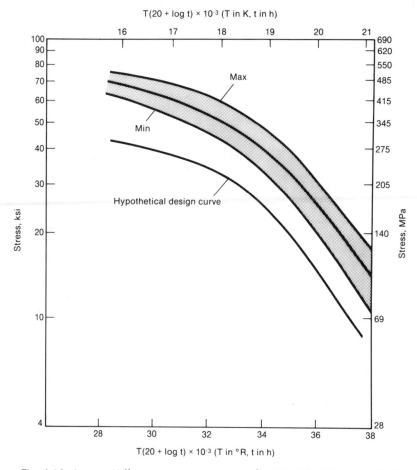

T(20 + log t) × 10⁻³ (T in K, t in h)

T(20 + log t) × 10⁻³ (T in °R, t in h)

Fig. 6.10. Larson-Miller stress-rupture curve for 1Cr-1Mo-¼ V rotor steel.

massiveness of the rotor, thermal gradients are set up in the rotor. The surface follows the ambient temperature change more closely than the interior, so that the surface would expand or contract relative to the interior but is prevented from doing so by the bulk of the rotor. Thermal strains thus result with every start-stop cycle and with load changes. This situation is illustrated in Fig. 6.11 (Ref 22).

Figure 6.11 describes a typical but simple cycle for a rotor in which a major load increase occurs, followed by steady operation at the high load and then by a major load decrease (Ref 22). The load variation with time is shown in Fig. 6.11(a). The patterns of temperature variation at the surface, middle, and bore of the rotor are shown in Fig. 6.11(b). The rotor surface stress varies with time, as shown in Fig.

6.11(c). The surface first tries to expand but is held in check by the bulk of the rotor, resulting in compressive stresses at the surface. If the load increase is sufficiently severe, compressive yielding occurs so that a residual tensile stress results when the loading cycle is completed. During steady operation, the residual tensile stress relaxes to a degree that depends on the temperature and time of operation at the steady load. When the load is decreased, the rotor surface goes into tension. This tensile stress is superimposed on the residual tensile stress. If tensile yielding occurs during a load decrease, a residual compressive stress results. This stress will not relax appreciably, however, because the temperature has reached a low value by now. The surface thermal strain variation with time is shown in Fig. 6.11(d). During repetition of this simple

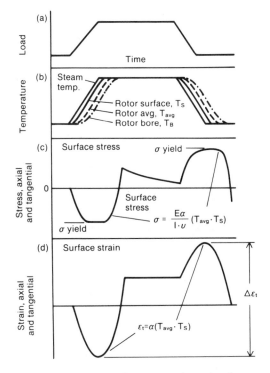

Fig. 6.11. Typical steam-turbine load-change cycle, showing variations in temperature, stress, and strain with time (Ref 22).

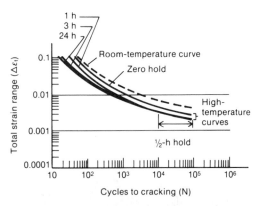

Fig. 6.12. Low-cycle-fatigue curves for Cr-Mo-V rotor steels at approximately 540 °C (1000 °F) (Ref 22).

cycle, at least three damage mechanisms can be operative: (1) fatigue due to the repeated cycles imposed by the strain range; (2) creep damage during stress relaxation at high temperature; and (3) creep damage under the steady operating loads. If the load changes are not severe and the thermally induced strains do not exceed the yield strain, then relaxation of residual stresses (item 2 above) does not become important. It is clear that the extent of damage depends on the strain range, the frequency of cycling, and the time and temperature under steady loading conditions. If the temperature is below the creep range, damage components 2 and 3 will be absent and damage will occur under simple low-cycle fatigue (LCF). At higher temperatures, the fatigue curves have to be modified to take into account frequency and "hold time" effects. Typical LCF design curves for HP rotor steels in the absence of creep effects are shown in Fig. 6.12 (Ref 22). The effects of frequency and hold time on the fatigue

behavior of Cr-Mo-V steels have been evaluated by several investigators. Some of their results as pertinent to life assessment of rotors are discussed in a later section.

Because the LCF damage arises primarily as a result of start-stop transients, the rates of temperature increase during start-up (ramp rate) and temperature decrease during shutdown must be carefully controlled. These rates are specified by turbine manufacturers in the form of cyclic life-expenditure (CLE) curves, as shown in Fig. 6.13 (Ref 22 and 23). These curves, which are derived by a procedure that will be described shortly, also can be used for computing LCF damage for various types of loading and unloading cycles.

The curves in Fig. 6.13 represent the percent life expended per cycle as a function of rotor diameter (packing diameter in the case of wheel-and-diaphragm-type rotors), surface stress-concentration factor, rate of steam temperature change, and magnitude of steam temperature change. For example, for a 635-mm- (25-in.-) diam rotor with a surface stress-concentration factor of 2.5, a 0.02% cyclic life expenditure would be caused by a 165 °C (300 °F) temperature change made at a rate of 110 °C (200 °F) per hour. In other words, only 50 such cycles could be withstood before fatigue-crack initiation would occur.

The CLE curves in Fig. 6.13 portray cyclic life expenditures and bore limits for rotors having packing diameters of 380,

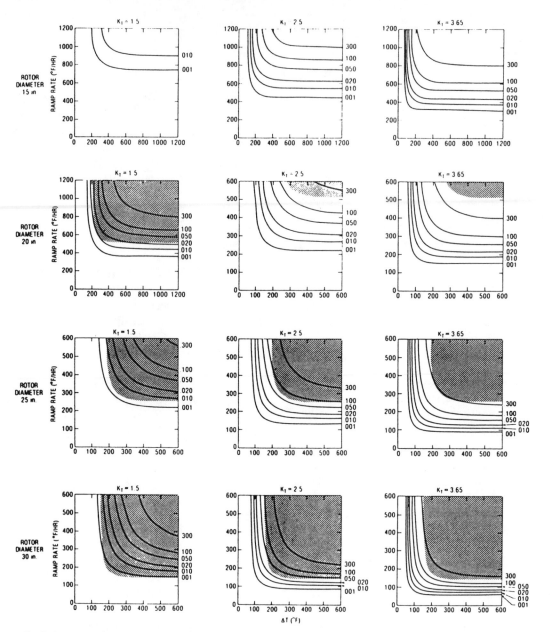

Shaded regions of curves represent bore-stress limits and define regions which should not be entered during temperature increases.

Fig. 6.13. Cyclic life-expenditure (CLE) curves (Ref 22 and 23).

510, 635, and 760 mm (15, 20, 25, and 30 in.) with surface stress-concentration factors of 1.5, 2.5, and 3.65. Cyclic life expenditures for other cases can be obtained by interpolation or extrapolation.

The curves in Fig. 6.13 are strictly valid only for *symmetrical* temperature transients, where the rate and magnitude of temperature increase are equal to the rate and magnitude of temperature decrease. For nonsymmetrical cycles, results of sufficient accuracy may be obtained by using a "pseudosymmetrical" range derived by averaging the ramp rates and ΔT values for the upramp and downramp.

The probability that a crack will initiate

and propagate at the center of a rotor must be kept as low as practicable because, for a cold rotor, bursting could result if crack initiation and propagation in this region are not detected. In order to minimize this probability, rotor design and operating instructions are aimed toward limiting the combined centrifugal and thermal bore stress to some fraction (say 90%) of the yield strength. This is achieved by avoiding the shaded area in Fig. 6.13. This region should not be entered *during start-ups or load increases*, when the thermal and centrifugal stresses are additive at the bore.

Crack growth in rotors can occur by fatigue, creep, or a combination of the two. There are two major sources of fatigue stresses. The first and larger of the two is the transient stress due to start-stop cycles, which is highest in the tangential direction at the bore and assists in the propagation of radial-axial bore cracks. In addition, the rotors may undergo an alternating bending stress superimposed on a high, steady-state mean stress during each revolution. This contributes to fatigue-crack propagation of tangential-radial cracks from the surface. The rate of fatigue-crack growth is generally given by the Paris law (Ref 24). At temperatures in excess of about 480 °C (900 °F), growth of cracks can also occur by creep or by a combination of creep and low-cycle fatigue (Ref 25). The phenomenology of crack growth under these conditions has been described in earlier chapters. A case study of crack-growth analysis may be found in Ref 26. Fatigue- and creep-crack-growth data are presented in Fig. 6.14 and 6.15.

Remaining-Life-Assessment Methods for Rotors

Remaining-life assessment can be based on a crack-initiation or a crack-propagation criterion. For surface cracks due to thermal fatigue and for cracks in the blade-attachment areas due to creep, crack initiation is used as the failure criterion. For bore cracks, crack initiation was used as the failure criterion until a few years ago, but with the emergence of clean steel technology, the advent of fracture mechanics, and the increasing need for extending the lives of rotors, application of crack-growth considerations has become common in recent years. When crack initiation is used as the failure criterion, history-based calculational methods are often used to estimate life expenditure. These methods are then supplemented with inspection and fracture-mechanics analyses in the case of the bore.

Analytical Methods for Crack Initiation. In the analytical methods, the operating history, rotor geometry, and heat-transfer properties are used to calculate the stress, strain, and temperature distribution in the rotor. This information is then used in combination with the standard creep-rupture data and low-cycle-fatigue data for the steel to estimate the creep, fatigue, or creep-fatigue life expended as appropriate to the situation.

Creep-Life Expenditure. A simplistic estimation of the creep life expended can be made by assessing the relaxed long-term bore stresses and rim stresses against the standard rupture data (e.g., Fig. 6.10) using the life-fraction rule. To be conservative, the lower-bound values in the stress-rupture plot are used. However, several problems arise when this method is used to calculate the life expenditure at blade-groove walls. It is well known that Cr-Mo-V rotor steels are subject to long-term degradation of creep properties due to strain-induced thermal softening. Rupture data such as those in Fig. 6.10 are based on short-term tests, and the extrapolation of these data to long service times is questionable. A more serious objection, however, is that the smooth-bar creep-rupture data are not applicable to the behavior at blade-groove walls, which have built-in stress concentrations. Data from notched bars which contain similar stress concentrations have to be used instead of smooth-bar rupture data. Because notch-weakening or notch-strengthening behavior is detected only in low-stress (low-strain-rate) tests, very-long-time tests on notched bars will have to be conducted. Extrapolation of short-term data will be nonconservative. The behavior of rotor

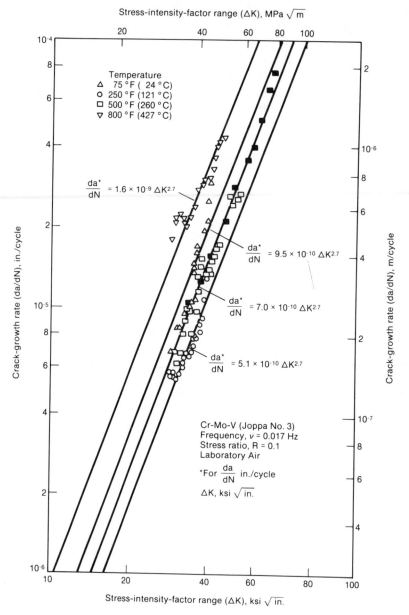

Fig. 6.14. Fatigue-crack-growth rates in Cr-Mo-V steel tested at 0.017 Hz (Ref 25).

steels in the presence of notches can be widely divergent depending on steel composition and heat treatment. Hence, any notched-bar data used for life estimation must be rotor-specific. Manjoine and Goldhoff have shown that for steels which tend to be weakened by notches, the notched-bar rupture life can decrease with increasing stress concentration at the notch root and with decreasing specimen size (Ref 27 and 28). Hence, any notched-bar data generated in the laboratory must be obtained on specimens simulative of the geometry and section size at the blade groove or the bore. Rupture testing of very large specimens with stress-concentrating features simulative of blade grooves is done by turbine manufacturers, but the data are proprietary. Notched-bar rupture tests on specific rotors to estimate creep-life consumption have been reported in a few instances, but in all cases testing was done after the rotors had

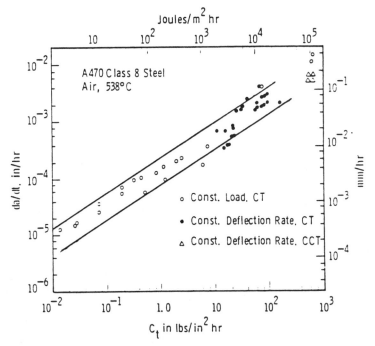

Fig. 6.15. Creep-crack-growth-rate behavior of type A470, class 8 steel at 538 °C (1000 °F) in air (Ref 26).

been taken out of service (Ref 27, 29, and 30). Generating the necessary data for in-service rotors is not practical. For these reasons, stress-based rupture-life evaluations have only limited value.

In the presence of stress concentrations, failure is not generally governed by rupture strength under relaxed stresses but rather by the accumulation of strain by repeated stress relaxations. For instance, at a rotor blade groove, the concentrated stress at the root of the groove relaxes with time and approaches the nominal stress. This local stress relaxation is accompanied by a gradual increase in local plastic strain. This strain must not exceed the strain capability of the material during its lifetime. Hence, calculation of strain accumulation with service and assessing it against a critical failure strain might be a more appropriate procedure for life assessment. Unfortunately, the strain capability (related to rupture ductility) is a function of time (strain rate) and temperature, and attempts to predict its long-term behavior on the basis of short-term tests have been unsuccessful, as described in Chapter 3. Recently, however,

Timo (Ref 22) has devised a test procedure for predicting the "crossover time" between notched- and smooth-rupture-specimen behavior, although details of this procedure have not been published.

Schlottner and Seeley have proposed a damage criterion based on the premise that the failure strain at a given temperature decreases with increasing time to rupture (Ref 31). This was discussed extensively in the section on rupture ductility in Chapter 3. Another observation is that specimens accumulating creep strain at higher strain rates fail at shorter times but have higher strain capabilities, as illustrated in Fig. 6.16. This indicates that the amount of strain accumulation as well as the rate at which the strain accumulates are important in a strain-based damage criterion. A material property that encompasses both of these concepts is the average creep rate to rupture, $\dot{\epsilon}_{avg}$, which is defined as ϵ_r/t_r, where ϵ_r is simply the strain obtained by extending the secondary-creep-rate curve up to t_r, and t_r is the rupture time. Plots of $\dot{\epsilon}_{avg}$ vs t_r for Cr-Mo-V steels were found to yield a linear relationship, regardless of test con-

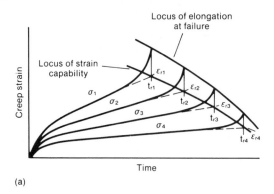

(a)

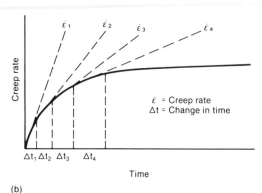

(b)

(a) Variation in strain capability with time to rupture.
(b) Strain-rate damage process.

Fig. 6.16. Schlottner-Seeley procedure for life assessment (Ref 31).

ditions, and the general relationship was found to be

$$t_r = P\dot{\epsilon}_{avg}^Q \qquad \text{(Eq 6.1)}$$

where P and Q are constants. This relationship is similar to the Monkman-Grant relationship described in Chapter 3 except that Schlottner and Seeley have extended the concept to include even the primary creep stage so that $\dot{\epsilon}_{avg}$ is not simply the minimum creep rate but any arbitrary value of instantaneous creep rate along the creep curve, as illustrated in Fig. 6.16. If the creep-strain-vs-time curve is known, it can be broken into time intervals of Δt_i, during which a given value of $\dot{\epsilon}_i$ obtains. The time to rupture, t_{ri}, corresponding to each creep rate can be calculated from Eq 6.1 by substituting $\dot{\epsilon}_i$ for $\dot{\epsilon}_{avg}$. The fractional damage D_i during each time increment along the creep curve can be

expressed as $\Delta t_i/t_{ri}$. The total damage during n time increments is assumed to be the sum of the incremental damage, or

$$D = \sum_{i=1}^{n} \left(\frac{\Delta t_i}{t_{ri}} \right) \qquad \text{(Eq 6.2)}$$

The time to failure is designated as the time at which D = 1.

The unique quality of the strain-rate damage-calculation method is that it can be applied to any loading condition that results in the accumulation of creep strain. In general, two steps are required. In the first step, the creep-strain-vs-time behavior must be established so that the creep rates are known at any time. Then the creep rates are used to predict damage using Eq 6.2.

The method of estimation of creep behavior must be specific to the component being analyzed. In some cases simple equations can be used, whereas in other cases a complex finite-element analysis may be required. The Schlottner and Seeley method provides a unique way of estimating damage based on time life fractions using a strain-exhaustion criterion.

LCF Life Expenditure in Pure Fatigue.
For pure-fatigue conditions of continuous cycling, data similar to those shown in Fig. 6.12 can be used in conjunction with the life-fraction rule to calculate fatigue-life expenditure. Frequently, the universal slopes equation (Eq 4.11) is used to estimate the shape of the fatigue curve based on a knowledge of the ultimate tensile strength and the tensile ductility. This method can be directly applied to cold-start conditions, where the maximum fatigue strains occur at relatively low temperatures. Kramer *et al* also applied the method to hot-start conditions for rotor locations where the maximum temperature did not exceed about 425 °C (800 °F), because their low-cycle-fatigue tests showed no hold-time effects (i.e., creep) on fatigue endurance even at 425 °C (Ref 29). Timo (Ref 22) has used a slightly modified version of the universal slopes equation suggested by Tavernelli and Coffin (Ref 32), as follows:

$$\Delta\epsilon_t = 0.5DN_f^{-0.5} + \frac{\sigma_u}{E} \qquad \text{(Eq 6.3)}$$

where D, N_f, σ_u, and E are defined as for Eq 4.11. This equation describes the behavior of a wide variety of low-alloy rotor steels. From Eq 6.3, and a knowledge of the percent reduction in area and the ultimate tensile strength, the expected fatigue life for a given strain range, and hence the fatigue-life expenditure, can be estimated.

LCF Life Expenditure in Creep-Fatigue. For temperatures higher than 425 °C (800 °F), there is a large body of data indicating a reduction in fatigue life due to decreased frequency of cycling as well as due to hold time. Several approaches for estimation of cumulative damage under these conditions were described in Chapter 4. Validation of these approaches with specific reference to high-temperature rotor steels will be discussed here. This will then be followed by a description of a commonly employed method for damage assessment of rotors.

One of the most comprehensive examinations of the applicability of life-prediction methods to Cr-Mo-V rotor steels was performed by Leven (Ref 33). A series of tests at 540 °C (1000 °F) with hold times up to 1 h at constant tensile stress or constant tensile or compressive strain formed the basic data. Analyses were carried out using the linear damage-summation method, the frequency-modified strain-range equation, and the strain-range-partitioning method. Leven concluded that all three methods predicted the actual lives within a factor of two and that none of them was significantly better than the others. Similar conclusions have been reached by Kuwabara and Nitta (Ref 34). Batte performed strain-controlled fatigue tests with hold times ranging from 0.5 to 15 h at 540 °C (1000 °F) and concluded that all three methods provided the same degree of predictive capability with respect to the 0.5-h-hold tests (Ref 35). For the low-strain-range, 15-h-hold tests, however, the linear damage-summation procedure provided the most accurate description of the data. An attempt to examine the creep-fatigue behavior of Cr-Mo-V rotor steels under mixed loading conditions was made by Curran and Wundt (Ref 36). In their tests, the first part of the package consisted of a 23-h hold at constant tensile load and was followed by fully reversed strain-controlled fatigue cycling. The results of tests up to 500 h in duration at 480 to 540 °C (895 to 1000 °F) indicated that the linear damage-summation method was generally nonconservative. Melton has investigated the low-cycle fatigue behavior of Cr-Mo-V rotor steels. The effects of waveshape and frequency on lifetime were studied, and the results were analyzed using the strain-range-partitioning, frequency-separation, and energy-damage-function methods (Ref 37). A comparison was made between the prediction capabilities of the analytical methods, and it was concluded that the best fit to the data was obtained using the frequency-separation model. A creep component was found to be more damaging during the tensile part of the cycle than during the compressive part of the cycle.

Extensive low-cycle-fatigue characterization of Cr-Mo-V rotor steels has also been performed by Bisego, Fossati, and Ragazzoni (Ref 38). Strain-controlled fatigue tests were carried out at 480 and 540 °C (895 and 1000 °F) with hold times up to 24 h and at different strain rates. It was concluded that the strain-range-partitioning method resulted in a better life-prediction capability than the linear damage-summation rule. The results of Ostergren, showing the absence of significant frequency and hold-time effects, have already been described in Chapter 4 (Ref 39). Ostergren concluded that the energy damage function without any frequency modifications provided a good fit to the data.

A review of the creep-fatigue data and the recommended analytical procedures for life estimation of Cr-Mo-V steels shows widely divergent views. The problem seems to lie in the fact that the effect of hold time on fatigue life is a function of the type of strain cycle employed, the strain range, and the temperature. The need for conducting laboratory tests which are realistic and representative of the strain cycles and conditions obtaining in HP rotors has been emphasized by Thomas and Dawson (Ref 40) and has been discussed in Chapter 4. From the data of Thomas and Dawson, it appears that under realistic conditions pertinent to HP rotor surface cracking and at low strain ranges, hold-time effects may be marginal. This seems to agree with the view of Timo (Ref 22), as will be apparent from the following discussion of their recommended procedure for remaining-life assessment.

In applying any of the damage-summation methods to rotors, the thermal strains corresponding to the various transients, as well as the strain-concentration factors corresponding to the stress-concentration factors at the critical regions, need to be determined. The magnitude of the thermal strain produced is dependent on the magnitude and rate of temperature change, the surface heat-transfer coefficient, the massiveness (as measured by diameter or thickness) of

the component, and the thermal properties of the component. The thermal strains and stresses throughout the rotor can be calculated in a number of ways, including finite-element analysis.

The approximate relationship among thermal strain, rotor size, and material properties is illustrated in Fig. 6.17 (Ref 22). Use of the relationship illustrated in this figure to calculate the thermal strains at the rotor surface requires material-property data specific to a given rotor. Timo has suggested that approximate calculations may be performed by using average values of rotor material properties, evaluated at an intermediate temperature (Ref 22). The following property values may be used for this purpose:

$$d = \text{Thermal-expansion coefficient}$$
$$= 8.5 \times 10^{-6} \text{ in./in.} \cdot {}^{\circ}F$$

$$\frac{k}{\gamma c} = \text{Thermal diffusivity} = 0.3 \text{ ft}^2/\text{h}$$

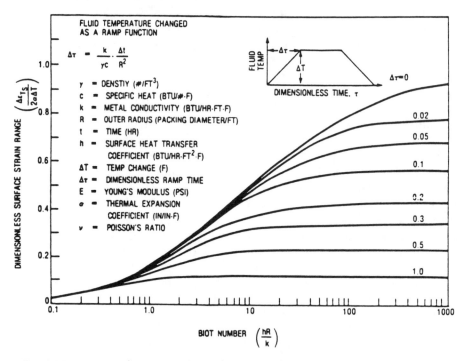

Fig. 6.17. Dimensionless nominal cylinder-surface thermal strain, used for calculating nominal thermal strain range on surfaces of turbine rotors (Ref 22).

ϵ_y = Cyclic yield strain

$\quad$ = 2×10^{-3} in./in.

$\dfrac{hR}{k}$ = Biot number = 100

Having determined the nominal thermal strain for a given transient, it is then necessary to estimate the concentrated strain produced in regions of strain concentration. It is well known that when the yield strain is exceeded, the strain-concentration factor increases above the elastic stress-concentration factor K_t. Values of an effective strain-concentration factor K_ϵ have been derived (see Fig. 6.18) for semibiaxial nominal surface stresses which occur at the rotor periphery and bore due to temperature transients (Ref 41). The true stress-strain curve of the steel must be used in calculating strain-concentration factors. Because the rotor strain-softens in service, the cyclic true stress-strain curve for the steel must be used.

Once the effective strain for a given transient is known, the fatigue life for that transient can be determined by entering the "appropriate" $\Delta\epsilon_t$-vs-N_f curve, which incorporates the hold-time effects for the steel. For low strain ranges where no yielding occurs, there is no residual stress and therefore no "hold time" as such. For such cases, use of data for a nominal 1/2-h hold time has been recommended (Ref 22). The data applicable to Cr-Mo-V rotor steels as provided by Timo are given in Fig. 6.12. In actual operation of rotors, several types of transients, such as cold starts, warm starts, and hot starts, occur. The number of each type of transient event is known from the operating records. The fractional fatigue life expended for each type of transient can be calculated and summed to determine the cumulative fatigue damage. The creep-damage fraction can be calculated as the ratio of time in service divided by the time to rupture at the operating temperature and stress. The cumulative creep-fatigue damage is obtained by summation of the fa-

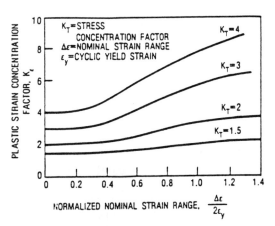

Fig. 6.18. Plastic strain-concentration factors for low-alloy steels (Ref 41).

tigue-life consumption and the creep-life consumption. This procedure is very similar to the "elastic route" suggested in ASME Code Case N-47 and described under "Design Rules for Creep-Fatigue" in Chapter 4. Alternatively, the "inelastic route" suggested in Code Case N-47 may be employed by using pure LCF curves (without hold time) and the bilinear creep-fatigue-damage curve suggested by Fig. 4.28. The choice of the appropriate LCF curves, the safety factors applied, and the value of D can vary from one investigator to another.

Example:
A Cr-Mo-V steel rotor has been in baseload service for 100,000 h at a nominal temperature of 540 °C (1000 °F). During this time it has been subjected to 105 cold starts and about 183 warm starts. Based on finite-element analysis, the $\Delta\epsilon_t$ values corresponding to the cold-start and warm-start conditions at a surface groove (K_t = 2.8) are estimated to be 0.003 and 0.002, respectively. The cyclic yield strain ϵ_y for the steel is 0.002. The creep-rupture life under the nominal operating conditions is 300,000 h. Calculate the cumulative life expended at the groove root.

Answer:

Step 1: Calculate the nominal thermal strain. In this case, nominal thermal strain is given as 0.003 for cold-start conditions and 0.002 for warm-start conditions.

Step 2: Calculate the effective strains. For cold-start conditions, $\Delta\epsilon/2\epsilon_y = 0.003/0.004 = 0.75$, $K_\epsilon = 4.2$ (from Fig. 6.18), and thus effective $\Delta\epsilon = 4.2 \times 0.003 = 0.0126$. For warm-start conditions, $\Delta\epsilon/2\epsilon_y = 0.002/0.004 = 0.5$, $K_\epsilon = 3$ (from Fig. 6.18), and thus effective $\Delta\epsilon = 3 \times 0.002 = 0.006$.

Step 3: Determine N_f from appropriate LCF curves that include hold-time effects at the temperatures where the strains peak in the two types of starts. Let us assume that for cold-start conditions at $\Delta\epsilon = 0.0126$, $N_f = 200$ cycles, and that for warm-start conditions at $\Delta\epsilon = 0.006$, $N_f = 2000$ cycles.

Step 4: Calculate cumulative fatigue-life fraction expended. Fatigue-life fraction expended = $105/200 + 183/2000 = 0.62$.

Step 5: Calculate cumulative creep-life fraction expended. Creep-life fraction expended = $100,000/300,000 = 0.33$.

Step 6: Calculate total life fraction expended. Total life fraction expended = $0.62 + 0.33 = 0.95$.

Carlton, Gooch, and Hawkes have described an alternative procedure in which the cumulative damage is expressed as the sum of the three damage components—i.e., pure fatigue, stress-rupture, and stress relaxation (Ref 42). On a per-cycle basis, the life fraction consumed is expressed as

$$\frac{D}{D_r} = \frac{1}{N_0} + d + \frac{t}{t_r} \qquad \text{(Eq 6.4)}$$

where D is the damage per cycle; D_r is the damage at crack initiation, assuming endurance to be determined by initiation; N_0 is the lower-bound continuous cycling (pure fatigue) endurance; d is the damage fraction per cycle accumulated by relaxation of the thermal stress during the hold time and is taken as the greater of those calculated for ductility-exhaustion and life-fraction models; t is the hold time; and t_r is the

lower-bound creep-rupture life at the primary membrane stress. This procedure differs from that of Timo mainly in the sense that it breaks down the creep-fatigue damage into two components—i.e., pure fatigue and stress relaxation. The procedure outlined by Timo combines these two components by using fatigue curves that incorporate hold-time effects.

Kubawara and Nitta have described a procedure (Ref 34 and 43) that is similar to that of Carlton, Gooch, and Hawkes except that, instead of using the pure fatigue curve, they have used Leven's frequency-modified fatigue-life equation to estimate the fatigue-life expenditure. For each type of transient where residual tensile stresses were calculated to be present, they have calculated creep-life consumption using the life-fraction rule for the steady-state relaxed stress as well as for the various stress decrements during stress relaxation. The cumulative damage so calculated was somewhat inconsistent with observations on actual rotors containing groove cracks; nevertheless, their study is among the very few ever reported on validation of damage rules against actual field experience. The only other published work relating to field experience is that of Kramer, Randolph, and Weisz (Ref 29). Using a modified version of Leven's frequency-modified fatigue-life correlation, they showed that creep-assisted low-cycle fatigue initiating at manganese sulfide inclusions at the bore may have been a probable cause of the bursting of the HP-IP rotor at the TVA Gallatin Station.

Nondestructive Methods For Damage Evaluation. Nondestructive methods that have been demonstrated to have potential for assessing rotors with respect to crack initiation include strain measurements, studies of creep cavitation using replicas, hardness measurements, and x-ray measurements.

Strain Measurements. Measurement of the deformation and distortion of rotor forgings during service has been an integral part of life estimation for some turbine manufacturers (Ref 44). Lateral differences in creep rate in horizontally heat treated rotors manufactured in the 1960's had led to unacceptable distortions. Since that time, vertical heat treatment is the general practice. Most high-temperature rotors in large units in the United Kingdom have been installed with provision for accurate determination of creep deformation at scheduled outages. An example of the variation in creep strain at the rotor bore with service time, calculated from measurements at the rim, for several 500-MW HP and IP rotors is shown in Fig. 6.19 (Ref 44). Such data, coupled with a proper choice of failure strain, could provide a suitable estimate of creep-life expenditure. Unfortunately, there is very little published data on the subject, and it is not clear how widespread this practice is in industry.

Creep-Cavitation. The progress of creep-cavitation as a function of creep-rupture life expended has been investigated on the basis of interrupted creep tests of two different forgings at different temperatures and stresses by Goto (Ref 45) and by Tanemura *et al* (Ref 46). The relationship between the number fraction of cavitating grain boundaries, A, and the fractional creep life expended, t/t_r, is shown in Fig. 6.20. These data could be adequately described by the model expressed in Eq 5.9. Some additional data have also been pub-

lished by Carlton, Gooch, and Hawkes (Ref 42). If a larger database comprising more heats and test conditions could be established, correlations of this type could be used for estimating creep-life expenditure. By using surface replication techniques, the procedure could be made completely nondestructive. The main limitation in applying replication would be accessibility to critical locations in the dovetail (blade-attachment) area and in the bore. Kadoya *et al* have developed a remote replication technique for use inside the rotor bore (Ref 47). Creep-cavitation has been observed in the blade-attachment areas, but there are no reported instances of creep-cavitation in rotor bores.

Hardness Evaluations for Creep. It has been recognized for some time that in materials where the principal damage under creep-exposure conditions consists of thermal or strain-induced softening, room-temperature hardness can provide an index of creep life expended.

The first attempt to develop hardness as an index of creep damage was that of Goldhoff and Woodford (Ref 48). In their study, the effect of prior creep exposure for times up to 60,000 h at four temperatures in the range 482 to 593 °C (900 to 1100 °F) on the subsequent rupture life in a standard test at

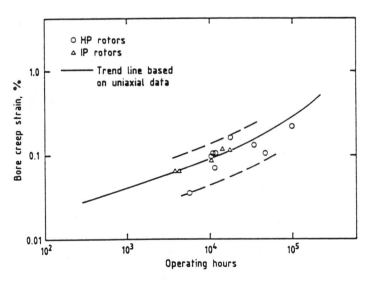

Fig. 6.19. Progress of creep deformation at the bore, calculated from measurements at the rim, for Parsons 500-MW HP and IP rotors in CEGB units (Ref 44).

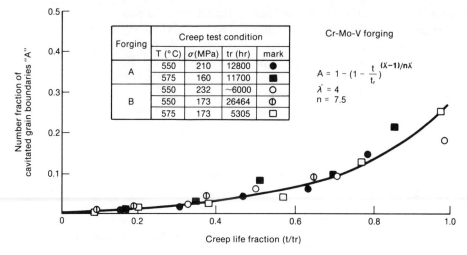

Fig. 6.20. Evolution of creep-cavitation with creep-life fraction expended for Cr-Mo-V rotor steels (Ref 46).

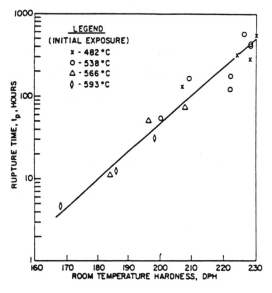

Fig. 6.21. Correlation between post-exposure rupture time in the standard test at 538 °C and 240 MPa and room-temperature hardness for Cr-Mo-V rotor steel (Ref 48).

538 °C and 240 MPa (1000 °F and 35 ksi) was examined. Some of the exposed specimens were interrupted at various strains. Postexposure specimens were cut from the uniform-strain sections of exposed bars. A good correlation was observed between room-temperature hardness measured on the exposed creep specimens and the post-exposure rupture life, as shown in Fig. 6.21.

If similar calibrations could be established between prior creep life expended or the remaining life fraction in the postexposure test and the hardness values for a range of Cr-Mo-V rotor steels, this method could be applied to estimation of remaining life. Unfortunately, more systematic data of this nature are not available.

Recently, Goto attempted to use the hardness technique as a stress indicator (Ref 49) and observed that the application of stress accelerated the softening process and shifted the hardness to lower parameter values compared with the case of simple thermal softening on a plot of hardness vs a modified Larson-Miller-type parameter, as shown in Fig. 6.22. He designed a new parameter defined by

$$G' = G + \Delta G \qquad \text{(Eq 6.5)}$$

$$G = \log[T(20 + \log t)] \qquad \text{(Eq 6.6)}$$

and

$$\Delta G = 0.000217(\sigma - 108) \text{ [for } \sigma > 108 \text{ MPa]}$$
$$\text{(Eq 6.7)}$$

$$\Delta G = 0 \qquad \text{[for } \sigma < 108 \text{ MPa]}$$
$$\text{(Eq 6.8)}$$

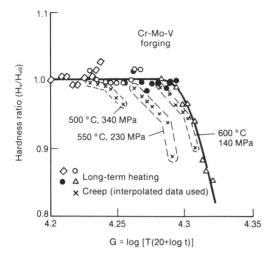

Fig. 6.22. Plot of hardness ratio vs G parameter for long-term heating and creep of Cr-Mo-V rotor steel (Ref 49).

where T is temperature in K, t is time in hours, and σ is stress in MPa. G is the parameter describing the thermal-softening behavior and ΔG is the parameter that incorporates the effect of stress. The parameter G' thus includes the effects of temperature and stress. When plotted in terms of G', all of the hardness values, regardless of stress, could be normalized into a single curve. By comparing hardness values at unstressed and stressed locations in a rotor, if ΔG could be determined, then the local stresses could be estimated. The values of time to rupture, t_r, at this stress and the known temperature could then be estimated from the rupture data for the steel. Because the initial hardness values of different rotors are likely to be different, Goto normalized all of the hardness results in terms of the initial hardness, as H_v/H_{v0}. By measuring H_v at a nonstressed location, any rotor could be "placed" with respect to the master plot of G vs H_v/H_{v0}. By subsequently measuring H_v at a stressed location, ΔG could be determined and used as described. This method has been successfully applied to estimation of the creep-life consumption at the T-root corner of a rotor disk by Tanemura *et al* (Ref 46). The following example will serve to illustrate, step by step, the use of the procedure.

Example:

A rotor has been operating for 150,000 h. The temperature in the first-row blade-attachment regions is estimated to be 538 °C (811 K). Hardness measurements at a groove and at a nonstressed region near the groove give values of $H_v = 225$ and 230, respectively. Calculate the creep life expended.

Answer:

Step 1: "Place" the rotor on the plot of H_v/H_{v0} vs G, based merely on long-term heating. $G = \log[T(20 + \log t)] = 4.307$. The corresponding $H_v/H_{v0} = 0.95$ from a plot of G vs H_v/H_{v0} (Fig. 6.22). Hence, $H_{v0} = 230/0.95 = 242$.

Step 2: $H_v/H_{v0} = 225/242 = 0.93$ at the stressed location and, corresponding to this H_v/H_{v0}, $G = 4.3085$. In other words, although the time and temperature of exposure at the stressed location were the same as at the unstressed location, accelerated softening has occurred as if the time-temperature conditions corresponded to higher values of G.

Step 3: Estimate the stress. $\Delta G = G_{stressed} - G_{unstressed} = 4.3085 - 4.3070 = 0.0015$. $\Delta G = 0.0015 = 0.000217(\sigma - 108)$. $\sigma = 115$ MPa.

Step 4: Calculate the expected rupture life. From a Larson-Miller rupture curve for Cr-Mo-V steel (Fig 6.10), find that at $\sigma = 164.7$ MPa, the Larson-Miller parameter = 20,555. $T(20 + \log t) = 20,555$. Substituting for T = 811 K (538 °C), we get $t_r = 220,000$ h. Remaining life = 70,000 h.

Kimura *et al* have used a very similar technique wherein hardness changes are related to time, temperature, and stress (Ref 50). They show an excellent correlation between predicted rupture lives and actual rupture lives for rotor steels tested at various hardness levels. Unfortunately, they have only reported the general form of the correlation without publishing the correlation constants.

Hardness changes as functions of time

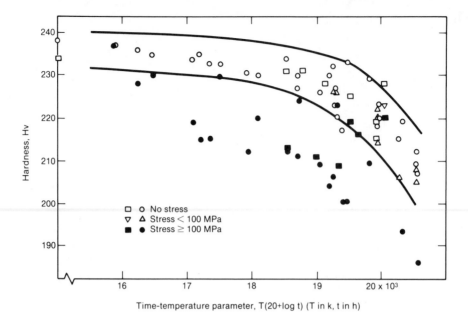

Fig. 6.23. Hardness changes in 1Cr-Mo-V rotor forging steel during thermal exposure and creep testing at 450 to 550 °C (840 to 1020 °F) (Ref 44).

and temperature of exposure for numerous heats of Cr-Mo-V rotor steels have also been described by Batte and Gooch, as shown in Fig. 6.23 (Ref 44). Their results confirm the finding of Goto that the effect of stress on aging becomes significant only above approximately 100 MPa. For situations where stresses are lower, Fig. 6.22 or similar data can be used to estimate the local metal temperature because the service time is known. Based on the temperature estimate, the time to rupture can be estimated from standard rupture data, provided that the operating stresses are known. The creep life expended, from a crack-initiation point of view, can be calculated.

Electrical Resistivity Measurements for Creep Damage. It has been found that long-term heating of Cr-Mo-V steels results in a reduction of the electrical resistivity and that applied stresses lead to even greater reductions (Ref 51). Figure 6.24(a) shows the decrease in resistivity expressed as a ratio of the resistivity value for the thermally exposed condition to that for the unexposed condition, as a function of a Larson-Miller-type exposure parameter; the constant C in the parameter has not been defined. Data obtained on samples from a rotor disk and bore are plotted in Fig. 6.24(a) to show the

effect of stress (Ref 46). The decrease in the resistivity ratio, ΔR_p, can then be converted to a creep-life fraction consumed using calibration plots of the type shown in Fig. 6.24(b). A linear relationship is observed between the creep-life expended and the decrease in the resistivity ratio up to t/t_r values of 0.7. At larger values of t/t_r, where microstructural damage due to strain softening yields to cavitation-type damage, the linear relationship becomes invalid. Based on the limited data presented by Tanemura *et al* (Ref 46), this technique appears to be very promising.

Hardness and X-Ray Evaluations for Fatigue. Kadoya *et al* have found that low-cycle-fatigue damage in Cr-Mo-V steel rotors results in strain softening, the extent of which can be determined by hardness measurements and by x-ray line-width measurements (Ref 47). Figure 6.25(a) and (b) illustrate the relationship among hardness change, x-ray line width, and fatigue-life fraction consumed based on laboratory LCF tests at 500 °C (930 °F). In Fig. 6.25(a), H_v/H_{v0} denotes the ratio of the Vickers hardness values after and before the test. Kadoya *et al* propose to use the hardness correlation for life estimation of bores and the x-ray correlation for life estimation at

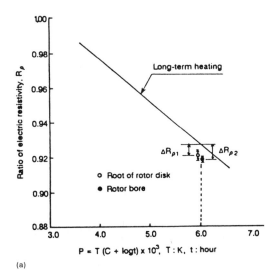

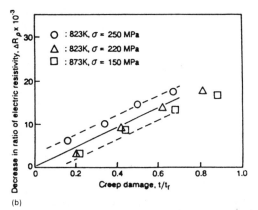

(a) Resistivity ratio as a function of exposure condition. (b) Decrease in resistivity ratio with expended creep life.

Fig. 6.24. Resistivity technique for estimating life fraction, t/t_r, expended in creep (Ref 46).

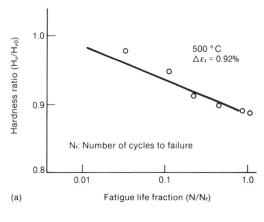

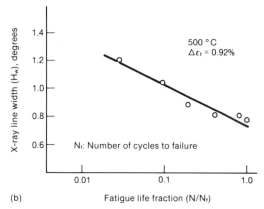

Fig. 6.25. Variation of (a) hardness and (b) x-ray line width with LCF life fraction for Cr-Mo-V forgings (Ref 47).

surface groove locations. Once again, these techniques are in their infancy but hold much promise as nondestructive tools for fatigue-life estimation.

Nondestructive Evaluation of Toughness.
A major task, particularly with respect to rotor-life assessment in the presence of bore cracks, is to estimate the current toughness at the critical location. This toughness, generally expressed in terms of the plane-strain stress intensity for fracture, K_{Ic}, determines the critical crack size for failure. Although cracks may initiate and propagate at higher temperatures by creep or by low-cycle fatigue, the risk of final fracture is greatest at

low temperature under transient conditions. Turbine manufacturers may sometimes have a record of the preservice FATT of the rotor from core bar samples or other samples. Unfortunately, the FATT (and hence K_{Ic}) is degraded in service due to temper embrittlement, and the kinetics of the phenomenon are not sufficiently well established. A limited amount of FATT data on rotors retired after extended service has become available in recent years. These data may be used as a guide in choosing appropriate K_{Ic} values for life estimation. Such procedures, however, tend to be overly conservative. Several nondestructive as well as relatively nondestructive tests involving removal of very small samples have been investigated in recent years (Ref 52 to 55). Of these, eddy-current examination, analytical electron microscopy, and secondary ion mass spectroscopy (SIMS) have been unsuccessful. Compositional correlations, Auger electron spectroscopy, chemical etching, electrochemical polarization, and use of single Charpy specimens have shown considerable promise. In these studies, Ni-Cr-Mo-V low-pressure turbine steels have been investigated more extensively than have the Cr-Mo-V steels used in HP rotors. Because the basic techniques are easily extendable to Cr-Mo-V steels, they are described in this section.

Compositional Correlations. Several compositional correlations have been suggested for Ni-Cr-Mo-V steels, but the body of data pertaining to Cr-Mo-V steels is much more limited. The relationship between the J factor [i.e., $(Sn + P)(Mn + Si) \times 10^4$] and the shift in the FATT due to temper embrittlement is expected to be similar to that for 2¼Cr-1Mo steels, shown in Fig. 7.9 and 7.10 in Chapter 7.

Auger Electron Spectroscopy. Because temper embrittlement is caused primarily by segregation of certain impurity and alloying-element species to grain boundaries, considerable effort has been focused on the quantitative analysis of the grain-boundary composition, which can then be related to the ΔFATT. Zhe *et al* systematically investigated the temper-embrittlement behavior of Cr-Mo-V steels in which the concentrations of manganese, silicon, phosphorus, and tin were varied in a controlled fashion (Ref 56). Hardness and grain size were also varied. Specimens were aged at 520 °C (970 °F) for times up to 6000 h. Auger analysis of grain-boundary fractures even in tin-doped steels did not reveal the presence of tin but only that of phosphorus (Ref 56). On the other hand, clear evidence of segregation of both phosphorus and tin has been found in samples from three actual rotors retired from service (Ref 20 and 57). Presumably, this difference is due to the difference in the exposure temperatures for the laboratory and field samples. The data indicate that segregation of tin may occur only at lower temperatures and therefore could not be reproduced under the accelerated aging conditions in the laboratory tests. When the whole body of available data is

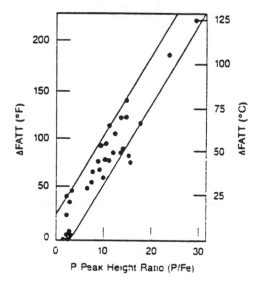

Fig. 6.26. Correlation of ΔFATT with phosphorus segregation, based on Auger analysis, for Cr-Mo-V steel (Ref 55).

taken together, a correlation between phosphorus segregation and ΔFATT is observed, as shown in Fig. 6.26. This relationship could be used to estimate FATT for in-service rotors based on Auger analysis of small samples.

Single Charpy Tests. Strong justification for the use of single Charpy specimens can be found in the early work of Newhouse, who found that the FATT of the steels could be estimated from the percentage fibrosity or impact energy of a single specimen tested at a given temperature (Ref 58). In a recent study, a model has been developed for estimating FATT for Cr-Mo-V rotor steels (Ref 59). Over 700 test data points from core bar samples from 18 rotors, aged in the laboratory for up to 95,000 h, were included in the analysis. The best fit was found to be:

$$(T - FATT) = 81.14 + 160 \log$$
$$\times [-\log(1 - Fib/100)]$$

$$(Eq\ 6.9)$$

where Fib is the percent fibrosity in a single Charpy test and T is the test temperature in °F. The correlation between excess

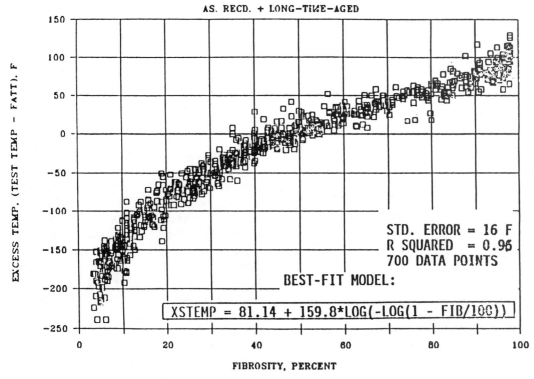

Fig. 6.27. Correlation of FATT with fibrosity for Cr-Mo-V rotor steel (Ref 59).

temperature and fibrosity for Cr-Mo-V steels is shown in Fig. 6.27. Validation of this relationship has also been carried out on a number of ex-service rotors. Values of FATT predicted on the basis of single specimens agreed with the values actually determined from multiple specimens to an accuracy of ±25 °F.

Small Punch Test. The use of small disk-like specimens subjected to bending loads in a punch has been described in the literature (Ref 60 to 62). According to this procedure, a thin plate specimen measuring approximately 10 by 10 by 0.5 mm (0.4 by 0.4 by 0.02 in.) is subjected to punch deformation with a 2.4-mm- (0.09-in.-) diam steel ball in a specially designed specimen holder. The test is performed at various temperatures in an Instron tensile-testing machine. From the load-deflection curves obtained at the various temperatures, the fracture energy is calculated and plotted as a function of test temperature to determine the ductile-to-brittle transition temperature.

Baik, Kameda, and Buck observed a unique correlation between the transition temperature determined in the small punch test, T_{sp}, and the FATT as determined from standard Charpy V-notch specimens (Ref 61 and 62). These correlations, however, were specific to the impurities present. This technique has been extended to Cr-Mo-V rotor steels by Takahashi et al (Ref 63). These authors characterized T_{sp} values using samples with different degrees of temper embrittlement which had been removed from a Cr-Mo-V steel rotor after 22 years of service. The correlations between the FATT and T_{sp} values (see Fig. 6.28) were found to be in between those for different impurities investigated earlier by Baik, Kameda, and Buck. For the Cr-Mo-V rotor steels, the T_{sp} values were about 0.57 of the FATT values. Based on these encouraging results, Takahashi et al have advocated the use of the small punch test as a semiquantitative tool for characterizing the degree of embrittlement of rotors.

Chemical Etching. Based on observations reported in the literature that grain bound-

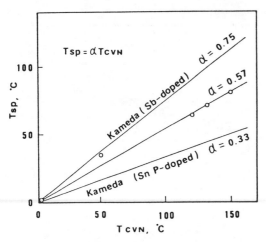

$T_{sp} = \alpha T_{CVN}$

Kameda (Sb-doped) $\alpha = 0.75$

$\alpha = 0.57$

Kameda (Sn P-doped) $\alpha = 0.33$

Fig. 6.28. Relationship between changes in ductile-to-brittle transition temperature obtained from small punch tests and Charpy tests for Cr-Mo-V steels (Ref 63).

aries of embrittled steels are attacked preferentially by picric acid solutions, chemical corrosion tests have been performed on Ni-Cr-Mo-V steel samples in a saturated picric acid solution with an addition of 1 g of tridecyl benzene sulfonate (per 100 ml of aqueous picric acid). Reasonable correlations between the grain-boundary groove depth as measured metallographically and the ΔFATT of the sample due to prior temper embrittlement were observed (Ref 52 and 53). Further work has shown that the depth of the grooves after etching can be successfully measured even from plastic replicas, thus making the technique very attractive for field use (Ref 54). Because this technique has yet to be extended to Cr-Mo-V rotor steels, it has been described only briefly.

Electrochemical Tests. The sensitivity of picric acid etch-test solutions to attack in segregated phosphorus regions has prompted its use for electrochemical tests. Kimura *et al* and Shoji and Takahashi (Ref 50 and 64) have shown that a reactivation scan from the passive to the active region of the electrochemical polarization curve could separate heats with different segregation levels (i.e., embrittlement). In this method, the material of interest is dipped into an elec-

trolytic solution and electrochemically polarized, and the polarization curve is recorded. Typically, an aqueous solution containing 2×10^{-2} mole/litre of picric acid and 1×10^{-2} mole/litre of sodium trimethyl benzene sulfonate was used as the electrolytic solution. As shown in Fig. 6.29(a), the minimum current I_r in the reactivation process flows in the embrittled material sampled from the high-temperature portion of the rotor, whereas no current flows in the nonembrittled samples. The relationship between I_r and the normalized ΔFATT with respect to the initial FATT is shown in Fig. 6.29(b). Shoji and Takahashi have suggested that the difference in behavior between the embrittled and nonembrittled samples is due to differences in their repassivation behavior. In the nonembrittled samples, a repassivation film is formed readily, whereas in the embrittled samples formation of the repassivation film is inhibited due to grain-boundary segregation of phosphorus, facilitating current flow.

The electrochemical polarization technique has been applied to Ni-Cr-Mo-V steels, but with little success (Ref 52). The reasons for this are not clear at present. The exact chemical composition of the electrolytic solution and the details of the technique used by Shoji and Takahashi have not been published. Whether their technique is sensitive to embrittlement by tin as well as by phosphorus is not known. The shape of the ΔFATT-vs-I_r curve (Fig. 6.29b) indicates that ΔFATT values of as much as 50% of the initial FATT may go undetected, because the curve is flat in this region. Despite these limitations, this technique has been extended and widely applied in field use (Ref 64). Its successful application to the characterization of temper embrittlement of a service-exposed rotor has been described by Tanemura *et al* (Ref 46).

Application to Service Rotors. Among the techniques described above, only the single-specimen Charpy technique and the small punch technique give the current FATT directly. All the other techniques only allow estimation of ΔFATT due to

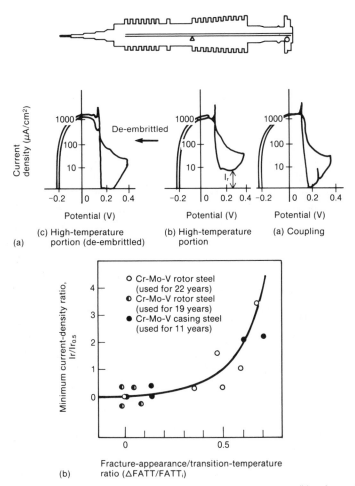

(a) Polarization curves from embrittled and nonembrittled locations in a rotor. (b) Relationship of polarization behavior to FATT ($I_r = 0.5$; 3.1 $\mu A/cm^2$; initial FATT$_i$ = 148 °C, or 298 °F).

Fig. 6.29. Electrochemical methods for detecting embrittlement (Ref 50 and 64).

service-induced embrittlement. These latter values, therefore, have to be used in conjunction with the FATT of the material in the virgin condition to calculate the current value of FATT. The preservice (virgin) value of FATT may be obtained from the manufacturer's records. It may also be estimated from empirical correlations such as

$$FATT = 6.9 + 2316P - 43Cr$$
$$- 108Mo + 0.3UTS \quad (Eq\ 6.10)$$

where FATT is expressed in °C, the compositions in wt%, and the ultimate tensile strength in MPa. The above correlation is based on regression analysis of data pertaining to 35 production rotors (Ref 65).

Once the current value of FATT at the critical location is known, it is then converted into a K_{Ic} value using any of the methods described previously. This value of K_{Ic} is then used in the remaining-life analysis. It is now becoming common practice to take ring samples during overboring or bottle boring of old rotors being subjected to life assessment. In these cases the limitation is only in terms of the number and size of samples available for FATT determination. Auger analysis, the single-specimen Charpy method, and the use of subsize Charpy specimens may be employed to circumvent this problem. Frequently, there is a need to estimate the FATT at the bore or the surface of a rotor when it is known at only one of these locations. It has been reported that, for numerous rotors forged in the 1970's, the average FATT at the bore is about 24% higher than the average surface FATT (Ref 30). Empirical relationships of this type are often used to estimate FATT values at inaccessible locations on the basis of tests on specimens from accessible locations. The limited data available in the literature have shown that in current Cr-Mo-V rotors the most severe embrittlement occurs at locations exposed to temperatures from 370 to 425 °C (700 to 800 °F) during operation (Ref 20 and 46).

Remaining Life Assessment Based on Crack Growth. The methodology for crack-growth analysis includes (1) time-independent fatigue, (2) time-dependent creep-crack growth, and (3) creep-fatigue interactions.

Time-Independent Fatigue-Crack Growth. At low temperatures where creep processes do not occur (below about 480 °C, or 900 °F) or at high frequency of thermomechanical cycling, crack growth is generally fatigue-dominated. The fatigue-crack-growth rate can be expressed by the Paris law (Eq 4.44):

$$\frac{da}{dN} = C'\Delta K^n \qquad \text{(Eq 6.11)}$$

in the region where $\Delta K > \Delta K_{Th}$. The ΔK value corresponds to the maximum value of K at the tip of a crack at start-up. The minimum stress is assumed to be zero in all cases. The value of C' varies with temperature but is relatively constant at a given value of T. Several investigators have characterized the Paris-law fatigue-crack-growth region in Cr-Mo-V rotor steels (Ref 20, 21, 29, 66, and 67). The constant C' is a function of both temperature and frequency. At a given temperature, C' increases with decreasing frequency (Ref 25). For instance, at 400 °C (750 °F), changing the frequency of fatigue from 1 Hz to 0.0017 Hz results in a change of C' from 8.2×10^{-10} to 30×10^{-4}. Values of n generally range from 2.7 to 3.7. Data from a variety of sources have been statistically analyzed by Ammirato *et al*, and the resultant curves at various temperatures and the mean curves based on the analysis are shown in Fig. 6.30 (Ref 68). At

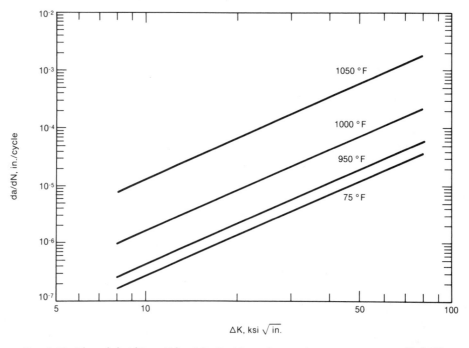

Fig. 6.30. Plot of da/dN vs K for 1Cr-Mo-V steel at various temperatures (Ref 68).

540 °C (1000 °F), the Paris-law relationship was found to be

$$\frac{da}{dN} = 6.57 \times 10^{-9} \, \Delta K^{2.35} \quad \text{(Eq 6.12)}$$

where ΔK is in ksi $\sqrt{\text{in.}}$ and da/dN is in in./cycle.

Time-Dependent Crack Growth. The time-dependent creep-crack-growth rate in Cr-Mo-V steels has been shown to depend on the crack-tip driving force C^* (identical to C_t under steady-state creep conditions) through the relationship

$$\frac{da}{dt} = bC^{*m} \quad \text{(Eq 6.13)}$$

Reference 69 provides a summary of a wide variety of da/dt-vs-C^* data for turbine rotor steels at 565 °C (1050 °F). Using these data and other data from Ref 70 and 71, Ammirato *et al* have suggested values

of m = 0.67 and b = 0.1294 (see Fig. 6.31; Ref 68), resulting in the relationship

$$\frac{da}{dt} = 0.1294C^{*0.67} \quad \text{(Eq 6.14)}$$

As described in Chapter 3, C^* is a function of the stress σ, crack size a, crack geometry, and creep rate $\dot{\epsilon}$. Ammirato *et al* reported that the stress and temperature dependence of $\dot{\epsilon}$ in Cr-Mo-V steels could be expressed by the relationship

$$\dot{\epsilon} = 2.74 \times 10^{-3} \, \exp(-63{,}600/T)\sigma^{10.5}$$

$$\text{(Eq 6.15)}$$

where $\dot{\epsilon}$ is in h^{-1}, T is in °R, and σ is in ksi. This relationship resulted in the following expression for C^*:

$$C^* = 0.006489 \, \exp(-63{,}600/T)\sigma^{11.5}a$$

$$\text{(Eq 6.16)}$$

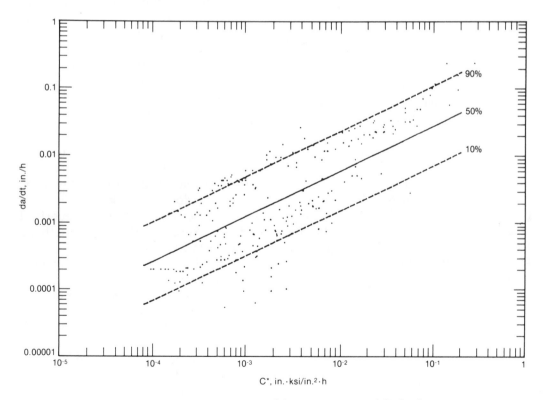

Fig. 6.31. Plot of da/dt vs C^* for 1Cr-Mo-V steel (Ref 68).

This equation was applicable to an edge crack in a half plane with uniform stress and was presumed to provide a conservative estimate for cracks of finite surface length in a stress gradient. The same C^* solution could also be used for buried cracks with an even greater degree of conservatism. In Eq 6.16, C^* is expressed in $in. \cdot ksi/in.^2 \cdot h$. The stress used in calculating C^* in the procedure followed by Ammirato *et al* was the steady-state stress in the uncracked rotor at the crack-tip location closest to the bore (i.e., maximum stress). The temperature used in the C^* calculation was the value at the crack-tip location farthest from the bore (i.e., maximum temperature). These precautions ensured that sufficient conservatism would be retained. Because C^* could be calculated analytically, with the stress, crack size, temperature, and creep behavior being known, and because the relationship between C^* and the crack-growth rate was known, the crack size as a function of operating time could be readily derived. In the computer code actually developed by Ammirato *et al*, the coefficients and exponents in the creep-rate equation (Eq 6.15) are used as default values. If these parameters specific to a rotor could be determined by creep tests, the actual values could then be used to derive a more accurate relationship for C^*. A case study of crack-propagation analysis at the blade-groove walls of a HP-IP rotor is described by Swaminathan *et al* (Ref 26).

Creep-Fatigue-Crack Growth. For turbine rotor steels, Saxena *et al* (Ref 72 and 73) and Swaminathan *et al* (Ref 74) have provided a comprehensive set of data on crack-growth rates with hold times ranging from 5 s to 24 h at 540 °C (1000 °F). Utilizing their data, Ammirato *et al* (Ref 68) have derived the following expression for crack growth under creep-fatigue conditions:

$$\frac{da}{dN} = 6.57 \times 10^{-9} \, \Delta K^{2.35}$$

$$+ 5.0 \times 10^{-7} \, \Delta K^{1.34} t_h^{0.33}$$

$$+ 0.13 C^{*0.67} t_h \qquad \text{(Eq 6.17)}$$

By substituting the appropriate ΔK and C^* expressions and the hold time, and by stepwise integration of the crack size from an initial size to the final size, the remaining life can be determined.

Life-Assessment Procedure. The life-assessment procedure under crack-growth conditions essentially consists of the following steps:

1. NDE inspection to determine the initial size a_i and the locations of flaws
2. Finite-element analysis to determine the stress and temperature distributions within the rotor for steady-state and transient conditions
3. Use of the applicable K expression for the flaw to determine the critical crack size a_c for the worst-case conditions
4. Separation of variables and integration of Eq 6.12 between the limits of crack sizes a_i to a_c to determine the number of cycles to failure, N_f, in fatigue
5. Computation of C^* or C_t using the expression appropriate to the crack geometry and substitution of the values of σ, A, and n
6. Separation of variables and integration of da between the limits a_i to a_c to determine the time to failure under creep conditions using Eq 6.16 or under creep-fatigue conditions using Eq 6.17.

Fracture-mechanics-based computer programs that include some or all of the above ingredients have been developed by many rotor-inspection organizations. One such program, known as Stress and Fracture Evaluation of Rotors (SAFER), is briefly described below (Ref 68).

A logic diagram of the major components of the SAFER code is shown in Fig. 6.32. The code integrates the inputs from ultrasonic bore NDE examination, stress analysis, and material properties to calculate the remaining life of the rotor. The rotor-failure mode addressed by the code is essentially growth of cracks by low-cycle fatigue or creep from flaws near the bore. The code does not address crack initiation, pri-

Inputs

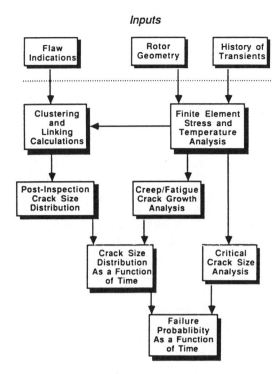

Fig. 6.32. Logic diagram for SAFER code for rotor-life prediction (Ref 68).

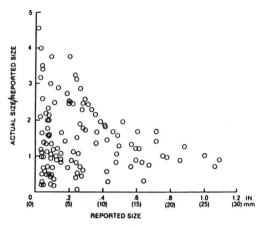

Fig. 6.33. Uncertainty in indication size as reported by NDE (Ref 21).

marily reflecting the dominant problem of link-up and crack extension among small indications. The program performs a cluster analysis by searching the nondestructive examination data and locating defects predicted to link together by either ligament yielding or stress rupture. SAFER calculates the temperature and stress distributions in the rotor, determines the crack growth from flaws under steady state or from repeated cycling, and computes the number of cycles or time required for the cracks to reach critical size on the basis of the fracture toughness of the rotor material.

To perform rotor analysis using SAFER, the engineer must first obtain the rotor geometry from drawings or actual measurements. The duty cycle is obtained from plant records where available. Material properties such as fracture toughness, yield strength, fatigue-crack growth, creep-crack growth, and creep-rate coefficients obtained from literature surveys, vendor's data, or supplemental tests on bore ring samples can be used, or the default values stored in the program for Cr-Mo-V material can be used. SAFER also contains steam properties and default heat-transfer boundary conditions. Any default values can be overriden by user input. To incorporate uncertainties in NDE, stress analysis, and material properties, a probabilistic version of the code is also available.

A major uncertainty in life assessment is the one arising from interpretation of NDE results. Regardless of how sophisticated the NDE procedure and equipment are, questions inevitably remain in regard to the exact size, shape, and type of the indications that are detected. In addition, NDE results cannot define the effects of material discontinuities on mechanical properties. Both of these aspects have been investigated by Schwant and Timo (Ref 21). The first aspect was investigated by comparing hundreds of sonic indications carefully mapped in rotors with the results of subsequent metallographic examination (see Fig. 6.33). These results show uncertainty factors as high as 5, particularly for small flaw sizes. Mechanical tests of regions containing defects (see Fig. 6.34) show that apparently innocuous and small defects can sometimes lead to severe reductions in fatigue life and associated scatter (Ref 21). Additional research programs, intended to build on this experience and to define the uncertainties in NDE results and mechanical properties using more retired rotors, are underway.

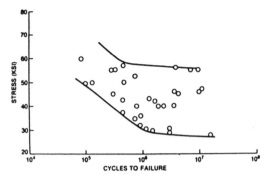

Fig. 6.34. Scatter in fatigue strength of LP rotor material due to forging tears (Ref 21).

Integrated Methodology. The various life-assessment methods previously described can be conveniently integrated into a phased approach consisting of the three levels of assessment described in Chapter 1 (Ref 75). An initial assessment is made on the basis of service history to identify rotors that require more detailed evaluation and to provide an equipment priority. This is followed by the application of methods that reduce conservatism of the initial assessment. This evaluation is described in more detail below.

The first step in level I assessment is the gathering of service information, such as unit running hours; number of hot, warm, and cold starts; past inspection results, failure history, if any; pertinent reports; past repairs and modifications; design parameters; temperature records; and vibration history. Significant questions to be answered are (1) whether the rotor has exceeded design temperature, load, or speed for long periods of time; (2) whether the foreseen service in life extension will exceed the design conditions; and (3) whether the failure history of the rotor has been significant. With the aid of the service history, the life exhausted under creep, fatigue, or creep-fatigue conditions, as appropriate, should be computed using the calculational procedures previously outlined, the standard low-cycle-fatigue, creep, and rupture curves, and the design or estimated values of temperatures and loads. If the remaining life fraction is equal to or greater than the ex-

pected life-extension period, no further assessment is required. The normal outage schedule for NDE examinations should be continued along with the maintenance of accurate logs. If the remaining life fraction is equal to or less than the life-extension period in future service, a level II assessment is required.

For level II assessment, two activities are ordinarily required: measurement of metal temperatures during transients and steady running; and inspection of the rotor. Metal temperatures will provide refined cycle-life-expenditure estimations and inspection will provide information about existing flaws. As a minimum, inspection should consist of visual examination, magnetic-particle testing, and ultrasonic testing of the bore. Peripheral and axial ultrasonic tests are also desirable. The use of "creep pips" attached to the rotors, which are markers by which the periodic strain can be measured directly (Ref 76), is an additional example of a level II monitoring technique. The inspection results are conservatively analyzed and applied to the assumed number of start-up cycles and load changes for the extended life period. The elements of conservatism are typically as follows:

1. A multiplier of 2 is applied to the sizes of all indications to account for inspection uncertainty.
2. Indications are assumed to be cracks in the radial-axial plane.
3. "Crack" growth caused by typical turbine operating cycles (average severity of about 0.02% surface life expenditure per cycle) is calculated by fracture-mechanics methods.
4. A critical crack size is calculated for a severe cycle—i.e., at the "bore limit" loading rate. The measured temperature-time profile and the combined thermal and centrifugal stresses are the data inputs. Fracture toughness (K_{Ic}) is estimated from correlations with impact energy or fracture-appearance transition temperature (FATT); the current FATT is derived from a calculated original FATT (from the

rotor composition) plus the embrittlement brought about by prolonged exposure at high temperature. Most of the information about changes in FATT with time resides in the manufacturer organizations. If that information cannot be obtained, rough estimates can be made through comparisons of composition and service time with measurements made on retired rotors or on rotors that are candidates for life extension (Ref 20, 30, 57, and 77).

The above calculations will provide an estimate of the rate of damage accumulation from thermal fatigue. Creep damage is evaluated as in level I, except that the measured temperatures are used. Expended life is the sum of the fatigue and creep components, and remaining life is calculated as before. If the remaining life is equal to or less than the life-extension period, a level III assessment is indicated.

The increase in accuracy for level III assessments is accomplished by way of refined stress analysis and measurement of the actual material properties, to establish where in the data population the specific rotor should be positioned. Measurements on trepanned samples, at locations specified by the refined stress analysis, will typically include toughness (K_{Ic}) or impact properties (FATT), fatigue-crack-growth rate (da/dN), and isostress creep-rupture properties (t_r). One overall scheme for remaining-life assessment based on these kinds of information is shown in Fig. 6.35. However, the evaluator can greatly accelerate the level III assessment by relying on the extensive work on rotor integrity carried out in developing the SAFER code (Ref 68). An example of level III analysis is given in Fig. 6.36 (Ref 78).

In spite of the large amount of research and development devoted to rotor evaluation in recent years, numerous uncertainties still remain; examples are the significance of creep-fatigue interactions at the rotor bore, flaw size and shape uncertainties, the ways in which flaws interact with one another, and the variability of material properties within a given rotor. The alternative to deterministic "best case/worst case" analysis is the probabilistic approach. A postprocessor code incorporating probabilistic considerations has been developed for use with SAFER or with any transient thermomechanical stress-analysis program. However, data from service-exposed material are lacking. Further improvements are now underway to resolve other shortcomings in the SAFER code (Ref 68).

Advanced Manufacturing Technologies

Prior to and during the 1950's, ingots for rotors were made in open-hearth furnaces, either acidic or basic, in air. Over the years, several major advances have been made in steelmaking technology including vacuum pouring and degassing, vacuum-carbon deoxidation, electric-furnace steelmaking, and secondary refining processes such as electroslag remelting (ESR). Such advances have significantly reduced the hazards of hydrogen cracking, major inclusions, segregation, and high residual-element contents.

Improved conventional practices include ladle desulfurization and vacuum-carbon deoxidation (VCD). In ladle desulfurization, molten steel is desulfurized in a ladle by use of calcium- and magnesium-base reagents. This results in very low contents of sulfur and associated manganese sulfide inclusions. In the VCD process, molten steel is deoxidized in the vacuum ladle after vacuum degassing, utilizing the carbon content of the steel itself for deoxidation. Because neither silicon nor aluminum is required for deoxidation, inclusion content and the risk of temper embrittlement due to silicon are reduced. The VCD process also results in reduced segregation at the center of the ingot.

Advanced melting and refining practices include electroslag remelting (ESR), electroslag hot tapping (ESHT), and central-zone remelting (CZR). In the ESR process, the ingot is used as a consumable electrode and is melted off through a refining slag. The ESHT process is a modification of ESR with the difference that molten slag is

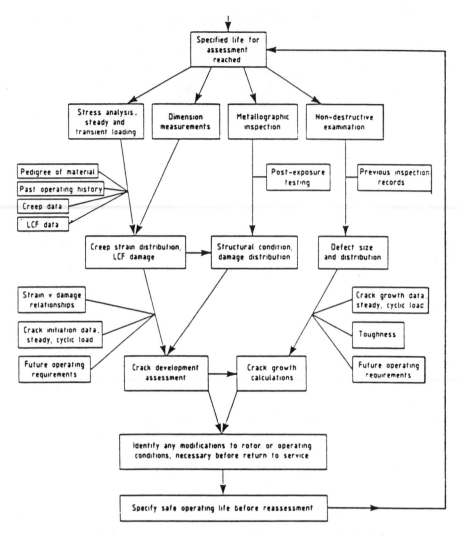

Fig. 6.35. Metallurgical aspects of remanent-life assessment for high-temperature rotor forgings (Ref 44).

added to the melt surface of the ingot electrode, which provides an ideal hot top. In the CZR process, the core of a conventional ingot is removed by trepanning. Using the cored ingot as a mold, the center is filled with high-quality electroslag metal by consumable-arc melting of an electrode under a refining slag.

Trends in impurity levels achieved in Cr-Mo-V rotor forgings over the last few decades are shown in Fig. 6.37 (Ref 20). To demonstrate the effectiveness of advanced steelmaking technologies, Swaminathan and Jaffee evaluated three full-size forgings of Cr-Mo-V rotors produced by low-sulfur

silicon deoxidation, vacuum-carbon deoxidation, and electroslag remelting processes (Ref 79). All these processes reduced sulfur to very low levels. As a result, excellent bore quality and cleanness were obtained. Compared with conventional forgings, the fracture toughness, creep-rupture strength, rupture ductility, and LCF strength were found to be superior, as shown in Fig. 6.38 (Ref 79). Considering the fact that, in general, improvements in rupture strength and ductility and in rupture strength and fracture toughness are mutually opposed, the simultaneous improvements achieved with respect to all of these properties must be

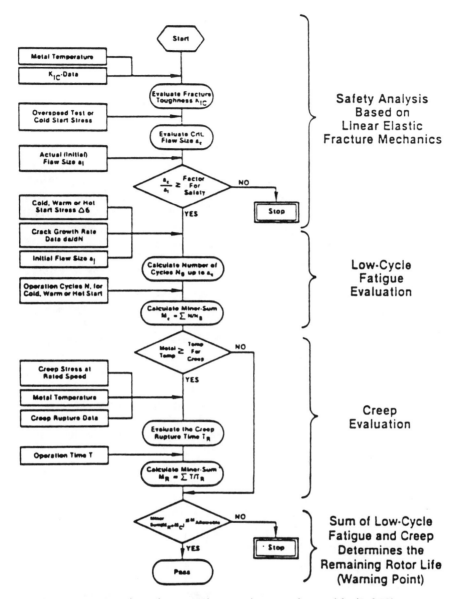

Fig. 6.36. Phased approach to evaluation of rotor life (Ref 78).

considered a major milestone in rotor technology.

Modifications of Heat Treatment and Alloy Content

In the past, one of the constraints on development of Cr-Mo-V steels with improved toughness has been the inability to make changes in heat treatment and alloy content without adversely affecting the creep and rupture strengths. It has been reported that as the rupture strength increases, FATT also increases. The best creep-rupture properties are achieved through a microstructure consisting of upper bainite. Unfortunately, lower bainite and martensite structures are preferable from a toughness point of view. Different rotor manufacturers have made their own trade-offs between toughness and rupture strength. The Europeans traditionally have oil quenched their rotors whereas the U.S. practice has been to air cool rotors. Systematic work by Berger has recently shown that oil quenching can improve

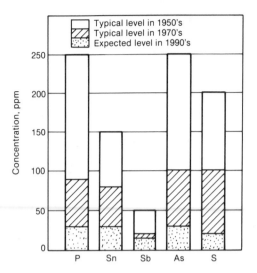

Fig. 6.37. Trends in impurity levels in Cr-Mo-V rotor steels (Ref 20).

toughness as well as the creep strength of rotor steels (Ref 80). Data from the oil-quenching experiments are included in Fig. 6.38 for comparison (Ref 79). Oil quenching coupled with minor alloy modifications has also been suggested as a way of maintaining both creep strength and toughness (Ref 81).

Casings

Steam-turbine casings are massive steel castings that encase the internal stationary and rotating components of the turbine. Casings have two critical functions: (1) containing the steam pressure and (2) maintaining support and alignment of the internal components.

Cracking of the casing can lead to steam leaks and, in extreme situations, to bursting. Casing distortion can cause damage by allowing contact between the stationary and rotating parts. Concerns about thermal distortion as well as thermal-fatigue cracking have increased in recent years due to the increasing use of older machines for cyclic operation. There are a number of additional reasons for the concern over the older casings made in the 1950's. They have accumulated longer service time and more start-stop cycles than more recent casings. The older casings were designed as pressure ves-

sels and had thick-wall sections that were more prone to thermal fatigue. They were manufactured from such alloys and heat treated in such a way as to increase strength at the expense of ductility. The steelmaking processes of the 1950's and early 1960's were not as refined as today's processes. Consequently, the steels made then contain much larger levels of impurities and inclusions than modern steels and are more susceptible to temper embrittlement and stress-relief cracking. Modern casings have the benefit of optimized steel compositions and improved foundry practices and design.

Damage Mechanisms in Casings

Damage in casings and remedial actions have been discussed in several recent papers (Ref 82 to 88). Damage generally consists of distortion and/or cracking. Distortion of casings can be either creep distortion or thermal distortion. Of these, the latter is the more severe. In design of casings, pressure containment requires thick, rigid walls, whereas thermal-gradient considerations require thin, flexible walls. The geometry therefore has to be optimized. Thermal distortions still occur due to rapid start-stop cycles, water induction, and other transient events. These generally can be prevented by strict adherence to the turbine manufacturer's operating guidelines. Distortion may exist at horizontal and vertical joint flanges, seal surfaces, nozzle and stationary diaphragm fits, etc. Creep distortion has been observed in a few high-temperature/high-pressure casings. In general, these dimensional changes can be accommodated by changes in alignment. In many instances of thermal distortion, the casing can be restored to its original condition by mechanical working followed by heat treatment (Ref 82 and 83).

Cracks in casings are typically located at the inlets of the HP and IP turbine sections, where the local thermal stresses are higher. Cracks have also been observed, although to a much smaller degree, at the inlet sections of LP hoods. Cracking in HP and IP sections is typically found on the in-

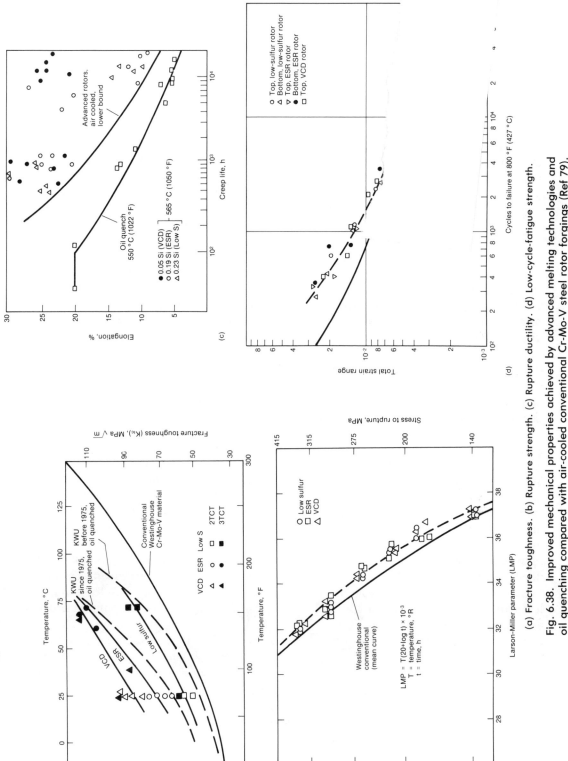

Fig. 6.38. Improved mechanical properties achieved by advanced melting technologies and oil quenching compared with air-cooled conventional Cr-Mo-V steel rotor forgings (Ref 79).

(a) Fracture toughness. (b) Rupture strength. (c) Rupture ductility. (d) Low-cycle-fatigue strength.

terior surfaces of steam chests, valve bodies, nozzle chambers, seal casings, diaphragm fits, and bolt holes (Fig. 6.39). Cracking in LP sections is typically found on the inner surfaces of inlet bowls, support struts, bolt holes, and diaphragm fits. Locations of cracking found in older casings have also been illustrated by Vogan and Morson (Ref 84). The causes of cracking in the HP and IP section casings and the relative percentages of these incidents according to Rasmussen (Ref 83) are as follows:

- Low-cycle/thermal fatigue, 65%
- Brittle fracture, 30%
- Creep, 5%

Some of the failures are combinations of the three modes listed above. Clearly, the most common type is low-cycle-fatigue cracking, resulting from repeated tensile and compressive thermal stress cycles. The cracks are transgranular and may be concentrated near stress raisers or spread over a larger area, such as the inside surface of a valve body. These cracks generally tend to grow slowly and can be found during the five-year inspection cycles, before they grow to critical size. If cracks are allowed to grow to critical size, then catastrophic brittle fracture can result. It is not uncommon to leave casings with low-cycle-fatigue cracks in continued service. If the casing material has sufficient fracture toughness, the fatigue cracks may never grow to critical size, because they may be arrested well before that, due to the decreasing through-wall thermal stresses.

Brittle cracks result from rapid crack growth occurring during transients, in casings with inherently low toughness or those whose toughness has been severely degraded due to temper embrittlement in service. The crack-growth rate during each cycle is related to the thermal stress level and hence the severity of the transient. In essence, these cracks are low-cycle-fatigue cracks but are growing at a much faster rate than the earlier category. These cracks can be transgranular or intergranular. They may be found as isolated cracks or in association with the slow-growing low-cycle-fatigue cracks. In view of the rapid crack-growth rates, the brittle form of low-cycle-fatigue cracking is the most severe form of cracking and, if found, requires immediate remedial action. The propensity of this form of cracking is clearly related to the FATT (K_{Ic}) of the steel. It has been reported that casings made in the 1950's which have been in service for more than 30 years can have FATT values as high as 230 °C (450 °F).

In most instances of cracking, the cracks are ground out and local repairs are performed. Repair procedures include grinding, welding, and metal stitching. The applicability of these procedures is based on metallurgical as well as economic considerations, as discussed in detail by Rasmussen (Ref 83). Weld-repair procedures vary widely in terms of filler metal, weld preparation, life expectancy, and postweld and preweld heat treatments. Because the entire field of service experience and repair technology for casings is reviewed elsewhere (Ref 82 to

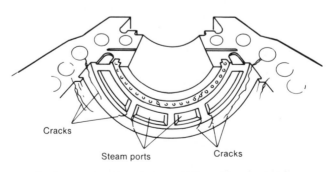

Cracks at various locations are indicated by wiggly lines.

Fig. 6.39. Crack locations in first-stage nozzle area of a steam-turbine casing (Ref 83).

88), only the essentials have been presented above.

Materials for Casings

Compositions and temperature ranges of application for casting-grade steels used for casings are listed in Table 6.4. The evolution of these steels with increasing unit size, and its implications for the foundry industry, have been discussed by Crombie (Ref 89). The thrust in the early years was toward improving creep strength to accommodate steadily increasing temperatures, which led to progressive changeovers from C-0.5Mo to 1Cr-0.5Mo steel and from 0.5Cr-Mo-0.25V to 1Cr-1Mo-0.25V steel. In the light of numerous instances of reheat cracking (stress-relief cracking) in weld heat-affected zones in the late 1960's and early 1970's, the importance of rupture ductility was realized. The material 0.5Cr-0.5Mo-0.25V was standardized by many manufacturers, but only after implementation of stringent specifications relating to control of residual elements (particularly phosphorus, antimony, tin, copper, aluminum, and sulfur), deoxidation practices, and welding procedures. Casing designs were also modified to eliminate manufacturing and in-service cracks. As an alternative to 0.5Cr-0.5Mo-0.25V steel, other manufacturers opted to use 2.25Cr-1Mo steel in the belief that this material had higher creep ductility, higher low-cycle-fatigue resistance, and better weld repair-

ability. Current designs use one or the other of these last two materials, depending on the turbine manufacturer. In the heat treated condition, the microstructures of casing steels consist of ferrite/pearlite mixtures. An improved 1¼Cr-½Mo steel composition with titanium and boron additions was investigated as a candidate for casing applications in the early 1970's (Ref 90). This steel has significantly higher tensile strength, rupture strength, and rupture ductility than those of standard 1¼Cr-½Mo steel, but its development has not been pursued further.

Remaining-Life Assessment

Remaining-life assessment of casings can be based on crack-initiation or crack-propagation criteria. Because casings are not highly stressed components, have large section sizes, and are made of relatively ductile materials, remaining-life assessments generally are based on crack-growth considerations. Crack-initiation considerations are useful at the design stage. For computing the initiation-based life, LCF data with appropriate correlations for hold time and strain concentrations are used in conjunction with the damage rules described in Chapter 4. A more common situation, however, is one in which a casing is found to contain many cracks and the plant owner has to decide how long to wait before repairing the casing. This situation involves crack-growth analysis. The material-prop-

Table 6.4. Typical compositions of steels commonly used for manufacture of turbine casings (Ref 89)

Material	Composition, wt %						Range of application, °C
	C	Si	Mn	Cr	Mo	V	
Flake-graphite gray cast iron(a)	3.00	1.80	0.70	. . .	. . .	. . .	400 max
Ferritic SGI(a)	3.22	2.94	0.25	. . .	. . .	. . .	. . .
Carbon steel	0.20 max	0.50 max	1.10 max	. . .	. . .	. . .	. . .
C-½Mo	0.25 max	0.20-0.50	0.50-1.00	. . .	0.50-0.70	. . .	480 max
Cr-½Mo	0.15 max	0.60 max	0.50-0.80	1.00-1.50	0.45-0.65	. . .	525 max
2¼Cr-1Mo	0.15 max	0.45 max	0.40-0.80	2.00-2.75	0.90-1.10	. . .	538 max
Cr-Mo-V	0.15 max	0.15-0.30	0.40-0.60	0.70-1.20	0.70-1.20	0.25-0.35	565 max
½Cr-Mo-V	0.10-0.15	0.45 max	0.40-0.70	0.40-0.60	0.40-0.60	0.22-0.28	565 max

(a) Typical composition.

erty information needed in this case includes (1) the critical crack size based on K_{Ic} or J_{Ic}, taking into account temper-embrittlement degradation specific to the casing material; and (2) fatigue-crack-growth-rate data with the appropriate hold-time correction. Frequently, casings are found to be quite ductile and to have rather large values of K_{Ic}. In such cases, the size of the remaining ligament may govern failure rather than the critical crack size based on fracture-mechanics considerations.

Logsdon *et al* have described an excellent case study of remaining-life analysis for an SSTG (ships service turbine generator) steam-inlet upper casing which contained numerous cracks in the inlet steam passageways and the steam chest (Ref 91). The casing was made of a 2¼Cr-1Mo steel casting and had been in service for nearly 27 years, at temperatures up to 540 °C (1000 °F). Detailed mechanical tests were carried out on samples removed from the casing. Some of the highlights from this mechanical test program were as follows. (1) The steel was found to cyclically harden at all test temperatures up to 540 °C (1000 °F). (2) The creep-rupture properties of the steel were similar to those reported in the literature for low-carbon (0.009%) grades of 2¼Cr-1Mo steel. (3) The fracture toughness of the steel was high, with a tearing modulus greater than 100 even at 24 °C (75 °F). Hence, the critical crack size was taken to be either the ligament size between holes or the crack size at which K_p, the stress intensity due to steam pressure, started to increase suddenly, depending on the crack orientation. (4) Fatigue-crack-growth rates under load control and displacement control were comparable. The crack-growth rates increased with temperature, and could be described by the Paris law, $da/dN = C'\Delta K^n$, where the values of C' and n were as given in Table 6.5. (5) The effect of hold time t_h on the fatigue-crack-growth rate in steam at 540 °C (1000 °F) could be described by the equation

$$\frac{da}{dN} = 4.87 \times 10^{-11}(\Delta K)^{3.8}$$

$$+ 3.81 \times 10^{-6}(\Delta K)^{0.94}t_h^{0.53}$$

$$(\text{Eq } 6.18)$$

where da/dN is expressed in in./cycle and ΔK in ksi$\sqrt{\text{in}}$. Figure 6.40 shows the actual data and the fitted curve. The remaining life N from an inspection crack size to a final crack size a_f could be obtained by integration of the above equation. This procedure is very similar to one illustrated in Chapter 4. The essential steps in remaining-life analysis are: (1) determination of initial crack size a_i by inspection; (2) stress analysis to determine the distribution of stresses and temperatures for steady operation and for various transients; (3) computation of ΔK corresponding to the peak stresses for the various transients, crack size, and geometry; (4) estimation of the tolerable final crack size a_f based on FATT and K_{Ic} tests on samples removed from the casing, or

Table 6.5. Values of C' and n in fatigue-crack-growth relation for 2¼Cr-1Mo steel (da/dN = C'ΔKⁿ) (Ref 91)

Source	Temperature °C	Temperature °F	C' (cm/cycle, MPa$\sqrt{\text{m}}$)	C' (in./cycle, ksi$\sqrt{\text{in.}}$)	n
Logsdon *et al* (Ref 91)	24	75	1.5×10^{-10}	8.1×10^{-11}	3.3
	427	800	9.5×10^{-10}	5.0×10^{-10}	2.9
	538	1000	3.2×10^{-9}	1.7×10^{-9}	2.8
Gudas (Ref 92)	24	75	4.5×10^{-10}	2.5×10^{-10}	3.3
Hackett (Ref 93)	24	75	1.0×10^{-10}	5.8×10^{-11}	3.6
Logsdon (Ref 94)	24	75	1.5×10^{-10}	8.5×10^{-11}	3.3
Brinkman *et al* (Ref 95)	24	75	5.1×10^{-11}	2.9×10^{-11}	3.7
	510	950	1.4×10^{-8}	7.0×10^{-9}	2.2

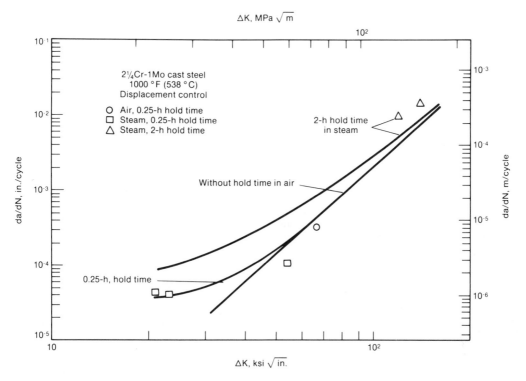

Fig. 6.40. Effect of hold time on fatigue-crack-growth-rate properties of 2¼Cr-1Mo cast steel (Ref 91).

based on ligament fracture or other criteria depending on crack size, crack location, and the toughness of the steel; and (5) integration of the ΔK expression (Eq 6.18) within limits of a_i and a_f. Data on service-exposed casings and additional case studies of remaining-life analysis may be found in Ref 96 and 97.

A generic methodology based on the phased three-level approach has been described by McNaughton *et al* (Ref 75). Stresses in turbine casings or cylinders are generated by the internal pressure of the steam, and by the transient thermal gradients that accompany start-ups, major load changes, and shutdowns. Thus, the assessment procedure must be capable of accounting for damage accumulation by low-cycle fatigue, creep damage sustained during steady running, the combined effects of creep and fatigue, and the degradation in toughness brought about by segregation during long-term exposure to high temperatures. Unlike rotors, casings are not axisymmetric rotating elements, and they

are castings rather than forgings. Therefore, evaluation does not lend itself so readily to bounding analysis. Furthermore, most of the experiential information from which correlations could be developed is held by manufacturers.

Level I assessment consists of acquiring and using the following information:

- Design drawings and as-built measurements
- Design-stress calculations
- Manufacturing records, particularly of deviations and repairs
- Records and drawings of modifications made during service
- Other service records of problems and repairs
- Starting and loading procedures throughout the service life
- Original material properties
- Steam-temperature records.

With this information it is possible to perform a conservative assessment of life expenditure in fatigue and creep. Whenever possible, such evaluations should be compared with the manufacturer's statistical records of problems and failures in the fleet of similar machines. Inasmuch as the methodology for life assessment at level I has not been established to the degree that it has for some other components, a utility embarking upon a life-extension program will find it advantageous to inspect casings at the earliest opportunity.

Level II in the generic scheme involves thorough NDE or inspection of the inner casings and outer shells, and measurements of metal temperatures during transients and at steady state. Stress calculation in closed form may not be appropriate at this stage, and finite-element methods may have to be invoked. If the remaining life calculated in this stage is inadequate to fulfill the intended life extension or if a crack is found, level III assessment must be conducted on the basis of actual material properties measured on samples taken from critical locations. Properties to be measured are strength, brittle-fracture resistance (K_{Ic} or FATT), isostress creep-rupture times, and minimum strain rates in creep. Examples of level III assessment for inner and outer cylinders are given in Ref 91 and 96 to 98.

Valves and Steam Chests

The principal damage mechanisms in valves and steam chests are low-cycle fatigue, creep, erosion, and wear. Valves and steam chests are classified as critical components. Valves must function reliably under severe service conditions or a catastrophic overspeed accident can occur. Steam chests and valve casings are pressure vessels operating at high temperature and pressure; the consequences of failure involve both safety and outage considerations. Low-cycle-fatigue damage is caused by thermal gradients during transients such as start-ups and load changes, whereas creep damage accumulates during steady-state conditions at full load. Damage to valve seats is usually the result of erosion by steam-borne particles; stems, bushings, and seal rings are degraded by wear brought about by relative motion between contacting surfaces.

Remaining-life assessment of valves and steam chests is conducted in very much the same way as for turbine casings. Manufacturing and service records are essential; the kinds of information required for creep and fatigue damage at each of the assessment stages have already been described. Erosion and wear damage cannot, in general, be predicted from design parameters, and the rate of damage accumulation can only be estimated from sequential inspections. Examples of valve and steam-chest evaluations are given in Ref 98.

Steam-Turbine Blades

The function of turbine blades (also called buckets) is to convert the available energy in steam into mechanical energy. Blades are designed to serve reliably for 30 years or more. The design system used for blades must take into account the material properties, the anticipated duty requirements, and the environment. There are significant differences in the design considerations used for short, first-row blades and for long, latter-stage blades. In LP turbines, blade problems arise primarily from corrosion fatigue due to impurities condensed from the steam. LP turbine blade problems have been described adequately elsewhere (Ref 99) and will not be considered here.

Design Considerations

The primary forces acting on a blade are centrifugal tensile forces due to rotation and bending forces due to the steam. In addition to the steady forces, the blades are also exposed to oscillating forces, nonuniform steam flow at the nozzle exit, and other sources of nonuniform flow that provide fatigue stimuli. First-stage blades are also exposed to a very complex stimulus during partial steam admission, and blades must withstand occasional overspeed stresses and fatigue stresses due to start-stop transients.

The components of a short-blade assembly can be broken down into four categories as shown in Fig. 6.41 (Ref 100). These categories are the cover, the tenons, the vanes, and the dovetails. The cover serves as the sealing surface for the radial steam seals and structurally couples the blades together. The tenons, which protrude through holes in the covers and are peened to form heads, serve the function of attaching the blades to the cover. The tenons must be ductile enough to withstand high plastic strains due to peening and must also withstand the centrifugal forces of rotation and steam bending forces acting on the blades. The blades themselves are designed to efficiently convert steam energy into mechanical energy. The blade dovetails serve the function of attaching the blades to the rotor disk. They also comprise the load path for the mechanical power generated in the steam path. Blade dovetails are exposed to high centrifugal loads and to steam bending loads. Two types of dovetails used for attaching the blades to the rotor disk are illustrated in Fig. 6.42.

Blade Materials and Properties

Steam-turbine blades are made from forged AISI types 403, 410, and 422 stainless steels containing 12% chromium. Addition of molybdenum, tungsten, and vanadium, such as in type 422, leads to increasing creep strength. These steels generally are austenitized, quenched, and tempered to achieve the desired strength. Types 403 and 410 generally are austenitized at 955 to 1010 °C (1750 to 1850 °F), whereas the more highly alloyed type 422 is austenitized at 980 to 1050 °C (1800 to 1925 °F). The steels are tempered in the range 565 to 705 °C (1050 to 1300 °F) to achieve hardness levels from 24 to 32 HRC. The microstructure of the fully heat treated product consists of virtually all tempered martensite.

The principal damage mechanisms of concern are high-cycle fatigue and creep-rupture. In particular, the stress-concentrating effects of the dovetail grooves are of great importance. Because creep-rupture at the grooves might be promoted by low rup-

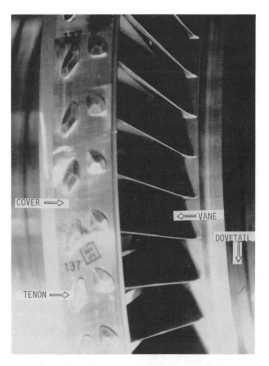

Fig. 6.41. Typical short turbine buckets after final assembly (courtesy of G.A. Cincotta, General Electric Co., Schenectady, NY) (Ref 100).

ture ductility, rupture ductility is a vital parameter. Design of the airfoils is based on HCF data of the type shown in Fig. 6.43 (Ref 101). Notched-bar testing as well as full-scale blade testing are invariably employed by the blade manufacturers to develop appropriate correction factors for the dovetail groove area. Because fatigue is of great concern, considerable caution is employed during manufacture to avoid any potential surface sites for crack initiation such as pits, dents, and inclusions. The sulfur content is stringently controlled to minimize inclusions. The ability to tolerate occasional and inadvertent surface defects as well as foreign-object damage is also important, and hence adequate toughness for the blades must be ensured. Freedom from in-service temper embrittlement must be ensured by control of the residual-element content. Electroslag refining (ESR) has been employed by some manufacturers for control of sulfur content and for

Fig. 6.42. Tangential-entry (left) and axial-entry (right) blade-attachment schemes (courtesy of G.A. Cincotta, General Electric Co., Schenectady, NY) (Ref 100).

improved toughness. More complete data regarding heat treatment and mechanical properties of types 403, 410, and 422 stainless steels may be found in Ref 101.

Remaining-Life Analysis of Blades

The major concern with respect to steam-turbine blades is the failure of blades in the last or next-to-last stage of the low-pressure turbine by corrosion fatigue. The only failure mechanisms applicable to blades in the high-pressure and intermediate-pressure turbines are creep, high-cycle fatigue, and temper embrittlement. In all instances, the initiation of cracks or cracklike defects is the failure criterion employed.

Rieger and McCloskey have summarized the various factors affecting the lives of blades (Ref 102). According to them, the basic parameters which affect blade life are:

1. Steady-state stresses resulting from centrifugal loads, steam bending loads, thermal gradients, assembly toler-

ances, geometric untwists, attachment tolerances, attachment constraints, and tenon cold working

2. Dynamic stresses caused by nonsteady steam forces, nozzle wakes, thermal transients, start-stop transients, sequential arc operation, per-revolution diaphragm harmonics, and flow instabilities. These stresses also include the effects of geometric stress raisers such as those occurring at root attachments, vane-platform fillets, tie-wire fillets or hole edges, and cover or attachment discontinuities, as well as the effects of residual preloads from riveting and assembly tolerances.

3. Structural stiffness and mass properties of the blade shape resulting in blade and blade-group material frequencies and the proximity of these frequencies to any strong exciting harmonics of rotational speed such as the nozzle-passing frequency (NPF)

4. Material properties such as ultimate

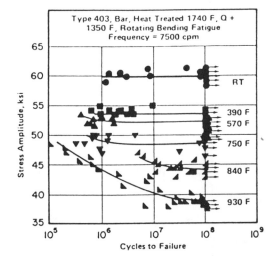

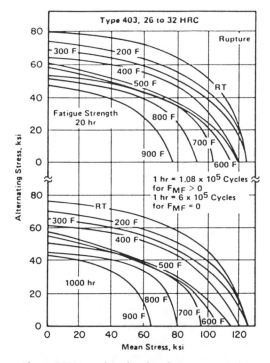

Above: S-N curves from bending-fatigue tests. Below: Stress-range diagram for AISI type 403 stainless steel.

Fig. 6.43. High-cycle-fatigue data for blade steels (Ref 101).

strength, fatigue strength, creep strength, toughness, system damping properties, and corrosion and erosion resistance

5. Load-history details resulting from centrifugal overloads, start-up overspeeds, and operational start-stop

cycles; and from the per-revolution excitation spectrum arising from the diaphragm flow distribution, including NPF excitation, sequential arc admission, and flow instabilities.

The combined effect of steady stresses and dynamic stresses can, in principle, be depicted via the Goodman-type diagram for type 403 stainless steel in Fig. 6.43 (Ref 101 and 103). Goodman diagrams, however, are unable to show how the combined effect of several exciting frequencies influences fatigue life. The cumulative damage under a spectrum of loading conditions can be more readily calculated using the linear damage-summation procedure of Palmgren and Miner described in Chapter 4.

The number of cycles to crack initiation has been found to obey the relationship (Ref 104 and 105)

$$\frac{\Delta\epsilon}{2} = \frac{\sigma_f'}{E}(2N_f)^b + \epsilon_f'(2N_f)^c \quad \text{(Eq 6.19)}$$

where $\Delta\epsilon/2$ is the total true strain amplitude (not engineering strain) for a given event at the crack location, σ_f' is the strength coefficient, E is the elastic modulus, ϵ_f' is the ductility coefficient, and b and c are exponents. Note that this equation is very analogous to the Coffin-Manson relationship as modified by Manson (see Eq 4.11, Chapter 4). The values of E, σ_f', b, ϵ_f', and c for type 403 stainless steel at 335 °C (635 °F) have been reported to be 28,000 ksi, 131 ksi, −0.083, 0.381, and −0.58, respectively (Ref 106). The elastic and plastic contributions of strain to fatigue life and the cumulative fatigue life based on this relationship are shown in Fig. 6.44.

The cumulative damage in blades occurs where the load history consists of more than one harmonic component. Special cycle-counting procedures are needed where two or more harmonic components occur simultaneously. The number of the various "events" leading to dynamic loading and the strain amplitude corresponding to each of these events must be sorted out from the loading history. The fractional life con-

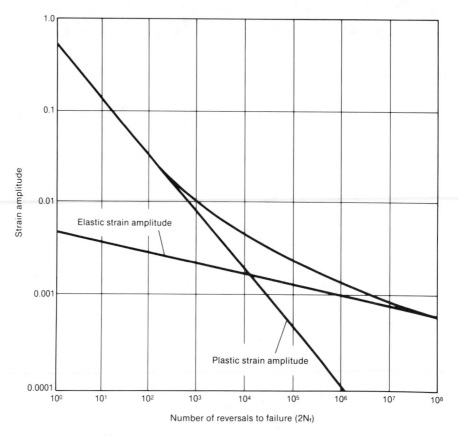

Fig. 6.44. Strain life vs number of cycles for AISI type 403 stainless steel (Ref 106).

sumed due to each type of event can then be determined as N/N_f by entering the low-cycle-fatigue curve at the appropriate value of $\Delta\epsilon$. Summation of the fractional damages due to the various types of events leads to the cumulative damage. A further refinement of the procedure consists of using a bilinear damage curve as shown in Fig. 6.45 (Ref 107). A formalized code for computing steady and dynamic stresses for various blade geometries and for estimating cumulative damage, known as BLADE (Ref 108), has been developed by Rieger *et al.* Due to uncertainties in the knowledge of material properties, operating history, and stresses, Rieger and McCloskey estimate that, with the current state of knowledge, blade-life estimations are possible only within a factor of 3 to 4 compared with the actual life. In view of this, detailed analyses of fatigue stresses, creep stresses, and their influences on crack initiation often are used

for optimization and relative evaluation of blade designs, but rarely for life prediction. This procedure, however, may have value in estimating inspection intervals in those instances where crack initiation occurs on the inside surfaces of the blade roots and NDE inspection is not possible without blade removal. Current procedures for blade replacement are based purely on inspection. The blades are inspected using magnetic-particle, dye-penetrant, eddy-current, or ultrasonic techniques. If any cracks are detected, the blade is replaced. An electrochemical technique similar to the one described earlier for rotors has been developed by Shoji and Takahashi (Ref 64) to detect temper-embrittlement degradation of blades. In cases of extreme degradation of toughness, the critical crack sizes may be smaller than those detectible by NDE techniques so that rapid brittle failure may occur even in the absence of detectible cracks.

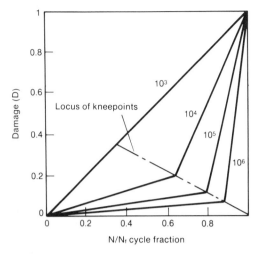

Fig. 6.45. Bilinear damage curves for failure at different cycles (Ref 107).

In such cases, the temper-embrittlement probe can be a valuable complementary NDE tool.

Diaphragms and Nozzle Boxes

Nozzles are stationary components whose essential function is to direct the flow of steam to the rotating blades at the appropriate angle (Ref 109). The nozzle airfoils are welded onto semicircular flat plates in older designs. The modern nozzle box consists of two semicircular pressure vessels with large integral steam paths, as shown in Fig. 6.46 (Ref 109). The nozzles themselves either are investment castings or are electrodischarge machined from solid forgings. The completed steam path is then welded into the box. Modern nozzle boxes are made from 12% Cr forgings. Nozzle diaphragms are manufactured in several ways depending on their size. The smaller ones are fabricated with nozzle partitions inserted in a punched band. Intermediate-size nozzle diaphragms are made of cast steel, and the latter-stage diaphragms of the turbine are fabricated by welding the airfoils directly to fabricated rings and webs. Maximum dimensional conformity to the design is essential in manufacturing the steam-path components. Whereas the diaphragm airfoils are made from 12% Cr steels (AISI type 403), ring materials are made from

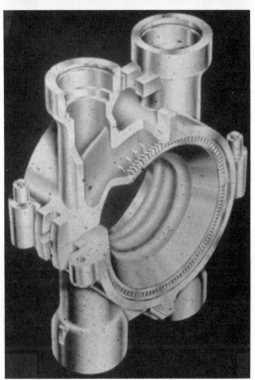

(Top) Typical diaphragm employed in older designs. (Middle) Nozzle plate. (Bottom) Double-flow nozzle box.

Fig. 6.46. Diaphragm and nozzle-box designs (courtesy of G.A. Cincotta, General Electric Co., Schenectady, NY) (Ref 109).

low-alloy steel plates, depending on the operating temperature.

It is critical to integrate the nozzle-diaphragm design with that of the bucket design to achieve maximum turbine efficiency. The important design parameters are the number of nozzles, nozzle setback, sidewall angle and location and partition contour. These factors also affect the reliability of the buckets. Each stage has to be carefully chosen so that a natural bucket resonant frequency is not coincident with damaging stimulus frequencies.

Remaining-Life Assessment

Because diaphragms and nozzle boxes are stationary components, they are tolerant to cracks. Hence, crack initiation is not the usual criterion for retirement, and these components often are allowed to continue in service with cracks present. The principal damage mechanisms are creep and thermal distortion of the nozzle chambers and cracking of the nozzle roots due to creep and thermal fatigue (Ref 109). Distortion problems are generally corrected by realignment or remachining. When cracks reach sufficient depth, weld repair is performed. When cracks are no longer repairable, the frequency of repair renders it uneconomical, or distortions become too severe, the nozzles are retired. Under these circumstances, no detailed remaining-life-assessment methodology is generally applicable.

Solid-Particle Erosion

A major service problem related to the first few stages of nozzles and blades in the HP-IP sections is the problem of solid-particle erosion. The principal erodent is the hard magnetite scale from the boiler tubes, which exfoliates during start-stop transients and enters the turbine. The damage to the nozzles and blades reduces turbine efficiency and warrants frequent repair/replacement of these components. The results of an EPRI study indicate that the average yearly cost of erosion damage in the United States is 70¢ per kilowatt (Ref 110). An ASME-ASTM-MPC study found that total SPE damage in the United States costs $100 million annually (Ref 111). The erosion occurs on the blade-vane leading edges, the cover, and the tenons, and is caused by exfoliation of scales from the boiler tubes, particularly during transient conditions. Approaches to solving this problem have included use of bypass systems that enable the steam to bypass the turbine during transients, use of chromizing and chromate treatments in the boiler to prevent exfoliation, and use of erosion-resistant coatings on the blades and vanes.

Results from laboratory evaluations of the corrosion resistance of a variety of coatings have been described by Quereshi, Levy, and Wang (Ref 112). They reported that four coating processes—detonation-gun coating, cladding using a method called Conforma-clad, chemical-vapor deposition, and boride diffusion coating—were among the best performers. A similar comparative study of several coating processes, including plasma spraying with Cr_3C_2, chromizing, VC diffusion coating, Cr_2O_3 coating, Cr_3C_2 diffusion coating, and boride diffusion coating, has been conducted by Watanabe, Ikeda, and Miyazaki (Ref 113). The boride diffusion coating appeared to be the best. They also subsequently installed boride-coated nozzles in many operating units and found the results to be promising. Sumner, Vogan, and Lindinger have described the General Electric Company's evolving experience with coatings (Ref 114). Plasma-sprayed WC coatings were applied to numerous first-stage and reheat-stage buckets in a number of units. Although the WC coatings reduced the erosion damage, the need for a more satisfactory coating led to further development of plasma-sprayed Cr_3C_2 coatings and boride diffusion coatings. Comparative performances of the coated specimens relative to the uncoated type 403 stainless steel specimens in laboratory tests are shown in Fig. 6.47. The diffusion coating was found to be about 30 times better, and the plasma-spray coating about 20 times better, than the uncoated steel (Ref 114 and 115). The plasma-spray coating of Cr_3C_2, which was 205 to 305 μm (8 to 12 mils) thick, was deemed to be the most

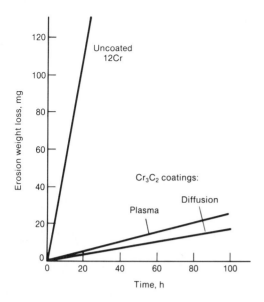

Note: At end of 100-h test
- 80% of original plasma coating present
- <10% of original diffusion coating present

Fig. 6.47. Results of laboratory solid-particle erosion tests (Ref 114 and 115).

applicable to all the steam-path components with the exception of the nozzle boxes and bolted-on nozzle plates used in the first stages of older designs. Because of limited access to the surface requiring the coating—i.e., the trailing-edge pressure surface—the line-of-sight plasma-spray coating could not be used for coating the nozzles. In these cases the boride diffusion coating was deemed desirable. Limited field experience with both types of coatings has shown considerable reduction in erosion damage (Ref 116). It appears that boride diffusion coatings applied by pack cementation and Cr_3C_2 coatings applied by plasma spraying will continue to be the industry's top choices, with minor variations in composition and processes depending on the vendor. Mechanisms of SPE and additional results from coating evaluations have also been described by Veerport *et al* (Ref 117).

High-Temperature Bolts

A steam turbine has many bolted joints that need to be separated for maintenance or repair. According to one estimate (Ref 118), in a typical 500-MW machine there are about 700 bolts and studs in diameters from 40 to 150 mm (1.6 to 5.9 in.) and lengths from 200 mm to 1 m (7.9 to 39.4 in.) operating in the creep range at temperatures up to 570 °C (1060 °F) in joints which have to withstand steam pressures up to 16 MPa (2.3 ksi). The main requirement of bolts in the creep range is to maintain steamtight joints without relaxing or fracturing.

Joint Tightness

From a joint-tightness point of view, the most important property for design of bolts is the stress-relaxation characteristics of the bolt material. For a given joint, the load required to exceed the steam load is applied to the flange area by tensile loading of bolts. In-service creep, however, causes relaxation of this load. The elastic strains produced by initial tightening of the bolts are progressively converted to creep strains, thereby reducing the effective load on the joint. The stress-relaxation characteristics must be such that even when the stresses are relaxed, they are still in excess of the design stress. Depending on the material and duty requirements, bolts are usually tightened to a predefined cold strain, say 0.15%, from which the initial load (or stress) on the bolt can be calculated. Based on stress-relaxation data at the operating temperature, the relaxed stress corresponding to the retightening intervals (i.e., at the end of about 30,000 h) is calculated. This stress should meet or exceed the design stress. With regard to joint life, an idealized situation is assumed in which overhauls, and therefore joint dismantling, take place at fixed intervals (say 30,000 h) between which no retightening is carried out. Ideally, to avoid replacement of bolts, at least five or six retightening operations would be carried out on each bolt during the life of the turbine. The creep relaxation of a bolt reused in this way must continue to meet the design requirement—i.e., the relaxation behavior for each subsequent tightening cycle of 30,000 h must be such that the residual stress on the bolt does not drop below the design stress at any time.

Typical compositions of bolt steels, their properties, and various national standards pertaining to them are summarized in Tables 6.6 and 6.7, from the work of Everson, Orr, and Dulieu (Ref 119). For convenience, the Central Electricity Generating Board of the United Kingdom has classified these compositions into groups numbered from 1 to 8. The stress-relaxation behavior of several of these groups of steels is illustrated in Fig. 6.48 (Ref 118). The stress values in Fig. 6.48 are the residual stresses at the end of 30,000 h following an initial cold prestrain of 0.15%. From these data alone, the nickel-base alloy seems to have the best resistance to stress relaxation, followed by a 12% Cr alloy and a 1Cr-1Mo-0.75V-Ti-B alloy. In actual selection of bolt materials, the compatibility of the thermal-expansion coefficient of the bolt with respect to the joint, as well as its susceptibility to various fracture mechanisms, have to be taken into account.

In the absence of sufficient stress-relaxation data, guidance can be obtained from the conventional creep-rupture data, because useful correlations exist between the two parameters, as illustrated in Fig. 6.49 for two bolt steels (Ref 118). The stress-rupture strengths of a number of bolting alloys are compared in Fig. 6.50. Their performance is in the same order as in terms of the stress relaxation data.

Bolt Fracture Mechanisms

The fracture of a bolt can lead to more drastic consequences than the relaxation of a bolt, which results only in leakage. Fracture of bolts can result from metallurgical causes, poor design, or poor tightening procedures. The two most important failure mechanisms for bolts are creep-rupture and brittle fracture. In plants which are subject to daily start-stop cycles, low-cycle fatigue can also result in crack initiation. These damage mechanisms are discussed below.

Creep-Rupture. Fracture of a bolt can occur if the local creep strain reaches the

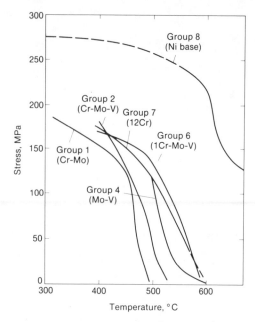

Fig. 6.48. Comparison of stress-relaxation behavior after 30,000 h as a function of temperature for various bolt materials subjected to a cold prestrain of 0.15% (Ref 118).

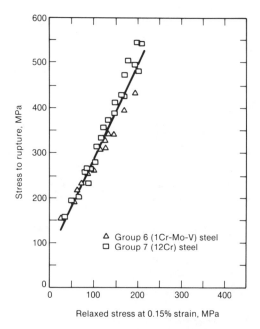

Fig. 6.49. Relationship between relaxed stress and rupture strength at the same duration for times from 1000 to 30,000 h and temperatures from 475 to 600 °C (885 to 1110 °F) (Ref 118).

Table 6.6. Commonly used ferritic alloy bolt steels (Ref 119)

Specification	Grade	Typical composition, % C	Cr	Mo	V	Others	Heat treatment, °C	RT tensile strength(a), MPa	Rupture(a)(b), 10^4 h, at temperature (°C) of: 500	550	600
United States: A193-84a											
	B7	0.4	1	0.2	...	...	T ≥ 593	≥790	NA	NA	NA
	B7M	0.4	1	0.2	...	...	T ≥ 620	≥690	NA	NA	NA
	B16	0.4	1	0.6	0.3	...	T ≥ 650	≥760	NA	NA	NA
Germany: DIN 17240											
	21CrMoV57	0.2	1.2	0.7	0.3	...	930 OQ + 720	700-850	271	170	...
	40CrMoV47	0.4	1.2	0.6	0.3	...	930 OQ + 700	850-1000	271	170	...
	X22CrMoV121	0.2	12	1	0.3	0.5 Ni	1050 AC + 680	900-1050	368	222	108
France: NF A 35 558											
	25 CD 4	0.25	1	0.2	...	...	880 OQ + ≥600	600-750	176	79	...
	20 CDV 5.07	0.2	1.3	0.7	0.25	...	930 OQ + ≥600	700-850	271	170	...
	Z20 CDNbV11	0.2	11	0.7	0.2	0.4 Nb	1120 AC + ≥670	880-1030	392	274	157
Russia	20Kh1M1F1TR	0.2	1.2	1	0.9	Ti, B	1000 Q + 700	NA	NA	NA	NA
Australia	Comsteel 029	0.2	1	1	0.6	Ti, B	990 OQ + 700	820-1000	NA	NA	NA

(a) NA = not available. (b) All values in MPa.

Table 6.7. United Kingdom ferritic bolt steel designations, heat treatments, and properties

Steel designations(a) BS 1506:1986	CEGB group	Nominal composition, % C	Cr	Mo	V	Others	Typical heat treatment, °C	RT tensile strength, MPa	Strength, MPa Rupture (c)	Rupture (d)	Relaxation(b) (c)	Relaxation(b) (d)
631-850	GP 1 (Cr-Mo)	0.4	1	0.5	...	...	870 OQ + 660	850-1000	234	97	81	...
671-850	GP 2 (Cr-Mo-V)	0.4	1	0.5	0.25	...	950 OQ + 700	850-1000	324	151	83	...
...	GP 3 (3Cr-Mo-V)	0.3	3	0.5	0.75	...	...	...	...	...	...	...
...	GP 4 (Mo-V)	0.2	...	0.5	0.25	...	...	...	...	...	...	...
...	GP 5 (1Cr-Mo-V)	0.2	1	1	0.75	Ti, B	(e)	(e)	(e)	(e)	(e)	(e)
681-820	GP 6 (1Cr-Mo-V)	0.2	1	1	0.75	...	980 OQ + 700	820-1000	418	280	141	70

(a) BS = British Standard. CEGB = Central Electricity Generating Board. (b) Stress relaxation for 0.15% strain. (c) Values at 10^4 h, 500 °C. (d) Values at 10^4 h, 550 °C. (e) Superseded by Durehete 1055.

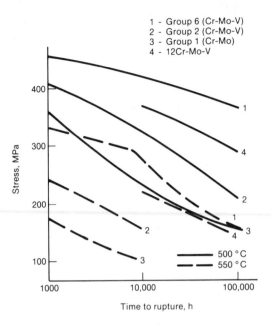

1 - Group 6 (Cr-Mo-V)
2 - Group 2 (Cr-Mo-V)
3 - Group 1 (Cr-Mo)
4 - 12Cr-Mo-V

Fig. 6.50. Rupture strengths of bolt steels (Ref 119).

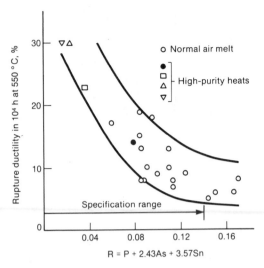

$R = P + 2.43As + 3.57Sn$

Fig. 6.51. Relationship between residual-element content and ductility (Ref 119).

creep ductility of the material from which the bolt is made. During every retightening cycle, stress relaxation can take place, leading to accumulated creep strain. Hence, the rupture ductility as measured in long-duration stress-rupture tests is a very important consideration. In addition, notched-bar tests are also carried out in the laboratory to ensure freedom from notch sensitivity. Obviously, low rupture ductilities lead to notch sensitivity, and the long-term values of neither of them can be predicted on the basis of short-term tests. Numerous failures of 1Cr-1Mo-V steel bolts were attributed to notch-sensitive failures which subsequently resulted in a new class of more ductile Cr-Mo-V steels containing titanium and boron. Unfortunately, improvements in rupture strength (and hence in relaxation strength) are often at odds with improvements in rupture ductility, and optimization of the two properties is not easy.

In addition to compositional modifications involving titanium and boron, major improvements in rupture ductility without compromises in creep strength have been achieved by reducing the impurity-element contents of the bolt steels. Figure 6.51 shows the degradation in rupture ductility of Cr-Mo-V steels at 550 °C (1020 °F) in 10,000 h of rupture with increasing phosphorus, arsenic, and tin contents. Similar effects due to bismuth and lead have also been demonstrated by Oakes, Bridge, and Judge (Ref 120). The residual-element content could be reduced by careful scrap selection, avoidance of air melting, and use of double vacuum melting (see Fig. 6.52).

Brittle fracture of bolts can occur if the fracture toughness of the bolt material is too low. Because high-strength steels are used for bolts, they have inherently low toughness. The toughness can be further degraded in service by temper embrittlement. A logical sequence of failure events might involve initiation of a creep-rupture crack at operating conditions followed by brittle fracture of the bolt during a transient. Fortunately, some of the solutions applied to improve the high-temperature rupture ductility of the steels, such as use of fine-grain steels and control of residual elements, also result in improved toughness and reduced susceptibility to temper embrittlement.

Low-Cycle Fatigue. Under repeated start-stop conditions, low-cycle fatigue can become an additional damage mechanism, contributing to crack initiation. Limited LCF data on a few bolt materials available in the published literature are shown in Fig. 6.53 (Ref 121). The highest resistance to

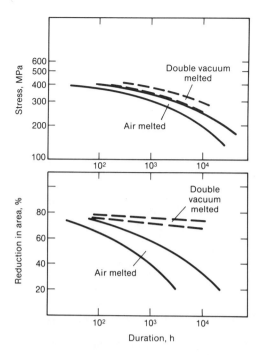

Fig. 6.52. Effect of double vacuum melting on stress-rupture properties of 1Cr-Mo-V steels at 550 °C (1020 °F) (Ref 120).

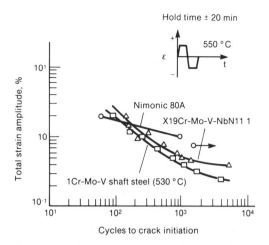

Fig. 6.53. Low-cycle-fatigue behavior of steam-turbine bolt materials at high temperature (Ref 121).

LCF is indicated for Nimonic 80, followed by a 10Cr-Mo-V-Nb steel.

The Metallurgy of Bolt Alloys

The basic compositions of bolt steels include Cr-Mo, Cr-Mo-V, Cr-Mo-V-Ti-B, and 12Cr-Mo-V nickel-base alloys. The evolution of these compositions was in step with the need for steels with increased creep strengths. The Cr-Mo steels used until the late 1940's had creep strengths adequate for service at temperatures up to about 480 °C (895 °F). With the increasing need for a higher-strength steel, a 1Cr-1Mo-¼V steel strengthened by stable V_4C_3 precipitates was developed. This alloy was found to be adequate for steam temperatures up to 540 °C (1000 °F). When steam temperatures reached about 565 °C (1050 °F) in the mid-1950's, a 1Cr-1Mo-¾V steel, in which vanadium and carbon had been stoichiometrically optimized to get the largest volume fraction of V_4C_3 and hence the highest creep strength, was developed. Unfortunately, this development had overlooked the importance of rupture ductility, and

many creep-rupture failures of bolts due to notch sensitivity occurred. The loss in rupture ductility was subsequently countered by grain refinement and by compositional modifications involving titanium and boron. Further developments aimed at providing additional creep strength have included development of 12Cr-Mo-V steels and René 80 superalloys. Reduction of impurity-element content has been another major development in recent years. The physical metallurgy of low-alloy ferritic steels has been discussed in Ref 119, and that of a new developmental 10% Cr steel has been described in Ref 121. Additional mechanical-property data on bolt steels may also be found in Ref 118 to 122.

Remaining-Life Assessment of Bolts

In addition to the various damage mechanisms discussed above, it is well known that ferritic steel bolts soften with service exposure, leading to a corresponding decrease in creep and stress-relaxation strengths. The deterioration in strength may lead to premature relaxation of stresses and consequent leaks. Monitoring of hardness or rupture strength using sacrificial bolts may provide a basis for bolt replacement under these conditions. In most cases, however, crack initiation is likely to precede creep-strength degradation as a failure mechanism. Hence, detection of cracks by NDE techniques during overhauls, or anticipated crack initiation based on calculational procedures, constitute the basis for bolt replacement. For conservatism, the latter procedure seems to be the more commonly used practice. Cumulative creep damage could be calculated by summing the time life fractions expended at various stress levels during each relaxation cycle and those under the steady-state stress, using the linear damage rule. Alternatively, the accumulation of creep strain during each relaxation cycle could be calculated and summed to get the cumulative creep damage. The Schlottner and Seeley procedure based on addition of time life fractions at various strain rates (discussed earlier for rotors) is also applicable. The cumulative fatigue damage can be calculated by entering the LCF curves (with hold time) at the appropriate strain levels, corresponding to the various transients, and summing the fatigue-life fractions consumed. Linear summation of creep- and fatigue-life consumptions indicates the total life fraction consumed. The procedure illustrated earlier for turbine-rotor surface grooves provides an example of the type of calculation that is required.

References

1. K.G. Reinhard, A Contribution to Higher Availability for Steam Power Plants, *Brown Boveri Review*, Vol 71, Mar/Apr 1984, Brown Boveri, Inc., Baden, Switzerland
2. K. Fujiyama *et al*, Life Diagnosis Experience and System Technology for Steam Turbine Components, in *International Conference on Life Assessment and Extension*, The Hague, Netherlands, June 13-15, 1988, Paper 2.1.3, Vol II, p 20-30
3. K. Kuwabara *et al*, Material Degradation and Life Assessment of Turbine Components After Long Term Service, in *International Conference on Life Assessment and Extension*, The Hague, Netherlands, June 13-15, 1988, Paper 2.5.2, Vol II, p 99-105
4. J. Ewald *et al*, Remaining Life Evaluation Measures on Turbines, in *International Conference on Life Assessment and Extension*, The Hague, Netherlands, June 13-15, 1988, Paper 3.7.1, Vol III, p 137-150
5. K.H. Mayer and H. Konig, Determination of Residual Life of Steam Turbine Components, in *International Conference on Life Assessment and Extension*, The Hague, Netherlands, June 13-15, 1988, Paper 1.10.1, Vol I, p 211-218
6. Y. Hirota *et al*, Changes of Material Properties and Life Management of Steam Turbine Components Under Long Term Service, *Mitsubishi Tech. Rev.*, Vol 19 (No. 3), 1982
7. R. Viswanathan and R.I. Jaffee, Metallurgical Factors Affecting the Reliability of Fossil Steam Turbine Rotors, *ASME J. Engg. Gas Turbines and Power*, Vol 107, July 1985, p 642-651
8. S.H. Bush, Failures in Large Steam Turbine Rotors, in *Rotor Forgings for Turbines and Generators*, R.I. Jaffee, Ed., Pergamon Press, New York, 1982, p 1-1 to 1-27
9. J.T.A. Roberts, *Structural Materials in Nuclear Power Systems*, Plenum Press, New York, 1981
10. D.P. Timo, Discussion of Axial Bores, in *Rotor Forgings for Turbines and Generators*, R.I. Jaffee, Ed., Pergamon Press, New York, 1982, p 5-127
11. K.H. Mayer, in *Rotor Forgings for Turbines and Generators*, R.I. Jaffee, Ed., Pergamon Press, New York, 1982, p 5-115 and 5-123
12. R.I. Jaffee, private communication, Electric Power Research Institute, Palo Alto, CA, 1988
13. R. Viswanathan, unpublished research, Electric Power Research Institute, Palo Alto, CA
14. D.P. Timo, R.M. Curran, and R.J. Placek, The Development and Evolution of Improved Rotor Forgings for Modern Large Steam Turbines, in *Rotor Forgings for Turbines and Generators*, R.I. Jaffee, Ed., Pergamon Press, New York, 1982, p 3-115 to 3-128
15. F.E. Werner, "Transformation Characteristics of CrMoV Rotor Steel," Research Memo 8-0102-M5, Westinghouse Research Laboratories, Mar 1956
16. I. Roman, C.A. Rau, A.S. Tetelman, and K. Ono, "Fracture and Fatigue Properties of 1Cr-Mo-V Bainitic Turbine Rotor Steels," Report NP 1023, Electric Power Research Institute, Palo Alto, CA, Mar 1979
17. S.J. Manganello and J.F. Boyce, United States Steel Corp., Monroeville, PA, circa 1955 (report title unavailable)
18. R. Viswanathan, Strength and Ductility of CrMoV Steels in Creep at Elevated Temperatures, *ASTM J. Test. and Eval.*, Vol 3 (No. 2), 1975, p 93-106
19. R. Viswanathan, Effects of Residual Elements on Creep Properties of Ferritic Steels, *ASM Metals Engg. Qtrly.*, Nov 1975, p 50-56
20. R. Viswanathan and R.I. Jaffee, Toughness of Cr-Mo-V Steels for Steam Turbine Rotors, *ASME J. Engg. Mater. Tech.*, Vol 105, Oct 1983, p 286-294
21. R.C. Schwant and D.P. Timo, Life Assess-

ment of General Electric Large Steam Turbine Rotors, in *Life Assessment and Improvement of Turbogenerator Rotors for Fossil Plants*, R. Viswanathan, Ed., Pergamon Press, New York, 1985, p 3.25-3.40

21a. G.T. Jones, *Proc. Institute of Mechanical Engineers*, Vol 186, 1972, p 31-32

22. D.P. Timo, "Design Philosophy and Thermal Stress Considerations of Large Fossil Steam Turbines," General Electric Fossil Steam Turbine Seminar, Schenectady, NY, Publication 84T16, 1984

23. J.R. Lindsey, M.R. Bishop, and R.L. Ullinger, Improving Steam Turbine Reliability, *Power*, Mar 1986

24. P. Paris and F. Erdogen, A Critical Analysis of Crack Propagation Laws, *Trans. ASME*, Dec 1963, p 528-534

25. T.T. Shih and G.A. Clarke, Effects of Temperature and Frequency on the Fatigue Crack Growth Rate Properties of a 1950 Vintage CrMoV Rotor Material, in *Fracture Mechanics*, G.V. Smith, Ed., STP 677, American Society for Testing and Materials, Philadelphia, 1979, p 125-143

26. V.P. Swaminathan, N.S. Cheruvu, A. Saxena, and P.K. Liaw, An Initiation and Propagation Approach for Life Assessment of an HP-IP Rotor, in *Life Extension and Assessment of Fossil Power Plants*, R.B. Dooley and R. Viswanathan, Ed., EPRI CS 5208, Electric Power Research Institute, Palo Alto, CA, 1987, p 659-676

27. M.J. Manjoine, Simulated Service Testing at Elevated Temperatures, Paper 19 in *Proceedings of the Joint International Conference on Creep*, Institute of Mechanical Engineers, London, 1963, p 7.37-7.48

28. R.M. Goldhoff, Stress Concentration and Size Effects in a CrMoV Steel, in *Proceedings of the Joint International Conference on Creep*, Institute of Mechanical Engineers, London, 1963, p 4.19-4.32

29. L.D. Kramer, D.D. Randolph, and D.A. Weisz, "Analysis of the Tennessee Valley Authority, Gallatin Unit No. 2 Turbine Rotor Burst," Winter Meeting of ASME, New York, Dec 5-10, 1976

30. N.S. Cheruvu, W.E. Howard, and W.O. Beasley, "Life Analysis of Oklahoma Gas and Electric Co. Mustang 3, HP-IP Rotor," proceedings of the Joint Power Generation Conference, Milwaukee, JPGC PWr-29, American Society of Mechanical Engineers, New York, 1985

31. G. Schlottner and R.E. Seeley, Estimation of the Remaining Life of High Temperature Steam Turbine Components, in *Residual Life Assessment, Non-Destructive Examination and Nuclear Heat Exchanger Materials*, C.E. Jaske, Ed., PVP Vol 98-1, American Society

of Mechanical Engineers, New York, 1985, p 35-45

32. J.F. Tavernelli and L.F. Coffin, Jr., "Experimental Support for Generalized Equation Predicting Low Cycle Fatigue," Paper No. 61-WA-199, American Society of Mechanical Engineers, New York, Nov 1961

33. M.M. Leven, *Experimental Mechanics*, Vol 353, Sept 1973

34. K. Kuwabara and A. Nitta, "Estimation of Thermal Fatigue Damage on Steam Turbine Rotors," Report E277001, Central Research Institute for the Electric Power Industry, Tokyo, July 1977

35. A.D. Batte, Creep-Fatigue Life Predictions, in *Fatigue at High Temperature*, R.P. Skelton, Ed., Applied Science Publishers, London, 1983

36. R.M. Curran and B.M. Wundt, "Interpretive Report on Notched and Unnotched Creep Fatigue Interspersion Tests in CrMoV, 2¼Cr-1Mo and Type 304 Stainless Steels," MPC 8-ASME, Metal Properties Council, New York

37. K.N. Melton, Strain Wave Shape and Frequency Effects on the High Temperature, Low Cycle Fatigue Behavior of a 1Cr-1Mo-V Ferritic Steel, *Mater. Sci. Engg.*, Vol 55, 1982, p 21-28

38. V. Bisego, C. Fossati, and S. Ragazzoni, "Low Cycle Fatigue Characterization of a HP-IP Steam Turbine Rotor," ASTM Symposium on Low Cycle Fatigue, Lake George, NY, American Society for Testing and Materials, Philadelphia, Oct 1985

39. W.J. Ostergren, Correlation of Hold Time Effects in Elevated Temperature Low Cycle Fatigue Using a Frequency Modified Damage Function, ASME-MPC Symposium on *Creep-Fatigue Interaction*, MPC-3, Metal Properties Council, New York, 1976, p 179-202

40. G. Thomas and R.A.T. Dawson, The Effect of Dwell Period and Cycle Type on High Strain Fatigue Properties of a 1Cr-Mo-V Rotor Forging at 500-550°C, in *Proceedings of International Conference on Engineering Aspects of Creep*, Vol 1, I Mech E Conf Pub 1080-5, Sheffield, Mechanical Engineering Publications, London, 1980

41. D.C. Gonyea, Method for Low Cycle Fatigue Design Including Biaxial Stress and Notch Effects, in *Fatigue at Elevated Temperatures*, STP 520, American Society for Testing and Materials, Philadelphia, 1973, p 678-687

42. R.G. Carlton, D.J. Gooch, and E.M. Hawkes, "The Central Electricity Generating Board Approach to the Determination of the Remanent Life of High Temperature Turbine Rotors," Paper C300/87, Institute of Mechanical Engineers, London, 1987

43. A. Nitta and K. Kuwabara, "Effect of Creep Fatigue Interaction on Life Expenditure of a

Turbine Rotor," Report E277002, Central Research Institute for the Electric Power Industry, Tokyo, July 1977

44. A.D. Batte and D.J. Gooch, Metallurgical Aspects of Remanent Life Assessment of High Temperature Rotor Forgings, in *Life Assessment and Improvement of Turbogenerator Rotors for Fossil Plants*, R. Viswanathan, Ed., Pergamon Press, New York, 1985, p 3.79-3.100

45. T. Goto, minutes of the Industry Advisory Group Meeting of the Electric Power Research Institute, San Antonio, TX, Dec 1988

46. K. Tanemura *et al*, Material Degradation of Long Term Service Rotor, in *International Conference on Life Assessment and Extension*, The Hague, Netherlands, June 13-15, 1988, Paper 1.8.4, Vol I, p 172-178

47. Y. Kadoya, K. Kawamoto, T. Goto, and H. Itoh, "Material Characteristics NDE System for High Temperature Rotors," Paper 85-JPGC-PWr-10, ASME/IEEE Joint Power Generation Conference, 1985

48. R.M. Goldhoff and D.A. Woodford, The Evaluation of Creep Damage in CrMoV Steel, in *Testing for Prediction of Material Performance in Structures and Components*, STP 515, American Society for Testing and Materials, Philadelphia, 1972, p 89-106

49. T. Goto, "Study on Residual Creep Life Estimation Using Nondestructive Material Properties Tests," Mitsubishi Technical Bulletin No. 169, Apr 1985

50. K. Kimura *et al*, Life Assessment and Diagnosis System for Steam Turbine Components, in *Life Extension and Assessment of Fossil Power Plants*, R.B. Dooley and R. Viswanathan, Ed., EPRI CS 5208, Electric Power Research Institute, Palo Alto, CA, 1987, p 677-685

51. S. Kirihara *et al*, Fundamental Study of Nondestructive Detection of Creep Damage for Low Alloy Steel, *J. Soc. Mater. Sci.*, Vol 33, 1984, p 1097

52. S.M. Bruemmer, L.A. Charlot, and B.W. Arey, "Grain Boundary Composition and Intergranular Fracture of Steels," Report RD 3859, Vol 1, Electric Power Research Institute, Palo Alto, CA, Jan 1986

53. R. Viswanathan and S.M. Bruemmer, In Service Degradation of Toughness of Steam Turbine Rotors, *ASME J. Engg. Mater. Tech.*, Vol 107, Oct 1985, p 316-324

54. R. Viswanathan, S.M. Bruemmer, and R.H. Richman, Etching Technique for Assessing Toughness Degradation of In-Service Components, *ASME J. Engg. Mater. Tech.*, Vol 110, Oct 1988, p 313-318

55. R. Viswanathan, Emerging Non-Destructive Techniques for Damage Assessment, in proceedings of conference on *Advances in Materials Technology for Fossil Power Plants*, Chicago, R. Viswanathan and R.I. Jaffee,

Ed., American Society for Metals, Metals Park, OH, 1987

56. Q. Zhe, Y.Q. Wing, S.C. Fu, and C.J. McMahon, "Impurity Induced Embrittlement of Rotor Steels," Report CS 3248, Vol 1, Electric Power Research Institute, Palo Alto, CA, Nov 1983

57. Q. Zhe, S.C. Fu, and C.J. McMahon, "Temper Embrittlement of CrMoV Turbine Rotor Steels," Report CS 2242, Electric Power Research Institute, Palo Alto, CA, Feb 1982

58. D.L. Newhouse, Relationships Between Charpy Impact Energy, Fracture Appearance and Test Temperature in Alloy Steels, *Weld. J. Res. Suppl.*, Mar 1963, p 1-10

59. S. Ganesh and A. Kaplan, unpublished work in progress for the Electric Power Research Institute, Project RP2481-3, General Electric Co., Schenectady, NY

60. M.P. Manahan, A.S. Argon, and O.K. Harling, *J. Nucl. Mater.*, 1981, p 103-104

61. J.M. Baik, J. Kameda, and O. Buck, *Scripta Met.*, Vol 17, 1983, p 1143-1147

62. J.M. Baik, J. Kameda, and O. Buck, STP 883, American Society for Testing and Materials, Philadelphia, 1986, p 95

63. H. Takahashi *et al*, Residual Life Assessment and Non Destructive Evaluation of Material Degradation by Means of Small Punch Test, in *International Conference on Life Assessment and Extension*, The Hague, Netherlands, June 13-15, 1988, Paper 2.7.2

64. T. Shoji and H. Takahashi, Non Destructive Evaluation of Materials Degradation During Service Operation by Means of Electrochemical Method, in *Life Extension and Assessment of Fossil Power Plants*, R.B. Dooley and R. Viswanathan, Ed., EPRI CS 5208, Electric Power Research Institute, Palo Alto, CA, 1987, p 745-759

65. L.K. Tu and B.B. Seth, Effect of Composition, Strength and Residual Elements on Toughness and Creep Properties of CrMoV Turbine Rotors, *Metals Technology*, Mar 1979, p 79

66. V. Berg, J.E. Bertilsson, K.H. Friedl, and R.B. Scarlin, On the Limits of the Design of Large Steam Turbine Rotors Operating at Elevated Temperatures, in *Rotor Forgings for Turbines and Generators*, R.I. Jaffee, Ed., Pergamon Press, New York, 1982, p 3-72 to 3-90

67. W. Engelke *et al*, Rotor Forgings for KWU Designed Turbine Generators, in *Rotor Forgings for Turbines and Generators*, R.I. Jaffee, Ed., Pergamon Press, New York, 1982, p 3-147 to 3-163

68. F.V. Ammirato *et al*, "Life Assessment Methodology for Turbogenerator Rotors," Report CS/EL 5593, Vol 1-4, Electric Power Research Institute, Palo Alto, CA, Mar 1988

69. A.M. Curzon and C.H. Wells, "Development

of a Crack Growth Algorithm for Time Dependent Analysis of Steam Turbine Rotors," Report NP 3187, Electric Power Research Institute, Palo Alto, CA, July 1983

70. A. Saxena, T.T. Shih, and H.A. Earnst, Wide Range Creep Crack Growth Behavior of A470 Class 8 (CrMoV) Steel, in *Fracture Mechanics: Fifteenth Symposium*, STP 833, American Society for Testing and Materials, Philadelphia, 1984, p 516-531

71. D.J. Smith and G.A. Webster, Estimates of the C^* Parameter for Crack Growth in Creeping Materials, in *Elastic Plastic Fracture: Second Symposium*, Vol I, *Inelastic Crack Analysis*, STP 803, American Society for Testing and Materials, Philadelphia, 1983, p 1654-1674

72. A. Saxena, R.S. Williams, and T.T. Shih, A Model for Representing and Predicting the Influence of Hold Times on the Fatigue Crack Growth Behavior at Elevated Temperatures, in *Fracture Mechanics: Thirteenth Conference*, STP 743, American Society for Testing and Materials, Philadelphia, 1981, p 86-99

73. A. Saxena and J.L. Bassani, Time-Dependent Fatigue Crack Growth Behavior at Elevated Temperature, in *Fracture: Interactions on Microstructures, Mechanisms and Mechanics*, AIME/ASME Conference, Los Angeles, 1984, p 357-383

74. V.P. Swaminathan, T.T. Shih, and A. Saxena, Low Frequency Fatigue Crack Growth Behavior of A 470 Class 8 Rotor Steel at 538°C, *Engg. Fract. Mech.*, Vol 16 (No. 6), 1982, p 827-836

75. W.P. McNaughton, R.H. Richman, C.S. Pillar, and L.W. Perry, "Generic Guidelines for the Life Extension of Fossil Fuel Power Plants," Report CS 4778, Electric Power Research Institute, Palo Alto, CA, Nov 1986

76. D.J. Gooch and R.D. Townsend, "CEGB Remanent Life Assessment Procedures," in "Fossil Plant Life Extension Conference," Report CS 4207, Electric Power Research Institute, Palo Alto, CA, Aug 1985

77. H.C. Argo and B.B. Seth, Fracture Mechanics Analysis of Ultrasonic Indications in CrMoV Steel Turbine Rotors, in *Case Studies in Fracture Mechanics*, T.P. Rich and D.J. Cartwright, Ed., Army Materials and Mechanics Research Center, MS 77-5, 1977, p 3.2.1-3.2.11

78. J. Ewald, B. Meuhle, K. Keinburg, and H. Termuhlen, in *Life Assessment and Improvement of Turbogenerator Rotors for Fossil Power Plants*, R. Viswanathan, Ed., Pergamon Press, New York, 1985, p 3.77-3.98

79. V.P. Swaminathan and R.I. Jaffee, Significant Improvements in Properties of CrMoV HP Rotors by Advanced Steel Making, in *Life Assessment and Improvement of Turbogenerator Rotors for Fossil Power Plants*, R. Viswanathan, Ed., Pergamon Press, New York, 1985, p 6.57

80. C. Berger, in *Proceedings of the 9th International Forging Conference*, Vol III, Dusseldorf, May 1981, p 30-33

81. H. Finkler and E. Potthast, Heat Treatment of 1%CrMoV Steel With Special Regard to the Effects of Quenching Rate on the Properties, in *Rotor Forgings for Turbines and Generators*, R.I. Jaffee, Ed., Pergamon Press, New York, 1982, p 5-83

82. D.B. Berrong, "Resolving Turbine Casing Cracking and Distortion Cracking," in "Workshop Proceedings: Life Assessment and Repair of Steam Turbine Casings," R. Viswanathan, Ed., Report CS 4676 SR, Electric Power Research Institute, Palo Alto, CA, July 1986, p 8.1-9.1

83. D.M. Rasmussen, "Steam Turbine Case Repairs To Extend Operating Life," Ref 82, p 6.1-7.1

84. J.H. Vogan and A. Morson, "Improvements in Shell Design and Maintenance Techniques for General Electric In-Service Shells," Ref 82, p 7.1-8.1

85. C. Perry and E. Hartsell, "Successful Field Welding of High Pressure Turbine Casing Cracks," Ref 82, p 2.1-3.1

86. C.N. Damman, C.A. Loney, and D. Randolph, "TVA's Experience With Casings," Ref 82, p 3.1-4.1

87. R.R. Richardson, A.G. Pard, L.J. Radak, and D. Kramer, "Remaining Life Assessment and Repair of Casings," Ref 82, p 4.1-5.1

88. D. Ackerman, "Survey of Weld Repair Procedures for High Pressure Steam Turbine Casings," Ref 82, p 9.1-10.1

89. R. Crombie, High Integrity Ferrous Castings for Steam Turbines—Aspects of Steel Development and Manufacture, *Mater. Sci. Tech.*, Vol 1, Nov 1985, p 986-993

90. R. Viswanathan, Effect of Ti and Ti + B Additions on the Creep and Rupture Properties of 1¼Cr-½Mo Steel, *Met. Trans.*, Vol 8A, p 57-61

91. Logsdon *et al*, Residual Life Prediction and Retirement for Cause Criteria for SSTG Upper Casings, Part I, Mechanical Properties, Part II, Fracture Mechanics Analysis, *Engg. Fract. Mech.*, Vol 25 (No. 3), 1986, p 259-303

92. J.P. Gudas, unpublished data, cited in Ref 91

93. E.M. Hackett, David Taylor Naval Ship Research and Development Center, unpublished data, cited in Ref 91

94. W.A. Logsdon, "Fatigue Crack Growth Properties of ASTM A296 12Cr and ASTM 217, 2¼Cr-1Mo Cast Steels in 520°F 1200 psi Distilled Water," Westinghouse Research Laboratories, unpublished work, cited in Ref 91

95. C.R. Brinkman *et al*, "Interim Report on the Continuous Cycling Elevated Temperature Fatigue and Subcritical Crack Growth Behavior of 2¼Cr-1Mo Steel," ORNL TM 4993, Oak Ridge National Laboratory, Dec 1975

96. J.D. Byron, S.R. Paterson, R.R. Proctor, and T.J. Feiereisen, Assessment of Remaining Useful Life of Ships Service Turbine Generator Steam Chests, *Naval Engg. J.*, May 1986, p 95

97. P.A. Coulon, Approaches to Life Extension for High Temperature Turbine Casings Beyond 100,000 Hours, in *Life Extension and Assessment of Fossil Power Plants*, R.B. Dooley and R. Viswanathan, Ed., EPRI CS 5208, Electric Power Research Institute, Palo Alto, CA, 1987

98. "Pacific Gas and Electric Co., Life Extension Strategy for Fossil Plants," Report CS 4782, Electric Power Research Institute, Palo Alto, CA, 1987

99. *Corrosion Fatigue of Steam Turbine Blade Materials*, R.I. Jaffee, Ed., Pergamon Press, New York, 1981

100. M.F. O'Connor, "Design Aspects of Short and First Stage Buckets for Large Steam Turbines," Paper 84 T4, General Electric Large Steam Turbine Seminar, General Electric Co., 1984

101. *Aerospace Structural Metals Handbook*, Mechanical Properties Data Center, Department of Defense, AFML TR 68115, 1975

102. N.F. Rieger and T. McCloskey, Turbine Blade Life Assessment, in *International Conference on Life Assessment and Extension*, The Hague, Netherlands, June 13-15, 1988, Paper 1.10.2, Vol I

103. J. Sohre, "Steam Turbine Blade Failures, Causes and Corrections," proceedings of Texas A & M Fourth Turbomachinery Symposium, College Station, TX, 1974

104. D.F. Socie and J.D. Morrow, Review of Contemporary Approaches to Fatigue Damage Analysis, Chapter 8 in *Risk and Failure Analysis for Improved Reliability*, J.J. Burke and V. Weiss, Ed., Plenum Publishing Corp., 1980

105. D.F. Socie, "Fatigue Life Prediction Using Local Stress-Strain Concepts," 1975 SESA Spring Meeting, Chicago, May 1975

106. J.W. Morrow, "Laboratory Simulation of the Low Cycle Fatigue Behavior of the Hook Regions of a Steam Turbine Blade Subjected to Start-Stop Cycles," proceedings of Fourth National Congress on Pressure Vessel and Piping Technology, ASME, Portland, June 1983

107. S.S. Manson and G.R. Halford, "Practical Implementation of the Double Linear Damage Rule and Damage Curve Approach for Treating Cumulative Fatigue Damage," Report 81517, National Aeronautics and Space Administration, Apr 1980

108. H.N. Rieger *et al*, "BLADE Code and Manual," prepared for the Electric Power Research Institute, Palo Alto, CA, 1986

109. F.D. Ryan, "The Design of Major Station-ary Components of Fossil Large Steam Turbines," Paper 84 T5, General Electric Large Steam Turbine Seminar, General Electric Co., 1984

110. R.G. Brown, F. Quilliam, and D.E. Leaver, "Evaluation of ASME Solid Particle Erosion Task Group Questionnaire," Document 2-R-83-02, Electric Power Research Institute, Palo Alto, CA, 1983

111. R.C. Spencer, "Solid Particle Erosion, What Can the Industry Afford To Spend To Eliminate SPE?", 1980 EPRI/ASME Workshop on Solid Particle Erosion, EPRI Report CS 3178, Electric Power Research Institute, Palo Alto, CA, 1983

112. J. Quereshi, A. Levy, and B. Wang, Characterization of Coating Processes and Coatings for Steam Turbine Blades, *J. Vacuum Sci. Tech. A*, Vol 4 (No. 6), Nov/Dec 1986, p 2638-2647

113. O. Watanabe, K. Ikeda, and M. Miyazaki, "Erosion Resistant Surface Treatment for Turbine Nozzles," in "Solid Particle Erosion of Utility Steam Turbines," Report CS 4683, Electric Power Research Institute, Palo Alto, CA, Aug 1986, p 4.61-4.78

114. W.J. Sumner, J.H. Vogan, and R.J. Lindinger, "Reducing Solid Particle Damage in Large Steam Turbines," Ref 113, p 4.1-4.18

115. S.T. Wlodek, "Development and Testing of Plasma Sprayed Coatings for Solid Particle Erosion Resistance," Ref 113, p 4.43-4.80

116. R.J. Ortolano, "Improving Steam Turbine Resistance to Abrasion Erosion," Ref 113, p 2.69-2.86

117. C. Veerport, C. Maggi, and H. Bartsch, "The Protection of Steam Turbines Against Solid Particle Erosion," Ref 113, p 4.19-4.42

118. G.D. Branch *et al*, High Temperature Bolts for Steam Power Plant, in *International Conference on Creep and Fatigue in Elevated Temperature Applications*, Sheffield and Philadelphia, Institute of Mechanical Engineers, London, Conference Publication 13, 1973, p 192.1-192.9

119. H. Everson, J. Orr, and D. Dulieu, Low Alloy Ferritic Bolting Steels for Steam Turbine Applications: The Evolution of Durehete Steels, in *Advances in Materials Technology for Fossil Power Plants*, R. Viswanathan and R.I. Jaffee, Ed., American Society for Metals, Metals Park, OH, 1988

120. G. Oakes, M.R. Bridge, and S. Judge, High Purity Cr-Mo-V Steels for Bolting Applications, in Ref 119

121. K.H. Mayer and H. Konig, Creep, Relaxation and Toughness Properties of the Bolt and Blade Steel X19CrMoVNbN111, in Ref 119

122. S.M. Beech *et al*, An Assessment of Alloy 80 A as a High Temperature Bolting Material for Advanced Steam Conditions, in Ref 119

7

Petroleum Reactor Pressure-Vessel Materials for Hydrogen Service

General Description

In the past, concern in the petroleum industry regarding selection of materials for pressure vessels has focused largely on safety and economic issues which dictate against unexpected equipment failures that could result in hazardous exposures and forced outages at an estimated cost of $50,000 per hour (Ref 1). More recently, however, the need for extension of the lives of current plants well beyond their originally anticipated durations has also become an important issue.

The ferritic steel reactor vessels used in modern petroleum refining and petrochemical processing are operated under conditions that can be as severe as metal temperatures of 565 °C (1050 °F) and pressures of 27.6 MPa (4000 psi) (Ref 2 and 3). Component-failure scenarios consider not only the conditions for steady operation but also those that apply during start-stop transients. The mechanical behavior considered includes such properties as fracture toughness, resistance to creep-rupture, and resistance to

thermal fatigue. In addition, the corrosion and environmental behavior of materials both in normal operation and during process upsets and shutdowns must be taken into account. Because fabrication involves extensive welding, the properties of weldments are of great importance. Reactor pressure vessels are designed, fabricated, and inspected in accordance with the ASME Boiler and Pressure Vessel Code. Material specifications listed in the code cover (1) the strengths necessary to guarantee allowable stresses, including room- and design-temperature tensile, creep, and fatigue strengths; (2) notch toughness at the lowest operating temperature; and (3) weldability. In addition to the minimum code requirements for the fabricated condition, steels for high-temperature, high-pressure hydrogenation service are required to withstand such environmental-degradation processes as temper embrittlement, hydrogen embrittlement, hydrogen attack, and creep embrittlement. Excellent reviews of materials problems in the refinery industry (Ref 4 to 8) and potential problems in coal conversion (Ref 9 to

329

11) have been published in the literature. This chapter will draw heavily upon these resources.

Thick-wall reactor pressure vessels are the hearts of the processing units in refineries. These large vessels range up to more than 905 t (1000 tons) in weight, and some have wall thickness of about 305 mm (12 in.). Some of the most critical requirements with respect to pressure-vessel size are those of hydroprocessors. The term "hydroprocessing" represents a wide range of refining processes including hydrocracking, hydrodesulfurization, and others in which oil is reacted with hydrogen in the presence of a catalyst at high pressures (up to 27.6 MPa, or 4000 psi) and high temperatures (345 to 455 °C, or 650 to 850 °F). The chemical reactions that occur in these processes include cracking of large hydrocarbon molecules, conversion of nitrogen compounds into ammonia (NH_3), and conversion of hydrogen compounds into hydrogen sulfide (H_2S). Such processes are important in production of low-sulfur fuels from high-sulfur crude oils and conversion of higher-boiling-point residual oils into lighter transportation fuels. In addition, production and upgrading of synthetic crude oils from liquefied coal, oil shale, and tar sands will require many hydroprocessing reactors. Because these processes involve exposure to hydrogen and hydrogen gases at high pressures, resistance to attack by these gases imposes major restrictions on materials selection. Because hydroprocessing represents the most severe case of hydrogen service, major emphasis is placed on the hydroprocessor reactor vessels discussed in the following sections. Most of the damage mechanisms described are equally applicable to reactor vessels in low- or intermediate-pressure hydrogen service (e.g., catalytic reformers). These vessels, made of 1Cr-½Mo and 1¼Cr-½Mo steels, operate at higher temperatures, and creep-damage mechanisms assume greater importance.

Materials of Construction

Pressure-vessel shells generally are made of low-alloy Cr-Mo steels. To prevent attack by H_2S and other sulfur-containing compounds, the shells are internally clad with austenitic stainless steels. (In this case, "cladding" is taken to mean roll cladding as well as weld-overlay processes.) Early reactors that had thinner walls were made of roll-clad plate, but for the thicker-wall vessels made subsequently, a weld-overlay process has been found to be more economical. A typical hydrodesulfurization vessel is made of 2¼Cr-1Mo steel overlaid internally with type 347 stainless steel. For plate-formed vessels, a typical grade of 2¼Cr-1Mo steel is ASTM A387, grade 22, class 2. The equivalent grade for ring and nozzle forgings is ASTM A336, grade F22. Sometimes a dual-layer overlay, such as a first layer of type 309 stainless steel and a second layer of type 347 stainless steel, is employed. In addition, reactor internals are made entirely from austenitic stainless steels for corrosion resistance. To support the stainless steel structures within the vessel, stainless steel brackets have in the past been arc welded to the cladding. The brackets are usually of either type 321 or type 347 stainless steel and are attached to the cladding by welding with electrodes coated with type 347. The internal structures are loosely bolted to the brackets in ways that prevent excessive stresses from differential thermal expansion and contraction. Several alternative schemes for attaching the internals to the cladding have been used in industry, and the procedure described above is meant to be illustrative.

Shells are fabricated using either seamless ring forgings or rolled plate subsequently seam welded to form rings. The ring sections are attached by girth welds made by submerged-arc welding (SAW). For longitudinal seam welding of rolled plate, either electroslag welding (ESW) or submerged-arc welding can be used. If ESW is used to form the seam welds, the complete ring sections are quenched and tempered. If SAW is used for the seam welds, an alternate procedure is to quench and temper the plates rolled to shape, perform the seam weld, and then postweld heat treat the ring. Nozzle sections are welded to the reactor body using shielded metal-arc welding (SMAW) for small nozzles and submerged-arc welding (SAW) for large nozzles.

After the ring sections have been fabricated, weld-overlay cladding is performed using an automatic welding process, typically the SAW process. In many cases, a two-layer overlay is specified, with the first layer being type 309 stainless steel to account for base-metal dilution, followed by a layer of type 347 stainless steel. Type 308L stainless steel has also been used occasionally in place of type 347. Cladding thicknesses vary from 2.5 to 6.4 mm (0.1 to 0.25 in.) depending on the corrosivity of the feedstock and the expected life. Weld cladding of nozzles is also performed using automatic gas tungsten-arc welding (GTAW) or shielded metal-arc welding (SMAW).

After application of the cladding and attachment of the nozzles and brackets, the final heat treatment for the vessel consists of a stress-relief treatment for 15 to 30 h. If the entire vessel cannot be heat treated in the shop, the ring sections are stress relieved in the shop, final fabrication by girth welding is performed in the field, and the girth welds are locally stress relieved in the field.

Currently, various ASTM plate specifications permit the use of 2¼Cr-1Mo steels over a range of room-temperature tensile strengths from 415 to 930 MPa (60 to 135 ksi):

- A387, grade D (annealed)–415 MPa (60 ksi) min UTS
- A387, grade D (normalized and tempered)–515 MPa (75 ksi) min UTS
- A542, class 4–585 MPa (85 ksi) min UTS
- A542, class 3–655 MPa (95 ksi) min UTS
- A542, class 2–795 MPa (115 ksi) min UTS
- A542, class 1–725 MPa (105 ksi) min UTS

Because the enhanced properties cannot be completely retained in the creep range, the design temperature should be carefully considered in determining the level of material strength that is required for safety. One can use either annealed material, normalized-and-tempered material, or quenched-

and-tempered material provided that the allowable stresses and fabrication procedures are met. Because the object of heat treatment is to produce a desired microstructure and the desired mechanical properties, the decision as to the type of heat treatment is based on the thickness and the hardability of the alloy. Each alloy steel has a critical cooling rate which must be exceeded to avoid lamellar products and proeutectoid ferrite in the transformed microstructure. As long as the critical cooling rate is exceeded, the manner of cooling is unimportant and the final strength properties are determined primarily by the tempering temperature. In recognition of this, ASTM and ASME allow for liquid quenching of plates over 100 mm (4 in.) thick to obtain microstructures and properties comparable to those obtained by air cooling of thinner plates. American hydrocracker reactors were fabricated from annealed 2¼Cr-1Mo steel in the early 1960's. As reactor sizes grew, the need for higher-strength quenched-and-tempered grades became apparent. Considerable characterization work on the properties of quenched-and-tempered 2¼Cr-1Mo steels was carried out during the early 1960's. The high hardenability of 2¼Cr-1Mo steel permits through-section hardening by water quenching of plates up to 305 mm (12 in.) thick while still retaining good mechanical properties. The resulting microstructures are fully bainitic. Hence, industry practice since the mid-1960's has been to require water quenching after austenitization for all section thicknesses greater than 125 mm (5 in.). Even for smaller sizes, the common practice is to use water-quenched steel, although this is not imposed as a requirement.

Once the designer has chosen the design stress values, the heat treatments are then prescribed to achieve the desired level of rupture strength in the steel. Because rupture strength is related to room-temperature tensile strength (Fig. 7.1; Ref 12), a target value for the latter is set. To achieve the desired tensile strength, the tempering temperature is chosen using the plots of tempering parameter vs tensile strength of the type shown in Fig. 7.2 (Ref 4). This procedure is also useful in setting optimized postweld

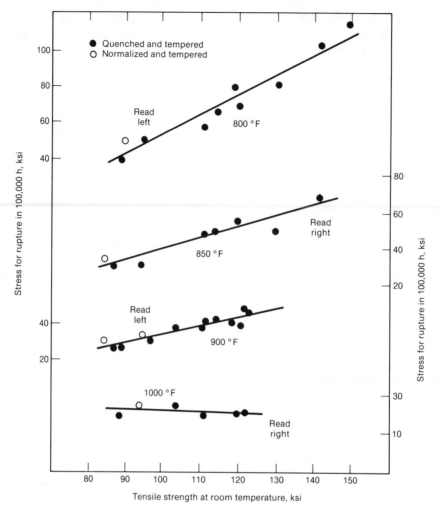

Fig. 7.1. Effect of room-temperature tensile strength on 100,000-h rupture strength of quenched-and-tempered and normalized-and-tempered 2¼ Cr-1Mo steel (Ref 12).

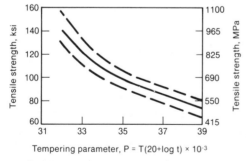

Dashed lines demarcate 95% confidence limits. For tempering parameter, temperature is expressed in °R, and time in hours.

Fig. 7.2. Tempering parameter vs tensile strength for quenched-and-tempered 2¼Cr-1Mo (ASTM A387, grade 22) steel plates and forgings (Ref 4).

heat treatments. Because PWHT at about 675 °C (1250 °F) may be performed both initially during fabrication and during subsequent repairs, the cumulative effect of these treatments in lowering the tensile strength (and hence the rupture strength) must be factored into the design.

Integrity Considerations for Pressure-Vessel Shells

Because a pressure vessel is a composite structure, its over-all integrity depends on the integrity of the steel shell, the austenitic cladding, and the interface between the two. The material-property requirements and the potential problems with respect to each of them will be considered in the following sections.

At the design stage, three material parameters are primarily taken into account in considering the integrity of shells of pressure vessels in hydrogen service: (1) the creep-rupture properties, which dictate the allowable stresses at various temperatures; (2) the fracture-toughness properties, which dictate inspection requirements and operating procedures for start-up and shutdown transients; and (3) resistance to hydrogen attack under operating conditions. From a life-assessment point of view, however, many in-service degradation phenomena have to be taken into account. In particular, temper embrittlement and hydrogen embrittlement can lead to progressive damage in service, requiring modifications of the inspection and operating procedures, and in extreme cases even to premature retirement of the reactor. Another factor not accounted for in design is subcritical crack growth under either static conditions (creep) or cyclic conditions (fatigue), as further exacerbated by environmental effects due to hydrogen and other corrosive media. Table 7.1 lists some of these damage phenomena that affect shell integrity. A detailed

Table 7.1. Potential problems for pressure-vessel shells (Ref 3)

Microstructure and phase stability
- Strength
- Toughness
- Hydrogen attack

Temper embrittlement (shutdown)
- Toughness
- Hydrogen embrittlement

Hydrogen embrittlement (shutdown)
- Sulfide stress cracking
- Toughness
- Slow crack growth

Environment-assisted crack growth
- Toughness
- Flow localization
- Fatigue-crack-growth rate

Corrosion

Hydrogen attack
- Strength and creep resistance
- Toughness

discussion of these phenomena is provided in the following sections.

Allowable Stress

Allowable design stresses for pressure vessels are normally based on the ASME Boiler and Pressure Vessel (unfired) Code, Section VIII, Division 1 or 2. Division 2 is more stringent in its design and inspection requirements, but Division 1 has a lower allowable stress. It should be emphasized that the code only specifies the upper limit for stress. Actual design stresses below the code allowable stress are invariably specified to build in a degree of conservatism. The degree of conservatism depends on designer and operator experience and varies from one company to another. For purposes of comparison, Section VIII, Division 1 allowable stresses as a function of temperature are shown in Fig. 7.3. The design allowable stresses for the Cr-Mo steels fall within a narrow band of 28 °C (50 °F), with higher-alloy steels lying on the high side of the band. The SA387 grades corresponding to 1¼Cr-½Mo, 2¼Cr-1Mo, and 3Cr-1Mo steels have stresses assigned up to 650 °C (1200 °F). It is clear from the narrow spread of the curves that allowable stress is a secondary consideration in selection of materials for hydrogen service. The temperature limit for operation is set more by a material's resistance to hydrogen attack, as dictated by the Nelson diagrams, than by its creep-rupture properties. The stress-rupture properties used in deriving the ASME allowable stress for 2¼Cr-1Mo steel (SA387, grade 22, class 2) are shown in Fig. 7.4. Most designers of hydroprocessor reactors limit design temperature to 455 °C (850 °F). This temperature is also the Nelson-curve limit for prevention of hydrogen attack, which sets a second temperature limit for operation. Other types of reactors that operate at lower hydrogen partial pressures can be made of lower-alloy steels and can operate at higher temperatures.

Fracture Toughness

The Charpy V-notch test is the test most commonly used to specify toughness levels

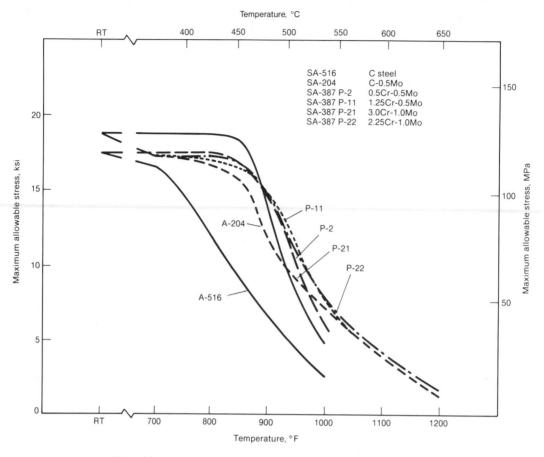

Fig. 7.3. Allowable stress as a function of temperature for commonly used Cr-Mo steels, from Section VIII, Division 1 of the ASME Boiler and Pressure Vessel Code.

for pressure-vessel steels. Although this test has many shortcomings, several investigations have successfully correlated Charpy test results with other fracture-toughness measurements such as K_{Ic} and J_{Ic}. The concern about toughness arises mainly in terms of potential failure occurring at low temperatures during start-up and shutdown transients. Because in the as-fabricated condition the Cr-Mo steels usually have excellent toughness even at low temperatures, generation of valid K_{Ic} data would require testing of very large specimens and would be prohibitively expensive.

ASME Section VIII, Division 2 rules specify toughness in terms of the minimum Charpy CVN energy at the lowest permissible vessel temperature from samples at the quarter-thickness location in the transverse orientation. The code requires an impact energy of 27 J (20 ft·lb) at the lowest temperature, although most companies involved in reactor design impose a more stringent requirement of 54 J (40 ft·lb) at 10 °C (50 °F), or the minimum service temperature. Using the Barsom-Rolfe correlation (Ref 13), this value of 54 J at 10 °C corresponds to a K_{Ic} value of 137 MPa$\sqrt{m}$ (125 ksi$\sqrt{in.}$) (Ref 4). This fracture toughness is believed to be well above the K level that will be reached during reactor start-up or shutdown conditions.

Temper Embrittlement. Although, on the basis of the 54 J/10 °C criterion, reactor vessels more than meet the toughness requirements in the as-fabricated condition, in-service temper embrittlement during service exposure in the range 345 to 455 °C (650 to 850 °F) can cause a significant degradation of toughness. Figure 7.5 illustrates

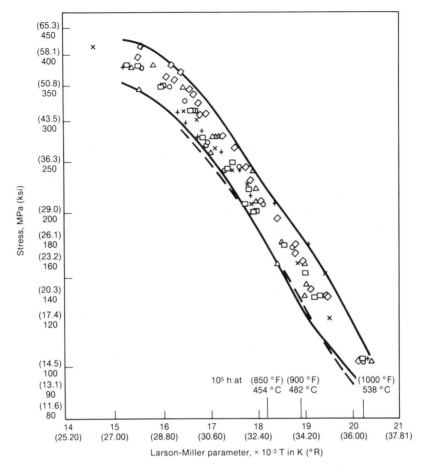

Fig. 7.4. Larson-Miller stress-rupture plot for 2¼ Cr-1Mo steel (Ref 10).

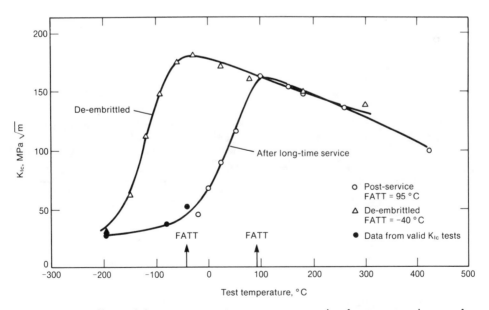

Fig. 7.5. Effect of long-term service exposure on the fracture toughness of 2¼ Cr-1Mo steel (Ref 6).

the change in the fracture toughness of a 2¼Cr-1Mo steel that was temper embrittled during approximately seven years of service in a hydrodesulfurizer unit (Ref 6). In addition to service embrittlement, some embrittlement from slow cooling after postweld heat treatment during vessel fabrication is also inevitable.

Temper embrittlement shifts the ductile-to-brittle transition temperature to higher values, resulting in corresponding reductions in fracture toughness and in the tolerable defect sizes at a given stress. The risk of failure, therefore, occurs not under operating conditions but during start-up and shutdown conditions. Brittle fractures in reactors that may have become temper embrittled can be prevented by reducing the stresses to values below those needed for crack propagation. This is achieved in industry by reducing the pressure to 25% of the design value when the active shell temperature is below 120 to 175 °C (250 to 350 °F) during the start-up and shutdown cycles. Vessels are generally preheated to this temperature before pressures are imposed during start-up. Similarly, they are depressurized during shutdown prior to reaching about 120 to 175 °C. These requirements lead to increased operating costs and reduced efficiency.

Field experience from older hydroprocessor reactors shows that, although transition-temperature shifts of 28 to 83 °C (50 to 150 °F) were typical of 2¼Cr-1Mo steel after long service exposures, shifts of over 110 °C (200 °F) could occur. Table 7.2 presents selected cases of extreme embrittlement (Ref 4). In one instance, a shift in transition temperature of as much as 159 °C (286 °F) has been reported. The shift in K_{Ic} in a 2¼Cr-1Mo steel taken from a hydrodesulfurizer after 3½ years of service is illustrated in Fig. 2.23 in Chapter 2. Similar data recently published on 2¼Cr-1Mo steel vessels tested after 30,000 hours and after 7 years of service exposure at about 450 °C (840 °F) also show significant drops in K_{Ic} due to temper embrittlement (Ref 14).

To evaluate the susceptibility of steels, either isothermal aging or accelerated step-cooling procedures can be employed. The principle of step cooling was discussed in Chapter 2. The step-cooling procedure commonly employed for pressure-vessel steels in the refinery industry is shown in Fig. 7.6 (Ref 6). A recent study by Shaw on behalf of the American Petroleum Institute has shown that the shift in transition temperature produced in such a step-cooling treatment is approximately one-third of what might be expected after 30 years of service in an actual reactor vessel (Ref 15). This factor of 3 is generally used in conjunction

Table 7.2. Examples of severely temper-embrittled 2¼Cr-1Mo steel removed from operating hydroprocessing reactors (Ref 4)

Year reactor built and operating location	Material	Exposure time, h	Exposure temperature °C	Exposure temperature °F	Transition temperature — De-embrittled °C	De-embrittled °F	Embrittled (as received) °C	Embrittled (as received) °F	Embrittlement shift °C	Embrittlement shift °F
1968/USA(a)	Plate(b)	21,000	425-455	800-850	−61	−78	59	138	120	216
	SAW weld metal(c)	...	...	...	−8	+17	151	303	159	286
	ESW weld metal	...	...	...	−90	−130	35	95	125	225
1969/Japan(d)	Plate(e)	30,000	350-450	660-840	−25	−13	82	180	107	193
ca 1967/Japan(d)	Plate(f)	60,000	350+	660+	−66	−87	65	149	131	236

(a) Transition temperatures are TT_{40} values. Reference, Chevron (unpublished). (b) Tensile strength, 615 MPa (89 ksi). (c) Quenched and tempered. (d) Transition temperatures are FATT values. Reference, Sawada *et al* (Ref 13a). (e) Tensile strength, 585 MPa (85 ksi). (f) Tensile strength, 650 MPa (94 ksi).

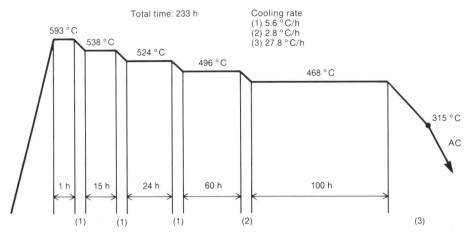

Fig. 7.6. Typical step-cooling cycle for temper-embrittlement studies (Ref 6).

with step-cooling treatments as a screening test for qualifying reactor materials and for specifying steels.

Long-Term Embrittlement Studies. Two long-term temper-embrittlement studies serve as landmark studies that have characterized the long-term isothermal temper-embrittlement susceptibilities of commercial-grade reactor materials. In the study by Shaw, 64 samples of Cr-Mo steels collected from various oil companies, reactor fabricators, and steelmakers, with an emphasis on 2¼ Cr-1Mo steels but also including a few samples of 1¼ Cr-½Mo steel and 3Cr-1Mo steel, were characterized (Ref 15). Plates, forgings, and welds were included. Isothermal embrittlement exposures extended up to 20,000 h at various temperatures ranging from 455 to 510 °C (850 to 950 °F). The salient conclusions from this study are listed below:

1. The maximum isothermal embrittlement occurred in the range 425 to 510 °C (800 to 950 °F).
2. Susceptibility to embrittlement varied widely. Some of the steels reached peak embrittlement at 425 or 470 °C (800 or 875 °F) after 20,000 h, whereas other steels had not reached peak embrittlement even after 20,000 h at 510 °C (950 °F). Embrittlement susceptibility was observed at temperatures as low as 345 °C (650 °F).

3. The 2¼ Cr-1Mo weld metal was found to have a higher susceptibility to embrittlement than base-metal plates or forgings.
4. Temper embrittlement was not found to affect the upper-shelf impact energy of the steel. Quenched-and-tempered steels (martensite-bainite) had lower transition temperatures than slow-cooled (ferrite-pearlite) steels prior to temper embrittlement. After 20,000 h of embrittlement at 455 °C (850 °F), however, the final transition temperatures for both conditions were the same.
5. Within a very broad band of scatter, a general trend of increasing embrittlement susceptibility with increasing manganese, silicon, phosphorus, and tin contents was observed.
6. The maximum value of the shift in the 54-J (40-ft·lb) transition temperature due to isothermal embrittlement, I_t, was found to be relatable to the shift in transition temperature due to step cooling, SCE, through the expression

$$I_t = 0.67(\log t - 0.91)SCE \text{ (Eq 7.1)}$$

where t is the time of exposure under isothermal conditions expressed in hours. This equation predicts that the maximum shift in transition temperature in a reactor after 30 years of service would be roughly three times the corresponding value measured from step-cooling tests.

7. Based on 25 heats subjected to isothermal embrittlement for times up to 20,000 h in the range 650 to 950 °F (343 to 510 °C), the following statistics were obtained for the 54-J (40-ft·lb) transition temperature (TT):

As-fabricated TT
    Average: −27 °F (−32 °C)
    Range: −170 to +80 °F (−112 to +27 °C)
Shift in TT
    Average: 76 °F (42 °C)
    Range: 5 to 196 °F (3 to 109 °C)
Embrittled TT
    Average: 49 °F (10 °C)
    Range: −55 to +160 °F (−48 to +71 °C)

Results of two extensive characterization studies at the Chevron Oil Co. have been reported by Erwin and Kerr (Ref 4). Their findings are generally in agreement with those of the API (Shaw) study. Some additional observations made by Erwin and Kerr are as follows. (1) Although the shifts in the 54-J (40-ft·lb) transition temperature and in the FATT were generally equal, ΔFATT could sometimes exceed the Δ54-J (Δ40-ft·lb) TT by 11 °C (20 °F). (2) A ranking of material classes in order of decreasing susceptibility to embrittlement gave the following order: submerged-arc welds (SAW), shielded metal-arc welds (SMAW), electroslag welds (ESW), and plate or forgings. The ranking was also observed to apply to the as-fabricated toughness prior to embrittlement. These differences between product forms were believed to be due not to differences in the fabrication procedure but merely to compositional differences associated with them.

A major problem in dealing with temper-embrittlement literature results from the use of the 54-J or 40-ft·lb transition temperature (TT_{40}) as a measure of toughness by some investigators, while others use a 50% ductile-to-brittle fracture transition temperature (FATT). Shaw chose to use the former, because measurement of FATT always involves subjective errors in reading the fracture surface. To facilitate conversion from one criterion to the other, the following correlations proposed by Sato *et al* may be used (Ref 16):

$$FATT\ (°C) = 1.14 \times TT_{40}\ (°C) + 21.6$$

$$(Eq\ 7.2)$$

$$\Delta FATT\ (°C) = 1.20 \times \Delta TT_{40}\ (°C) + 2.6$$

$$(Eq\ 7.3)$$

Compositional Effects. The temper-embrittlement susceptibility of various Cr-Mo steels used in reactors have been compared by Murakami, Nomura, and Watanabe (Ref 6). Their results are shown in Fig. 7.7. Among the steels, the 2¼Cr-1Mo and 3Cr-1Mo steels exhibited the highest susceptibilities. Simulated heat-affected-zone materials from various Cr-Mo steels have also been compared (Ref 17). The results from this study are shown in Fig. 7.8. It was observed that 1Cr-½Mo steels showed almost no susceptibility to temper embrittlement. In all cases, the simulated HAZ treatment was found to increase appreciably the embrittlement susceptibility, the most pronounced effect being observed in the 1¼Cr-½Mo steels. In all cases, increasing the phosphorus, tin, and silicon contents increased the susceptibility. Because the contents of these elements were not controlled, comparison between the various Cr-Mo steels was not possible. Nevertheless, the need to take into account weldment behavior rather than simply base-metal behavior was clearly brought out.

The recognition that the impurity elements phosphorus, tin, arsenic, and antimony play a major role in temper embrittlement led Bruscato (Ref 18) to propose a correlation between the shift in transition temperature and the compositional factor X, defined as

$$X = \frac{10P + 5Sb + 4Sn + As}{100} \quad (Eq\ 7.4)$$

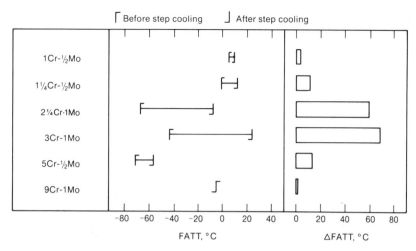

Fig. 7.7. Comparison of temper-embrittlement susceptibilities of Cr-Mo pressure-vessel steels (Ref 6).

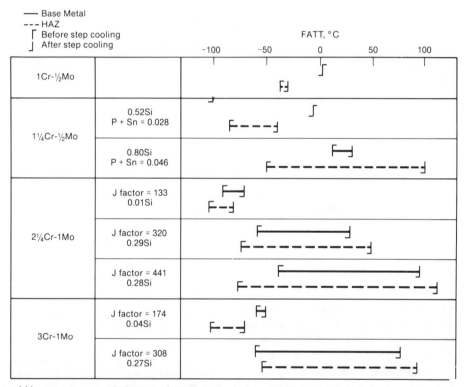

Postweld heat treatment: 1Cr-½Mo and 1¼Cr-½Mo, 20 h at 650 °C (1200 °F); 2¼Cr-1Mo and 3Cr-1Mo, 20 h at 690 °C (1275 °F). Silicon, phosphorus, and tin contents are in wt %.

Fig. 7.8. Ranges of FATT for base metal and synthetic HAZ material (peak temperature, 1350 °C, or 2460 °F) in various Cr-Mo steels before and after step cooling (Ref 17).

where X is the embrittlement factor and the concentration of each element is expressed in ppm. Further observations concerning the deleterious effects of manganese and sil- icon led Watanabe *et al* (Ref 19) to propose another embrittlement factor J, defined as

$$J = (Si + Mn)(P + Sn) \times 10^4 \quad \text{(Eq 7.5)}$$

where the concentrations of the various elements are expressed in weight percentages. This factor implies that the effects of silicon and manganese are additive; similarly, the effects of phosphorus and tin are treated as additive. Because the (Mn + Si) and (P + Sn) terms are interactive, no embrittlement can occur if one of them is zero. Results of a study by McMahon *et al* (see also Fig. 2.28, Chapter 2) are only in partial agreement with this implication (Ref 20). Their results show that the impurity combination (P + Sn) alone, in the absence of manganese and silicon, seems quite capable of causing appreciable embrittlement. Further, they showed that whereas tin exacerbates the embrittlement due to phosphorus, it is incapable of causing significant embrittlement by itself. Phosphorus, on the other hand, causes severe embrittlement regardless of the presence or absence of tin. For a given J-factor, phosphorus in combination with manganese and/or tin caused more severe embrittlement than tin. These results are inconsistent with the J-factor approach which implies an equivalent role for phosphorus and tin. The inconsistency can be resolved if the effect of tin with respect to phosphorus is recognized to be interactive rather than additive. Based on an analysis of these implications, Viswanathan and Jaffee have proposed that the J-factor should be modified as (Si + Mn + Sn) (P) with the appropriate coefficients (Ref 21). Development of more complete and accurate embrittlement equations may also have to recognize many other synergisms between other alloying elements and impurity elements, variability in strength levels, and other microstructural features. In the interim, however, application of the J-factor as a semiempirical approach to estimation of the potential embrittlement susceptibilities of steels has gained widespread industry acceptance. Because commercial steels always contain both phosphorus and tin, the inconsistencies arising from the presence of only one of the impurities as discussed earlier do not apply. Hence, within limits of broad scatter, the J-factor plots have been used successfully. For the population of

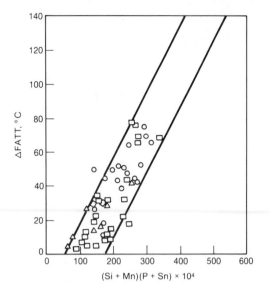

Fig. 7.9. Correlation between J factor and ΔFATT results from step-cooling studies on 2¼ Cr-1Mo steel (Ref 21).

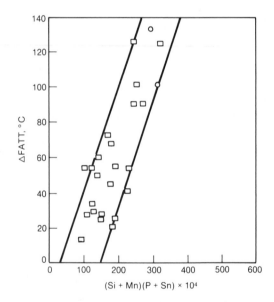

Fig. 7.10. Correlation between J factor and ΔFATT results from long-term studies of isothermal embrittlement of 2¼ Cr-1Mo steels (Ref 21).

steels of commercial composition, such as that of the API study, the J-factor correlation seems to be fairly satisfactory, as shown in Fig. 7.9 and 7.10. Figure 7.9 is based on the results of step-cooling studies, and Fig. 7.10 is based on long-term (20,000 to 60,000-h) isothermal studies. Clearly, the slope of the isothermal databand seems to be higher than that of the step-cooled databand. Additional correlations may be found in Ref 14.

In Fig. 7.9 and 7.10, data from several investigations have been included along with the data originally published by Watanabe *et al* to demonstrate the general usefulness of the J-factor approach (Ref 21). The data were obtained for different product forms, grain sizes, and heat treatment conditions, and yet the correlation between the J-factor and the degree of embrittlement is reasonably good. Based on these plots, a steel composition which will be relatively low in temper-embrittlement susceptibility can be defined.

Results from the study by McMahon *et al* (Ref 20) are consistent with other results in the literature. The deleterious effects of manganese, silicon, phosphorus, and tin in 2¼Cr-1Mo steels have been documented by several studies. Commercial steels invariably contain residual levels of nickel. In addition, they contain carbon at levels of 0.1 to 0.2%. To ascertain that the embrittlement behavior observed in the laboratory heats of steels are applicable to commercial grades, McMahon and coworkers have completed a series of studies in which the effects of carbon and nickel were investigated (Ref 22 and 23). Neither varying the carbon level in the range 0.1 to 0.2% nor addition of 0.3% Ni to the series of steels was found to result in a significant change in embrittlement behavior. The only effect of nickel was to lower the FATT of the steels in the nonembrittled condition.

Copeland and Pense investigated the effects of sulfur contents in the range 0.006 to 0.029% on the toughness of 2¼Cr-1Mo steels (Ref 24 and 25). The principal effect of sulfur was found to be a reduction of the upper-shelf Charpy V-notch energy, although at low temperatures the energy values approached each other. It was also observed that at sulfur contents greater than 0.03%, the steels did not always meet the minimum requirement of 54 J at 10 °C (40 ft·lb at 50 °F).

Effects of Microstructure. Although numerous studies have been conducted on the temper embrittlement of 2¼Cr-1Mo steels, only four have touched upon the microstructural aspects in even a cursory manner. These results have been reviewed by Viswanathan (Ref 26).

Results from the studies of Swift and Gulya (Ref 27) and of Emmanuel, Leyda, and Rozic (Ref 28) on 2¼Cr-1Mo steels exposed to various heat treatment schedules are summarized in Table 7.3 (Ref 26). The quenched-and-tempered material, containing a predominantly bainitic structure, has much higher strength and toughness than the normalized-and-tempered material. The shifts in FATT (ΔFATT) due to subsequent embrittlement are very small and of the same magnitude for the quenched-and-tempered and normalized-and-tempered materials. The effects of tensile strength on temper embrittlement are conflicting. The ΔFATT is found to increase or decrease with tensile strength for different heats.

The effects of variations in grain size also were evaluated by Swift and Gulya by varying the austenitizing temperature. The grain size was varied widely over a range from ASTM −0.4 to ASTM +7.5. The ΔFATT values ranged from 5 to 42 °C (9 to 76 °F), with the maximum value being obtained at an intermediate grain size. The range of grain sizes encountered in commercial practice is expected to be much narrower than the range covered in this study. For all practical purposes, therefore, the effect of grain-size variations on susceptibility to temper embrittlement for these steels can be concluded to be negligible.

Results of Emmanuel, Leyda, and Rozic on two weld metals (WA and WB) and on a base plate (BB) of 2¼Cr-1Mo steel showed no systematic and appreciable effect of tensile strength on temper-embrittlement susceptibility, as can be seen in Table 7.3 (Ref 28).

Temper-embrittlement studies by Kerr showed that when 2¼Cr-1Mo steel plates were either quenched and tempered to tensile strengths above 840 MPa (122 ksi) or normalized and tempered to tensile strengths below 600 MPa (87 ksi), susceptibility to

Table 7.3. The effect of heat treatment on the temper-embrittlement susceptibilities of 2¼Cr-1Mo steels embrittled by step cooling (Ref 26)

| Identification | Section size, cm | Heat treatment(a) | Tensile strength, MPa | Yield strength, MPa | Fracture appearance transition temperature, °C | | | Reference | Transformation product(b) |
					Non-embrittled	Embrittled	Shift		
B6730	2.5	FCT	470	294	24	32	8	27	Ferrite + bainite
B6730	2.5	QT	582	439	−73	−62	11	27	Bainite
			1076	969	−32	2	34		
A2766		QT	600	472	−37	−7	44	27	Bainite
			917	807	110	116	6		
WA (ES weld)	15	QT	649	518	−37	−12	25	28	Bainite + ferrite
			769	642	−1	21	20		
			922	797	99	121	22		
WB (SMAW)	15	SR	614	486	−18	24	42	28	Bainite + ferrite
			711	597	26	21	47		
			852	759	29	29	0		
BB	15	QT	714	607	−57	−46	11	28	Bainite + ferrite
			719	614	−62	−40	22		
			909	806	32	57	25		

(a) FCT, QT, and SR denote furnace cooled and tempered, quenched and tempered, and stress relieved, respectively.
(b) Not reported in the reference, but estimated on the basis of transformation curves.

temper embrittlement was negligible (Ref 29). The maximum susceptibility was encountered in plates quenched and tempered to tensile strengths in the intermediate range from 600 to 700 MPa (87 to 102 ksi).

Emmer, Clauser, and Low have reported that for both quenched-and-tempered and normalized-and-tempered section plates, the susceptibility to temper embrittlement is greater than that of annealed plates (Ref 30). For the susceptible steels, the FATT in the nonembrittled condition increased with increasing strength but the ΔFATT decreased with increasing strength in the tensile-strength range from 485 to 835 MPa (70 to 121 ksi).

Clear-cut comparisons between different transformation products tempered to the same strength level are unavailable. The range of microstructures investigated is also limited and consists mostly of bainite-ferrite aggregates (see Table 7.3). The limited data that have been published indicate a trend of slightly increasing temper-embrittlement susceptibility with products of faster cooling

rates from austenite. Variations of ΔFATT with variations in tensile strength are unsystematic and do not suggest any broad correlation between the two parameters. In general, the ΔFATT values reported are small as a result of either the inadequacy of the embrittling procedures used or the inherently lower susceptibility of the steel to embrittlement. In either case, the spread in ΔFATT due to structural variations is also small, making it difficult to distinguish between real microstructural effects and apparent effects due to scatter in the data. It is therefore reasonable to conclude that microstructural variations in the range encountered in commercial practice do not significantly affect the temper-embrittlement susceptibilities of 2¼Cr-1Mo steels. There is some evidence for 1¼Cr-½Mo steels, however, that microstructures produced by simulated heat-affected-zone heat treatments render the steel more susceptible to embrittlement than the base metal (Ref 17).

Control of Temper Embrittlement. It is

clear from the earlier discussion that control of phosphorus, tin, silicon, and manganese offers the best hope for controlling the temper-embrittlement problem with respect to new vessels. The feasibility of this can be illustrated as follows: The FATT (nonembrittled) of commercial 2¼Cr-1Mo steels is generally below room temperature. The average value, based on about 60 heats used in the API program, is −20 °C (−4 °F). This means that a maximum shift in FATT of about 45 °C (81 °F) due to temper embrittlement during the 30-year life of a pressure vessel may be allowable, if the FATT is always to be maintained below room temperature. In terms of a step-cooling test, the above ΔFATT corresponds to a value of 15 °C (27 °F) using the factor-of-three correlations suggested by Shaw. Taking a conservative approach, a J-factor of 100 may be selected as the permissible upper-bound value, based on Fig. 7.9. This target value can be met by controlling either the alloy content (Mn + Si) or the impurity content (P + Sn). Reduction of manganese to very low levels is expected to result in loss of hardenability and strength. On the other hand, steels with a maximum silicon content of 0.1% can be easily produced by vacuum carbon deoxidation (VCD). The VCD practice is widely employed in manufacture of Ni-Cr-Mo-V turbine rotor steels and 2¼Cr-1Mo pressure-vessel steels. On the other hand, low levels of silicon are claimed to result in reduced tensile strength and increased creep strength at elevated temperatures (Ref 31). Hence the choice of VCD steel may depend on the specific design criteria that are applicable. Average levels of phosphorus and tin in commercial steels in recent years are approximately 0.01 wt %. A 0.005 wt % level is also readily achievable with the current state of steelmaking technology. With these considerations in mind, it can be shown that a VCD grade containing 0.4 Mn, 0.1 Si, 0.01 P, and 0.01 Sn as well as a conventional grade containing 0.4 Mn, 0.3 Si, 0.006 P, and 0.006 Sn will meet the requirements from the temper-embrittlement point of view. Such compositions are both realistic and within current ASTM specifications. It can therefore be concluded that the temper-embrittlement problem *per se* can be readily avoided in pressure vessels by exercising proper control of composition. This would result not only in increased reliability but also in improved economy due to flexibility of operation.

In the last ten years, steelmakers have made great progress in reducing the embrittlement susceptibilities of 2¼Cr-1Mo steels. Today, phosphorus content is typically required to be below 100 ppm, which is readily achieved with electric furnace melting (Ref 4). Tin cannot be readily removed by conventional refining processes; hence, reductions in tin content have been achieved mainly through control of raw materials. Consideration is also given to restriction of silicon content to below 0.10%, which is effective in plates and forgings but not practical for weld metals. Use of fine-grain deoxidation techniques and restriction of sulfur to 0.10% have also led to increased initial toughness as well as reduced susceptibility to embrittlement in plates and forgings. Under today's steelmaking practices, temper embrittlement can virtually be eliminated in wrought products, but some concern remains with respect to weldments.

Another important step in controlling temper embrittlement consists of screening materials, particularly weld metals, by use of the accelerated step-cooling procedure. Susceptible steels and welds are thus readily screened out prior to use in reactor fabrication. Although some companies specify "J" values and control embrittlement by specifying compositions, others simply specify the 54-J (40-ft·lb) transition-temperature requirement subsequent to the step-cooling treatment. The particular procedure used varies from one organization to another.

For reactors already in service, whose weld toughness (54-J or 40-ft·lb TT) lies in the range of 50 to 350 °F, the procedure used to reduce the risk of brittle fracture during transients continues to reduce the pressure to one-fourth the design pressure in the temperature range from 10 to 175 °C (50 to 350 °F) (Ref 4). This is based on a recommendation issued by the API in 1974 to all refineries in the United States and Canada (Ref 32) and is cited as part of the ASME Boiler and Pressure Vessel Code covering pressure reduction to prevent massive brittle fracture (Ref 33). Pressure reduction on start-up and shutdown is standard practice in North America, Japan, and many other countries. This practice has paid off, and there has been no operational brittle fracture of a reactor due to temper embrittlement.

Hydrogen Embrittlement. In the presence of hydrogen or hydrogen-containing compounds, atomic hydrogen can dissolve in the steel and cause embrittlement. Such properties as tensile strength, yield strength, hardness, and impact energy remain unaffected. The principal effects of embrittlement are manifested as reduced ductility in tensile tests and as reduced threshold stress intensity for crack propagation. Because embrittlement involves diffusion of hydrogen atoms to crack-tip regions, this phenomenon is very strain-rate-dependent and is not detectable in impact tests. Constant-extension-rate tests (CERT) under low strain rates are often employed for relative evaluation of the susceptibilities of materials. Increased susceptibility is detected as a reduction in ductility, a reduction in time to failure, and an intergranular failure mode. Although the CERT is useful as a qualitative test, it gives no information that can be used in design or life assessment.

A test method commonly employed to derive a design-stress criterion is to conduct tensile tests in the hydrogen environment on smooth tensile bars at different stress levels. The curve for stress vs time to failure is generally similar to the S-N curve for fatigue, with a clearly defined threshold stress below which failure does not occur. This value of stress can be used as a basis for defining design-stress values and the allowable maximum tensile strength for the steel necessary for resistance to embrittlement. Because of the similarity in shape of stress-vs-time-to-failure curves and fatigue S-N curves, hydrogen embrittlement is often referred to as static fatigue.

An alternative approach, which yields a more quantitative measure of the effects of hydrogen embrittlement, is the use of fracture mechanics. Precracked compact-type (CT) or wedge-opening load (WOL) specimens are subjected to a rising applied load, and the load at which crack propagation starts is used to define a threshold value of K termed K_{IH}. Alternatively, a precracked specimen bolt-loaded to a given applied load is exposed to the environment. The load on the specimen relaxes as a result of crack propagation until it is reduced to a critical value at which the crack is arrested. This critical load is then used to determine the K_{IH} value. The procedures used are similar to those used for determining values of K_{Ic} in air, described in Chapter 2 (ASTM E399-72), with the difference being that lower strain rates are employed in the rising-load tests and crack growth occurs in the presence of hydrogen. The hydrogen may be either precharged into the specimen or present in the test environment. High-pressure H_2S environments sometimes have been employed because crack-growth rates are rapid under these conditions, although the K_{IH} values obtained have been found to be identical to those in pure hydrogen. This procedure therefore is helpful in rapid determination of K_{IH}. Regardless of the type of test employed, susceptibility to hydrogen embrittlement is indicated by the fact that K_{IH} values are lower than the corresponding K_{Ic} values in air. Because K_{IH} is related to the nominal stress and the defect size in the same way as K_{Ic} is related to them, it can be used quantitatively to determine the critical values of stress and defect size that will result in fracture.

Landes and McCabe have conducted an extensive investigation of two grades of $2\frac{1}{4}$Cr-1Mo steels—namely, normalized-and-tempered ASTM A387, class 2, grade 22 steel and ASTM A542, class 3 steel (Ref 34). The microstructures were predominantly bainitic and the average yield strengths for the two grades were 345 and 496 MPa (50 and 72 ksi), respectively. Tests were conducted on fatigue-precracked CT specimens 25 mm (1 in.) thick at room temperature in a H_2–6% H_2S gas mixture at pressures up

Table 7.4. Values of K_{Ic} and K_{IH} from rising-load tests on base metal (BM), weld metal (WM), and heat-affected-zone (HAZ) material in ASTM A387 and A542 steels (Ref 34)

Material	Pressure		K_{IH}		K_{Ic} from J_{Ic}	
	MPa	psi	MPa$\sqrt{m}$	ksi$\sqrt{in.}$	MPa$\sqrt{m}$	ksi$\sqrt{in.}$
A387 BM	10.3	1500	83.8	76.2	286	260
A387 BM	24.1	3500	76.9	69.9	286	260
A387 WM	24.1	3500	68.1	61.9	307	279
A387 HAZ	24.1	3500	45.8	41.6	297	270
A387 WM(a)	24.1	3500	48.2	43.8	. . .	. . .
A542 BM	5.5	800	96.8	88.0	295	268
A542 BM	10.3	1500	98.0	89.1	295	268
A542 BM	24.1	3500	67.9	61.7	295	268
A542 WM	24.1	3500	77.0	70.0	. . .	. . .

(a) Temper embrittled.

to 24 MPa (3.5 ksi). Table 7.4 provides a comparison of the K_{IH} values with K_{Ic} values estimated from J_{Ic} values. It is readily apparent that the K_{IH} values are considerably lower than the K_{Ic} values in air. The A542 grade has higher values of K_{IH} than the A387 grade despite its higher strength. In the A387 grade, the lowest values of K_{IH} were obtained in the simulated heat-affected-zone material and in the weld metal in the temper-embrittled condition. Results of a similar study on a number of pressure-vessel steels with yield strengths ranging from 585 to 795 MPa (85 to 115 ksi) tested in 21-MPa (3000-psi) hydrogen gas environments have also been reported by Loginow and Phelps (Ref 35).

Hydrogen-precharging-type fracture-mechanics experiments have been conducted by Kerr *et al* (Ref 36) and by Groeneveld and Elsea (Ref 37). Hydrogen contents in the specimens were typically 2 to 3 ppm or 6 to 7 ppm, corresponding to hydrogen concentrations that might be expected during typical reactor operation at 370 °C (700 °F) and a hydrogen pressure of 10.3 MPa (1500 psi) and at 455 °C (850 °F) and a hydrogen pressure of 17.2 MPa (2500 psi), respectively. Hydrogen-precharging-type experiments have also been reported by Iwadate, Watanabe, and Tanaka (Ref 14).

It has been pointed out that delayed-cracking experiments performed by precharging with hydrogen lead to lower values of K_{IH} than those measured in high-pressure gaseous hydrogenous environments (Ref 14). The basis for this claim is not readily apparent. Nevertheless, the former tests are more typical of reactor shutdown conditions and hence the data from such tests represent a more realistic assessment.

There are two sources of hydrogen embrittlement in reactor vessels. First, hydrogen can enter the steel through aqueous corrosion in the presence of H_2S. Alternatively, hydrogen dissolved in the steel during operation can be retained in the steel and cause embrittlement at low temperatures. In either case, the concern with embrittlement is at low temperatures (<150 °C, or <300 °F) during shutdown conditions and not at high temperatures during operation.

Hydrogen embrittlement arising from aqueous corrosion in the presence of H_2S is generally known as sulfide stress cracking (SSC). H_2S is a "poison" for the recombination reaction of atomic hydrogen to form molecular hydrogen. Hence it promotes dissolution of large quantities of atomic hydrogen, sometimes as high as 10 ppm. Fortunately, SSC generally is not a problem in reactor vessels because most of them are clad with stainless steels. Use of steels at tensile strengths below 690 MPa (100 ksi) also ensures freedom from cracking. The problem is of greater importance with respect to heat-exchanger bolting, valve parts, and oil-well tubulars made of high-strength steel.

A more imminent problem results from

the direct introduction of hydrogen into the steel during reactor operation. Despite the cladding, hydrogen can diffuse through the cladding into the base plate at elevated temperatures. If the reactor is cooled too quickly for hydrogen to diffuse out of the steel, it becomes supersaturated at ambient temperature.

The through-thickness steady-state distribution of hydrogen in the wall of a stainless-steel-clad reactor can be readily calculated using simple Fick's-law considerations. The parameters needed for the calculation are the temperature and pressure of operation, the wall thicknesses of the cladding and the shell, and the hydrogen solubility and diffusivity expressions applicable to the cladding and the base metal. An illustration of this procedure, based on the paper by Adams and Welland, is provided below (Ref 38).

According to Fick's law, the flux J (quantity of hydrogen which diffuses through a unit area per unit time) is related to the diffusion coefficient D and the concentration gradient dC/dx through the relationship

$$J = -D \frac{dC}{dx} \qquad \text{(Eq 7.6)}$$

For a clad system (shown in Fig. 7.11) with a cladding thickness t_c and a base-metal thickness t_b, the flux through the cladding and through the base metal can be equated to give

$$-J = D_c \frac{C_1 - C_2}{t_c} = D_b \frac{C_3 - C_4}{t_b} \qquad \text{(Eq 7.7)}$$

where C_1, C_2, C_3, and C_4 are the hydrogen concentrations at the locations shown in Fig. 7.11 and D_c and D_b are the diffusion coefficients for hydrogen in the cladding and base metal, respectively. The clad surface will have a hydrogen pressure $P_{H_{2(1)}}$ equal to the process hydrogen partial pressure, while the external partial pressure $P_{H_{2(4)}}$ is zero. Using Sievert's law, we get

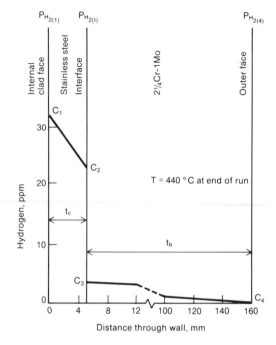

Fig. 7.11. Variation of hydrogen concentration through a reactor wall at steady state under a hydrogen pressure of 15 MPa (150 bars) at 440 °C (825 °F) (Ref 38).

$$C_1 = S_c \sqrt{P_{H_{2(1)}}} \quad \text{and} \quad C_4 = S_b \sqrt{P_{H_{2(4)}}} = 0$$
$$\text{(Eq 7.8)}$$

At the cladding/base metal interface, where the partial pressure of hydrogen is $P_{H_{2(I)}}$,

$$C_2 = S_c \sqrt{P_{H_{2(I)}}} \quad \text{and} \quad C_3 = S_b \sqrt{P_{H_{2(I)}}}$$

and hence

$$C_2/C_3 = S_c/S_b \qquad \text{(Eq 7.9)}$$

where S_c and S_b are the solubilities of hydrogen in the cladding and base metal, respectively. It can be shown that

$$C_2 = \frac{C_1}{1 + D_b S_b t_c / D_c S_c t_b} \quad \text{and}$$

$$C_3 = \frac{C_1}{S_c/S_b + D_b t_c / D_c t_b} \qquad \text{(Eq 7.10)}$$

where the values of S_b, D_b, S_c, and D_c are known from literature (Ref 38) and are expressed as

$$S_c = 12.88 \exp(-1078/T) \text{ ppm}$$

$$\text{(Eq 7.11a)}$$

$$D_c = 93.1 \exp(-6767/T) \text{ cm}^2/\text{h}$$

$$\text{(Eq 7.11b)}$$

$$S_b = 43.0 \exp(-3261/T) \text{ ppm}$$

$$\text{(Eq 7.11c)}$$

$$D_b = 5.04 \exp(-1600/T) \text{ cm}^2/\text{h}$$

$$\text{(Eq 7.11d)}$$

where T is expressed in K. By substitution of typical values of $t_b = 15$ cm (5.9 in.) and $t_c = 0.5$ cm (0.2 in.), C_2 and C_3 can be calculated at 440 °C (825 °F) and a pressure of 150 bars (2150 psi, or 15 MPa). The steady-state concentration profile thus obtained is plotted in Fig. 7.11. Similar calculations have been performed for a 200-mm- (7.9-in.-) thick reactor vessel with a cladding 10 mm (0.4 in.) thick operating at 425 °C (800 °F) and a hydrogen pressure of 13.8 MPa (2000 psi) by Johnson and Hudak (Ref 39). Their calculations show that at steady state a maximum of 3 to 5 ppm of dissolved hydrogen may be present in the steel shell at the cladding interface. If this level is below the safe hydrogen level permissible, hydrogen embrittlement will not be a problem. If, on the other hand, this level exceeds the safe hydrogen level permissible, the reactor must be degassed at the operating temperature until the actual hydrogen content is decreased to a value below the safe level. The safe hydrogen levels have been estimated to be about 8.5 ppm for a steel with a strength level below 690 MPa (100 ksi) and about 4.3 ppm for a steel with a strength level of 760 MPa (110 ksi).

Estimation of safe H_2 levels is based primarily on a knowledge of the effects of tensile strength and hydrogen content on K_{IH}. Based on extensive investigations by Kerr

et al (Ref 36) and by Groeneveld and Elsea (Ref 37), such information has already been gathered for base metal and weld metal in 2¼Cr-1Mo steels, as shown in Fig. 7.12 (Ref 4). The same results have been cross-plotted in Fig. 7.13 (Ref 4). The safe hydrogen levels can be calculated by setting the condition that the actual stress-intensity factor due to a crack at the applied stress should not exceed the K_{IH} value for the steel. The stress-intensity expression for surface flaws is given by

$$K = 1.94\sigma \left[\frac{a}{\phi - 0.212(\sigma/\sigma_y)^2} \right]^{1/2}$$

$$\text{(Eq 7.12)}$$

where σ is the applied stress, a is the crack size, and ϕ is a crack-shape factor. For a long surface crack with a length more than ten times its depth, $\phi = 2.46$. By assuming a crack depth of a = 25 mm (1 in.) and an applied stress of one-third the yield strength, the safe H_2 content as a function of the tensile strength of the steel can be calculated (see Fig. 7.14). These data show that for tensile strengths up to 690 MPa (100 ksi), the safe H_2 level exceeds the solubility of hydrogen and hence no degassing is required. For tensile strengths above 760 MPa (110 ksi), however, the safe levels of H_2 approach or fall below the expected level of dissolved hydrogen, thus requiring degassing of the reactor prior to shutdown. Refinery-industry practice is completely consistent with these guidelines. The assumptions regarding 25-mm- (1-in.-) deep pre-existing flaws as well as stresses approaching one-third the yield strength are both highly conservative and reflect stringent industry requirements in this regard. Some of this conservatism, however, is justified in view of the possible presence of unknown residual stresses and in view of temper embrittlement/hydrogen embrittlement interactions which may appreciably reduce the K_{IH} values, as discussed below.

Combined Effect of Hydrogen Embrittlement and Temper Embrittlement. Several

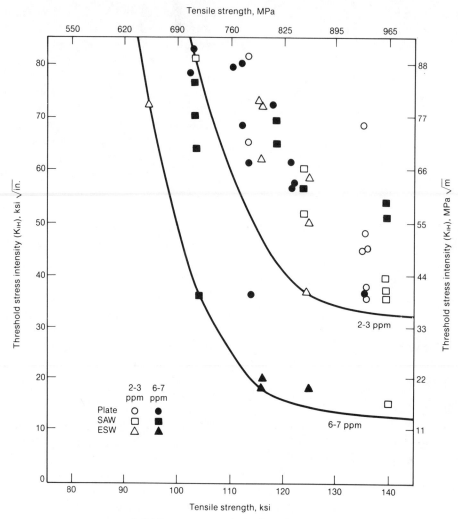

Solid lines represent lower-bound curves.

Fig. 7.12. Threshold stress intensity for cracking, K_{IH}, vs tensile strength for 2¼Cr-1Mo steel charged to two different bulk hydrogen concentrations (Ref 4).

studies have shown that the threshold stress intensity for cracking in hydrogen, K_{IH}, can be appreciably reduced in steels which have been subjected to prior temper embrittlement (Ref 34 and 40 to 43). The effect of temper embrittlement on the hydrogen-embrittlement behavior of commercial-purity 2¼Cr-1Mo steels has been investigated by the Japanese Pressure Vessel Research Council (Ref 43) and by Landes and McCabe (Ref 34). In both studies, K_{IH} was found to decrease systematically with ΔFATT due to temper embrittlement. These results recently have been augmented by more published data (Ref 14). Figure 7.15 presents the rela-

tionship between K_{IH} and the degree of temper embrittlement expressed by 50% FATT data from Charpy tests (Ref 14). Variation in fracture toughness, K_{Ic}, measured in air is also plotted for comparison. Measurements of K_{IH} were performed using two methods: the rising-load method with hydrogen-charged 25-mm-(1-in.-) thick specimens, and the bolt-loaded method with 25-mm (1-in.) WOL specimens in a 500-ppm H_2S solution. The absorbed hydrogen levels ranged from 2 to 4.2 ppm. The K_{Ic} values were estimates from J_{Ic} tests. The salient points that emerge from reviewing these data are (1) K_{IH} decreases rapidly

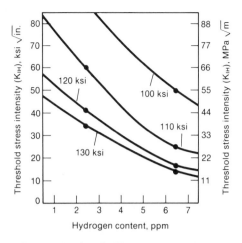

Fig. 7.13. Threshold stress intensity for cracking, K_{IH}, vs hydrogen concentration for 2¼Cr-1Mo steel at various tensile strengths (Ref 4).

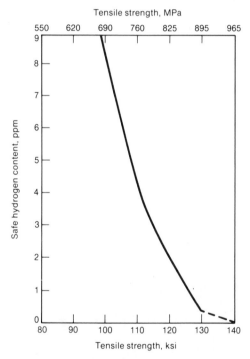

Fig. 7.14. Safe hydrogen concentration to avoid hydrogen-crack growth below 150 °C (300 °F), assuming a crack 25 mm (1 in.) deep and a stress equal to one-third of the yield strength (Ref 4).

as FATT increases from about −100 °C (−150 °F) to about +75 °C (165 °F) and then reaches a plateau at a level as low as 25 MPa√m̄ (23 ksi√in.) for the steel contain-

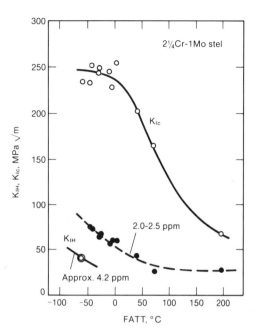

Fig. 7.15. Relationship between K_{IH} and FATT for 2¼Cr-1Mo steel at hydrogen contents of 2 to 2.5 ppm and 4.2 ppm (Ref 14).

ing 2 to 2.5 ppm of hydrogen; and (2) at a hydrogen level of 4.2 ppm, the same trend is apparent except that the K_{IH} values are reduced even further. The relationship between K_{IH} and FATT could be expressed in the form (Ref 14)

$$K_{IH} = 0.0014 \, FATT^2$$
$$- 0.421 \, FATT + 57.0 \quad (Eq \ 7.13)$$

for a steel containing 2 to 2.5 ppm of hydrogen, where K_{IH} is expressed in MPa√m̄, and FATT in °C.

The results presented in Fig. 7.15 have significant implications with respect to both operating practice for current vessels and specification of cleaner steels for future vessels. For vessels currently in operation, safe hydrogen levels may be appreciably reduced when temper embrittlement is present. It was previously shown that at an applied K of 44 MPa√m̄ (40 ksi√in.) — i.e., σ = 228 MPa (33 ksi), a = 25 mm (1 in.), and φ = 2.46 — the safe H₂ level was 8.5 ppm for a steel with a tensile strength of 690 MPa

(100 ksi), using Eq 7.13. If the same vessel had been subjected to severe temper embrittlement with a resultant FATT of 100 °C (212 °F), the K_{IH} value would have been reduced to 29 MPa$\sqrt{m}$ (26 ksi$\sqrt{in.}$), even at a hydrogen level of only 2 to 2.5 ppm. Hence, the safe hydrogen level would now be well below 2 ppm. Alternatively, at a K_{IH} value of 29 MPa$\sqrt{m}$ (26 ksi$\sqrt{in.}$), the tolerable flaw size at a hydrogen content of 2 to 2.5 ppm would be as low as 9.45 mm (0.372 in.). This means that either degassing procedures would now be required or tolerable crack sizes would have to be more restricted when temper embrittlement and hydrogen embrittlement occurred simultaneously. Additional studies are needed to address this very important issue, and revisions of current practice in terms of allowable hydrogen levels and defect levels may become necessary.

Results presented in Fig. 7.15 also show that decreasing the temper-embrittlement susceptibility of a steel (by control of the J factor) in the region where K_{IH} is insensitive to FATT will not result in any benefit from a hydrogen-embrittlement point of view. In the region where K_{IH} is very sensitive to FATT, even small reductions in the temper-embrittlement susceptibility of the steel may lead to large benefits. This aspect should be kept in mind during implementation of compositional controls for future vessels.

The correlation between hydrogen embrittlement and temper embrittlement can be expressed schematically as shown in Fig. 7.16. This figure has been adapted from Ref 6, with a minor modification to include the effect of strength level based on insight gained from other studies (Ref 44). At very low levels of embrittlement and at

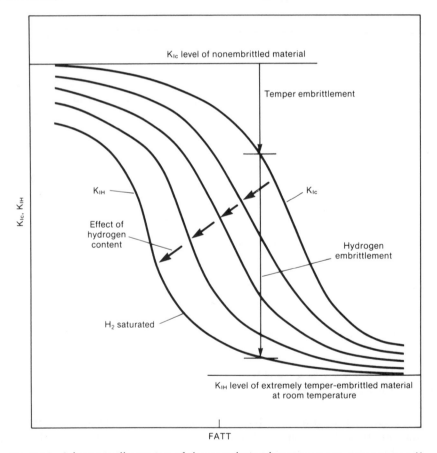

Fig. 7.16. Schematic illustration of the correlation between room-temperature K_{Ic}, K_{IH}, and FATT (Ref 6).

high degrees of embrittlement, the K_{Ic} and K_{IH} values approach each other. It is in the intermediate range of temper embrittlement that major interactive effects are observed.

Industry Experience With Failure Due to Embrittlement. Industry experience with respect to failure of pressure vessels due to hydrogen embrittlement and temper embrittlement has been reviewed by Kerr (Ref 4). This experience essentially includes that of Chevron at the Richmond Isomax complex and that of the Nippon Mining Co. at the Mizushima refinery. At the Richmond site, in 1966, cracking was observed in eight reactors made of $2\frac{1}{4}$Cr-1Mo steel after only a few months of operation. Cracks had developed in welds which were either nozzle welds in the top and bottom heads of the reactor or circumferential welds in the shell. The reactors were repaired by removing weld metal in the areas of the cracks, rewelding, and reheat treating of all the reactors at the site. Based on detailed studies, the cracking was attributed to excessive strength levels in the weld and base metal which had resulted in reduced toughness and a high susceptibility to hydrogen-induced cracking. This problem was overcome by tempering the repaired vessels to a lower tensile strength, below 760 MPa (110 ksi), which also resulted in increased toughness. The subsequent industry practice of limiting the tensile strength of ASTM SA387, grade 22, class 2 steel for hydroprocessing reactors to 690 MPa (100 ksi) and requiring an impact energy of 54 J at 10 °C (40 ft·lb at 50 °F) was the direct result of Chevron's experience.

The second major failure experience occurred in 1974 during field repair of a hydrodesulfurizer reactor at the Mizushima refinery of the Nippon Mining Co. The reactor had been in service for 3 to 4 years, after which it was removed from operation to be converted for use in some other process. This modification called for local weld repair and postweld heat treatment of the $2\frac{1}{4}$Cr-1Mo steel shell at stainless steel internal attachment welds. During the heat treatment, a brittle fracture of the reactor occurred around 270° of the reactor circumference. An extensive investigation of the failure was undertaken. The scenario for the failure of the shell was found to involve the following sequence of events. (1) Deep cracks initiating at attachment welds had formed due to a combination of high residual stress and degradation of the stainless steel welds during initial fabrication. The initial fabrication called for attachment of all stainless steel internals prior to final postweld heat treatment of the reactor. This practice had led to sigma-phase embrittlement of the stainless steel welds and very high residual stresses due to differential thermal expansion at these welds. The resulting cracks in the welds also penetrated the $2\frac{1}{4}$Cr-1Mo base metal. (2) The pre-existing cracks would have been within the tolerable critical crack size for the base metal except for the fact that, during service, severe temper embrittlement of the base metal had occurred, reducing its toughness. (3) Localized postweld heat treatment during the postservice field repair led to severe thermal stresses. The combination of temper embrittlement and the thermal stresses caused the actual crack size to exceed the critical crack size, resulting in rapid unstable crack propagation and final failure. The importance of this failure and various lessons learned from it have been published in a number of papers. The major changes in industry practice that have come about as a result of this failure are as follows. First, the heat treatment sequence has been modified such that, at all areas of high local stress, type 347 stainless steel is applied after final postweld heat treatment of the vessel. As an additional precaution, a buttering layer of sigma-free, ductile type 309 stainless steel is applied prior to postweld heat treatment. Second, improved field welding procedures and start-up and shutdown procedures have been implemented to reduce thermal stresses. Third, an improved awareness of the temper-embrittlement problem and the methods of controlling it has come about. Numerous publications by Watanabe *et al* have described this failure incident and the resultant remedial actions (Ref 45 to 51).

Hydrogen Attack

High-temperature exposure of carbon and low-alloy steels to high-pressure hydrogen leads to a special form of degradation known as hydrogen attack. In contrast to hydrogen embrittlement, which degrades toughness at low temperatures and imposes restrictions on start-up and shutdown procedures, hydrogen attack leads to a degradation of material properties at the operating temperature.

Hydrogen attack is basically a decarburization reaction. If the attack is confined to the surface, it is known as surface attack. If the attack occurs internally, the resultant product—i.e., methane—is unable to escape, forms bubbles, and leads to permanent internal damage. The reaction involved can be written as

$$2H_2 + C \rightleftarrows CH_4 \qquad \text{(Eq 7.14)}$$

where the carbon is in the form of carbides. The methane bubbles nucleate at carbides, grow under methane pressure, and link up to form fissures and cracks, as shown in Fig. 7.17. Applied stress aids the damage process but is not a necessary prerequisite. Surface decarburization is favored by lower partial pressures of hydrogen and higher temperatures. Under these conditions, the carbon in the steel diffuses to the surface to interact with the hydrogen. In internal attack, which is promoted by higher partial pressures of hydrogen and lower temperatures, hydrogen diffuses inward to react with the carbides. The most critical parameters affecting the susceptibility to attack are metal temperature, hydrogen partial pressure, applied stress, chemical composition, and heat treatment. Increasing the contents of strong carbide-formers in the steel, such as chromium, molybdenum, vanadium, tungsten, and niobium, decreases the susceptibility of the steel to attack.

Surface decarburization is characterized by a decrease in the carbon content of a shallow surface layer. The lower-carbon material is expectedly more ductile, but is also weaker and softer. Because there is no ac-

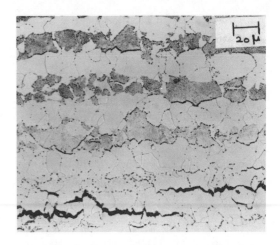

Fig. 7.17. Decarburization and fissuring in a pressure-vessel steel due to hydrogen attack (photo courtesy of M. Prager, Metal Properties Council, New York).

companying fissuring, surface decarburization *per se* is of little concern, except insofar as it serves as a forewarning of more serious internal damage.

Internal decarburization has been further divided into two categories: fissuring and blistering. The basic mechanism for both involves internal decarburization and formation of methane. The principal differences seem to be that the latter type of damage is more often associated with inclusion stringers and laminations, with less evidence of decarburization around grain boundaries. For practical purposes, such a distinction appears to be purely superficial. A fuller description of the mechanisms and manifestations of hydrogen attack can be found in the article by Nelson (Ref 52).

Damage to steels by high-temperature, high-pressure hydrogen is preceded by a period of time when no noticeable change in properties can be detected by the usual test methods. The length of time before hydrogen attack is detected is termed the incubation time. It is generally believed that the incubation time represents the time beyond which the methane pressure inside a cavity is sufficiently high to overcome the opposing surface-tension forces so that the cavity can exceed the critical nucleus size and become stable. A schematic illustration of the kinetics of hydrogen attack based on grain-

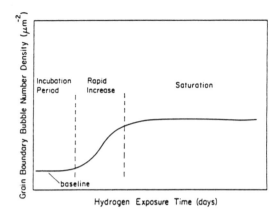

Fig. 7.18. Number density of grain-boundary cavities as a function of hydrogen exposure time (Ref 53).

boundary bubble-density measurements by Stone *et al* is shown in Fig. 7.18 (Ref 53). Beyond the incubation period, bubble density increases rapidly with time and then reaches a saturation point. Higher temperatures, higher stresses, higher hydrogen pressures, and prior cold work have the effect of reducing the incubation time and raising the saturation value of cavity density, as shown in Fig. 7.19 (Ref 53). In view of the importance of incubation time, the American Petroleum Institute has published incubation curves for carbon steels and ½Mo steels, based on both laboratory and field experience.

Several mechanistic studies have attempted to model the nucleation and growth of cavities in terms of the thermodynamic diffusion characteristics of the process. The superimposed effect of applied stresses has been taken into account by Stone *et al* (Ref 53). Based on their model, it is now possible to predict cavity density and size as a function of stress, temperature, and hydrogen partial pressure in 2¼Cr-1Mo steels. Further theoretical developments are needed before the rupture lives of components under operating conditions can be predicted.

Effects of Hydrogen Attack on Mechanical Properties. Hydrogen attack is manifested by losses in room-temperature tensile strength, ductility, impact energy, and density, as shown in Fig. 7.20 (Ref 54 and 55). In all cases, these effects start to occur only

when a critical temperature of prior exposure to hydrogen is exceeded. With increasing alloy content of the steel, the critical exposure temperature for susceptibility is shifted to increasingly higher temperatures. The effect of hydrogen attack on creep-rupture life at 540 °C (1000 °F) for a ½Mo steel is shown in Fig. 7.21 (Ref 56). A pronounced reduction in rupture life is observed. Figure 7.22 shows the relationships between stress-rupture and time for 1¼Cr-½Mo, 2¼Cr-1Mo, and 3Cr-1Mo steels at 550 °C (1020 °F) (Ref 57). In a hydrogen atmosphere at a pressure of 10 MPa (1400 psi), all the steels ruptured at shorter times in hydrogen than in argon. The difference was more pronounced in the lower-alloy steels.

The effect of hydrogen pressure on rupture time for 1Cr-½Mo steel at 540 °C (1000 °F) has been investigated by Holmes *et al* (Ref 58). Figure 7.23, based on their results, shows that rupture time is appreciably reduced by increasing the pressure of hydrogen.

The effect of temperature on the rupture strength of quenched-and-tempered 1¼Cr-½Mo steels in hydrogen has been investigated by Watanabe *et al* (Ref 59). Their results (see Fig. 7.24) show a significant drop in the rupture strength at all temperatures down to 500 °C (930 °F).

In view of the importance of 2¼Cr-1Mo steels, several studies have focused on their creep-rupture properties under hydrogen-exposure conditions (Ref 60 to 62). A detailed review of these studies can be found in Ref 4. The principal conclusions from these studies were as follows. (1) Losses in creep-rupture properties were apparent only in 2¼Cr-1Mo steels in the annealed (slow-cooled) condition and only at temperatures at or above 540 °C (1000 °F). For rapidly cooled and tempered microstructures, no deleterious effects were apparent. (2) In the range of temperatures of practical interest — i.e., below 480 °C (900 °F) — no deleterious effects were observed in any of the steels. (3) At all exposure temperatures up to 595 °C (1100 °F), attack of 2¼Cr-1Mo steels consisted primarily of surface decar-

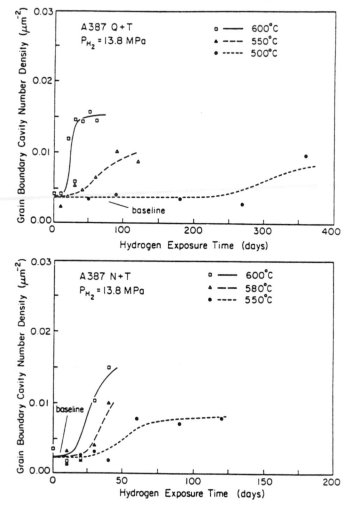

Fig. 7.19. Number density of grain-boundary cavities for ASTM A387 steels in the quenched-and-tempered (above) or normalized-and-tempered (below) condition as a function of hydrogen-exposure time and temperature at a hydrogen pressure of 13.8 MPa (2000 psi) (Ref 53).

burization with no evidence of internal fissuring. It is very important to note that all of these studies were short-time studies with rupture times extending up to only 2000 to 3000 h.

The most comprehensive and long-duration (up to 13,000 h) test results in the range of temperatures where 2¼Cr-1Mo steels are utilized are those of Erwin and Kerr (Ref 4). Quenched-and-tempered steel specimens were tested in argon at 103 kPa (15 psi) and in hydrogen at 13.8 MPa (2 ksi), at temperatures of 455, 510, and 595 °C (850, 950, and 1100 °F). Creep-rupture results from this study are shown in Fig.

7.25. At all the test temperatures, the creep-rupture strengths of the steels were actually found to be higher in hydrogen than in argon, presumably because of a strengthening effect of dissolved hydrogen. This result is in contrast with the results of Ishizuka and Chiba at 550 °C (1020 °F), which showed a degradation in creep-rupture strength due to hydrogen (Ref 57).

Factors Affecting Hydrogen Attack. The factors affecting hydrogen attack have been reviewed by Stone *et al* (Ref 53). The environmental variables known to influence hydrogen attack include temperature, pressure, and stress. The material variables in-

P_{H_2}, 30 MPa; holding time: 360 h

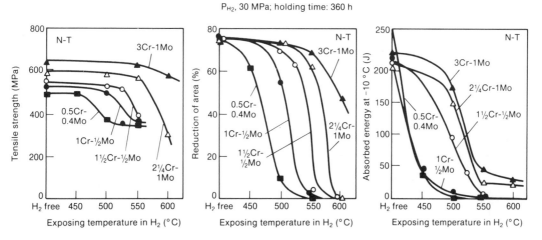

Fig. 7.20. Effects of hydrogen-exposure temperature on the mechanical properties of Cr-Mo steels with carbon contents of 0.12% (Ref 54 and 55).

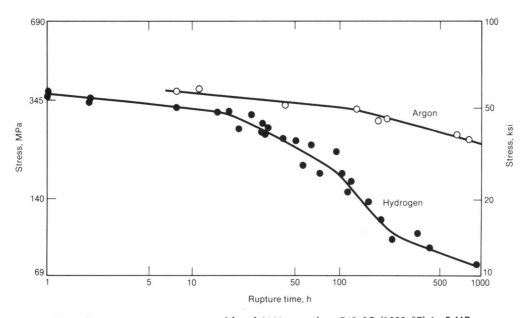

Fig. 7.21. Loss in creep-rupture life of ½ Mo steel at 540 °C (1000 °F) in 5-MPa (725-psi) hydrogen (Ref 56).

clude alloy content, impurity content, heat treatment, cold work, and grain size. The influences of these variables can be considered primarily in terms of their effects on the nucleation and growth of methane bubbles.

The methane-formation reaction given by Eq 7.14 is exothermic, so that the equilibrium methane pressure, which is the driving force for methane-bubble nucleation and growth, will increase with decreasing temperature for a given hydrogen pressure. On the other hand, diffusion rates and reaction rates will decrease with decreasing temperature. The balance between these opposing tendencies determines the net effect of temperature. Shih and Johnson have rationalized the shape and position of plain carbon steels in the Nelson diagram (discussed later) on this basis (Ref 63). In the range of hydrogen pressures and temperatures of practical interest, it is anticipated that the methane gas in equilibrium with hydrogen does not behave as an ideal gas; hence the

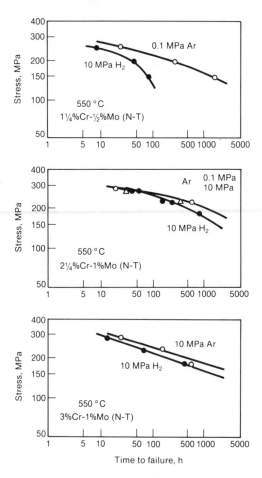

Fig. 7.22. Relationships between stress rupture and time for Cr-Mo steels in hydrogen and in argon at a pressure of 10 MPa (1400 psi) (Ref 57).

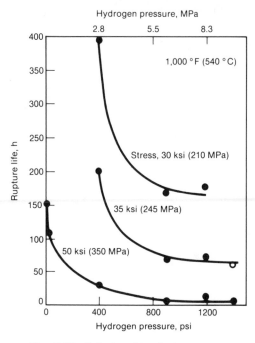

Fig. 7.23. Relationships between rupture life and hydrogen pressure for a 1Cr-½Mo steel at various applied stresses (Ref 58).

methane pressure would become insensitive to the hydrogen pressure above a few hundred MPa. Consistent with this trend, the Nelson curves become independent of pressure at low temperatures.

It has been observed that application of stress exacerbates hydrogen attack in steels. According to Stone *et al*, there are at least two reasons why a tensile stress should enhance hydrogen attack (Ref 53): first, tractions across the grain boundaries add to the driving force for bubble growth already provided by the internal methane pressure; and second, applied stresses can cause grain-boundary sliding and produce stress-concentration sites which are favorable for methane-bubble nucleation.

Alloying with carbide-stabilizing elements

is known to increase resistance to hydrogen attack. Several reasons for this beneficial effect can be cited: (1) reduction in the carbon activity, which in turn reduces the rate of methane production and the equilibrium methane pressure; (2) reduced supply of dissolved carbon due to the greater stability of the alloy carbides; (3) creep strengthening by the alloying elements (e.g., molybdenum), which may reduce bubble growth controlled by matrix deformation; and (4) grain-boundary strengthening and reduced grain-boundary sliding and a consequent reduction of favorable sites for bubble nucleation. The various elements which reduce hydrogen attack are chromium, molybdenum, tungsten, vanadium, titanium, niobium, zirconium, tantalum, thorium, manganese, phosphorus, and sulfur. On the other hand, hydrogen attack can also be enhanced by carbon, nickel, copper, and sulfur. Recently, pronounced detrimental effects due to aluminum have been reported in 1Cr-½Mo steels, as illustrated in Fig. 7.26 (Ref 64). In this figure, the critical embrittling temperature is defined as that

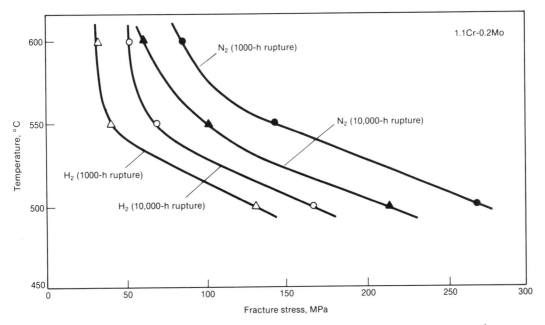

Fig. 7.24. Relationships between temperature and fracture stress in internal-pressure rupture tests (Ref 59).

temperature above which, if the steel is exposed, the reduction in area at room temperature will be reduced by more than 50% of its initial value. Control of aluminum content to levels below 0.010% was therefore indicated as being very desirable. The favorable effect of stress-relief (SR) treatments is also shown in this figure.

The effects of heat treatment on susceptibility to hydrogen attack have been reviewed in detail in Ref 55. It has been reported that excessively high austenitizing temperatures (greater than 1000 °C, or 1830 °F, for 1¼Cr-½Mo steel) lead to increased susceptibility. Furthermore, a quenched-and-tempered 1¼Cr-½Mo steel was reported to be more susceptible to attack than a normalized-and-tempered steel. The effects due to austenitizing and cooling rates have been attributed to grain-size effects and the distributions and natures of the carbide phases. Postweld heat treatments and tempering treatments have pronounced effects on the susceptibility to attack. The effects of postweld heat treatment are illustrated in Fig. 7.27 (Ref 65). Extended tempering and postweld treatments lead to alloy carbide phases which are increasingly stable. The se-

quence of reactions by which M_3C is transformed to M_7C_3 and eventually to M_6C and $M_{23}C_6$ is well known. The more well tempered the structure is, the more stable the carbides are and hence the more resistant the steel is to hydrogen attack. In general, heat-affected-zone material around welds has been shown to be more susceptible to attack than base metal (Ref 55). Once again, this may be due to the larger grain size, coupled with the possible existence of prior defects and residual stresses at these locations. Because pockets of methane will form at nucleation sites at grain boundaries and adjacent to carbides, inclusions, etc., it is helpful to achieve as uniform a structure as possible. Large nonmetallic inclusions are particularly harmful because even a small incipient crack will have a large effective crack length if it forms near an inclusion; strung-out inclusions from rolling and segregation streaks should be meticulously avoided. Excessive alloy segregation is also undesirable because it leads to alloy-lean areas which are susceptible to attack.

Cold working of steels during fabrication can increase susceptibility to hydrogen attack. Forming can introduce 5 to 10% cold

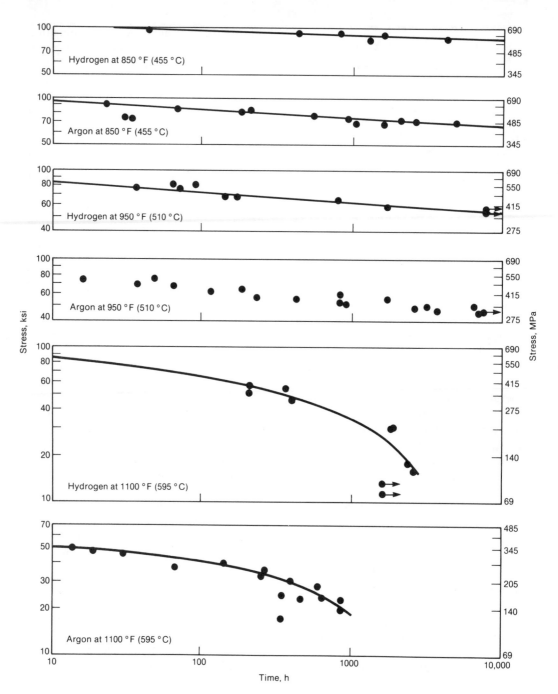

Fig. 7.25. Stress-rupture curves for 2¼Cr-1Mo steels in hydrogen at 2000 psi (14 MPa) and argon at 15 psi (105 kPa) (Ref 4).

work into the plates used in the fabrication of heavy-wall pressure vessels. Large strains can introduce fracture in and around carbides and inclusions as well as residual stresses in the surrounding matrix. The fracture process would increase the number of favorable sites for methane-bubble nucleation, and residual stresses would add to the driving force for nucleation and growth of the bubbles. This problem generally can be ameliorated, but not totally eliminated, by proper stress-relief treatments. It is also im-

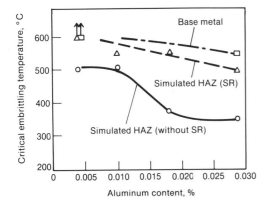

Fig. 7.26. Effect of aluminum content on the critical embrittling temperature of 1Cr-1Mo steel exposed to hydrogen for 100 h at 20 MPa (2900 psi) (Ref 64).

portant to eliminate stress-concentrating features in the vessels by careful design and fabrication.

Detection of Hydrogen Damage. Inspecting for and detecting hydrogen damage is usually more difficult than detecting ordinary high-temperature oxidation or sulfidation. There is no visible evidence of attack. Conventional NDE methods are of limited value. Surface decarburization can be detected by hardness measurements using por-

table hardness testers. Surface replication can also reveal decarburization. Light grinding of the surface is necessary before replication to eliminate corrosion products and residual decarburization from heat treatment. Depth of attack can be determined only by metallography and by microhardness traverses across the thickness.

Internal decarburization and fissuring can be detected only by destructive metallography and mechanical tests on through-the-wall samples. Internal decarburization can take place without surface decarburization so that surface replication techniques are ineffective. On a fracture surface, internal attack looks very similar to creep-rupture failure, with cavities present at the grain boundaries resulting in intergranular fracture. The trapped methane can be detected and measured after breaking open a specimen in a vacuum chamber. The simplest test is a bend-flattening test at room temperature. For instance, the normal ductility of low-carbon steel is such that a sample 6.4 mm (¼ in.) thick can be flattened through an angle of 180° without cracking. Hydrogen-damaged samples will crack at lower angles due to reduced ductility. Radiogra-

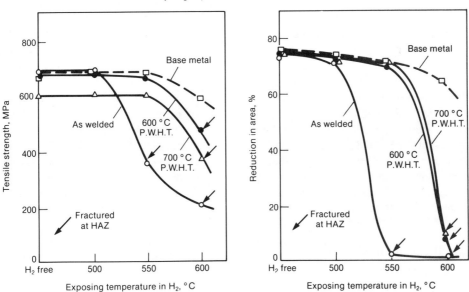

Fig. 7.27. Effect of postweld heat treatment (5 h) and temperature on the tensile properties of a welded joint in 2¼Cr-1Mo steel prepared by shielded metal-arc welding (hydrogen pressure, 30 MPa (4350 psi); hold time, 360 h) (Ref 65).

phy is incapable of detecting microfissures. Increased attenuation of an ultrasonic pulse in the presence of fissures has sometimes been observed. Based on this, ultrasonic techniques have been employed with mixed success (Ref 66). Because ultrasonic testing provides rapid and wide coverage, it is a desirable technique, but should be used in conjunction with conventional metallography.

Nelson curves, which are experience-based curves that define the safe operating limits in pressure-temperature space for various steels, are shown in Fig. 7.28 (Ref 67). These curves have traditionally been used in the chemical and petroleum industries for material selection as well as for design of vessels for hydrogen service. The Nelson curves have been plotted so as to pass below the minimum conditions of temperature and hydrogen partial pressure below which any damage has been detected regardless of the length of time in service. Satisfactory performance points were plotted only if the samples or equipment had been exposed for a minimum of one year. Unsatisfactory performance points were plotted regardless of the length of exposure. Data to support the curves were derived from a variety of commercial processes as well as from laboratory tests. It is important to note that the Nelson curve for a material does not imply that the material will not fail in the region below the curve, but only that such failures have not occurred in the past. The curves are subject to downward revision based on experience. It is also important to note that the Nelson curves provide only guidelines for reference conditions. Actual vessel design may follow the curve or be 14 to 28 °C (25 to 50 °F) below the curve, to ensure additional conservatism. The actual design practice is proprietary to the various companies designing the vessels.

In Fig. 7.28, the dashed curves represent tendencies for steels to decarburize at the surface, whereas the solid lines denote internal attack. Note that low temperatures and high pressures promote internal attack, whereas high temperatures and low pressures promote surface decarburization. The

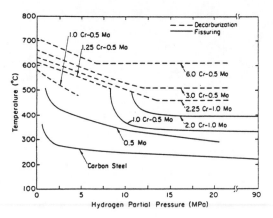

Fig. 7.28. Nelson curves for hydrogen attack of steels (Ref 67).

higher-alloy steels such as 2¼Cr-1Mo steel have shown only decarburization in service, whereas the lower-alloy steels seem to be subject to both types of attack. The resistance to damage is clearly found to increase with increasing alloy content. Among the alloying elements, molybdenum has been found to be highly effective, as can be readily seen in Fig. 7.29.

Several limitations of the Nelson-curve approach have been pointed out in the literature, as follows:

1. Because Nelson curves are based on prior experience, they are subject to continuing revision in the light of new failures. For example, recent Russian work (Ref 68 and 69) lowered the Nelson curves for plain carbon steel. Although the temperature was lowered by only about 25 °C (45 °F), the pressure was lowered by 5.5 MPa (800 psi). Nearly all the Nelson-curve changes over the years have involved lowering of the carbon steel and C-Mo steel curves in the hydrogen-pressure range below 10.3 MPa (1.5 ksi) absolute. This was, in fact, the principal modification in the latest revision, by which the curve for C-½Mo steel was displaced downward by approximately 55 °C (100 °F). Lowering of the curve was based on three failures documented by service reports to the API.

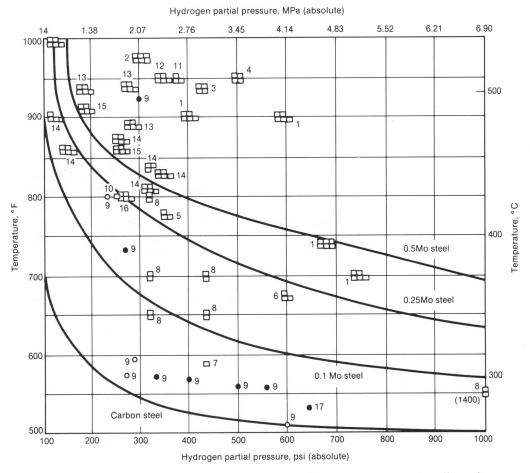

Fig. 7.29. Operating limits for steels in hydrogen service, showing the effect of molybdenum (Ref 67).

A detailed review of the circumstances leading to the revisions may be found in the paper by Bonner (Ref 70).

2. The Nelson curves do not accurately represent weld-metal or heat-affected-zone (HAZ) behavior. It has been shown that C-½Mo steels suffered hydrogen attack at temperatures 30 to 40 °C (70 to 104 °F) below the Nelson curves (Ref 71). Methane blisters were observed by Merrick and McGuire in C-½Mo and 1¼Cr-½Mo piping below their respective Nelson curves subsequent to the 1977 revision of the Nelson curves (Ref 72).

3. Potential effects of other gaseous components on hydrogen attack have not been taken into account. For instance,

it has been reported that in the presence of 10% H_2S along with hydrogen, hydrogen attack did not occur in an AISI 1020 steel under conditions well exceeding the Nelson-curve limits (Ref 73). In such instances the Nelson curves may represent an overly conservative approach.

4. The bulk of the data used in preparing the Nelson curves is based on steels used in accordance with earlier design codes that specified annealed steels. The curves may not be representative of quenched or normalized-and-tempered steels selected for higher-stress applications. Hence they may not be sufficiently conservative for ASME Section VIII, Division 2 re-

quirements. Furthermore, there are no actual data points for 2¼Cr-1Mo steels in any heat treated condition.

5. The effects of impurities, inclusions, cold work, residual stresses, and pre-existing flaws are not taken into account in drawing the curves. Because the curves are drawn through the minimum data points, it is only hoped that they cover the "worst case" situation with respect to all of these unknowns. Further conservatism is built into design by using design curves 14 to 28 °C (25 to 50 °F) below the Nelson curves.

6. The most important limitation is that the Nelson curves lack a foundation in terms of a basic understanding of hydrogen attack. Hence, prediction of hydrogen-attack behavior under conditions outside the envelope defined by the Nelson curves is nearly impossible. This limitation has imposed a major cost penalty due to adherence to conservative practices with respect to material selection, design, and process conditions.

7. A very important consideration that is often overlooked in applying Nelson curves to the design of stainless-steel-clad vessels is the fact that the ferritic steel shells of these vessels are not in direct contact with the process hydrogen. It is anticipated that the partial pressure of hydrogen available to cause hydrogen attack of the base metal is lower than the pressure of hydrogen inside the reactor. Unfortunately, with the exception of a study by Erwin and Kerr (Ref 4), there are no reported observations regarding hydrogen attack in clad steels. The results of this study showed that hydrogen attack of 2¼Cr-1Mo steels can occur even in the clad condition, but under more severe conditions than those suggested by the Nelson curves. It was concluded that stainless-steel-clad vessels would have higher threshold temperatures and longer incubation times for hydrogen attack than unclad vessels. The petroleum industry has not

yet taken advantage of this. These considerations, however, do not apply to design of pipes, unclad vessels, or refractory-clad vessels. In refractory-clad vessels, provision has to be made for failure of the refractory lining, allowing direct access of hydrogen to the base metal.

Industry Experience With Failure Due to Hydrogen Attack. During the past 35 years, thousands of pressure vessels and countless miles of piping have performed satisfactorily in a variety of high-pressure hydrogen processes. There have been very few instances of catastrophic failure, reflecting the conservatism of design and operating practices.

Some instances of catastrophic failure, and some near misses due to hydrogen-related cracking have been reported by Sorrell and Humphries (Ref 7). These instances included: (1) total fragmentation in service of a 305-mm- (12-in.-) diam carbon steel pipe at a German ammonia plant in 1974, attributed to erroneous use of carbon steel in a line intended to be made of alloy steel; (2) catastrophic rupture of a 305-mm- (12-in.-) diam pipe, once again due to inadvertent use of carbon steel in place of 1¼Cr-½Mo steel; (3) extensive damage found in the reactor vessels of three catalytic reforming units in 1955 due to inadvertent use of the steel at temperatures of 315 to 370 °C (600 to 700 °F), although the vessels were designed to operate at 150 °C (300 °F); (4) catastrophic failure of a C-½Mo steel reactor in 1970 during tightness testing, due to prior operation at excessive temperatures; (5) failure of a carbon steel pipe due to excessive temperatures; and (6) unspecified failures due to poor weld quality. Preventive actions for such failures include material verification prior to service, selection and maintenance of proper refractory linings (where applicable), and avoidance of hot spots (Ref 7).

Instances of methane blistering have been reviewed by Merrick and McGuire based on their own experience as well as on industry experience (Ref 72). Their experience per-tained to piping systems associated with catalytic reforming units. The industry experience consisted of nine instances reported to the NACE Group Committee T-8 on Refining Industry Corrosion. The steels involved were mostly C-½Mo and 1¼Cr-½Mo compositions. Out of a total of 13 instances reported, at least five occurred below the Nelson curves. Because none of these situations had led to leaks or major failures, the authors recommended that careful inspection and repair/replacement programs be adopted and that no revision of the Nelson curves was warranted (Ref 72).

Numerous instances of hydrogen attack in Japanese plants have been reviewed by the Japanese Pressure Vessel Research Council Subcommittee on Hydrogen Embrittlement (Ref 55). A summary of these instances as tabulated by the subcommittee is presented in Table 7.5. The damage mechanisms covered include fissuring, blistering, and decarburization. Out of a total of 18 instances reported, almost half had occurred during operation below the Nelson curves. Damage had been observed in the welds and/or the base metal. Damage in the welds was generally attributed to residual welding stresses and the presence of hard microstructures, emphasizing the need for proper postweld heat treatments. The reasons for the base-metal damage have not been discussed.

High-Temperature Creep

The concern over high-temperature creep properties arises in the following contexts. First, in reactors for hydrocracking service, creep and creep rupture *per se* are not a problem because the limiting operating temperatures are well below the creep range. However, long-term degradation of creep-rupture properties due to hydrogen attack is a potential concern. A lack of theoretical models and experimental data in the temperature range of interest, and possible counterbalancing effects lending conservatism to the design as discussed earlier, make it difficult to explicitly take the hydrogen effects into account. There are, however, no reported instances of creep-rupture failures in these reactors. In other types of reactors not involving hydrogen environments or involving much-lower-pressure hydrogen environments (e.g., catalytic-cracking and catalytic-reforming reactors), operating temperatures may be sufficiently high to warrant taking creep, creep-rupture, and creep-crack growth into consideration in the context of remaining-life assessment. Second, low creep ductility of the steels may be quite important in connection with both long-term creep-embrittlement-related cracking in service and stress-relief cracking during initial fabrication or subsequent weld repair. The characteristics of both of these phenomena were discussed in detail in Chapter 3.

Creep embrittlement generally has not been a problem in hydrocrackers, because the operating temperatures and steel strength levels (UTS limited to 690 MPa, or 100 ksi) in these units are lower than those normally required for embrittlement to occur. In addition, these reactors generally employ 2¼Cr-1Mo steel, which is more resistant to creep embrittlement than lower-alloy steels. The various factors affecting creep embrittlement of 2¼Cr-1Mo steels have been reviewed by Emmer, Clauser, and Low (Ref 30) and by Viswanathan (Ref 74).

Murakami, Watanabe, and Nomura (Ref 6) have shown that when simulated heat-affected-zone properties of 1¼Cr-½Mo and 2¼Cr-1Mo steels are compared at 550 °C (1020 °F), the former steel exhibits much lower creep-rupture ductility than the latter, despite its lower creep-rupture strength (see Fig. 7.30). The reduced rupture ductility renders the 1¼Cr-½Mo steel much more prone to notch-sensitive creep-rupture failure, as shown in Fig. 7.30. Accordingly, these investigators have observed creep cracking in several 1¼Cr-½Mo steel reactors operating at about 530 °C (985 °F) (Ref 6). A number of other instances of cracking due to creep embrittlement in nozzle-to-shell welds in 1¼Cr-½Mo steel reactors have also been reviewed recently by Bagnoli, Leedy, and Wada (Ref 75). The cracks generally are observed at stress

Table 7.5. Japanese experience with high-temperature hydrogen damage in petroleum refinery equipment (Ref 55)

Equipment	Material	Operating temperature, °C	Hydrogen partial pressure, MPa — Operating	Hydrogen partial pressure, MPa — Limit(a)	Duration in service, years	Damage — Type(b)	Damage — Location(c)
Pipe for thermocouple in ammonia plant	5Cr-0.3Mo	450-500	5-6	>30	2.5-3	C, DC, N	BM
Catalysis-cooling pipe in ammonia plant	Carbon steel	490	6	1.1	~5	C, DC	BM
	1Cr-0.4Mo	490	4.8	4	2.5	C, N	BM
	Carbon steel	490	4.8	1.1	2.5	C, DC	BM
	1Cr-0.2Mo	490	4.8	4(d)	0.8	C, N, DC	BM
Pipes in hydroformer plant	ASTM A106B (C-Si)	321-332(e)	1.21	0.7	4	C	W
		343-388(f)	0.82	0.7	4	C	W
Methanol converter	C-Si + SUS 304 lining	10-430	1.14	0.7	4	C	BM
Ammonia converter	1Ni-0.85Cr	80-580	30(g)	0.3-5(h)	10	B, C	BM
Ammonia conversion furnace	4.4Cr-0.76Mo	400-500	15	>30	0.5	C	W
	0.77Cr-0.20Mo	400-500	15	3.6	0.5	DC	BM
Liner of ammonia converter	C-Si	150-230	12-16	17	40	B	BM
Combined-feed exchanger	ASTM A204A (0.5Mo)	315-338	3.2	11	10	B	BM
Pipes in platformer plant	C-Si	Norm, 54.4; max, 340	2	0.8	12	C, DC	BM, W
Heat exchanger in ammonia plant	0.5Mo	270-300	8-10	18	2	C, DC	W
	0.5Mo	350	1.64	9.5	4.8	C	W
	0.5Mo	520	0.97	1.05	4.2	C	W
	0.5Mo	410-450	1.04	1.8	1.3	C	W
Pipes in platformer plant	1.1Cr-0.5Mo	545	2.62	5.6(j); 2(h)	9	DC	BM

(a) Estimated from Nelson curves (1977). (b) B–blistered; C–cracked; DC–decarburization; N–nitridation. (c) BM–base metal; W–weld metal. Details of postweld heat treatment can be found in original references cited in Ref 55. (d) For 1Cr-0.5Mo. (e) Design temperature, 368 °C. (f) Design temperature, 374 °C. (g) Total. (h) For 1.0Cr-0.5Mo. (j) For 1.25Cr-0.5Mo.

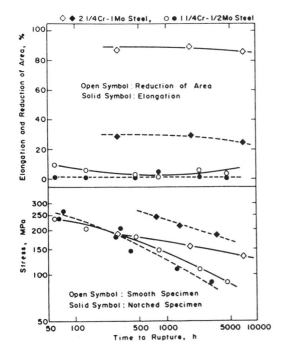

◇ ◆ 2 1/4 Cr-1Mo Steel,   ○ ● 1 1/4 Cr-1/2 Mo Steel

Open Symbol : Reduction of Area
Solid Symbol : Elongation

Open Symbol : Smooth Specimen
Solid Symbol : Notched Specimen

Fig. 7.30. Creep-test results at 550 °C (1020 °F) for smooth and notched specimens of synthetic HAZ material (peak temperature, 1350 °C, or 2460 °F) in 1¼Cr-½Mo and 2¼Cr-1Mo steel (theoretical stress-concentration factor for notched specimen, 1.9) (Ref 6).

raisers, such as nozzle-attachment welds, skirt-attachment welds, external lugs, and pads. Such cracking could be attributable to high operating temperatures, peak stresses at the weld toe, and poor creep ductility of the HAZ material (Ref 6). Precipitation of M_2C carbides has been cited as a contributing factor (Ref 75).

The susceptibility of steels to stress-relief cracking is governed by the same metallurgical variables that govern creep embrittlement, as reviewed in Chapter 3. Hence, experience with stress-relief cracking in reactors is expected to be very similar to experience with creep embrittlement. It is generally known that 2¼Cr-1Mo steel is much less susceptible to cracking than 1¼Cr-½Mo steel. Reduction of thermal stresses during welding, avoidance of stress raisers during design and fabrication, and control of strength level, impurities, and grain size are some of the preventive measures generally employed.

Subcritical Crack Growth

Subcritical crack growth in reactors may occur as a result of creep, low-cycle fatigue or hydrogen embrittlement. Creep-crack growth may not be very important in hydrocrackers, because the temperatures in these reactors are generally kept below 455 °C (850 °F). In other reactors, where lower partial pressures or the absence of hydrogen permits use of much higher temperatures, creep-crack growth can become an important factor. Frequent start-stop transients also can cause growth of low-cycle-fatigue cracks. The creep contribution to low-cycle fatigue depends on operating temperature. Crack growth also can occur by hydrogen-assisted mechanisms during the low-temperature periods from shutdown to start-up.

Extensive creep-crack-growth data applicable to Cr-Mo steels were discussed and illustrated in Chapters 3 and 5. Some additional creep-crack-growth data applicable to 2¼Cr-1Mo and 1¼Cr-½Mo heat-affected-zone material are shown in Fig. 7.31 based on the work of Iwadate, Watanabe, and Tanaka (Ref 14). These data showed that the creep-crack growth at 550 °C (1020 °F) could be described by the following expressions:

$$\frac{da}{dt} = 3.10 \times 10^{-21} K_I^{10.5} \text{ [for 2¼Cr-1Mo]}$$

(Eq 7.15)

$$\frac{da}{dt} = 6.80 \times 10^{-26} K_I^{14.5} \text{ [for 1¼Cr-½Mo]}$$

(Eq 7.16)

Crack-growth rates in 2¼Cr-1Mo steel were found to be two orders of magnitude lower than those in 1¼Cr-½Mo steel. In equations 7.15 and 7.16, da/dt is expressed in units of mm/h, and K is expressed in MPa√m.

Crack-growth rates in the presence of hydrogen at low temperatures have also been measured for 2¼Cr-1Mo steels by Murakami *et al* (Ref 6). The tests were conducted using WOL bolt-loaded specimens in a 500-

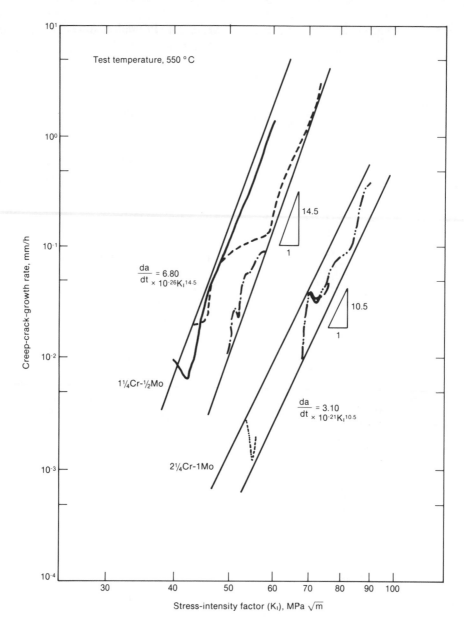

Fig. 7.31. Creep-crack-growth rates for HAZ materials in 1¼Cr-½Mo and 2¼Cr-1Mo steels (Ref 14).

ppm H_2S solution. The absorbed hydrogen levels in the specimen ranged from 2.0 to 4.2 ppm. Figure 7.32 shows the linear relationship between crack-growth rate and stress-intensity factor. The upper-bound relationship is expressed by the equation

$$\frac{da}{dt} = 2.40 \times 10^{-24}\, K_I^{11.7} \text{ mm/h} \quad \text{(Eq 7.17)}$$

where K_I is expressed in MPa$\sqrt{m}$. Sources of information on fatigue-crack growth in

hydrogen environments for 2¼Cr-1Mo steels include the work of Braziel *et al*, Suresh *et al*, and Murakami *et al* (Ref 76 to 78).

The crack-growth laws for creep and for hydrogen-assisted conditions have the general form

$$\frac{da}{dt} = CK^m \quad \text{(Eq 7.18)}$$

Thus, the remaining life t of the structural component can be defined as the time inter-

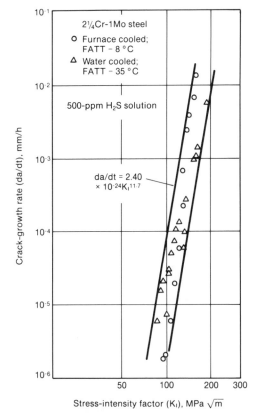

Fig. 7.32. Crack-growth rate for 2¼Cr-1Mo steel in a 500-ppm H₂S solution (Ref 6 and 14).

Table 7.6. Potential causes of cladding degradation (Ref 3)

Microstructural or phase changes including sensitization:
   Embrittlement (for example, σ or χ phases)
   Loss of corrosion resistance

Low-cycle-fatigue cracking (thermally induced): fatigue cracks into base metal

Compositional changes due to environment: carburization and subsequent sensitization

Loss of adherence

Hydrogen embrittlement of the weld overlay during shutdown

Corrosion

val required for the crack to grow from size a_i, detected by nondestructive examination (NDE), to the critical flaw size a_c for fracture of the component, as given by

$$t = \int_0^t dt = \int_{a_i}^{a_c} \frac{da}{CK_I^m}$$

$$= \frac{2}{(m-2)CM^{m/2}\sigma^m}$$

$$\times \left[\left(\frac{1}{a_i} \right)^{(m-2)/2} - \left(\frac{1}{a_c} \right)^{(m-2)/2} \right]$$

(Eq 7.19)

where $M = 1.21\ \pi/Q$ for surface cracks, $M = \pi/Q$ for embedded cracks.

Cladding Integrity

Because the principal function of the stainless steel cladding in reactor vessels is to protect the underlying steel from corrosive environments, degradation phenomena that affect the integrity of the cladding need to be considered. The concern is twofold: (1) loss of cladding or cracks in the cladding can permit access of the aggressive environment to the base metal; and (2) cracks in the cladding can penetrate the base metal, as was evidenced in a hydrodesulfurizer reactor (Ref 45 to 51). Several of the degradation phenomena that affect the cladding are listed in Table 7.6 (Ref 3). These phenomena are briefly reviewed below.

Corrosion

Among the principal corrodents that may cause cladding degradation are H₂S/H₂ mixtures, naphthenic acid, chlorides, and polythionic acid. Corrosive attack in H₂S/H₂ environments can occur in the hot portions of hydrodesulfurizers and other hydrotreating processes—specifically, in preheating furnaces, transfer lines, reactor outlet piping, and feed/effluent exchangers. Corrosion in such environments can cause damage by metal loss as well as by plugging of the reactor by the sulfide corrosion products. Austenitic stainless steels are generally resistant to this form of attack. However, even in these materials, corrosion rates can exceed 0.254 mm/year (10 mils/year) at 425 °C (800 °F) if the H₂S partial pressure exceeds 69 kPa (10 psi) (Ref 79). A typical corrosion design curve for H₂S/H₂ mixtures is shown in Fig. 7.33 (Ref 9).

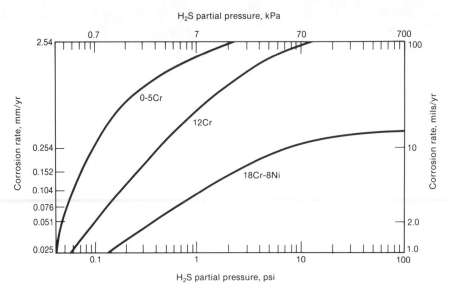

Fig. 7.33. Typical corrosion design curve for H_2S/H_2 service at 425 °C (800 °F) (Ref 9).

Condensed naphthenic acid is a potentially strong corrosive agent for stainless steels in the temperature range from 220 to 400 °C (430 to 750 °F) (Ref 80). Adequate precautions have to be taken to combat this form of corrosion.

Stress-corrosion cracking (SCC) due to chlorides and polythionic acid is a potential concern in all stainless steel parts, principally during shutdown periods. Chloride-induced SCC is generally prevalent below the dew point of water. Both forms of attack are facilitated if the stainless steel has been sensitized during postweld heat treatment. Use of stabilized grades of steel, such as types 347, 321, ELC 304, etc., and controlled shutdown procedures to minimize residual stresses are among the principal preventive measures used to ameliorate the SCC problem.

Embrittlement

Embrittlement of stainless steel cladding has in the past led to cracking in the cladding, followed by penetration of the base metal during shutdown repairs (Ref 45 to 51). Embrittlement has been attributed to formation of a brittle sigma phase in the ferrite phase which increased the susceptibility of the steel to hydrogen embrittlement. A minimum ferrite content of 3 to 4% is normally specified for the weld metal in manual submerged-arc welds in type 347, to prevent hot cracking. Ferrite content is limited to 8 to 10% to minimize sigma-phase formation. Hence, the ferrite content of the cladding may range from 3 to 10%. The ferrite phase serves to nucleate the sigma phase during postweld heat treatment.

Some of the general features associated with cracking of cladding in reactors have been summarized by Johnson and Hudak (Ref 39), as follows:

- The cracks are circumferential rather than longitudinal. Many occur in the "attachment" welds, where triaxial stresses are likely to be present.
- The cracks are not present when the vessel is placed in service but are found during shutdowns after several months or years of service.
- It has not been established whether the cracks form during service or during cooling accompanying shutdown.
- The cracks initiate at the cladding or attachment-weld surface and grow inward toward the dissimilar-metal interface.
- Metallography reveals that the cracks

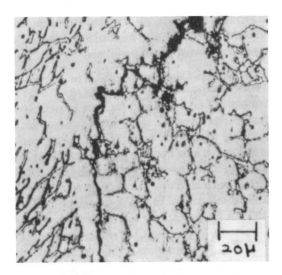

Fig. 7.34. Crack propagation through delta ferrite and sigma phases in type 347 stainless steel weld-metal cladding (Ref 39).

propagate through the ferrite/sigma networks between austenite grains (Fig. 7.34).

- The cracks are prevalent in type 347 weld metal and often are arrested at the interface with the underlying type 309 weld-metal layer.
- On rare occasions, a crack penetrates the type 309 layer and exposes the underlying Cr-Mo steel to the highly corrosive (sulfidizing) environment, thus permitting sulfide corrosion as well as enhanced hydrogen embrittlement of the structural Cr-Mo steel. In the case of the Mizushima hydrodesulfurizer reactor, several cracks had penetrated the 2¼Cr-1Mo base metal.

Based on the general cracking pattern observed, Johnson and Hudak have suggested several conclusions. From the fact that the cracks are circumferential rather than longitudinal, it can be concluded that they are not the result of the hoop stress of normal vessel service. Instead, the stress primarily responsible is probably the internal stress due to differential thermal expansion and contraction of the respective dissimilar metals (austenitic cladding and ferritic base material). These effects are such that the cladding undergoes maximum tensile stress during cooling to room temperature. On cooling from 425 to 24 °C (800 to 75 °F), it is estimated that the differential strain is on the order of 0.2% (Ref 39). The corresponding biaxial differential shear stress between the ferritic and austenitic alloy materials is on the order of the 0.2% yield strength of the austenitic material. This estimate was recently confirmed by the experimental measurements of Watanabe *et al* (Ref 46) to the effect that the maximum residual tensile stress in the cladding is 345 MPa (50 ksi). In view of the existence of this level of stress, it has seemed reasonable to assume that the cladding cracks form only during cooldown periods. During normal service, on the other hand, the tensile stress in the cladding is much lower, and therefore little if any cracking during service is to be expected.

The fact that the cracks propagate through the ferrite and sigma phase between the austenite grains indicates that attention must be given to the properties of these constituents. The sigma phase is essentially an intermetallic compound having the approximate composition FeCr. This phase is stable only below a certain temperature (approximately 900 °C, or 1650 °F) and forms by a diffusion mechanism during isothermal treatment in a relatively narrow temperature range. In reactor vessel claddings, the sigma phase is believed to form during the stress-relief treatment at about 690 °C (1275 °F), prior to vessel service.

In reactor vessel welds in type 347 stainless steel containing over 10% delta ferrite, sigma-phase formation can be severely embrittling. The kinetics of sigma-phase formation have been summarized in a review article by Willingham and Gooch (Ref 81). In reactor vessels containing less than 10% ferrite, moderate amounts of sigma phase can form during the normal postweld heat treatment of 20 h at 690 °C (1275 °F), but the amount is tolerable except at regions of high strain or stress concentration.

In the as-deposited condition, type 347 stainless steel weld metal has a Charpy V-notch toughness of 54 to 102 J at 10 °C (40 to 75 ft·lb at 50 °F), but this value is decreased by postweld heat treatment and can be further reduced by the presence of hydrogen. The combined effect of hydrogen and sigma phase has been studied by Watanabe *et al* (Ref 46 and 82). Ductility losses due to both factors have been quantified. Subsequently, fracture toughness investigations have been carried out by Johnson and Hudak (Ref 39). They have reported that even in weld metal containing a medium amount of ferrite (6%), a postweld heat treatment of 30 h at 690 °C (1275 °F) followed by exposure to hydrogen at 14.5 MPa (2100 psi) and 425 °C (800 °F) resulted in room-temperature K_H values as low as 33 MPa$\sqrt{m}$ (30 ksi $\sqrt{in}$.). Reactor shutdown procedures that reduce hydrogen content by degassing and reduce thermal stresses are the best available remedies for this problem. In addition, many fabricators have rescheduled welding of internal attachments and welding at other high-stress areas from before to after final postweld heat treatment.

Debonding

Recently, a new form of hydrogen-induced cracking has occurred in weld-clad steel in reactors (Ref 83). This cracking occurs at the boundary between the cladding and the base metal and is termed debonding. Debonding occurs several hours after shutdown following operation in high-temperature, high-pressure hydrogen, indicating that a combination of residual thermal stresses and hydrogen is involved in the cracking mechanism. Temperature/pressure conditions that lead to debonding have been defined by Ishiguro *et al* (Ref 84), as shown in Fig. 7.35.

Several studies have been undertaken to determine the nature, causes, and preven-

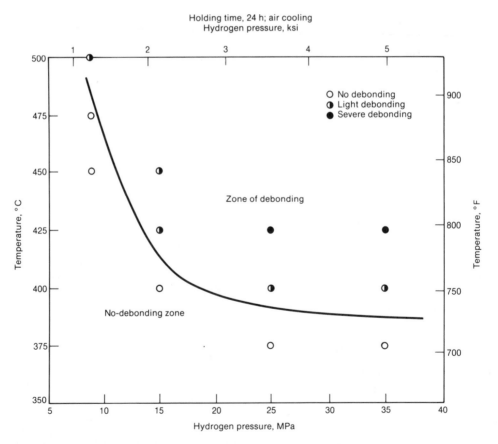

Fig. 7.35. Effects of hydrogen pressure and temperature on debonding of type 309/347 stainless steel weld overlay (Ref 84).

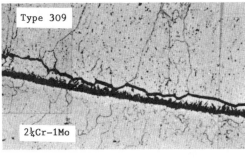

Fig. 7.36. Typical appearance of hydrogen-induced debonding of cladding (photo courtesy of M. Prager, Metal Properties Council, New York).

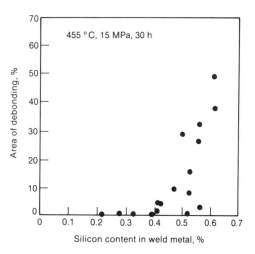

Fig. 7.37. Effect of silicon content on debonding for 30-h test at 455 °C in hydrogen at a pressure of 15 MPa (Ref 89).

tion of debonding (Ref 85 and 86). The characteristics of debonding that have emerged are:

1. Cracking occurs at grain boundaries in the cladding, adjacent to the fusion line (see Fig. 7.36) (Ref 87).
2. Cracking can be reproduced in the laboratory by charging of samples simulative of reactor operation followed by rapid cooling to room temperature.
3. Debonding is delayed cracking in the presence of hydrogen and residual stresses from shutdown.

From these characteristics it is evident that the problem of debonding could be mitigated by (1) slow cooling from the operating temperature to minimize thermal stresses (Ref 87), (2) reducing the hydrogen content of the steel by degassing at high temperatures prior to shutdown, and (3) reducing the hydrogen content of the steel by degassing at low temperatures above 100 °C (212 °F) (Ref 87). Improved weld-metal compositions and deposition processes have also been evaluated by investigators at Kawasaki Steel (Ref 88 and 89). They observed a pronounced effect of silicon content in the weld metal on the susceptibility to debonding, as shown in Fig. 7.37 (Ref 89). Based on this behavior, lowering the silicon content to levels below 0.4% was recommended. They also observed that increasing the ferrite content of the weld metal at the cladding/base

metal interface resulted in the elimination of coarse austenite grain boundaries and the introduction of elongated dendritic ferrite boundaries parallel with the interface. The latter were less susceptible to cracking. In comparison with submerged-arc welding, electroslag welding (ESW) was found to lead to less debonding. Based on all these observations, a weld process which optimizes the weld-metal composition and process variables, known as the "MAGLAY" process, has been advocated (Ref 88). This process utilizes the ESW process that employs Joule's heating of slag instead of an arc as the heat source and results in less weld dilution and improved bead smoothness.

Application of Refinery Experience to Coal-Liquefaction Reactors

The diminishing crude-oil reserves and the need for developing alternative liquid fuel resources have prompted worldwide interest in coal-liquefaction processes over the past decade. Most of the development effort has been concentrated on direct coal-liquefaction processes, also known as hydroliquefaction. In these processes, coal is dissolved in a hydrocarbon solvent and the resulting coal slurry is heated in the presence of high-pressure hydrogen. After the reaction is complete, the reactor is depressurized and the slurry is sent to product fractionation and solids separation. The typical operating conditions and some of the characteristics of the reactor vessel are shown in Fig. 7.38 (Ref 3). As can be seen, there are many similarities between the materials requirements for coal-liquefaction reactors and the reactors of conventional hydroprocessors. These similarities are as follows. (1) At least at the pilot-plant stage, construction of the coal-liquefaction vessel involves a 2¼Cr-1Mo steel shell with a two-layer cladding of type 309 and type 347 stainless steels. (2) Reactor environments are somewhat similar in both cases. (3) The material-damage mechanisms, and issues such as hydrogen attack, hydrogen embrit-

tlement, creep, low-cycle fatigue, subcritical crack growth, and toughness, are identical. The principal differences between the two types of processes arise in the following areas. (1) The erosivity of the reactants is expected to be more severe in coal liquefaction. (2) In view of the higher sulfur contents in coal, coal-liquefaction pressure vessels may be subjected to more exaggerated degradation by sulfur-base compounds than vessels used in oil refining. (3) Coal-liquefaction process streams contain NH_3, COS, and particulate-matter components which normally are not present in oil refining. (4) The upper limit of design temperature for coal liquefaction in commercial reactors is expected to be around 480 °C (900 °F), which is at least 28 °C (50 °F) higher than in hydroprocessing reactors.

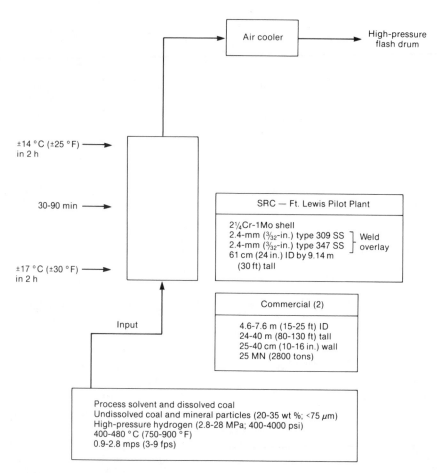

Fig. 7.38. Conditions for and characteristics of a coal-liquefaction reactor vessel (Ref 3).

This necessitates higher creep strength and increased resistance to hydrogen attack. (5) Reactor sizes are expected to be larger for coal liquefaction. Coupled with item 4, this necessitates the use of thicker-wall shell materials (wall thicknesses from 305 to 405 mm, or 12 to 16 in.), which constitutes a more severe requirement with respect to the hardenability of the ferritic steel.

It is anticipated that coal-liquefaction pressure vessels will also be designed on the basis of the ASME Boiler and Pressure Vessel Code, Section VIII. It has been suggested that the criticality of these vessels would dictate lower allowable stresses, as permitted by Division 1, in combination with the more stringent Division 2 requirements for fabrication. The number and type of postweld heat treatment cycles (minimum of 6.75 h at 675 °C, or 1250 °F, for a wall thickness of 305 mm, or 12 in.) are expected to reduce the tensile- and rupture-strength values to below those needed to meet Division 2 requirements (Ref 10). Development of improved alloys with higher creep-rupture strengths would allow Division 2 requirements to be met and wall-thickness requirements to be reduced. Some of these developmental efforts are reviewed in the following section. Specific problems in extending refinery technology to coal liquefaction have been reviewed in several publications (Ref 3, 10, and 11).

Improved Alloys for Pressure Vessels

The need for developing improved alloys is driven by the economics of fabrication and operation. With respect to oil-refinery processes, such as in hydrocrackers and hydrodesulfurizers, increasing the upper limit of operation from its current value of 455 °C (850 °F) to 480 °C (900 °F) and above would lead to major improvements in process efficiency. The major need here is for alloys higher in creep strength and resistance to hydrogen attack. With respect to coal-liquefaction processes, the large wall thicknesses required demand alloys with improved hardenability and cleanness so that

through-thickness toughness requirements can be met. In addition, development of more creep-resistant materials would permit reduced wall thicknesses, leading to smaller vessels, which are easier to fabricate and more economical to operate.

Ishiguro *et al* have conducted extensive studies on a low-silicon $2\frac{1}{4}$Cr-1Mo-$\frac{1}{4}$V-Ti-B steel (Ref 84 and 90). This modified steel has been found to have sufficient hardenability to produce bainitic structures at plate thicknesses up to 400 mm (16 in.) and creep-rupture strength about 50% greater than that of conventional $2\frac{1}{4}$Cr-1Mo steel, as shown on a Larson-Miller plot in Fig. 7.39 (Ref 84). The allowable design stresses based on the improved strength properties are shown in Fig. 7.40 (Ref 84). Assuming an operating temperature of 480 °C (900 °F), estimated allowable design-stress values were 15 and 37% greater than those for SA336, grade F21 steel in the Division 1 and Division 2 criteria, respectively. Remarkably enough, the improved rupture strength was achieved in combination with improved ductility. The lower silicon content and the additions of titanium and boron also led to increased resistance to temper embrittlement in comparison with $2\frac{1}{4}$Cr-1Mo steel. The modified steel also exhibited higher resistance to hydrogen-assisted cracking than that of $2\frac{1}{4}$Cr-1Mo steel. The properties of this steel have also been characterized by Klueh and Swindeman, and the claims made by the earlier investigators have been verified independently (Ref 91).

Several organizations are developing modified 3Cr-1Mo alloys with compositional changes intended to improve creep strength (Ref 92 to 94). One example of the strength attainable in modified 3Cr-1Mo compositions is illustrated in Fig. 7.41 (Ref 94), which compares the properties of a vanadium-enhanced 3Cr-1Mo steel with those of conventional $2\frac{1}{4}$Cr-1Mo steels. Other efforts to develop Cr-Mo steels for heavy-gage applications include carbon and nickel variations (Ref 95), manganese and nickel additions to $2\frac{1}{4}$Cr-1Mo steel (Ref 96), and minor modifications of 9Cr-1Mo steel (Ref 97). Considerable effort is needed to amass

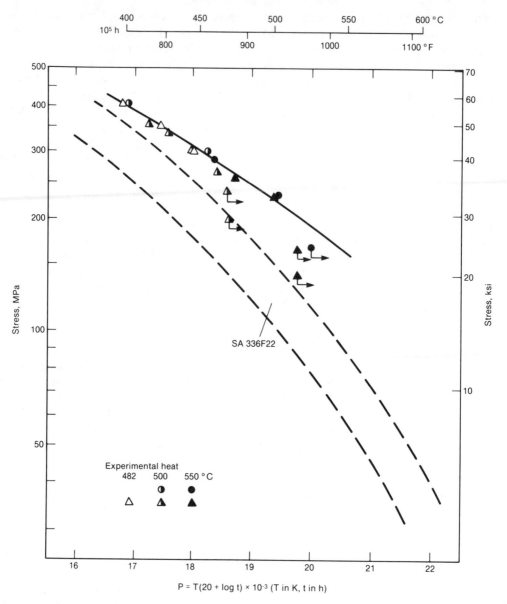

Fig. 7.39. Creep-rupture strength of a low-silicon 2¼Cr-1Mo-¼V-Ti-B developmental steel (Ref 84).

the data required to gain code approval for the new alloys for pressure-vessel construction. This is now being done under the umbrella of a Metal Properties Council (MPC) program.

Life-Assessment Techniques

Reactor vessels are thick-section components, operate at relatively low stresses, and are made of ductile, low-strength steels. In vessels used for high-pressure hydrogenation service, the upper limit of operating temperature set by hydrogen-attack considerations is lower than that permitted by creep considerations. These temperatures are below 455 °C (850 °F), and hence creep and creep-fatigue are not important mechanisms of crack initiation. Cracking due to other reasons such as environment and fatigue are invariably surface-originated and occur at predictable locations of stress

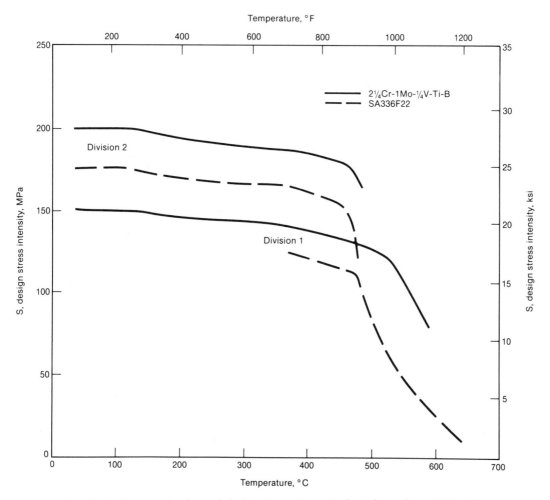

Fig. 7.40. Estimated values of design stress intensity for a low-silicon 2¼Cr-1Mo-Ti-B developmental steel and SA336, grade F22 steel (Ref 84).

concentrations, where they can be readily detected by conventional inspection techniques. For these reasons, crack-initiation-based methods have received little attention. In all but exceptional cases where combinations of severe thermally induced residual stresses and embrittlement (e.g., the hydro-desulfurizer at the Mishuzima refinery, described earlier) can decrease the crack tolerance of the material to very low levels, remaining-life methods have been based mainly on crack-growth analysis.

In vessels used for hydrogenation service, where hydrogen pressures are sufficiently low, the operating temperatures permitted by the Nelson diagram approach the creep range (e.g., catalytic reformers). Many of these vessels made of 1Cr-½Mo and 1¼Cr-½Mo steels operate at temperatures up to about 540 °C (1000 °F), where creep can also become an important damage mechanism. The applicable failure considerations in this case involve (1) creep rupture during operation, (2) creep-crack growth during operation followed by brittle fracture during a start-up/shutdown transient, or (3) environmentally assisted crack growth during shutdown followed by brittle fracture during a start-up/shutdown transient.

In vessels used in thermal-conversion processes where hydrogen is not an issue, the failure scenario primarily involves creep rupture during operation. These vessels are nominally made of carbon and C-Mo steels and operate at temperatures up to about 480 to 540 °C (900 to 1000 °F). Temper embrit-

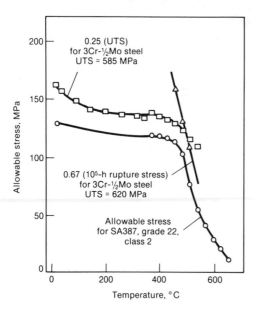

Fig. 7.41. Estimated allowable stress for a vanadium-modified 3Cr-1½Mo steel based on Section VIII, Division 1 of the ASME Boiler and Pressure Vessel Code (Ref 94).

tlement, hydrogen-assisted crack growth, and hydrogen attack are not significant considerations. In these cases, loss of rupture strength and internal creep damage can occur in the form of carbide spheroidization, carbide coarsening, and creep cavitation. These changes cannot be detected by conventional NDE techniques, because the damage occurs internally and uniformly, and without formation of manifest cracks until very late in life. In these cases, detection of incipient damage prior to cracking is essential. A large number of vessels built 30 to 40 years ago fall in this category and have caused concern in the industry. Several recent studies have reported on the spheroidization (transformation of lamellar carbides in the pearlite into carbide spheroids) and loss of creep-rupture properties of in-service vessels (Ref 98 to 101).

In general, potential methods for determining the expended life from a crack-initiation point of view are identical to those described for boiler headers in Chapter 5. For creep damage, these include calculations using time life fractions, strain and dimensional measurements, cavitation and carbide-coarsening measurements on replicas or metallographic sections, hardness measurements to estimate local temperatures, use of microstructural reference catalogs, and isostress-rupture tests. For pure fatigue and creep-fatigue damage, the calculational procedure would be similar to that described for headers in the section on calculational methods in Chapter 5 and is based on linear addition of fatigue and creep damage. The feasibility of extending the creep-fatigue design rule of Code Case N-47 (see the section on design rules for creep-fatigue in Chapter 4) to 2¼Cr-1Mo steel petroleum vessels has been recently reviewed by Jaske (Ref 102). This procedure is expected to be useful for design but is believed to be overly conservative for remaining-life assessment. Because the basic techniques and methods have been described elsewhere, they will not be repeated here.

The general principles of remaining-life prediction based on crack-growth analysis have been described in previous Chapters. Case histories and empirical material-property correlations specific to petroleum pressure vessels have been described recently by Iwadate, Watanabe, and Tanaka (Ref 14) and by Iwadate, Namura, and Watanabe (Ref 103). Slightly different failure scenarios apply to hydrogenation vessels and to vessels not involving large hydrogen pressures and operating in the creep regime. In the first case, a crack of initial size a_i is postulated to grow by a hydrogen-assisted mechanism until it reaches a critical size a_c, after which fracture occurs during a start/stop transient. In the second case, crack growth occurs primarily by creep under operating conditions; final failure can occur by leaking at the operating temperature or by fracture during a subsequent start/stop transient. The relative magnitudes of the critical crack size for brittle fracture, a_c, and the size of the remaining ligament will determine whether leak or fracture occurs.

The general procedure for life assessment (based on Ref 14; see also Fig. 7.42) involves the following steps:

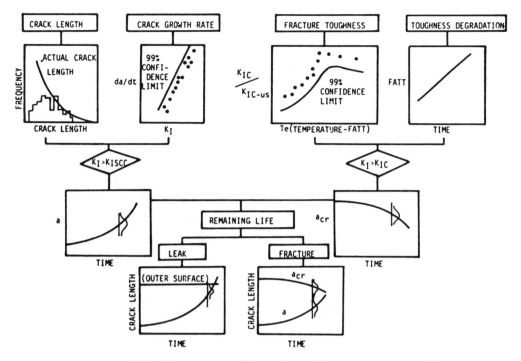

Fig. 7.42. Procedure for remaining-life prediction for a pressure vessel (Ref 14).

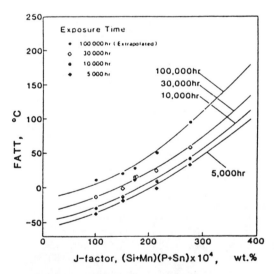

Fig. 7.43. FATT at most embrittling exposure temperature vs J factor for up to 100,000 h of service (Ref 103).

1. The actual initial crack sizes (a_i) are estimated from the flaw sizes detected by NDE.

2. The values of stress-intensity factor K_I are calculated using appropriate equa-

tions (e.g., Fig. 2.5) taking into account the total stresses due to residual thermal stresses and pressure stresses if present.

3. For the material in the current condition, the K_{Ic}-vs-temperature curve is estimated on the basis of (a) J_{Ic} tests followed by conversion to K_{Ic} values, (b) FATT tests followed by conversion to K_{Ic} values using procedures outlined in Chapter 2, or (c) estimates of current FATT using J-factor compositional correlations such as those in Fig. 7.10 or in Fig. 7.43 (Ref 103) followed by conversion of the FATT values to K_{Ic} values using procedures outlined in Chapter 2.

4. The critical crack size for fracture a_c for the "worst case" is calculated using peak stress values and the K_{Ic} values at the temperature where the peak stress is reached during a transient.

5. The propagating cracks are selected on the basis of the criterion that for these cracks $K_I \geq K_{IH}$, where K_{IH} is the threshold value of K for crack propa-

gation in a hydrogen environment. The K_{IH} value is estimated on the basis of data of the type presented in Fig. 7.15 (see also Eq 7.13).

6. Because the initial crack sizes a_i and the final crack size (a_c or remaining ligament, whichever is smaller) are known, the remaining life can be determined using known crack-growth-rate relationships such as Eq 7.15 to 7.17. For creep-crack-growth analysis, the C_t-based methodologies described in Chapters 3 and 5 can also be used.

References

1. T. C. Bauman, *J. Metals*, Vol 12, Aug 1977, p 8-11
2. *Petroleum Processing Handbook*, W.F. Bland and R. Davidson, Ed., McGraw-Hill, New York, 1967
3. T.E. Scott, Pressure Vessels For Coal Liquefaction—An Overview, in *Application of 2¼Cr-1Mo Steel For Thick Wall Pressure Vessels*, STP 755, American Society for Testing and Materials, Philadelphia, 1982, p 7-25
4. W.E. Erwin and J.G. Kerr, "The Use of Quenched and Tempered 2¼Cr-1Mo Steel For Thick Wall Reactor Vessels in Petroleum Refinery Processes: An Interpretive Review of 25 Years of Research and Application," Bulletin 275, ISSN 0043-2326, Welding Research Council, New York, Feb 1982
5. C.C. Clark, "Engineering Considerations in the Selection of Cr-Mo Steels For High Temperature Hydrogen Environments," Paper 128, presented at the meeting of the National Association of Corrosion Engineers, March 6-10, 1978
6. Y. Murakami, T. Nomura, and J. Watanabe, Heavy-Section 2¼Cr-1Mo Steel For Hydrogenation Reactors, in *Application of 2¼Cr-1Mo Steel For Thick Wall Pressure Vessels*, G.S. Sangdahl and M. Semchyshen, Ed., STP 755, American Society for Testing and Materials, Philadelphia, 1982, p 383-417
7. G. Sorrell and M.J. Humphries, High Temperature Hydrogen Damage in Petroleum Refinery Equipment, *Mater. Performance*, Aug 1978
8. G.R. Prescott, Material Problems in the Hydro Carbon Processing Industries, *Met. Progress*, July 1981, p 24-30
9. A.R. Ciuffreda and P.E. Krystow, "Petroleum Industry Materials Selection as Related to Coal Conversion Processes," in "Materials Problems and Research Opportunities in Coal Conversion," Report No. CR-402, workshop sponsored by the National Science Foundation and the Office of Coal Research, Columbus, OH, Apr 1974, p 469-479
10. A.M. Imgram and R.A. Swift, Pressure Vessel, Piping, and Welding Needs For Coal Conversion Systems, *ASM J. Mater. Energy Systems*, Vol 7 (No 3), Dec 1985, p 212-221
11. G. Sorrell, M.J. Humphreys, E. Bullock, and M. Van Der Voorde, Materials Technology Constraints and Needs in Fossil Fuel Conversion and Upgrading Processes, *Int. Met. Rev.*, Vol 31 (No 5), 1986, p 216-243
12. G.V. Smith, "Supplemental Report on the Elevated Temperature Properties of Chromium-Molybdenum Steels (An Evaluation of 2¼Cr-1Mo Steel)," ASTM Data Series Publication DS 6S2, American Society for Testing and Materials, 1971
13. J.M. Barsom and S.T. Rolfe, Correlations Between K_{Ic} and Charpy V-notch Test Results in the Transition Temperature Range, in *Impact Testing of Metals*, STP 466, American Society for Testing and Materials, Philadelphia, 1970, p 281-302
13a. S. Sawada *et al*, Temper Embrittlement Characteristics of 2¼Cr-1Mo Steels Used in Hydrogenation Units for 30,000 and 60,000 Hours at About 660 to 840F, in *Ductility and Toughness Considerations in Elevated Temperature Service*, ASME/MPC, New York, 1978, p 167-186
14. T. Iwadate, J. Watanabe, and Y. Tanaka, Prediction of the Remaining Life of High-Temperature Pressure Reactors Made of Cr-Mo Steels, *Trans. ASME, J. Pressure Vessel Tech.*, Vol 107, Aug 1985, p 230-238
15. B.J. Shaw, "Characterization Study of Temper Embrittlement of Chromium-Molybdenum Steels," proceedings of the American Petroleum Institute, Division of Refining, Vol 60 (Mid-Year Meeting, Chicago, May 11-14, 1981), API, Washington, 1981, p 225-247
16. S. Sato, S. Matsui, T. Enami, and T. Tobe, Strength and Temper Embrittlement of 2¼Cr-1Mo Steel, in *Application of 2¼Cr-1Mo Steel for Thick Wall Pressure Vessels*, G.S. Sangdahl and M. Semchyshen, Ed., STP 755, American Society for Testing and Materials, Philadelphia, 1982, p 363-382
17. Y. Murakami, Y. Nitta, T. Nomura, and Y. Nakajo, National Meeting of Japan Welding Society, Autumn 1979, No. 25, p 40-41 (also cited in Ref 6)
18. R. Bruscato, Temper Embrittlement and Creep Embrittlement of 2¼Cr-1Mo Shielded Metal Arc Welded Deposits, *Weld. J. Res. Suppl.*, Vol 49, Apr 1970, p 148s
19. J. Watanabe *et al*, "Temper Embrittlement of 2¼Cr-1Mo Pressure Vessel Steel," Presented at 29th Annual ASME Petroleum-Mechanical

Engineering Conference, Sept 15-18, 1974, Dallas

20. C.J. McMahon *et al*, The Effect of Composition and Microstructure on Temper Embrittlement in 2¼Cr-1Mo Steels, *Trans. ASME, J. Engg. Mater. Tech.*, Vol 102, 1980, p 369

21. R. Viswanathan and R.I. Jaffee, 2¼Cr-1Mo Steels For Coal Conversion Pressure Vessels, *Trans. ASME, J. Engg. Mater. Tech.*, Vol 104, July 1982, p 220-226

22. C.J. McMahon, Jr., and D.H. Gentner, "The Effect of Carbon Content on the Temper Embrittlement of 2¼Cr-1Mo Steel," Metal Properties Council Report, 1981

23. J.C. Murza, D.H. Gentner and C.J. McMahon, Jr., "The Effects of a High Residual Nickel Content on the Temper Embrittlement of 2¼Cr-1Mo Steels," Metal Properties Council Report, 1981

24. J.F. Copeland and A.W. Pense, "Factors Influencing the Notch Toughness and Transition Temperature of ASTM 542 Steel," presented at the ASME Petroleum-Mechanical Engineering Conference, Los Angeles, Sept 16-20, 1973

25. J.F. Copeland, Influence of Sulfur Content on the Fracture Toughness Properties of a 2¼Cr-1Mo Steel, *Trans. ASME, J. Pressure Vessel Tech.*, Vol 98, May 1976, p 135-142

26. R. Viswanathan, Influence of Microstructure on the Temper Embrittlement of Some Low Alloy Steels, in *MICON 78: Optimization of Processing, Properties and Service Performance Through Microstructure Control*, H. Abrams, G.N. Maniar, D.A. Nails, and H.D. Solomon, Ed., STP 672, American Society for Testing and Materials, Philadelphia, 1979, p 169-185

27. R.W. Swift and J.A. Gulya, *Weld. J. Res. Suppl.*, Feb 1973, p 575-675

28. G.N. Emmanuel, W.E. Leyda, and E.J. Rozic, in *2¼ Chromium 1 Molybdenum Steels in Pressure Vessels and Piping*, A.O. Schaefer, Ed., American Society of Mechanical Engineers, New York, 1971, p 78-122

29. J. Kerr, "Temper Embrittlement of Low Alloy Reactor Steels," Technical Report W.0.8025.43, Standard Oil of California, 1969

30. L.G. Emmer, C.D. Clauser, and J.R. Low, "Critical Review of Embrittlement in 2¼Cr-1Mo Steel," Bulletin 183, Welding Research Council, New York, May 1973

31. T. Sekine *et al*, "Development of Impurity-Reduced 2¼Cr-1Mo Heavy Section Steel Plates of Low Temper Embrittlement Susceptibility," Kawasaki Steel Corp. Report, May 1980

32. Letter dated July 30, 1974, from American Petroleum Institute, Division of Refining, to managers of all refineries in the United States and Canada, cited in Ref 4

33. Paragraph AD 155, Section VIII, Division 2, ASME Boiler and Pressure Vessel Code, American Society of Mechanical Engineers, New York

34. J.D. Landes and D.E. McCabe, Design Properties of Steels Used in Coal Conversion Vessels, in *Application of 2¼Cr-1Mo Steel For Thick Wall Pressure Vessels*, G.S. Sangdahl and M. Semchyshen, Ed., STP 755, American Society for Testing and Materials, Philadelphia, 1982, p 68-92

35. A.W. Loginow and E.H. Phelps, Steels For Seamless Hydrogen Pressure Vessels, *Corrosion* Vol 31, Nov 1975, p 404-412

36. J.G. Kerr, J.W. Coombs, P. Purgolis, and R.J. Olsen, "Environmental Behavior of 2¼Cr-1Mo Hydrocracker Reactor and Piping Materials," proceedings of the American Petroleum Institute, Division of Refining, Vol 52 (Mid-Year Meeting, New York, May 8-11, 1972), API, Washington, p 890-905

37. T.P. Groeneveld and A.R. Elsea, "Effect of Hydrogen on Properties of Reactor Vessel Steels at Temperatures Below 400 °F," Final Report to the American Petroleum Institute from Battelle Memorial Institute, Columbus, OH, Dec 31, 1973; published as "Hydrogen Assisted Crack Growth in 2¼Cr-1Mo Steel," Report 956, American Petroleum Institute, Washington, 1978

38. N.J.I. Adams and W.G. Welland, "The Practical Application of Fracture Mechanics to Hydrocarbon Reactor Vessels," Fifth International Conference on Pressure Vessel Technology, San Francisco, 1984

39. E.W. Johnson and S.J. Hudak, "Hydrogen Embrittlement of Austenitic Stainless Steel Weld Metal with Special Consideration Given to the Effects of Sigma Phase," Bulletin 240, ISSN 0043-2326, Welding Research Council, New York, Aug 1978

40. K. Yoshino and C.J. McMahon, Jr., *Met. Trans.*, Vol 5, 1974, p 363

41. R. Viswanathan and S.J. Hudak, Jr., *Met. Trans. A*, Vol 8A, Oct 1977, p 1633

42. C.L. Bryant, W.C. Feng, and C.J. McMahon, Jr., *Met. Trans. A*, Vol 9A, May 1978, p 625

43. Japanese Pressure Vessel Research Council Cooperative Study Program, Subcommittee on Hydrogen Embrittlement, H. Gondo, Chairman, Progress Report No. 1, Jan 1980

44. R. Viswanathan and R.I. Jaffee, Clean Steels to Control Hydrogen Embrittlement, in *Proceedings of the Fifth International Conference on Current Solutions To Hydrogen Problems in Steels*, Washington, 1982, American Society for Metals, Metals Park, OH, 1982

45. J. Watanabe *et al*, "A Fracture Safe Analysis of Pressure Vessels Made of 2¼Cr-1Mo Steels," Paper No. 126, CORROSION/76, Houston, Mar 22-26, 1976, National Association of Corrosion Engineers, 1976

46. J. Watanabe, K. Ohnishi, M. Murai, and R. Chiba, "On the Cracks in Stainless Steel Internal Attachment Welds for High Temperature, High Pressure Hydrogen Service," Paper No. 104, CORROSION/77, San Francisco, Mar 14-18, 1977, National Association of Corrosion Engineers, 1977

47. K. Kawakami, "A User's View on the Desulfurization Reactor," paper presented at the Japan Petroleum Institute, Dec 6, 1974, cited in Ref 4

48. S. Ishigaki *et al*, "On the Reactor for Direct Desulfurization of Heavy Oil," paper presented to the Japan Society of Mechanical Engineers, Oct 12, 1975, cited in Ref 4

49. J. Watanabe *et al*, "Investigation of the Failure of a Direct Desulfurization Reactor During Its Field Repair Work," proceedings of the Joint International Petroleum-Mechanical Engineering Conference, Mexico City, Sept 19-24, 1976, American Society of Mechanical Engineers, New York, 1976, p 7-14

50. S. Mima *et al*, "Design and Materials Problems of Direct Desulfurization Reactors," Ref 49, p 15-24

51. J. Watanabe *et al*, "Field Inspection of and Repairs to Heavy Wall Refinery Vessels," CORROSION/78, Houston, Mar 6-10, 1978, National Association of Corrosion Engineers, 1978

52. G.A. Nelson, "Action of Hydrogen on Steel at High Temperatures and Pressures, Bulletin 145, Welding Research Council, New York, Oct 1969, p 33-42

53. D. Stone *et al*, "Hydrogen Attack in Cr-Mo Steels at Elevated Temperatures," report to the U.S. Department of Energy covering the period Sept 1, 1980 to Aug 31, 1983, CU/MSE 07963-01, Cornell University, Ithaca, NY, 1983

54. H. Ishizuka and R. Chiba, Japan Steel Works Research Laboratories Report, 1967, p 43-66, cited in Ref 46

55. Japan Pressure Vessel Research Council, Subcommittee on Hydrogen Embrittlement, Temper Embrittlement and Hydrogen Attack in Pressure Vessel Steels, Report No. 2, May 1979

56. F.H. Vitovec, *Proc. Amer. Petroleum Inst.*, Vol 44 (No. III), 1964, p 179

57. H. Ishizuka and R. Chiba, Report of Welding Research Subcommittee on Chemical Plant Apparatus, The Japan Welding Engineering Society, July 1971

58. E. Holmes, P.C. Rosenthal, P. Thoma, and F.H Vitovec, Progress Report to Subcommittee on Corrosion, American Petroleum Institute, 1964 (No. 3), cited in Ref 46

59. J. Watanabe, H. Ishizuka, K. Onishi, and R. Chiba, proceedings of the 20th National Symposium on Strength, Fracture and Fatigue, Japan, 1975, p 101

60. D.L. Sponseller and F.H. Vitovec, Resistance to Hydrogen Attack at 1000°F of 2¼ Cr-1Mo and 0.5Mo-0.0006B Steels, in *Symposium on Heat Treated Steels for Elevated Temperature Service*, New Orleans, Sept 19-20, 1966, American Society of Mechanical Engineers, New York, 1966, p 74-85

61. J.A. Mullendore and J.P. Tralmer, "Hydrogen Attack in Cr-Mo Steels," proceedings of the American Petroleum Institute, Division of Refining, Vol 47 (Mid-Year Meeting, Los Angeles, May 15-17, 1967), API, New York, 1967, p 429-446

62. E.A. Sticha, Tubular Stress Rupture Testing of Chromium-Molybdenum Steels With High Pressure Hydrogen, *Trans. ASME, J. Basic Engg.*, Vol 91, Dec 1969, p 590-592

63. H.M.S. Shih and H.H. Johnson, *Acta Met.*, Vol 30, 1982, p 537

64. Effect of Aluminum Content on Hydrogen Attack of Simulated HAZ in 1Cr-½Mo Steels, Japan Pressure Vessel Research Council, Subcommittee on Hydrogen Embrittlement, H. Gondo, Chairman, May 18, 1981

65. H. Ishizuka and R. Chiba, *Tetsu-To-Hagane*, Vol 56 (No. 1), 1970, p 93

66. J. Bland, Ultrasonic Inspection Used To Detect Hydrogen Attack, *Petroleum Refiner*, Vol 37 (No. 7), July 1958, p 115-118

67. Nelson curves published in "Steels For Hydrogen Service at Elevated Temperatures and Pressures in Petroleum Refineries and Petrochemical Plants," API Publication 941, 2nd Ed., American Petroleum Institute, Washington, June 1977

68. Y.I. Arkharao, I.D. Grebshkova, and V.V. Dubovski, *Zashchita Metallov*, Vol 5 (No. 1), 1969, p 70

69. G.R. Odette, S. Vagaralli, and W. Oldfield, "Analysis of Hydrogen Attack on Pressure Vessel Steels," Quarterly Progress Report No. 3 to the U.S. Department of Energy, prepared at University of California, Santa Barbara, May 15, 1979

70. W.A. Bonner, Revision to the Nelson Curves, *Proc. Amer. Petroleum Inst.*, Vol 36, 1977, p 3-6

71. N. Bailey, Bulletin No. 18, Welding Research Institute, 1977, p 2-33

72. R.D. Merrick and C.J. McGuire, "Methane Blistering of Equipment in High Temperature Hydrogen Service," Paper No. 30, CORROSION/79, National Association of Corrosion Engineers, 1979

73. D. Eliezer and N.G. Nelson, *Corrosion*, Vol 35 (No. 1), 1979, p 17

74. R. Viswanathan, Effect of Impurities on the Creep Properties of Ferritic Steels, *ASM Metals Engg. Qtrly.*, Nov 1975, p 50-55

75. D.L. Bagnoli, J.W. Leedy, and T. Wada, "Embrittlement of 1Cr-½Mo and 1¼Cr-

½Mo Alloys After Long Time Service," Paper No. 160, CORROSION/88, St. Louis, Mar 88, National Association of Corrosion Engineers

76. R. Braziel, G.M. Simmons, and R.P. Wei, Fatigue Crack Growth in 2¼Cr-1Mo Steel Exposed to Hydrogen Containing Gases, *Trans. ASME, J. Engg. Mater. Tech.*, Vol 101, July 1979, p 199-204

77. S. Suresh, G.F. Zamiski, and R.O. Ritchie, Fatigue Crack Propagation Behavior of 2¼Cr-1Mo Steels for Thick Wall Pressure Vessels, in *Application of 2¼Cr-1Mo Steel for Thick Wall Pressure Vessels*, G.S. Sangdahl and M. Semchyshen, Ed., STP 755, American Society for Testing and Materials, Philadelphia, 1982, p 49-67

78. Y. Murakami, T. Nomura, and K. Ohnishi, unpublished results cited in Ref 14

79. A.R. Ciuffreda and P.E. Krystow, "Petroleum Industry Materials Selection as Related to Coal Conversion Processes," in "Materials Problems and Research Opportunities in Coal Conversion," Report No. CR-402, workshop sponsored by the National Science Foundation and the Office of Coal Research, Columbus, OH, Apr 1974, p 469-479

80. J. Gutzeit, *Mater. Performance*, Oct 1977, p 24

81. D.C. Willingham and T.G. Gooch, "Sigma Formation in Austenitic Steel Weld Metal," Welding Research Institute Bulletin, 1971

82. Y. Yoshino, J. Watanabe, and R. Chiba, Hydrogen Embrittlement of 300 Series Stainless Steel Weld Metals for Hydrodesulfurization Reactors, international conference on *The Effects of Hydrogen on Behavior of Materials*, AIME, New York, 1975

83. J. Watanabe *et al*, Hydrogen Induced Disbonding of Stainless Weld Overlay Found in Desulfurizing Reactor, in *Performance of Pressure Vessels With Clad and Overlaid Stainless Steel Linings*, American Society of Mechanical Engineers, 1981, p 1-21

84. T. Ishiguro *et al*, "Current Status of R & D Program of a Heavy Section Pressure Vessel Steel for High Temperature and High Pressure Hydrogenation Service," Report MR 81-6, The Japan Steel Works, July 1981

85. K. Naitoh *et al*, Study in Hydrogen Embrittlement of Pressure Vessels Overlaid With Stainless Steel — Part 2, Hydrogen Embrittlement of Transition Zone Between Weld Overlay and Base Metal, *Pressure Engg., JHPI*, Vol 18 (No. 5), 1980, p 263-270

86. K. Naitoh *et al*, Study in Hydrogen Embrittlement of Pressure Vessels Overlaid With Stainless Steel — Part 3, De-Bonding at Boundary Layer Between Weld Overlay and Base Metal, *Pressure Engg., JHPI*, Vol 18 (No. 5), 1980, p 272-276

87. N. Morishige, R. Kume, and H. Okabayashi, Influence of Low Temperature Hydrogen Degassing on Hydrogen-Induced Disbonding of Cladding, *Trans. Japan Weld. Soc.*, Vol 16 (No. 1), Apr 1985, p 12-18

88. S. Nakano *et al*, "MAGLAY Process — Electro-Magnetic Controlled Overlay Welding Process With ESW," Technical Report No. 2, Kawasaki Steel Corp., Japan, Mar 1981

89. S. Nakano *et al*, "On the Disbonding Characteristics of Weld Overlay by MAGLAY Process," Kawasaki Steel Corp., Japan, 1981

90. T. Ishiguro *et al*, A 2¼Cr-1Mo Pressure Vessel With Improved Creep Rupture Strength, in *Application of 2¼Cr-1Mo Steel for Thick Wall Pressure Vessels*, G.S. Sangdahl and M. Semchyshen, Ed., STP 755, American Society for Testing and Materials, Philadelphia, 1982, p 129-148; see also Report MR 80-13, Japan Steel Works, May 1980

91. R.L. Klueh and R.W. Swindeman, "Mechanical Properties of a Modified 2¼Cr-1Mo Steel for Pressure Vessel Applications," ORNL 5995, Oak Ridge National Laboratory, Dec 1983

92. T. Wada and T.B. Cox, in *Advanced Materials for Pressure Vessel Service With Hydrogen at High Temperatures and Pressures*, M. Semchyshen, Ed., American Society of Mechanical Engineers, New York, 1982, p 111

93. S.J. Manganello, in *Advanced Materials for Pressure Vessel Service With Hydrogen at High Temperatures and Pressures*, M. Semchyshen, Ed., American Society of Mechanical Engineers, New York, 1982, p 153

94. T. Ishiguro *et al*, in "Research in Chrome-Moly Steels," R.A. Swift, Ed., MPC-21, American Society of Mechanical Engineers, New York, 1984, p 77-93

95. T. Wada and G.T. Eldis, in *Application of 2¼Cr-1Mo Steel for Thick Wall Pressure Vessels*, G.S. Sangdahl and M. Semchyshen, Ed., STP 755, American Society for Testing and Materials, Philadelphia, 1982, p 343-361

96. R.J. Kar and J.A. Todd, in *Application of 2¼Cr-1Mo Steel for Thick Wall Pressure Vessels*, G.S. Sangdahl and M. Semchyshen, Ed., STP 755, American Society for Testing and Materials, Philadelphia, 1982, p 228-254

97. V.K. Sikka, in *Ferritic Alloys for Use in Nuclear Energy Technologies*, J.W. Davies and D.J. Michael, Ed., The Metallurgical Society of AIME, Warrendale, PA, 1984, p 317

98. W.B. Bedesem, Assessment of Refinery Materials in Elevated Temperature Service, in *Evaluation of Materials in Process Equipment After Long Term Service in the Petroleum Industry*, ASME-MPC Conference, New Orleans, MPC 12, 1980, p 1-12

99. E.L. Creamer, Metallurgical Condition of Several Carbon Steels After Long Term Elevated Temperature Exposure, Ref 98, p 13-22

100. S. Ibarra, On-Site Metallurgical Evaluation of an FCCU Reactor, Ref 98, p 23-33

101. W.R. Warke and G.P. Coker, The Effect of Long Time Exposure on Charpy Properties of Cr-Mo Pressure Vessel Steels, Ref 98, p 47-58

102. C.E. Jaske, "Fatigue Curve Needs for Higher Strength 2¼Cr-1Mo Steel for Petroleum Process Vessels," MPC Symposium on Fatigue Initiation, Propagation and Analysis for Code Construction, ASME Winter Annual Meeting, Chicago, Nov 1988

103. T. Iwadate, T. Nomura, and J. Watanabe, Hydrogen Effect on Remaining Life of Hydroprocessor Reactors, *Corrosion*, Vol 44 (No. 2), Feb 1988, p 103-112

8

Materials for Advanced Steam Plants

General Requirements

Increasing the efficiency of power plants can result in significant cost savings in terms of reduced fuel costs and reduced need for added capacity. The historical developments leading to the current need for improved coal-fired power plants have been reviewed by Armor (Ref 1). Two of the important parameters that have been identified and which also have major impacts on materials technology are the temperature/pressure configuration of the steam cycle and the number of reheat stages. The increased efficiency achievable in a steam plant by increasing the pressure and temperature of the steam and by adding a double-reheat feature has been well-documented in the literature (Ref 2 to 4). An extensive assessment of the feasibility of concepts, economics, and research and development efforts needed to improve the efficiency of pulverized-coal plants has been carried out in two independent studies sponsored by the Electric Power Research Institute (Ref 3 and 4). Based on these studies and on additional surveys, it was concluded that the equipment manufacturers are presently capable of providing

units with design steam conditions of 31.0 MPa (4500 psi) and 565/565/565 °C (1050/1050/1050 °F) (Ref 5). The triple temperatures given here are those of the throttle steam and the steam in the two subsequent reheat cycles. It was also concluded that with modest development efforts in materials and design, the industry would be capable of providing advanced-cycle units with steam conditions of 31.0 MPa (4500 psi) and 595/595/595 °C (1100/1100/1100 °F) (Ref 5).

The use of advanced cycles to improve operating efficiency requires different degrees of development. Figure 8.1 illustrates the improvements in heat rate that can be achieved by use of the various advanced cycles. For the three design levels shown, the development plan can be structured into three phases (Ref 5). This will initially provide the industry with an option of constructing an advanced plant with hardware designs already available (phase 0), followed by a plant which requires developments but that has an excellent chance of successful completion in a short time period with the potential to provide economical operation (phase I). The efforts in the last

383

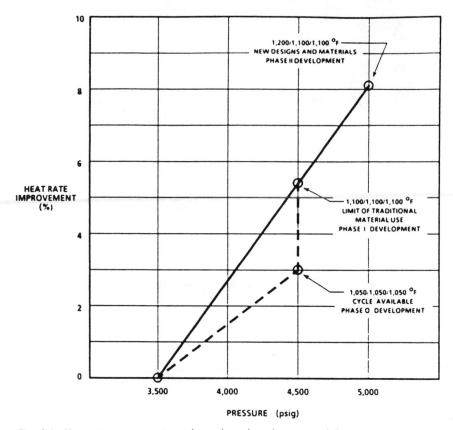

Fig. 8.1. Heat-rate improvement through cycle selection and the conceptual three-phase development program (Ref 5).

phase of the development plan are expected to require extended time periods for completion and cannot confidently be judged to be economical until initial developments have shown satisfactory in-service performance (phase II). The three-phase approach visualized by Hottenstine *et al* (Ref 5) is summarized in Table 8.1. Programs similar to the EPRI program are being implemented in Germany and Japan (Ref 6 to 9).

In the United States, there has been substantial experience based on two high-efficiency plants built in the 1950's. Philo 6, a 125-MW plant owned by the Ohio Power Company, has been operational since 1957 under design steam conditions of 31 MPa (4500 psi) and a 620/555/555 °C (1150/1050/1050 °F) double-reheat temperature cycle. Eddystone 1, owned by Philadelphia Electric Company, was designed to operate under steam conditions of 34.5 MPa (5000 psi) and 650/565/565 °C (1200/

1050/1050 °F), and has been operational since 1959. Although the Eddystone 1 plant has operated under derated conditions of 32.4 MPa (4700 psi) and 605 °C (1125 °F) for most of its service life, because of mechanical and metallurgical problems, the actual operating conditions have been more severe than the phase-I conditions of the EPRI program. A variety of materials problems and the research and development efforts needed to solve these problems have been identified as a result of all this experience.

A major challenge in constructing the advanced plants described above is in the area of materials technology. Research and development, as well as demonstrations of the capabilities of alternative materials, need to be carried out. This chapter will present a brief review of the anticipated materials problems and solutions, based on published literature (Ref 2 and 10 to 14).

Table 8.1. Steam conditions for coal-fired plants in EPRI program phases 0, I, and II (based on Ref 10)

Program phase	Pressure		Temperature(a)		Comments
	MPa	ksi	°C	°F	
0 . . . . 31.0		4500	565/565/565	1050/1050/1050	Available now. Verification of high-pressure pump and FWH performance recommended.
I 31.0		4500	595/595/595	1100/1100/1100	No major limitation. No new materials needed. Development of reliable designs for superheaters, reheaters, turbine forgings, and casings needed.
II 34.5		5000	650/595/595	1200/1100/1100	Significant research and development needed on austenitic rotor forgings and coal-ash-corrosion-resistant boiler tubing. Full-size test facilities needed.

(a) Temperatures given are for main steam and first and second reheats.

Boiler Materials

A variety of considerations dictate the selection of materials for use in the boiler. In the case of superheater/reheater tubes, headers, and steam piping, creep strength and rupture strength are the foremost considerations. In addition, for tubes, fire-side corrosion resistance and steam-side oxidation resistance are necessary. For thick-section components, such as headers and pipes, fabricability, weldability, fracture toughness, and resistance to thermal fatigue are needed.

Boiler tubing and piping alloys potentially capable of meeting the requirements of the advanced steam plant are listed in Table 8.2 (Ref 13 and 15 to 34). They fall roughly into the two categories of 9 to 12% Cr steels and austenitic alloys. It is anticipated that the 9 to 12% Cr steels will meet the requirements up to steam temperatures of 595 °C (1100 °F)—i.e., phases 0 and I in Table 8.1—as possible alternatives to conventional 18Cr-8Ni 300 series austenitic stainless steels such as TP304H, 321, 316H, and 347H, and that conventional and improved austenitic steels will be needed for steam temperatures above 595 °C (phase II). Remarkable progress has been made during the last decade in developing steels containing 9 to 12% Cr which are superior

to low-alloy ferritic steels and can be cost-effective substitutes for 300 series austenitic stainless steels in the range 540 to 595 °C (1000 to 1100 °F). Many developments have also taken place in improving the creep strength and hot corrosion resistance of austenitic steels.

Figure 8.2 illustrates the general concepts employed in developing the various improved alloys for use in boilers (Ref 13). Both solution-strengthening methods and precipitation-strengthening methods have been utilized. Improvements in the 9 to 12% Cr steels have been mainly in the area of optimization of composition with respect to solid-solution strengtheners such as molybdenum, tungsten, and carbon and with respect to precipitation strengtheners such as vanadium, niobium, and titanium. The development of austenitic stainless steels has followed a similar path. For instance, the base composition for 18Cr-8Ni steels is that of type 304 stainless steel; solid-solution strengthening with molybdenum results in type 316, whereas precipitation hardening with titanium or niobium, or both, results in type 321, type 347, and Tempaloy-A1, respectively. Further improvements have consisted of increases in chromium content, development of coextruded tubing, and grain-size refinements to optimize the cor-

Table 8.2. Nominal chemical compositions of boiler tubing alloys (Ref 13)

Steel	C	Si	Mn	Ni	Cr	Mo	W	V	Nb	Ti	B	Others
9%Cr Steels												
T9	0.12	0.6	0.45	...	9.0	1.0	...	...	...	...	...	...
HCM9M (Ref 15)	0.07	0.3	0.45	...	9.0	2.0	...	...	...	...	...	...
NSCR9 (Ref 16)	0.08	0.2	0.90	...	9.0	2.0	...	0.15	0.05	...	...	...
Tempaloy F-9 (Ref 17)	0.06	0.5	0.60	...	9.0	1.0	...	0.25	0.40	...	0.005	...
EM12 (Ref 18)	0.10	0.4	0.10	...	9.0	2.0	...	0.30	0.40	...	...	...
T91 (Ref 19)	0.10	0.4	0.45	...	9.0	1.0	...	0.20	0.08	...	...	...
9Cr-Mo-W (Ref 20)	0.07	0.06	0.45	...	9.0	0.5	1.8	0.20	0.05	...	0.004	...
9Cr-Mo-W TB9 (Ref 21)	0.08	0.05	0.50	0.1	9.0	0.5	1.8	0.20	0.05	...	...	0.05 N
9Cr-Mo-W NF616 (Ref 22)	≤0.15	≤0.50	≤1.0	...	9.0	0.5	1.8	0.20	0.05	...	0.0002	0.037 N
12%Cr Steels												
HCM12 (Ref 23)	0.10	0.3	0.55	...	12.0	1.0	1.0	0.25	0.05	...	...	...
AMAX12Cr (Ref 24)	0.07	0.3	0.60	...	12.0	1.5	1.0	0.20	0.05	...	...	...
HT9 (Ref 25)	0.20	0.3	0.55	...	12.0	1.0	...	0.25	...	...	...	...
JETHETE M154	0.14	0.2	0.70	2.4	12.0	1.8	...	0.35	...	...	...	0.05 N
TB12	0.08	0.05	0.50	0.1	12.0	0.5	1.8	0.20	0.05	...	0.003	0.05 W
18Cr-8Ni Steel												
TP304H	0.08	0.6	1.6	8.0	18.0	...	...	...	...	...	...	...
Tempaloy A-1 (Ref 26)	0.12	0.6	1.6	10.0	18.0	...	...	...	0.10	0.08	...	...
TP321H	0.08	0.6	1.6	10.0	18.0	...	...	...	...	0.5	...	...
TP316H	0.08	0.6	1.6	12.0	16.0	2.5	...	...	...	...	...	...
TP347H	0.08	0.6	1.6	10.0	18.0	...	...	...	0.8	...	...	...
15Cr-15Ni Steel												
17-14 CuMo (Ref 27)	0.12	0.5	0.7	14.0	16.0	2.0	...	...	0.4	0.3	0.006	3.0 Cu
Esshete 1250 (Ref 28)	0.12	0.5	6.0	10.0	15.0	1.0	...	0.2	1.0	...	0.006	...
Tempaloy A-2	0.12	0.6	1.64	14.1	17.8	1.57	...	...	0.24	0.10	...	...
20-25Cr Austenitic steel												
TP310S	0.08	0.6	1.6	20.0	25.0	...	...	...	...	...	...	...
NF709 (Ref 29)	0.15	0.5	1.0	25.0	20.0	1.5	...	...	0.2	0.1	...	...
HR3C (Ref 30)	0.06	0.4	1.2	20.0	25.0	...	...	...	0.45	...	...	0.2 N
NF707 (Ref 31)	0.08	0.5	1.0	35.0	22.0	1.5	...	...	0.2	0.1	...	...
800H	0.08	0.5	1.2	32.0	21.0	...	...	...	...	0.5	...	0.4 Al
Mod. 800H (Ref 32)	0.08	0.4	0.8	34.0	22.0	1.25	...	...	0.4	...	...	...
Incoloy 807	0.06	0.43	0.97	39.0	20.5	...	4.60	...	0.95	0.30	...	7.6 Co, 0.4 Al
HK4M	0.23	0.56	1.23	25.0	25.0	...	...	...	...	0.39	...	0.35 Al
High-Cr, high-Ni steel												
CR30A (Ref 33)	0.06	0.3	0.2	50.0	30.0	2.0	...	...	...	0.2	...	0.03 Zr
HR6W (Ref 34)	0.08	0.4	1.2	43.0	23.0	...	6.0	...	0.18	0.08	0.003	...
Inconel 617	...	0.40	0.40	54.6	22.0	8.5	...	...	...	0.40	...	12.5 Co, 1.2 Al
Inconel 671 (cladding)	0.05	...	...	51.5	48.0	...	...	...	...	...	...	...

rosion behavior on the fire side and the steam side.

Superheater Tubing

The superheater tubes in the boiler are likely to undergo the most severe service conditions and must meet stringent requirements with respect to fire-side corrosion, steam-side oxidation, creep-rupture strength, and fabricability. In addition to meeting these requirements, the alloy chosen must be cost-effective. There is a clear "layering" in terms of two classes of alloys, with each class finding its own niche with respect to temperature capability and cost-effectiveness. The approximate limiting temperature for optimum utilization of 9 to 12% Cr steels appears to be 595 °C (1100 °F), with austenitic materials being needed for higher-temperature applications. Hence, in discussing

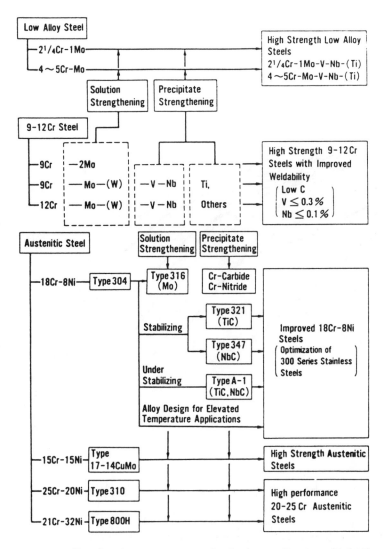

Fig. 8.2. Alloy-development concept for boiler applications (Ref 13).

material properties, each class of material will be treated separately in the following sections.

Creep-Rupture Strength. The chemical compositions of several 9 to 12% Cr steels are listed in Table 8.2. Among the 9% Cr steels shown in Table 8.2, 9Cr-1Mo-W (Ref 20 to 22) is the most recently developed, and consequently accounts for the smallest amount of long-time creep-rupture data. HCM9M (Ref 15) is a low-carbon 9Cr-2Mo steel which is already established in Japanese industry and is known as STBA27 according to MITI standards. Freedom from V and Nb coupled with low levels of C, contribute to the superior weldability and form-

ability of this alloy. The other 9% Cr alloys, NSCR9 (Ref 16), Tempaloy F9 (Ref 17), EM12 (Ref 18), and T91 (Ref 19)*, are modified 9Cr steels whose strength is achieved by precipitation strengthening due to vanadium and niobium additions.

Steels containing 12% Cr, such as HT9 (Ref 25), have been widely used in Europe for boiler tube and pipe applications (Ref 35). Application of HT9 in the United States and Japan has been limited by serious concerns about its weldability. Re-

*T91 and P91 both refer to the same alloy composition, but as designated for tubing or piping, respectively. The alloy is also referred to as "Super 9Cr alloy."

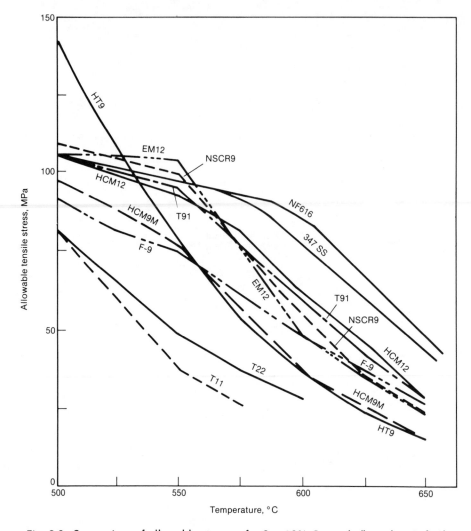

Fig. 8.3. Comparison of allowable stresses for 9 to 12% Cr steels (based on Ref 13).

cently, modified 12% Cr steels for boiler tubing have been developed. These steels, designated as AMAX 12Cr (Ref 24), HCM12 (Ref 23, 36, and 37), and TB12, exhibit improved weldability and creep-rupture strength compared with HT9. They have a duplex microstructure of ferrite and tempered martensite.

Apart from the extensive commercial application of HT9, several years of limited field experience is available on HCM9M (Ref 13), HCM12M (Ref 36 and 37), and T91 (Ref 38). Production of alloy NF616 on a commercial scale has been demonstrated (Ref 39). A comparison of allowable stresses for 9 to 12% Cr steels, based on ASME Section VIII, Division 1 rules, is shown in Fig. 8.3 (Ref 13). The highest allowable stresses are obtained for the 9Cr-Mo-W steel NF616, followed by HCM12M and T91. The data for AMAX 12Cr alloy are not plotted, because the long-time test data needed for establishing code allowable stresses are not available. Based on creep-strength considerations alone, it would appear, therefore, that HCM12M, T91, and NF616 are leading candidates for use up to steam temperatures of 595 °C (1100 °F). If longer-term creep-rupture data on NF616 confirm its superiority, it may be possible to use this alloy at temperatures up to 650 °C (1200 °F).

For steam temperatures exceeding 595 °C (1100 °F), improved austenitic steels are needed. For convenience, they can be classified as those containing less than 20% Cr and those containing more than 20% Cr. Alloy modifications based on the 18Cr-8Ni steels, such as TP304H, 316H, 347H, and Tempaloy A-1, and alloys with lower chromium and higher nickel contents, such as 17-14 CuMo steel, Esshete 1250, and Tempaloy A2, fall into the classification of steels with less than 20% Cr. The allowable tensile stresses for steels in this class are compared in Fig. 8.4. Tempaloy A1, Esshete 1250, and 17-14 CuMo steel are found to offer major improvements over the 300 series stainless steels. It has been reported that grain-size modifications of AISI type 347H stainless steel can in some instances lead to rupture properties somewhat better than those of Tempaloy A1 (Ref 40).

Several high-creep-strength alloys containing more than 20% Cr, such as NF707, NF709, and HR3C, have been developed, and offer low-cost alternatives to Incoloy 800 for use in the temperature range from 650 to 700 °C (1200 to 1290 °F). A comparison of the ASME Code allowable stresses for the high-chromium alloys is shown in Fig. 8.5. Clearly, NF709 and HR3C are leading candidates for use in the highest-temperature applications. The highest creep strength is achieved in Inconel 617, which contains 50% Cr, but it is also likely to be the most expensive alloy to use.

A comparison of allowable temperatures at a constant allowable stress of 49 MPa (7 ksi), as a function of chromium content, is shown in Fig. 8.6. With increasing chromium, a discontinuity is seen in the allowable metal temperatures of austenitic steels, rising about 50 °C (90 °F) above those of ferritic steels. In terms of increasing temperature capability, stable austenitic alloys offer the highest capability, followed by metastable austenitic steels, and then by ferritic steels. The fully enhanced, stable austenitic alloys are clearly capable of operating under phase II temperature conditions (650 °C, or 1200 °F).

Fire-side corrosion results from the presence of molten sodium-potassium-iron

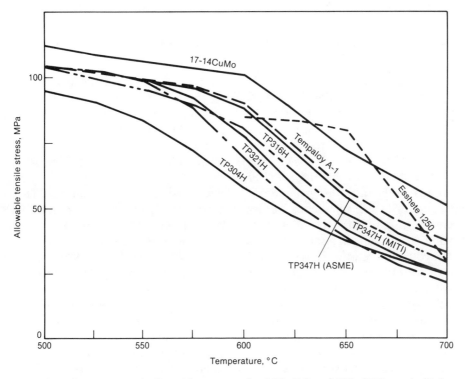

Fig. 8.4. Comparison of allowable stresses for 18Cr-8Ni and 15Cr-15Ni steels (Ref 13).

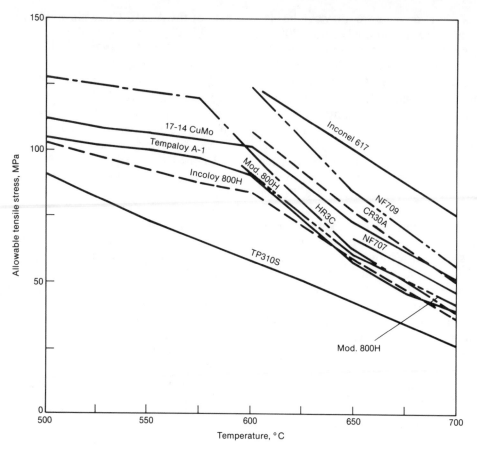

Fig. 8.5. Comparison of allowable stresses for austenitic alloys containing more than 20% Cr (based on Ref 13).

trisulfates. Because resistance to fire-side corrosion increases with chromium content, the 9 to 12% Cr ferritic steels are more resistant than the 2¼ Cr-1Mo steels currently used. The 12% Cr steel in turn shows better corrosion resistance than 2¼ % Cr steel and 9% Cr steel, as shown in Fig. 8.7 (Ref 41). Stainless steels and other superalloys containing up to 30% Cr represent a further improvement. Increasing the chromium content beyond 30% results in a saturation effect on the corrosion resistance, as shown in Fig. 8.8 (Ref 42). For practical purposes, when corrosive conditions are present, fine distinctions between ferritic steels may be academic, and it may be necessary to use austenitic steels containing chromium in excess of 20%.

A ranking of the performances of various austenitic alloys in the presence of trisulfates has been provided by Ohtomo *et al*

(Ref 43) on the basis of short-term laboratory tests (see Fig. 8.9). The plots of weight loss versus temperature exhibit a bell-shape curve. At temperatures below 600 °C (1110 °F), corrosion is believed to be low because the trisulfate exists in solid form. Above 750 °C (1380 °F), corrosion rates are once again low, as the trisulfates vaporize. The worst corrosion problem is in the range 600 to 750 °C (1110 to 1380 °F). The data indicate that the high-chromium alloys such as type 310 stainless steel and Incoloy 800H are superior to the other alloys tested, and that Inconel 671 (Ni-50Cr) is virtually immune to attack. Lower-chromium stainless steels, such as type 316H, type 321H, and Esshete 1250, show considerable susceptibility to attack. The alloy most susceptible to attack seems to be the 17-14 CuMo alloy used in the Eddystone 1 plant. However, chromizing was found to significantly improve the corrosion

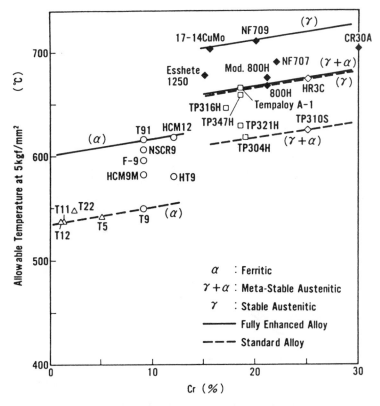

Fig. 8.6. Allowable metal temperatures at constant allowable stress of 5 kgf/mm^2 as a function of chromium content for various alloys (Ref 13).

resistance of this alloy. Recent results of field probe studies confirm the following ranking of alloys in increasing order of corrosion resistance: T91, HCM12, chromized T91, type 347 stainless steel, Incoloy 800, and Inconel 617 (Ref 44). In addition to alloy selection, other "fixes" to minimize fireside corrosion, such as shielding of the tubes and operating above the peak temperature of the bell-shape curve, may also be applied, if economical (Ref 11).

Steam-side oxidation of tubes and exfoliation of the oxide scale and its consequence in terms of solid-particle erosion damage to the turbine have already been discussed (Chapter 6). This problem is expected to be more severe in advanced steam plants, because the much higher steam temperatures employed are likely to cause more rapid formation of oxide scale.

Very limited data are available regarding the steam-side scale-growth characteristics of the ferritic tubing alloys. In the study by

Sumitomo Metal Industries (Ref 45), the oxide growth in steam for alloys T22 (2¼Cr-1Mo), T9, HCM9M, and the modified 9Cr-1Mo (T91) were compared, based on 500-h exposures. Results showed the superiority of the T91 alloy over the other alloys. Masuyama *et al* compared alloys HCM12, HCM9M, 321H, and 347H in field tests in the temperature range 550 to 625 °C (1020 to 1155 °F) over a period of one year (Ref 37). Samples were inserted in the tertiary and secondary superheaters and reheaters. From the results, they concluded that the resistance to steam oxidation of HCM12 is superior to those of 321H and HCM9M and comparable to that of fine-grain 347H for exposure to the high-temperature region of the reheater. Subsequent monitoring over a period of three years (see Fig. 8.10) has borne out their earlier conclusions (Ref 36). In addition to the inherent resistance of HCM12M steel to steam-side oxidation, Masuyama *et al* suggest that the

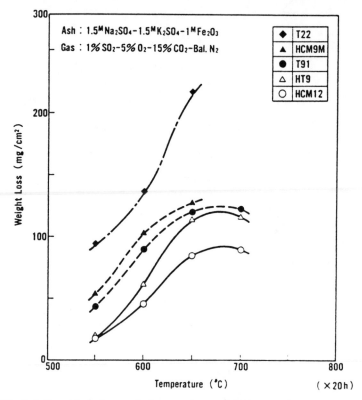

Fig. 8.7. Relationship between hot-corrosion weight loss and temperature for ferritic steels (Ref 41).

tendency toward exfoliation of oxide scale would also be less for this alloy than for austenitic steels (Ref 36 and 37). Additional improvements in 9 to 12% Cr steels may be possible by extending the chromizing (Ref 46 and 47) and chromate conversion treatments (Ref 48) that currently are applied to lower-alloy steels. In the case of 300 series stainless steels, grain refinement during heat treatment has been shown to be clearly beneficial (see Fig. 8.11). Internal shot blasting is also known to improve the steam oxidation resistance of 300 series stainless steels by enhancing chromium diffusion. It is therefore anticipated that these steels would be used in the fine-grain and shot-peened conditions. Results of steam oxidation tests at 650 °C (1200 °F) for times up to 2000 h have been reported for several austenitic steels (Ref 49).

Summary of Tube-Material Status. For phase 0 steam conditions (steam temperature of 565 °C or 1050 °F and metal temperature of 595 °C or 1100 °F), alloys T91,

HCM12M, and AISI type 304 stainless steel are viable candidates for superheater and reheater tubing, provided that fire-side corrosion is not a major problem. Under corrosive conditions, however, alloy HR3C and chromized type 347 stainless steel may be the most cost-effective options.

For intermediate-temperature applications corresponding to phase I steam conditions (595 °C, or 1100 °F), Tempaloy A1 and type 347 stainless steel are deemed to be adequate in the absence of corrosive conditions. Under corrosive conditions, alloy HR3C, chromized type 347 stainless steel, and alloy NF709 may offer the best combinations of creep strength and corrosion resistance.

For the highest-temperature application corresponding to phase II steam conditions (650/595 °C, or 1200/1100 °F), the creep strength requirements are met by Inconel 617, 17-14 CuMo steel, Esshete 1250, CR30A, and NF709. Among these alloys, 17-14 CuMo steel and Esshete 1250 have in-

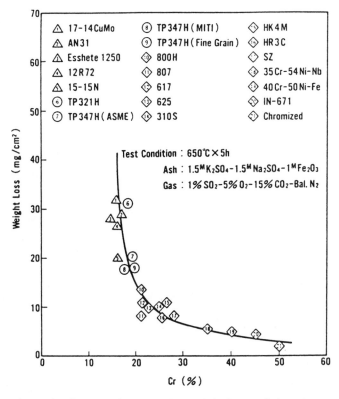

Fig. 8.8. Relationship between hot-corrosion weight loss and chromium content for various alloys (Ref 42).

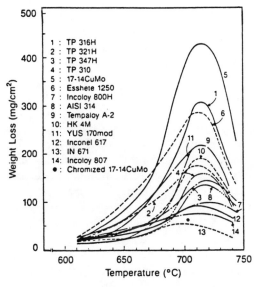

Fig. 8.9. Comparison of fire-side corrosion resistance of various alloys (Ref 43).

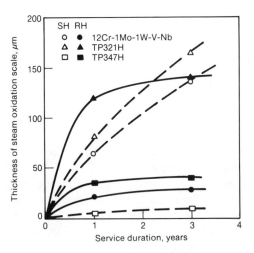

Fig. 8.10. Changes in thickness of steam oxidation scale with time in field trials of superheater (SH) and reheater (RH) sections (Ref 36).

adequate corrosion resistance and will have to be chromized or clad with corrosion-resistant claddings of Inconel 671 or type 310 stainless steel if corrosive conditions are present. Alternatively, NF709 and CR30A could be used without any corrosion protection.

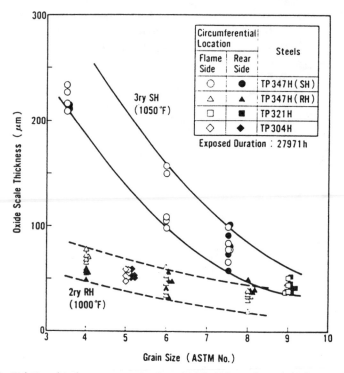

Fig. 8.11. Relationship between grain size of 18Cr-8Ni steels and thickness of steam oxidation scale after exposure for 27,971 h in superheater (SH) and reheater (RH) sections (Ref 13).

Fabrication and formability evaluation studies have been carried out by Haneda *et al* using 17-14 CuMo chromized tubes and 17-14 CuMo/310 stainless steel coextruded tubes (Ref 50). The weldability and bendability of both types of tubes were found to be satisfactory. Welded joints made by GTA welding of the chromized and the clad tubes showed superior mechanical properties.

Many of the alloys described here have been developed very recently and have not yet been sufficiently qualified on the basis of long-duration creep and rupture tests. Accurate prediction of the tubing costs, once the alloys have been commercialized, is also difficult. Final selection of tube alloys will rest to a large degree on economic considerations in addition to material properties. The materials-selection criteria described here should, therefore, be regarded as purely tentative.

Headers and Steam Pipes

The choice of materials for heavy-section components is governed not only by creep strength and fabricability but also by resistance to thermal fatigue (LCF) and by fracture toughness. Although the austenitic stainless steels are advantageous from a creep-strength point of view, their lower thermal conductivities and higher coefficients of thermal expansion make them less attractive than ferritic steels from a thermal-fatigue point of view. Unfortunately, the creep strengths of ferritic steels restrict their use to temperatures below 595 °C (1100 °F).

The thermal-conductivity and thermal-expansion behavior of a number of ferritic alloys are compared with the properties of austenitic alloys in Fig. 8.12 and 8.13 (Ref 51). Clearly, the thermal conductivities of the ferritic steels are higher than those of the austenitic alloys and the coefficients of thermal expansion are lower than those of the austenitic alloys. Both of these factors serve to decrease thermal stresses and increase the thermal-fatigue resistance of the ferritic alloys as compared to AISI type 316 stainless steel and alloy 800H. Among the ferritic steels, the 9 to 12% Cr steels offer the best possibilities of reducing the wall thickness

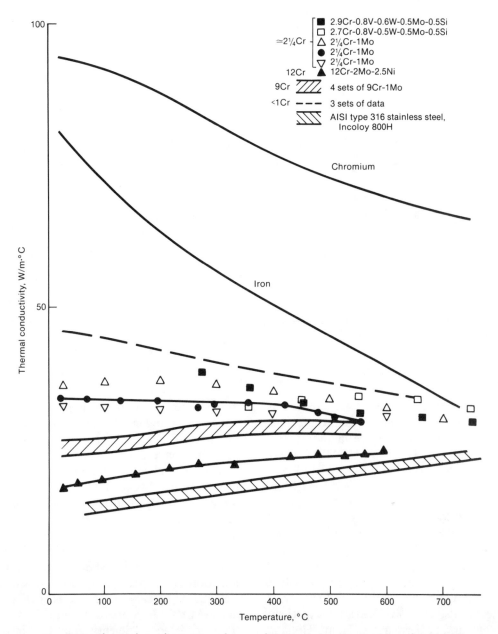

Fig. 8.12. Thermal conductivities of iron, chromium, ferritic steels, and austenitic alloys (Ref 51).

of piping for a given loading, because they have much higher creep strengths than the lower-alloy steels. This, again, makes them attractive candidates for use under cyclic conditions, even in conventional plants, where thermal fatigue might be a problem.

It is anticipated that, for steam temperatures up to 595 °C (1100 °F), ferritic steels containing 9 to 12% Cr, 1 to 2% Mo, and various levels of vanadium, niobium, and tungsten will be utilized. To ensure good fracture toughness and freedom from temper embrittlement, stringent control will have to be exercised over the levels of phosphorus, tin, antimony, arsenic, sulfur, copper, aluminum, manganese, and silicon. Low sulfur levels on the order of 20 to 50 ppm can be achieved using argon-oxygen decarburization (AOD) or modern ladle refining processes. Despite the high strength

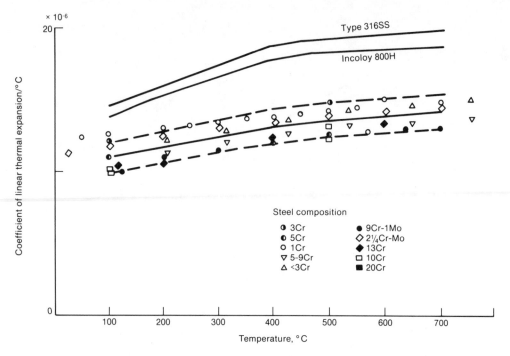

Fig. 8.13. Thermal-expansion data for various chromium steels (Ref 51).

levels expected for these steels, they must possess good weldability. As a result of extensive developments, three candidate steels have emerged. These are HCM9M, developed by Sumitomo and Mitsubishi Heavy Industries, the 12Cr-Mo-V steel HT9, and the modified super 9Cr alloy developed by ORNL/CE (P91) (Ref 11).

Alloy HT9 (DIN X20CrMoV121) is a well-established steel with an extensive stress-rupture database which exceeds 10^5 h at temperatures in the range 500 to 600 °C (930 to 1110 °F) for all product forms. There is also extensive operating experience (>20 years) in Germany, Belgium, Holland, South Africa, and Scandinavia for steam temperatures up to 540 °C (1000 °F) and some limited experience on a few small units with steam temperatures from 560 to 580 °C (1040 to 1075 °F). This experience generally has been satisfactory. Difficulties have, however, been reported during fabrication and particularly during welding and post-weld heat treatment. This arises because the relatively high carbon content of the steel (0.2%) and the correspondingly low M_s temperature promote the possibility of

austenite retention after welding, high residual stresses, and cracking prior to and during stress relief. It is reported that these problems have been overcome by careful control of preheat treatment and postweld heat treatment backed up by vigorous quality control. Difficulties have also been reported when the material has been given inadequate solution heat treatment. Due to these concerns, this alloy has not found much favor in the United States, the United Kingdom, or Japan. Alloys with improved weldability characteristics, such as HCM12M and AMAX 12Cr, are of recent origin and have not been sufficiently qualified in the laboratory to warrant immediate use in boiler piping. HCM12M has been adequately characterized for tubing applications but has not yet been applied to large-diameter, thick-wall pipes.

In view of its superior creep-rupture strength, the 9Cr-Mo-W steels can be considered for piping applications up to 595 °C (1100 °F). Additional long-time rupture data, demonstrations of fabricability, and field trials are needed before this alloy can qualify as a leading candidate. With regard to the 9Cr-2Mo steel (HCM9M), the feasibility of fabrication of large-diameter, thick-wall piping and application to in-plant header and main steam piping have already been demonstrated (Ref 52). The practical use of this material has been easy because its simple composition lends fabricability and weldability comparable to those of low-alloy steels. The toughness of large-diameter pipes has been found to be over 102 J/cm^2 (460 ft·lb/in.^{2}) at 0 °C (32 °F). Allowable stresses are comparable to those for the HT9 alloy (see Fig. 8.3), but lower than those for P91.

The modified 9Cr alloy, P91, appears to be quite superior to HT9 and to HCM9M in terms of creep-rupture strength and is, hence, the most promising candidate for use in header and steam piping for temperatures up to 595 °C (1100 °F). Fabrication and evaluation of large-diameter piping made of this alloy have been reported upon by Sumitomo investigators (Ref 53) and by Haneda, Masuyama, Kaneko, and Toyoda (Ref 54). Haneda *et al* have conducted field trials on pipes of T91 (Ref 54). Chubu Electric Power Co. (Kawago Power Station, No. 1 and No. 2 units: 31 MPa or 4500 psi; 565/565/565 °C or 1050/1050/1050 °F—steam phase 0 in Table 8.1) is to be the fore-runner in experience on uses of large quantities of large-diameter, thick-wall piping of P91 steel. Header and main piping for this plant are being fabricated from P91 steel.

Mechanical properties of laboratory and commercial heats of P91 steel have been extensively characterized by Sikka *et al* (Ref 55-59). Base metal and welds from full-scale pipes have also been tested (Ref 53 and 54). The 50% ductile-brittle transition temperatures have been found to be in the range −40 to 0 °C (−40 to +32 °F) and the 68-J (50-ft·lb) transition temperature around −40 °C for base metal. The impact toughness of the weld metal, however, has tended to be inferior to those of the base metal and the heat-affected-zone material, but adequate for the application (Ref 54).

Available low-cycle-fatigue data on a variety of 9Cr steels and 12Cr steels have been summarized by Skelton and Beckett (Ref 60). Within the scatterbands, the endurance of all materials was found to be similar, although the T91 steel appeared to be slightly better at 600 °C (1110 °F) (Ref 60). Figure 8.14 provides a comparison of the fatigue-endurance curves for a variety of structural alloys (Ref 61). In the low-cycle-fatigue region (high strain range), the fatigue behavior of the super 9Cr steel (P91)

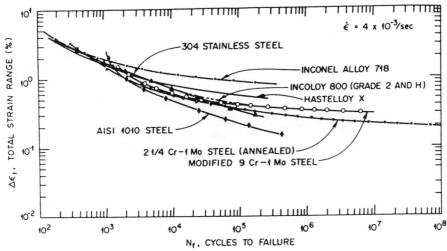

Lines represent best-fit values of actual data. Data for type 304 stainless steel include tests conducted at 540 and 565 °C (1000 and 1050 °F).

Fig. 8.14. Comparison of fatigue behavior at 540 °C (1000 °F) for several materials (Ref 61).

is similar to that of 2¼Cr-1Mo steels. In the high-cycle-fatigue region, however, the super 9Cr steel seems to be superior to the 2¼Cr-1Mo steels (Ref 61). There is little published data relating to creep-fatigue (hold-time effects) for either the 9Cr or 12Cr materials. This is a major limitation that should be addressed in future research programs.

Available fatigue-crack-growth data in a high-strain fatigue regime, as well as in the Paris-law regime, have been reviewed by Skelton and Beckett (Ref 60). In the Paris-law regime, the following fatigue-crack-growth-rate law is obeyed by the T91 material tested at 595 °C (1100 °F):

$$\frac{da}{dN} = 4.5 \times 10^{-7} \Delta K^{2.01} \, \text{mm/cycle}$$

(Eq 8.1)

where ΔK is expressed in MPa$\sqrt{\text{m}}$. Creep-crack-growth-rate studies on T91 at 540 and 595 °C (1000 and 1100 °F) by Jaske and Swindeman have shown that the da/dt-vs-C_t behavior is identical to that of low-alloy Cr-Mo and Cr-Mo-V steels and falls within the scatterband of behavior shown in Fig. 3.24 in Chapter 3 (Ref 62).

A potential disadvantage of ferritic steels in general is their apparent susceptibility to type IV cracking, which occurs at the edges of HAZ material adjacent to unaf-

fected parent material (Ref 14). Cracking is generally attributed to localized creep deformation in a "soft" zone present at the intercritical region, under the action of bending stresses. Cracking of this type has been observed in ½Cr-Mo-V pipework systems after extensive operating time (60,000 h) at 565 °C (1050 °F) and also in 2¼Cr-1Mo steels in laboratory tests. Similar failures have been observed in crossweld rupture specimens in ½Cr-Mo-V, 2¼Cr-1Mo, and 9Cr-1Mo steels. In T91 steel, type IV cracking has been observed in laboratory tests at several different laboratories (Ref 14). Townsend has concluded that the data at 600 °C (1110 °F) indicate a shortfall in weld properties of approximately 20 to 25% in terms of stress in comparison with the parent material, and in this respect the drop in properties is of similar magnitude to that noted in ½Cr-Mo-V crossweld samples. Failures of this type have not been reported in service for HT9 steel, but most operating experience with this material has been at temperatures below 550 °C (1020 °F). Limited laboratory data indicate that welds in HT9 can suffer type IV failures when the test temperature is 600 °C (1110 °F). Creep-rupture test results of Haneda *et al* are somewhat at variance with the conclusions of Townsend inasmuch as no significant reduction of rupture strength for the weldment was noted at 600 °C (1110 °F) for P91 steel (see Fig. 8.15) (Ref 54). At 650 °C (1200 °F), however, the weldment was

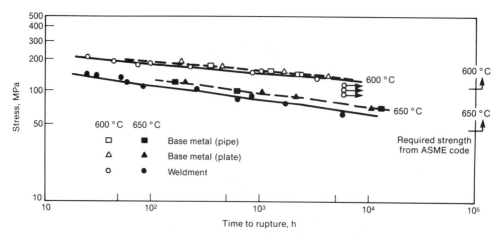

Fig. 8.15. Comparison of creep-rupture properties of modified 9Cr-1Mo steel weldment and base metal, illustrating effect of type IV cracking (Ref 54).

found to have lower rupture strength than that of the base metal. Haneda *et al*, however, concluded that this drop in rupture strength would not pose a problem because even the lower-strength material met the code allowable stress requirements.

Improved heat treatments to avoid type IV cracking in P91 steel have been investigated by Roberts *et al* (Ref 38). A partial heat treatment procedure in which pipe sections are normalized and partially tempered at 675 °C (1250 °F) instead of the conventional tempering at 760 °C (1400 °F), followed by girth welding of the sections and stress relieving of the entire component assembly at 745 °C (1375 °F) has been found to be successful. It avoids formation of the "soft" zone at the heat affected zone/base metal interface and eliminates the propensity for type IV cracking. A major limitation of this procedure is that it requires furnace facilities capable of implementing the stress-relief treatment of the entire component as the last operation in the fabrication sequence. This can be done in shop fabrication of headers and steam pipes but is difficult to do in field fabrication. The only alternative in such cases is to incorporate a 10 to 20% safety margin in terms of stress into the design.

For temperatures above 595 °C (1100 °F), use of austenitic steels would be essential. In existing plants, problems have occurred with type 316 steel at Drakelow "C" Power Station due to distortion under conditions of thermal ratchetting (Ref 14 and 63) and at Eddystone Power Station (Ref 64), where cracking was due to thermal embrittlement and creep-fatigue interaction after 140,000 h at 620 °C (1150 °F). Experience at Drakelow "C" (unit 12) with Esshete 1250 has been good (>80,000 h at 600 °C, or 1110 °F) with considerably less distortion than with type 316, probably because its high tensile strength reduced the susceptibility to thermal ratchetting (Ref 14 and 63). Austenitic steels in general, however, are more likely to be susceptible to creep-fatigue interactions than ferritic steels, due to their high coefficients of expansion and low thermal conductivities. For any new plant operating above 600 °C, it will be

essential to assess the risk to austenitic steels arising from creep-fatigue.

One of the embrittlement phenomena to contend with in using austenitic steel is the formation of an Fe-Cr intermetallic compound, known as sigma phase. Formation of this phase in service is known to have promoted creep-rupture failure in power-plant steam piping at the Eddystone 1 plant (Ref 64). Optimized compositions which also emphasize control of residual elements have been developed to avoid this problem. It is anticipated that performance of sigma-safe steel compositions designed and operated to minimize thermal stresses would be quite satisfactory in the advanced plants. Material-property data relating to creep and low-cycle fatigue of austenitic steels have been presented and discussed in previous chapters.

Material-property requirements for headers and steam pipes are likely to be similar, and hence they have been grouped together in the above discussion. Some minor differences exist which may affect material selection. The steam temperature is likely to be much more uniform in steam pipes, but subject to time-dependent and location-dependent fluctuations in headers. Hence, the thermal-fatigue-strength requirements are greater for headers than for steam pipes. Self-weight-induced stresses are less important for headers than for steam pipes, permitting heavier-wall construction and an attendant higher temperature/pressure capability for a given material when used in headers. One of the most important differences is that headers have many welded attachments to inlet stub tubes from reheaters and superheaters and intersections of outlet nozzles connecting pipework. Depending on the selection of materials for the superheater/reheater tubes and the header piping, dissimilar-metal welded joints may be required. The integrity of such austenitic-to-ferritic welds when 9 to 12% Cr steels form the ferritic components needs to be more thoroughly investigated.

Materials Selection for Boilers

Based on the discussion so far, a tentative list of promising candidates is presented in

Table 8.3. Candidate boiler materials for advanced plants (Ref 65)

Component	Phase 0: 31.0 MPa (4500 psi); 565/565/565 °C (1050/1050/1050 °F)	Phase I: 31.0 MPa (4500 psi); 595/595/595 °C (1100/1100/1100 °F)	Phase II: 34.5 MPa (5000 psi); 650/595/595 °C (1200/1100/1100 °F)
Furnace wall	1Cr-½Mo (T12) 1¼Cr-½Mo (T11)	1Cr-½Mo (T12) or 1¼Cr-½Mo (T11) for lower wall; Super 9Cr (T91) or HCM12 for upper wall	Same as phase I
Finishing superheater			
Noncorrosive	Super 9Cr (T91) HCM12 304 SS	Tempaloy A-1 347 SS	17-14 Cu-Mo Esshete 1250

Corrosive	HR3C
	Chromized 347 SS
		HR3C
		Chromized 347 SS
		NF709
			NF709
			CR30A
			Inconel 617
			17-14 Cu-Mo chromized
			  or coextruded with
			  Inconel 671 or 310 SS

Finishing reheater	Same as above	Same as above	Shot-blasted 347H
Headers and steam pipes	2¼Cr-1Mo (P22)	316H Super 9Cr (P91) 9Cr-2Mo (HCM9M) 12Cr-Mo-V (HT9) 9Cr-Mo-W (NF616)	316H
Separator or safety valve	A302 C-Mn-Mo-Ni WC9 casting 2¼Cr-1Mo	2¼Cr-1Mo 9Cr-2Mo (HCM9M)	2¼Cr-1Mo 316 casting 9Cr-2Mo (HCM9M)
Boiler recirculating pump	WC6 casting 1¼Cr-½Mo	WC6 casting	WC6 casting

Table 8.3, with reference to the phase 0, I, and II steam conditions described in Table 8.1. Table 8.3 is based on a similar description of materials selection for Japanese plants in a paper by Jaffee (Ref 11), which was modified in the light of later developments, as reviewed by Bakker (Ref 65). Materials selection for headers and for steam pipes are likely to be similar and, therefore, have been grouped together. In the piping systems, the average metal temperature is expected to be identical to the steam temperature; on the other hand, in superheater and reheater tubes, the metal temperature can be as high as 28 °C (50 °F) above the steam temperature. Hence, the 9 to 12% Cr steels, which have a creep-strength capability extending up to 595 °C (1100 °F), can be used for piping under phase I conditions but are relegated to phase 0 conditions when used for tubing. An excellent review of the state of the art with respect to boiler materials development can be found in Ref 66.

Steam-Turbine Materials

Candidate materials for use in the steam turbines of advanced plants are listed in Table 8.4. The rationale behind these selections on a component-specific basis is described in the following sections.

HP Rotors

It is well-recognized that 1Cr-1Mo-¼V steels currently in wide use are limited by their creep strength for service up to 540 °C (1000 °F). A useful database of industry experience is available pertaining to the use of

Table 8.4. Candidate materials for steam turbines in advanced steam plants (Ref 7)

Turbine part(s)	Steam temperature		
	565 °C (1050 °F) or lower	595 °C (1100 °F)	620 °C (1150 °F) or higher
SP, VHP, HP, VHP-HP, IP rotor	Cr-Mo-V steel (forging)	12%Cr steel (forging)	Austenitic superalloy or equivalent (forging)
Inner cylinder	Cr-Mo steel (casting)	9%Cr steel (casting)	316 austenitic steel (casting or forging)
Blade	Heat-resisting superalloy, if needed (rolled or forged)		
Steam valve	Cr-Mo steel or 9%Cr steel (forging)	316 austenitic steel (forging)	
Nozzle box	Cr-Mo steel or 9%Cr steel (casting)	316 austenitic steel (casting or forging)	
Inlet steam pipe at high temperature	Cr-Mo steel (forging) without dissimilar-metal weld	316 austenitic steel (forging) with dissimilar-metal weld	
Outer cylinder	Cr-Mo steel (casting)		
LP rotor	In case of reheat temperature of 565 °C or lower, Ni-Cr-Mo-V steel (forging)	In case of reheat steam temperature of 595 °C or higher, improved material or advanced material free from temper embrittlement (forging)	

12% Cr tempered martensitic steels at 565 °C (1050 °F). An excellent combination of strength, ductility, and toughness, as well as good experience with 12% Cr HP/IP rotors at 565 °C (1050 °F) and below, have made this type of steel a prime candidate for HP/IP rotors at up to 595 °C (1100 °F) in advanced plants. However, there is considerable doubt that currently available commercial versions of 12% Cr steel can be used at 595 °C without steam cooling the rotor down to 565 °C, because of their inadequate creep strength. Nevertheless, there is a strong incentive to use the most creep-resistant version of 12% Cr steel such that it will have a creep strength at 595 °C equivalent to that of currently used 12% Cr steels at 565 °C, in order to provide an additional margin of safety (Ref 10 and 11). For developing improved alloys, a tentative goal for rupture strength has been established at 100 MPa (14.5 ksi) for a rupture life of 100,000 h at 595 °C.

Large steam-turbine rotors made of 12% Cr steels have been produced and used in HP/IP turbines operating at up to 565 °C

(1050 °F) since about 1955 in Europe, about 1960 in the United States, and more recently in Japan. This experience has been reviewed by Newhouse (Ref 67). Typical compositions employed (see Table 8.5) have included the German alloy X21CrMoV 121, an 11Cr-Mo-V-Nb-N alloy developed in the United States, and an 11Cr-Mo-V-Ta-N alloy developed in Japan. In Western Europe and the United States, the standard practice has been to use electric-furnace melting with vacuum stream degassing (VSD), tap degassing, or ladle degassing to cast electrodes which are then electroslag remelted (ESR). In Japan, the ingots are produced by basic electric-furnace melting followed by vacuum carbon deoxidation (VCD). Reducing the silicon and sulfur contents and the contents of impurity elements has been a very important aspect of the development of this process. Worldwide, there are a number of developments of stronger materials for HP rotors. The EPDC program in Japan, the EPRI program in the United States, and the COST 501 program in Europe on modified 12Cr-Mo-V steels are noteworthy.

Table 8.5. Nominal chemical compositions of 12%Cr rotor steels

Alloy	Developed in	C	Mn	Si	Ni	Cr	Mo	W	V	Nb	Ta	N	Al
In use													
X21CrMoV121	Germany	0.23	0.55	0.20	0.55	11.7	1.0	...	0.30	...	...	...	...
11Cr-Mo-V-Nb-N	United States	0.16	0.62	0.25	0.38	11.1	1.0	...	0.22	0.57	...	0.05	...
11Cr-Mo-V-Ta-N	Japan	0.17	0.60	0.06	0.35	10.6	1.0	...	0.22	...	0.07	0.05	...
Developmental													
10Cr-Mo-V-Nb-Ta-N (alloy 3.1)	Japan (Ref 67)	0.09	0.29	0.06	0.30	10.13	0.98	...	0.20	0.1	0.07	0.07	...
10Cr-Mo-V-W-Nb-N (alloy 4.2)	Japan (Ref 67)	0.13	0.49	0.04	0.96	10.40	1.01	0.82	0.20	0.07	...	0.06	...
10Cr-Mo-V-Nb-N (TR1100, TMK1)	Japan (Ref 21 and 68)	0.14	0.50	0.05	0.6	10.20	1.5	...	0.17	0.06	...	0.04	0.002
10Cr-Mo-V-W-Nb-N (TR1150, TMK2)	Japan (Ref 21 and 68)	0.13	0.50	0.05	0.7	10.2	0.4	1.8	0.17	0.06	...	0.05	0.005
11Cr-Mo-V-W-Nb-N (TR1200)	Japan (Ref 21)	0.12	0.50	0.05	0.8	11.2	0.3	1.8	0.20	0.06	...	0.06	0.005
Superclean X21CrMoV121	France (Ref 73)	0.20	0.06	0.03	0.75	11.5	1.0	...	0.30	...	...	...	...
High-nitrogen steels	Germany/ Switzerland (Ref 74)	0.06	0.55	0.20	0.55	10.0	1.45	...	0.22	0.07	...	0.16	...
		0.08	0.55	0.20	0.55	12.6	2.0	...	0.10	0.06	...	0.33	...

Several developmental alloys (see Table 8.5) with improved creep strength have been reported in the literature. These alloys offer the potential of operating at steam temperatures higher than 595 °C (1100 °F) and have rupture-strength capabilities in excess of the currently used commercial versions of 12% Cr steel. Available mechanical-property data on these alloys are rather limited at this time.

In Fig. 8.16, the 100,000-h rupture-stress values for currently available commercial 12Cr-Mo-V steels are compared with those of some of the developmental alloys. At a typical design stress of about 125 MPa (18.1 ksi), the current alloys are limited to 565 °C (1050 °F) in the uncooled condition (i.e., metal temperature). The TR1100, TR1150, and TR1200 alloys appear to be capable of operation at 595 °C (1100 °F), 620 °C (1150 °F), and 640 °C (1185 °F), respectively. The TR1100 composition, also referred to as TMK1, has been utilized for actual production of an HP/IP rotor for use at the EPDC Wakamatsu station, where the steam temperature will be 595 °C (1100 °F) (Ref 68 and 69).

Figure 8.17 is a composite of the mean

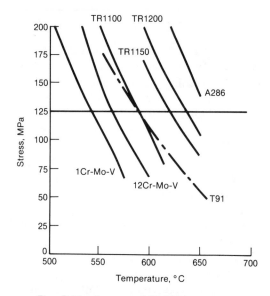

Fig. 8.16. Average 100,000-h rupture-stress values for commercial and developmental rotor steels.

stress-rupture curves for the three commercial alloys currently in use (Ref 67). Data for the 11Cr-Mo-V-Nb-N alloy (U.S.) are plotted as a scatterband based on 2 standard deviation. The data for the 11Cr-Mo-V-Ta-N alloy (Japan) fall within the scatterband.

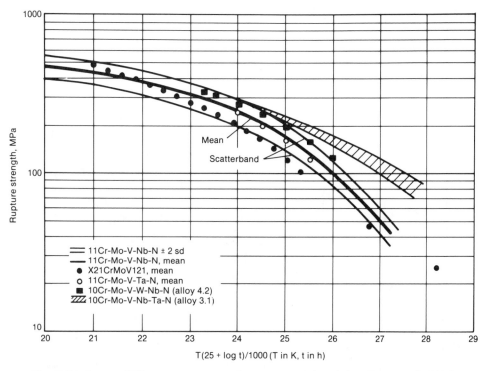

Fig. 8.17. Larson-Miller rupture curves for commercial and developmental 12% Cr rotor steels (based on Ref 67).

The 12Cr-Mo-V alloy X21CrMoV121 (Europe) is found to have lower rupture-strength values than those of the other two alloys at Larson-Miller parameter values in excess of about 24,000. Developmental alloys 10Cr-Mo-V-W-Nb-N (designated as alloy 4.2 in Ref 67) and 10Cr-Mo-V-Nb-Ta-N (designated as alloy 3.1 in Ref 67) appear to offer significant improvement over the commercial alloys, especially higher temperatures and longer times. Properties of alloy 4.2 are very similar to those of the TMK1 (TR1100) rotor composition, and properties of alloy 3.1 are similar to those of the TMK2 (TR1150) composition. Data for TMK1 and TMK2, therefore, are not plotted separately. Rupture data for TMK1 and TMK2 have been reported in a recent paper (Ref 70).

Figure 8.18 shows fracture-toughness behavior for several different grades of 12% Cr rotor steels in comparison with conventional Cr-Mo-V steels and low-sulfur Cr-Mo-V steels (Ref 67). The different grades of 12% Cr steel, as indicated by different

symbols include 12Cr-Mo-V steels made by ESR and, 11Cr-Mo-V-Ta-N steel made by VCD. The lower-bound curves for the 12% Cr steels and for the Cr-Mo-V steels are based on published results of Engelke *et al* (Ref 71). The lower-bound curve for 12% Cr steels is appreciably higher than that for conventional Cr-Mo-V steels at all temperatures and higher than that for low-sulfur Cr-Mo-V steels at temperatures above 0 °C (32 °F). No obvious distinctions among the different grades of 12% Cr steels are apparent.

Figure 8.19 depicts the high-cycle-fatigue design curves for the 12Cr-Mo-V German alloy for temperatures from room temperature to 550 °C (1020 °F) based on mean fatigue-strength data (Ref 71). Data for an 11Cr-Mo-V-Ta-N Japanese commercial rotor alloy at different temperatures are shown in Fig. 8.20 (Ref 67). In the figure, a developmental alloy in which tantalum was replaced by tungsten and niobium (alloy 4.2, Table 8.5) is also included. The developmental alloy appears to have higher

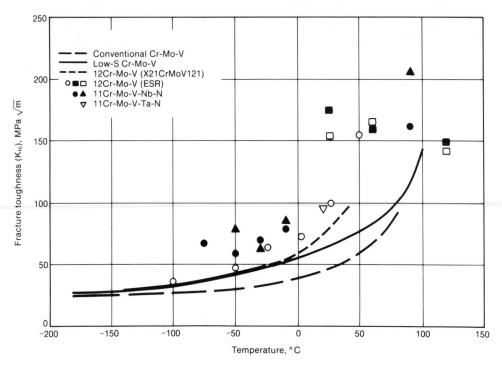

Fig. 8.18. Fracture-toughness data for Cr-Mo-V and 12%Cr steels from various sources (Ref 67).

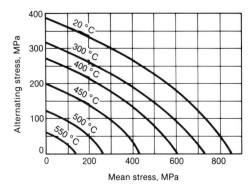

Fig. 8.19. High-cycle-fatigue design curves for 12Cr-Mo-V rotor steels (Ref 71).

fatigue strength than the commercial 11Cr-Mo-V-Ta-N alloy.

Figure 8.21 is a plot of the mean data for cycles to crack initiation versus strain range for 12Cr-Mo-V (X21CrMoV121) and 1Cr-Mo-V rotors, for temperatures below and within the creep range, nominally the same for the two materials (Ref 71). Additional data on an electroslag-remelted 12Cr-Mo-V steel are also included in the figure. Data on 11Cr-Mo-V-Nb-N steel have also been

reviewed by Newhouse (Ref 67). The available database is too limited, refers to different test conditions, and does not include information on the range of variability of the results. Hence, direct comparisons between different alloys are not possible.

Figure 8.22 presents upper-limit fatigue-crack-growth data for 12Cr-Mo-V rotor steel (X21CrMoV121), from Engelke *et al* (Ref 71), compared with actual data for an ESR rotor of the same steel. The curves are comparable (Ref 67).

Creep-crack properties of several modified 12% Cr rotor steels at 630 °C (1165 °F) have been evaluated by Kuwabara, Nitta, Ogata, and Sugai (Ref 72). Their results, shown in Fig. 8.23, led them to conclude that the relationships between da/dt and C* for these steels are identical to those for low-alloy steels and austenitic stainless steels.

Significant progress in developing improved 12% Cr steels has also been reported based on work in Europe (Table 8.5). Clean-steel versions of 12% Cr HP/IP rotors with low silicon and manganese levels

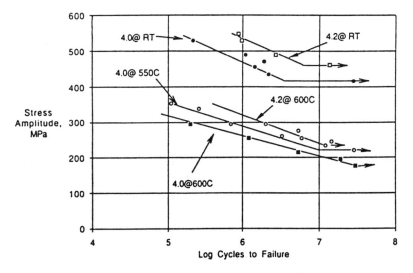

Fig. 8.20. Comparison of high-cycle-fatigue data for 11Cr-Mo-V-Ta-N commercial steel (designated 4.0) and 11Cr-Mo-V-W-Nb-N developmental rotor steel (alloy 4.2) (Ref 67).

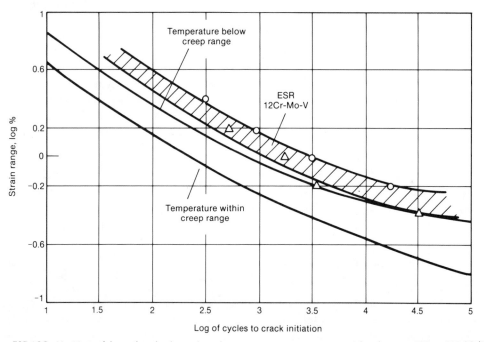

For ESR 12Cr-Mo-V steel (crosshatched area), o denotes room temperature and △ denotes 550 to 600 °C (1020 to 1110 °F).

Fig. 8.21. Mean low-cycle-fatigue curves for 1Cr-Mo-V and 12Cr-Mo-V rotors compared with data on ESR 12Cr-Mo-V steel from Kobe Steel (based on Ref 67 and 71).

and strengthened by niobium and nitrogen additions have been reported by Pisseloup, Poitrault, and Badeau (Ref 73). High-nitrogen, low-carbon steels developed by Uggowitzer *et al* (Ref 74) are reported to yield a hundredfold improvement in rupture life compared with the commercial 12Cr-Mo-V (X21CrMoV121) rotor steels. These alloys are made by ESR melting in a high-pressure nitrogen environment. Based

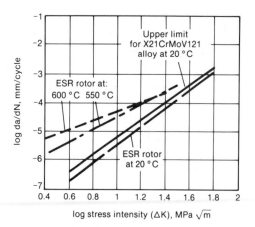

Fig. 8.22. Fatigue-crack-growth-rate data for ESR 12Cr-Mo-V rotor steel at 20, 550, and 600 °C compared with standard X21CrMoV121 alloy at 20 °C (Ref 67).

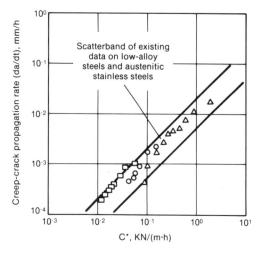

Fig. 8.23. Variation of creep-crack-propagation rate with C* integral for modified 12% Cr rotor steels at 630 °C (1165 °F) (Ref 72).

on extrapolation of short-time data, these authors have expressed the belief that the 10% Cr steel containing 0.16% N_2 might have rupture strength exceeding that of the best Japanese alloy, TR1200 (Ref 74). Investigation of laboratory heats of boron-modified 10% Cr steels has also shown that the target rupture strength of 100 MPa (14.5 ksi) for rupture in 10^5 h at 595 °C (1100 °F) can be met with modified compositions containing 100 ppm boron (Ref 75). Results from European research programs

have been reviewed by Scarlin and Schepp (Ref 76).

It is clear from the literature that several developmental alloys have shown the capability for rotor operation at 595 °C (1100 °F) in the uncooled condition. More fabricating experience and long-term property data are being gathered.

Austenitic steels are required for service temperatures above 595 °C (1100 °F). The EPDC research and development program on rotor steels for operation at 650 °C (1200 °F) have concentrated on alloy A286. Some experience with a similar austenitic steel, Refractaloy, is available from the Eddystone 1 power station in the United States. The technical problem in producing large forgings of A286 and Refractaloy is freezing segregation. This problem has been overcome through the use of ESR melting, where freezing segregation is minimized through the small molten pool and favorable solidification pattern. The austenitic steels are disadvantageous in large section sizes because of their low thermal conductivity and high coefficient of thermal expansion compared with ferritic steels. Thus, a high-temperature power plant with austenitic steel rotors and other heavy-section components will be less able to thermal cycle than one equipped with ferritic steel components. Also, design allowances must be made for the higher expansion of the austenitic components through sliding seals.

Blades for HP/IP Rotors

The principal requirements for blades are high creep strength, high-cycle-fatigue strength, toughness, and scaling resistance. Current materials for HP/IP turbines are 12Cr-1Mo, 12Cr-Mo-V, and 12Cr-Mo-V-W steels. All the commercial 12% Cr rotor forging compositions listed as being in use in Table 8.5 can also be used as blade forgings. The commercially available 12% Cr steel forgings are expected to be adequate for phase 0 conditions (i.e., steam at 565 °C, or 1050 °F). For phase I (595 °C/1100 °F) conditions, it is anticipated that the "developmental" 12% Cr steel rotor forging compositions or Nimonic 80A would be suitable candidates for the control-stage blading. In the long blades used in the re-heat turbine, it may be necessary to use nickel-base superalloys such as Nimonic 80A, M252, or Refractaloy 26 (Ref 77). For phase II conditions (650 °C/1200 °F), nickel-base superalloy blading would be required throughout.

LP Rotors

Currently, a 3.5Ni-Cr-Mo-V steel is widely used for rotors in the low-pressure section. Allowable metal temperatures are limited to about 345 °C (650 °F), primarily to avoid the risk of temper embrittlement at higher temperatures. It is anticipated that steam temperatures at the LP inlet for the advanced designs would be in the range 400 to 455 °C (750 to 850 °F). This means that either the LP rotors have to be cooled or that improved materials need to be used. Because the first approach would lead to reduced efficiencies, the latter approach is preferred. To eliminate the susceptibility of Ni-Cr-Mo-V steels to temper embrittlement altogether, a J factor—i.e., $(P+Sn)$ $(Mn+Si) \times 10^4$—of less than 10 would be required. It is believed that such low values of J can be achieved by employing a combination of basic oxygen or electric-furnace steelmaking, vacuum carbon deoxidation (to avoid silicon), and ladle desulfurization (to reduce sulfur and the need for adding manganese).

Progress has been made in producing high-purity steel in production-size ingots and forgings using modern secondary refining technology (Ref 78). Silicon or aluminum deoxidation can be replaced by vacuum carbon deoxidation, leaving low residual silicon contents of 0.02 to 0.05%. The molten steel may be desulfurized in ladle-refining furnaces to levels of 0.001 to 0.002% S, thus avoiding the need to use manganese to tie up the sulfur as MnS. Several trial rotors and production rotors weighing from 34 to 120 metric tons have been produced with values of J in the range 3 to 8. This is to be contrasted with typical J values of about 72 for conventional Ni-Cr-Mo-V rotors. A critical overview of the extensive and recent developments with respect to superclean rotor steels has been published by Jaffee (Ref 79). Highlights of some of the mechanical-property evaluations from this article are reported below.

Trial rotors from the superclean Ni-Cr-Mo-V steel have been found to have excellent toughness and resistance to temper embrittlement. The FATT values in the as-received condition are in the range −70 to −60 °C (−95 to −75 °F) compared with values of −40 to −10 °C (−40 to +14 °F) for conventional rotors. The upper-shelf energy is about 200 J, about 50 to 100 J higher than that for conventional rotors. The fracture-toughness (K_{Ic}) values are significantly higher than those of conventional rotors, even at very high yield strengths (see Fig. 8.24) (Ref 79 and 80). The variation in fracture toughness with temperature for a trial rotor forging of the superclean steel is shown

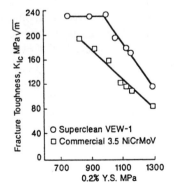
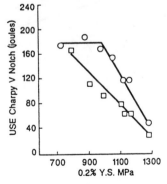

Fig. 8.24. Fracture toughness and Charpy V-notch upper-shelf energy (USE) of super-clean and commercial 3.5Ni-Cr-Mo-V steels at room temperature as a function of yield strength (Ref 79 and 80).

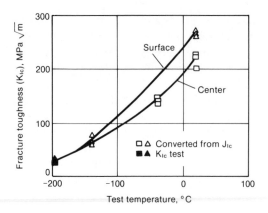

Fig. 8.25. Variation of fracture toughness with temperature for a trial rotor forging of superclean Ni-Cr-Mo-V steel (Ref 81).

perclean steel are plotted in Fig. 8.27. Using a stress-rupture strength of 10^5 h at 100 MPa (14.5 ksi) as the minimum requirement, the clean steels seem capable of operation at temperatures as high as 500 °C (930 °F). This means that, under the expected operating conditions at 455 °C (850 °F), adequate resistance to creep is completely ensured.

Turbine Stationary Components

Heavy-section stationary components employed at high temperature include inner and outer casings, valves, nozzle blocks, and inlet pipes. These parts must have high creep strength (and ductility), resistance to thermal fatigue, and good fracture toughness. Due to the complex shapes and large sizes, castings will be preferred to forgings for many of these applications. Thermal-fatigue considerations and the need for compatibility with other ferritic steel components such as 12% Cr rotors and 9 to 12% Cr steam inlet piping, in terms of the coefficient of thermal expansion, make castings of 9 to 12% Cr steels attractive for steam temperatures of 565 and 595 °C (1050 and 1100 °F)—i.e., phases 0 and I.

in Fig. 8.25 (Ref 81). Isothermal-embrittlement studies at 350, 425, and 480 °C for times up to 17,000 h show total immunity to temper embrittlement (Ref 82).

Figure 8.26 presents the results of low-cycle-fatigue testing of samples from two trial rotors of the superclean Ni-Cr-Mo-V steel (Ref 82). In comparison with the scatterband of data for the normal-purity (commercial) steels, the superclean steels represent a clear improvement.

Results of creep-rupture tests of the su-

Several casting-grade 9 to 12% Cr steels

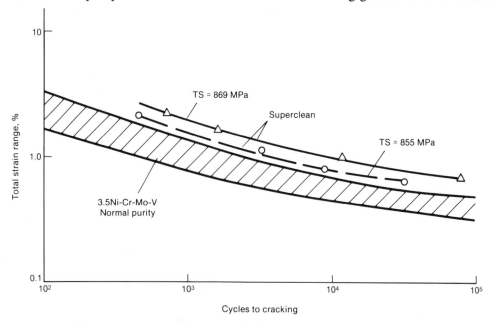

Fig. 8.26. Comparison of low-cycle fatigue behavior of 3.5Ni-Cr-Mo-V rotor steel in superclean and normal conditions (Ref 82).

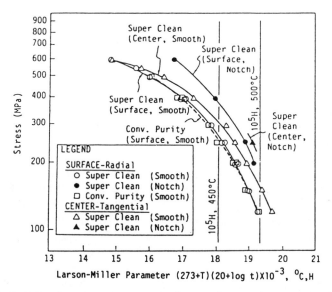

Fig. 8.27. Results of stress-rupture tests of superclean and conventional-purity LP rotor forgings (Ref 81).

Table 8.6. Compositions of 9 to 12% Cr alloy castings (Ref 77)

Source(a)	Composition, %									
	C	Si	Mn	Ni	Cr	Mo	V	W	Nb	N
Specification										
T91	0.08 / 0.12	...	...	0.4 max	8.0 / 9.5	0.85 / 1.05	0.18 / 0.25	...	0.04 / 0.10	0.03 / 0.07
Toshiba (Kawagoe)	0.12 / 0.16	0.2 / 0.6	0.4 / 0.8	0.3 / 0.7	9.5 / 11.0	0.7 / 1.0	0.2 / 0.3	...	0.05 / 0.12	0.03 / 0.06
MHI (Wakamatsu)	0.08 / 0.12	0.7 max	0.8 max	0.4 / 0.7	9.1 / 10.0	0.65 / 1.0	0.13 / 0.20	...	0.03 / 0.07	0.03 / 0.07
Typical										
T91										
MAN/GF	0.11	0.4	0.4	0.2	9.0	0.9	0.21	...	0.08	0.05
IHI/Okano	0.12	0.36	0.51	0.07	9.0	0.9	0.22	...	0.10	0.03
TSB 12 Cr (Kawagoe)	0.12	0.50	0.48	0.66	10.0	0.8	0.27	...	0.06	0.05
MHI 12 Cr (Wakamatsu)	0.12	...	...	0.5	10.0	0.8	0.25	...	0.06	0.05
Hitachi 12 Cr	0.13	0.28	0.58	0.58	10.23	1.1	0.22	0.23	0.06	0.045

(a) Names in parentheses refer to power stations where the steels will be used.

have been described by Jaffee (see Table 8.6) (Ref 77). Efforts are in progress to qualify and use modified 9Cr-1Mo steel (grade 91) in the form of castings. The Japanese turbine manufacturers have also developed several casting grades. They all have practically the same composition (9 to 12% Cr, 1 Mo, 0.2 V strengthened by niobium and nitrogen). It is anticipated that for steam temperatures of 565 and 595 °C (1050 and 1100 °F), the 9 to 12% Cr casting-grade steels will be utilized. For higher-temperature service, austenitic stainless steels will be used.

If stationary parts made of austenitic steel are used in conjunction with martensitic 12Cr-Mo-V rotors, incompatibility of thermal expansion becomes a major problem. Choice of materials and designs should take this into account. The problem of differential thermal expansion is also of concern in joining austenitic steam inlet piping to casings made of 2¼Cr-1Mo or other ferritic steels. In these cases, an alloy with an intermediate coefficient of expansion, such as Incoloy 800H, may be used as a transition piece, and welding of stainless steel to Incoloy 800H, as well as Incoloy 800H to 2¼Cr-1Mo steel, would be done using Inconel filler metal (Ref 10 and 11).

The bolting used to hold the high-temperature casing together must be selected to have sufficient elevated-temperature strength and make a good thermal-expansion match with the casing material. When the rotor and casing are made of ferritic steel, modified 12% Cr bolts work well up to 565 °C (1050 °F), but nickel-base superalloys are needed at 595 °C (1100 °F) or higher. Based on a critical survey of worldwide experience and based on relaxation tests, Mayer and Konig (Ref 83) have identified Nimonic 80A, Nimonic 90, and Refractaloy-26 as the best candidates for use up to 600 °C (1110 °F).

References

1. A.F. Armor, Worldwide View of Improved Coal-Fired Power Plants, in *Proceedings of the First International Conference on Improved Coal-Fired Power Plants*, A.F. Armor, W.T. Bakker, R.I. Jaffee, and G. Touchton, Ed., Report CS-5581-SR, Electric Power Research Institute, Palo Alto, CA, 1988
2. P.L. Rittenhouse *et al*, "Assessment of Material Needs for Advanced Steam Cycle Coal-Fired Plants," Report ORNL TM 9735, Oak Ridge National Laboratories, Oak Ridge, TN, Aug 1985
3. G.M. Yarenchak *et al*, "Engineering Assessment of an Advanced Pulverized Coal Power Plant," Report CS-2555, Electric Power Research Institute, Palo Alto, CA, Aug 1982
4. S.B. Bennett *et al*, "Engineering Assessment of an Advanced Pulverized Coal Power Plant," Report CS-2223, Electric Power Research Institute, Palo Alto, CA, Jan 1982
5. R.D. Hottenstine, N.A. Phillips, and R.A. Dail, "Development Plans for Advanced Fossil Fuel Power Plants," Report CS-4029, Electric Power Research Institute, Palo Alto, CA, May 1985
6. W. Nesbit, Japanese Challenge in Coal Combustion Technology, *EPRI Journal*, Vol 7 (No. 4), May 1982, p 20-23
7. H. Toda, H. Haneda, and A. Hizume, "Ultra Supercritical Pressure Power Generation," Mitsubishi Power Systems Bulletin MBB 82113E, Mitsubishi Heavy Industries, Ltd., Japan, 1982
8. T. Suzuki, "The Development of Coal-Firing Power Unit With Ultra High Performance in Japan," Electric Power Development Co., Ltd., presented at the Electric Power Research Institute Workshop on Fossil Plant Heat Rate Improvement, Charlotte, NC, Aug 26-28, 1981
9. H. Haas, W. Engelke, J. Ewald, and H. Termuehlen, Turbines for Advanced Steam Conditions, in *Proceedings of the American Power Conference*, Vol 44, 1982, p 330-338
10. M. Gold and R.I. Jaffee, Materials for Advanced Steam Cycles, *ASM J. Mater. Energy Systems*, Vol 6 (No. 2), 1984, p 130-145
11. A.F. Armor, R.I. Jaffee, and R.D. Hottenstine, Advanced Supercritical Power Plants — the EPRI Development Program, in *Proceedings of the American Power Conference*, Vol 46, 1984, p 70
12. R.W. Swindeman *et al*, "Alloy Design Criteria and Evaluation Methods for Advanced Austenitic Alloys in Steam Service," Report ORNL 6274, Oak Ridge National Laboratories, May 1986
13. F. Masuyama, H. Haneda, and B.W. Roberts, Update Survey and Evaluation of Materials for Improved Coal-Fired Power Plants, in *Proceedings of the First International Conference on Improved Coal-Fired Power Plants*, A.F. Armor, W.T. Bakker, R.I. Jaffee, and G. Touchton, Ed., Report CS-5181-SR, Electric Power Research Institute, Palo Alto, CA, 1988, p 5-85 to 5-108
14. R.T. Townsend, CEGB Experience and U.K. Developments in Materials for Advanced Fossil Plants, in *Advances in Materials Technology for Fossil Fuel Power Plants*, R. Viswanathan and R.I. Jaffee, Ed., American Society for Metals, Metals Park, OH, 1987, p 11-20
15. T. Yukitoshi, K. Nishida, T. Oda, and T. Daikoku, *ASME J. Pressure Vessel Tech.*, Vol 98 (No. 5), 1976, p 173-178

16. T. Zaizen *et al*, in *The 6th International Symposium on Heat Resistant Metallic Materials*, Czechoslovakia, 1981, p 66
17. K. Kinoshita, Nippon Kokan Technical Report, Vol 62, 1973, p 123
18. M. Ivenel, *Revue Generale Term*, 1964, p 555
19. V.K. Sikka, C.T. Ward, and K.C. Thomas, in *Production, Fabrication, Properties, and Application of Ferritic Steels for High-Temperature Applications*, American Society for Metals, Metals Park, OH, 1981
20. K. Oda, T. Fujita, and Y. Otoguro, "Report of the 123rd Committee on Heat Resistant Metals and Alloys," Japan Society for the Promotion of Science, Vol 26 (No. 2), 1985, p 261-269
21. T. Fujita, Advanced High Chromium Ferritic Steels for High Temperatures, *Met. Prog.*, Aug 1986, p 33-40
22. "Quality and Mechanical Properties of NF616 for Power Plant Boilers," Nippon Steel Corp., June 1987
23. F. Masuyama *et al*, presented at the International Conference on High Temperature Alloys, Preprint Paper No. 21, Petten, The Netherlands, Oct 15-17, 1985
24. E.J. Vinebert, T.B. Cox, C.C. Clark, and P. Boussel, presented at the International Conference on High Temperature Alloys, Preprint Paper No. 24, Petten, The Netherlands, Oct 15-17, 1985
25. P.G. Kalwa, K. Haarmann, and J.K. Janssen, presented at the Topical Conference on Ferritic Alloys for Use in Nuclear Energy Technology, AIME, Snowbird, UT, June 19-23, 1983
26. Y. Minami, H. Kimura, and M. Tanimura, presented at the International Conference on New Development in Stainless Steel Technology, ASM, Detroit, Sept 17-21, 1984
27. C.L. Clark, J.J.B. Rutherford, A.B. Wilder, and M.A. Cordovi, *Trans. ASME, J. Engg. Power*, Vol 82, 1960, p 35
28. J.D. Murray, *Weld. Met. Fabrication*, 1962, p 350
29. M. Kikuchi *et al*, presented at the International Conference on High Temperature Alloys, Preprint Paper No. 22, Petten, The Netherlands, Oct 15-17, 1985
30. Y. Sawaragi *et al*, *Sumitomo Kinzoku*, Vol 37, 1985, p 166
31. Nippon Steel Corp. (NSC) Technical Report, Dec 1984
32. H. Doi, C. Asano, M. Sukekawa, and S. Kirihara, in *International Conference on Creep*, Tokyo, Apr 14-18, 1986
33. Nippon Kokan K. (NKK) in-house data, cited in Ref 13
34. Y. Sawaragi and K. Yoshikawa, *Tetsu to Hagane*, Vol 72, 1986, p 672
35. H. Fricker and B. Walser, "Experience With High Chromium Ferritic Steels in Fossil-Fired Boilers," presented at BNES Conference on Ferritic Steels for Fast Reactor Steam Generators, May 30 to June 2, 1977
36. F. Masuyama, H. Haneda, K. Yoshikawa, and A. Iseda, Three Years of Experience With a New 12%Cr Steel in Superheater, in *Advances in Materials Technology for Fossil Fuel Power Plants*, R. Viswanathan and R.I. Jaffee, Ed., American Society for Metals, Metals Park, OH, 1987, p 259-266
37. F. Masuyama, H. Haneda, T. Daikoku, and T. Tsuchiya, "Development and Applications of a High Strength, 12%Cr Steel Tubing With Improved Weldability," Technical Review, Mitsubishi Heavy Industries, Ltd., Japan, Oct 1986, p 229-237
38. B.W. Roberts and D.A. Canonico, Candidate Uses for Modified 9Cr-1Mo Steel in an Improved Coal-Fired Plant, in *Proceedings of the First International Conference on Improved Coal-Fired Power Plants*, A.F. Armor, W.T. Bakker, R.I. Jaffee, and G. Touchton, Ed., Report CS-5181-SR, Electric Power Research Institute, Palo Alto, CA, 1988, p 5-58 to 5-82
39. H. Masumoto *et al*, Development of a 9%Cr-Mo-W Steel for Boiler Tubes, in *Proceedings of the First International Conference on Improved Coal-Fired Power Plants*, A.F. Armor, W.T. Bakker, R.I. Jaffee, and G. Touchton, Ed., Report CS-5581-SR, Electric Power Research Institute, Palo Alto, CA, 1988, p 5-203
40. K. Yoshikawa *et al*, *Therm. Nucl. Power*, Vol 12 (No. 12), 1985, p 1325-1339
41. H. Teranishi *et al*, presented at the International Conference on High Temperature Alloys, Preprint Paper No. 21, Petten, The Netherlands, Oct 15-17, 1985
42. T. Ikeshima, *Bull. Japan Inst. Metals*, Vol 22 (No. 5), 1983, p 389
43. Ohtomo *et al*, High Temperature Corrosion Characteristics of Superheater Tubes, Ishikawajima Harima Industries, *Engg. Rev.*, Vol 16 (No. 4), Oct 1983
44. A.L. Plumley and W.R. Roczmiac, Coal Ash Corrosion Field Testing of Advanced Boiler Tube Materials, in *Proceedings of the Second International Conference on Improved Coal-Fired Power Plants*, Electric Power Research Institute, Palo Alto, CA, Nov 1-4, 1988
45. "Properties of Super 9Cr Steel Tube (ASTM) A 213-T 91," Report 803 F–No. 1023, Sumitomo Metal Industries, July 1983
46. A.J. Blazewicz and M. Gold, "Chromizing and Turbine Solid Particle Erosion," ASME Paper No. 78, JPGC PWr-7, Joint ASME/IEEE/ASCE Power Generation Conference, Dallas, Sept 1978

47. P.L. Daniel *et al*, "Steamside Oxidation Resistance of Chromized Superheater Tubes," CORROSION 80, NACE Conference, Chicago, Mar 1980

48. I.M. Rehn *et al*, "Controlling Steamside Oxide Exfoliation in Utility Boiler Superheaters and Reheaters," Paper No. 192, CORROSION 80, NACE Conference, Chicago, 1980

49. K. Kubo, S. Murase, M. Tamura, and T. Kanero, Application of Boiler Tubing Tempaloy Series to the Heat Exchanger of Advanced Coal-Fired Boilers, in *Proceedings of the First International Conference on Improved Coal-Fired Power Plants*, A.F. Armor, W.T. Bakker, R.I. Jaffee, and G. Touchton, Ed., Report CS-5581-SR, Electric Power Research Institute, Palo Alto, CA, 1988, p 5-237 to 5-254

50. H. Haneda *et al*, Fabrication and Formability Evaluation of Thick-Wall Chromized and Co-extruded Superheater Tubings for Advanced Cycle Power Applications, *ASM J. Mater. Energy Systems*, Vol 8 (No. 4), 1987, p 426-441.

51. V.K. Sikka, "Status of Development and Commercialization of Modified 9Cr-1Mo Steel," Oak Ridge National Laboratories, May 24, 1979

52. H. Haneda *et al*, in *Proceedings of the 3rd International Conference on Steel Rolling*, Tokyo, Sept 2-6, 1985, p 669-676

53. "Properties of Super 9Cr Large Diameter Pipe," Report 803 F–No. 1024, Sumitomo Metal Industries, Apr 1983

54. H. Haneda, F. Masuyama, S. Kaneko, and T. Toyoda, Fabrication and Characteristic Properties of Modified 9Cr-1Mo Steel for Header and Piping, in *Advances in Materials Technology for Fossil Fuel Power Plants*, R. Viswanathan and R.I. Jaffee, Ed., American Society for Metals, Metals Park, OH, 1987, p 231-242

55. V.K. Sikka, C.T. Ward, and K.C. Thomas, Modified 9Cr-1Mo Steel—An Improved Alloy for Steam Generator Application, in *Ferritic Steels for High-Temperature Applications*, A.K. Khare, Ed., American Society for Metals, Metals Park, OH, 1983, p 65

56. V.K. Sikka, R.E. MacDonald, G.C. Bodine, and W.J. Stelzman, "Effects of Tempering Treatment on Tensile, Hardness and Charpy Impact Properties of Modified 9Cr-1Mo Steel," Report ORNL/TM-8425, Oak Ridge National Laboratories, Oct 1982

57. "A New Chromium-Molybdenum Steel for Commercial Applications," ORNL Technology Transfer Meeting, Knoxville, TN, Apr 6-8, 1982

58. C.D. Lundin, B.J. Kruse, and M.W. Richey, "Transformation, Metallurgical Response, and Behavior of the Weld Fusion and HAZ in Cr-Mo Steels for Fossil Energy Applications," Advanced Research and Technology Development Fossil Energy Materials Program, Fourth Quarter Progress Report for the period ending 12-31-82, Report ORNL/FMP-83/1, Oak Ridge National Laboratories, p 93-107

59. V.K. Sikka, M.G. Cowgill, and B.W. Roberts, Creep Properties of Modified 9Cr-1Mo Steel, Proceedings of Topical Conference on *Ferritic Alloys for Use in Nuclear Energy Technologies*, AIME, June 1983, p 413-423

60. R.P. Skelton and B.J.E. Beckett, Thermal Fatigue Properties of Candidate Materials for Advanced Steam Plant, in *Advances in Materials Technology for Fossil Fuel Power Plants*, R. Viswanathan and R.I. Jaffee, Ed., American Society for Metals, Metals Park, OH, 1987, p 359-366

61. C.R. Brinkman, J.P. Strizak, M.K. Booker, and V.K. Sikka, "A Status Report on Exploratory Time Dependent Fatigue Behavior of 2¼Cr-1Mo and Modified 9Cr-1Mo Steel," Report ORNL TM7699, Oak Ridge National Laboratories, Oak Ridge, TN, June 1981

62. C.E. Jaske and R.W. Swindeman, Long-Term Creep and Creep-Crack-Growth Behavior of 9Cr-1Mo-V-Nb Steel, in *Advances in Materials Technology for Fossil Fuel Power Plants*, R. Viswanathan and R.I. Jaffee, Ed., American Society for Metals, Metals Park, OH, 1987

63. B. Plastow, B.I. Bagnall, and D.E. Yeldham, *Weld. Fabr. Nucl. Industry*, British Nuclear Engineering Society (BNES), London, 1979

64. J.E. Bynum *et al*, "Failure Investigation of Eddystone Main Steam Piping," presented at the 65th Convention of the American Welding Society, Dallas, Apr 1984

65. W.T. Bakker, personal communication, Electric Power Research Institute, Palo Alto, CA, Nov 1988

66. W.T. Bakker, EPRI Perspective on Boiler Materials and Components, in *Proceedings of the First International Conference on Improved Coal-Fired Power Plants*, A.F. Armor, W.T. Bakker, R.I. Jaffee, and G. Touchton, Report CS-5581-SR, Electric Power Research Institute, Palo Alto, CA, 1988, p 5-287 to 5-300

67. D.L. Newhouse, "Guide to 12Cr Steels for High- and Intermediate-Pressure Turbine Rotors for the Advanced Coal-Fired Steam Plant," Report CS-5277, Electric Power Research Institute, Palo Alto, CA, July 1987

68. A. Hizume *et al*, The Probability of a New 12%Cr Rotor Steel Applicable to Steam Temperature Above 593 °C, in *Advances in Materials Technology for Fossil Fuel Power Plants*, R. Viswanathan and R.I. Jaffee, Ed., American Society for Metals, Metals Park, OH, 1987, p 143-153

69. A. Hizume *et al*, An Advanced 12Cr Steel Ro-

tor Applicable to Elevated Steam Temperature 593 °C, *ASME J. Engg. Mater. Tech.*, Vol 109, Oct 1987, p 319-325

70. K. Furuya *et al*, Advanced 12Cr Steel Rotors Developed for EPDC Wakamatsu's Ultra-High Temperature Turbine Project, in *Proceedings of the Second International Conference on Improved Coal-Fired Power Plants*, Electric Power Research Institute, Palo Alto, CA, Nov 1-4, 1988

71. W. Engelke *et al*, Rotor Forgings for KWU Designed Turbine-Generators, in EPRI Workshop Proceedings, *Rotor Forgings for Turbines and Generators*, WS-79-235, R.I. Jaffee, Ed., Electric Power Research Institute, Palo Alto, CA, Sept 1981, p 3-147 to 3-163

72. A. Kuwabara, A. Nitta, T. Ogata, and S. Sugai, Improved 12Cr Steels for Advanced Steam Turbine Rotors, in *Advances in Materials Technology for Fossil Fuel Power Plants*, R. Viswanathan and R.I. Jaffee, Ed., American Society for Metals, Metals Park, OH, 1987, p 153-162

73. J. Pisseloup, J. Poitrault, and J. Badeau, Manufacture of High Temperature Rotors Using an Optimized 12%Cr Steel, in *Advances in Materials Technology for Fossil Fuel Power Plants*, R. Viswanathan and R.I. Jaffee, Ed., American Society for Metals, Metals Park, OH, 1987, p 163-172

74. P.J. Uggowitzer, G. Stein, B. Anthamatten, and M.O. Speidel, Development of Nitrogen Alloyed 12%Cr Steels, in *Advances in Materials Technology for Fossil Fuel Power Plants*, R. Viswanathan and R.I. Jaffee, Ed., American Society for Metals, Metals Park, OH, 1987, p 181-186

75. R.B. Scarlin *et al*, Materials Development for Alternative Rotor Designs for Improved Fossil-Fired Power Plants, in *Advances in Materials Technology for Fossil Fuel Power Plants*, R. Viswanathan and R.I. Jaffee, Ed., American Society for Metals, Metals Park, OH, 1987, p 51-59

76. B. Scarlin and P. Schepp, State of the European Cost Activities on Improved Coal-Fired Power Plant, in *Proceedings of the Second International Conference on Improved Coal-Fired Power Plants*, Electric Power Research Institute, Palo Alto, CA, Nov 1-4, 1988

77. R.I. Jaffee, Keynote Address: Advances in Materials for Fossil Power Plants, in *Advances in Materials Technology for Fossil Fuel Power Plants*, R. Viswanathan and R.I. Jaffee, Ed., American Society for Metals, Metals Park, OH, 1987, p 1-10

78. R.I. Jaffee, Materials and Electricity, *Met. Trans. A*, Vol 17A, p 755-775

79. R.I. Jaffee, "Overview of Superclean Rotor Steel Development," in "Superclean 3.5% NiCrMoV Steels and Improved Cr-Mo-V Steels," proceedings of special meeting organized by Verein Deutscher Eisenhüttenleute, Dusseldorf, May 31 to June 1, 1988

80. P.J. Uggowitzer and R.M. Magdowski, New Stahle Hochster Reinheit, in *Moderne Stahle Schweiz Akad der Wissenschaften*, Zurich, 1987, p 41–66

81. Y. Yoshiyoka, O. Watanabe, M. Miyazaki, and R.C. Schwant, Superclean 3.5NiCrMoV for Low Pressure Rotors, in *Proceedings of the Second International Conference on Improved Coal-Fired Power Plants*, Electric Power Research Institute, Palo Alto, CA, Nov 1-4, 1988

82. K.H. Mayer, unpublished results cited in "Superclean 3.5% NiCrMoV Steels and Improved Cr-Mo-V Steels," proceedings of special meeting organized by Verein Deutscher Eisenhüttenleute, Dusseldorf, May 31 to June 1, 1988

83. K.H. Mayer and H. Konig, High Temperature Bolting of Steam Turbines for Improved Coal-Fired Power Plants, in *Proceedings of the Second International Conference on Improved Coal-Fired Power Plants*, Electric Power Research Institute, Palo Alto, CA, Nov 1-4, 1988

9

Life-Assessment Techniques for Combustion Turbines

General

The combustion turbine in its simplest form consists of a compressor, a combustor, and a turbine. In a common configuration, the compressor and the turbine are mounted on a single shaft which is also connected to the generator. Ambient air is drawn into the compressor and is pressurized to 10 to 14 atm (typical). The compressed air flows into the combustion section, where fuel is injected and burned.

The hot combustion gases are expanded through the turbine section and rejected to the atmosphere. The typical configuration of a turbine is shown in Fig. 9.1 (Ref 1).

Although a single-shaft configuration is most common in large utility turbines, aircraft derivatives employ two shafts such that one shaft is attached to the compressor and to one or two turbine stages, whereas the other shaft is coupled to the remaining stages of the turbine and to the electric generator. This arrangement increases opera-

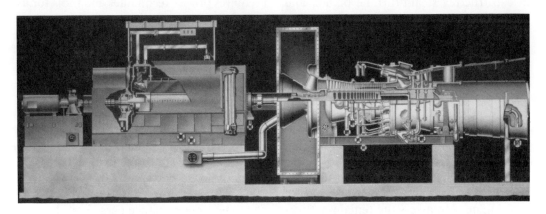

Fig. 9.1. Typical configuration of a combustion turbine (General Electric Model MS7001F) (Ref 1; courtesy of G.A. Cincotta, General Electric Co., Schenectady, NY).

415

tional flexibility by removing the need to run the compressor turbine and the power turbine at the same speed.

The simple open-cycle turbine operates at thermal efficiencies up to about 34% and at heat rates as low as 10,000 Btu/kW·h. The efficiency can be improved by utilizing the heat energy from the exhaust gases which exit the turbine at about 480 to 595 °C (900 to 1100 °F). This can be done in two ways. A regenerative heat exchanger placed in the exhaust stream can be used to heat the compressed air from the compressor prior to admission to the combustor. Alternatively, the waste heat can be used to generate steam in a waste-heat boiler to operate a steam turbine-generator. The latter configuration, known as the combined-cycle plant, can lead to thermal efficiencies approaching 50% and heat rates as low as 7500 Btu/kW·h.

Combustion turbines in utility service most commonly burn No. 2 fuel oil or natural gas; many turbines have dual-fuel capability. Some success has also been demonstrated with heavier oils and crudes, and currently there is much interest in burning gas and liquid fuels derived from coal conversion. The attractiveness of combustion turbines as viable power-generation options in today's context can be attributed to a number of factors. They are relatively inexpensive and quick to install. This gives the utilities enormous flexibility in the face of the high cost of borrowing, long lead times in constructing steam-turbine plants, and uncertainties in projected load growth. The increased efficiency of combined-cycle plants is also a big factor in fuel cost savings. It is anticipated that the use of combustion turbines in utilities will grow significantly during the next decade. In the meantime, there is an urgent need for techniques for assessing the conditions of components in turbines currently in service. Combustion turbines constitute approximately 7% of the total power-generation capacity of the U.S. electric utility industry. There has been a growing need to assess the remaining lives of aging components. The need for life-assessment technology is driven by many economic factors. Several turbines currently in operation have been in service for more than 50,000 h and are likely to perform reliably even after exceeding the original "design life." A properly managed life-assessment and refurbishment program may help extend the lives of these turbines by another 10 to 20 years, thus avoiding major capital expenditures for the utilities (Ref 2).

In most utilities, the turbine manufacturer's recommendations constitute the principal basis for replacement of components. Often, these recommendations are based on the manufacturer's past experience and are not machine-specific. The methodologies and criteria by which they are arrived at are unclear to the utilities. Utilities would like the capability for independent assessment of the conditions of their components. This concern has become increasingly serious as the cost of replacement parts has escalated. The cost of superalloy components alone may constitute as much as 15 to 20% of the total cost of the turbine. Operating practices also vary widely among utilities and can be at variance with manufacturers' recommended practices. Both above-design operation (e.g., rapid cycling, excess temperature) and below-design operation (derating) are common. In such cases, utilities need the technology to assess the penalties associated with above-design operation, as well as the excess life to be derived from below-design operation, on a unit-specific basis.

Intelligent scheduling of inspections also requires proper techniques and tools for assessing life expenditure of components. Lengthening of inspection intervals on the basis of judicious life assessment can lead to significant cost savings. The correct choice of schedules can also help avoid unforeseen outages.

One often-ignored reality of turbine operation in the utility context is the possibility of foreign-object damage to the hot-section components. The foreign objects are frequently loose parts, nuts, bolts, and tools left inside the turbine after overhauls by negligent maintenance personnel. In order to withstand such damage and have tolerance to pre-existing or service-induced flaws, it is essential that blade materials have adequate impact toughness. Because impact toughness is subject to in-service degradation, periodic assessment of this property is an essential part of any life-assessment strategy.

Future trends in turbine usage also dictate the need for improved life-assessment techniques. There is an increasing trend toward use of combustion turbines in cogeneration and combined-cycle plants operating in base-load or intermediate-load configurations. These types of duty cycles require high-reliability components with greater longevity than is expected of machines operating under peaking duty. There is also a trend toward use of higher turbine inlet temperatures to increase turbine efficiency. Time-dependent damage mechanisms, such as creep, can become more important in this context.

A last but not least consideration is the safety of plant equipment and plant personnel. Failure of blades, disks, and other rotating components can lead to consequential damage of other components downstream, and to catastrophic failure of the turbine. Accurate life-assessment techniques can help avoid such occurrences.

Components which operate in the creep regime include combustor baskets (liners), transition pieces, turbine nozzles (vanes), blades (buckets), and turbine disks (wheels). These are the components in the hot-gas path and hence the most subject to various forms of high-temperature damage. This chapter will review the materials of construction, damage mechanisms, and life-assessment techniques for nozzles and buckets. A detailed review of the literature and a utility survey pertaining to combustion-turbine materials problems has been published by Cialone and coworkers (Ref 3). Key information from that review will be presented here.

Materials of Construction

Materials developments for industrial turbines have generally followed in the trails of developments for aircraft turbines. In both types of engines, superalloys are used for the hottest blades and vanes. Both employ advanced cooling systems. There are a few major differences in terms of requirements, however, and they have led to slightly divergent development paths. These differences are as follows. (1) The design lives of industrial turbines are about ten times longer than those of aircraft turbines, and hence the operating temperatures and stresses are lower. (2) Industrial turbines are designed to operate on lower-grade fuels, and hence their resistance to sulfidation has been a greater concern. In aircraft engines, oxidation is the primary environmental damage mechanism. (3) Industrial turbines are designed for less-severe cyclic operation, and hence creep is more of a concern than fatigue. (4) Components of industrial turbines are much larger, and hence technological innovations are often constrained by processing limitations.

The development of new materials and coatings for industrial combustion turbines has been driven by several factors, including increased turbine inlet temperatures, the trend toward increased output and efficiency, and responses to problems encountered in service. Figure 9.2 illustrates the increase in firing temperature over the years and the corresponding developments in bucket alloys (Ref 4 and 5). It can be seen from this figure that the increase in alloy high-temperature strength, which accounted for the majority of firing-temperature increases until about 1970, slowed during the 1970's. This occurred as the result of two factors. First, emphasis was placed on the use of air cooling to increase firing temperatures. Second, as the metal temperatures approached 870 °C (1600 °F), hot corrosion of buckets became a more life-limiting factor. Together with the increased use of more contaminated fuels, this required that material developments be directed more toward improvement of hot corrosion resistance and the use of long-life corrosion-resistant coatings. The 1980's have seen the emphasis swing back toward development of stronger alloys in an effort to fill part of the need for machine uprating. It is anticipated

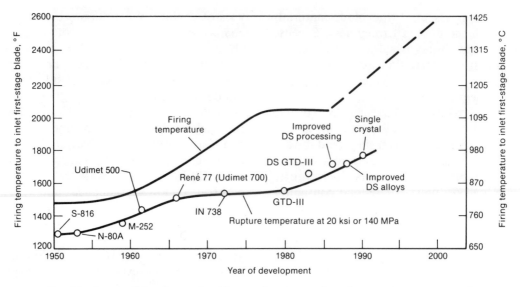

Fig. 9.2. Past and future trends of heavy-duty gas-turbine firing temperatures and corresponding blade-material developments (Ref 4 and 5).

that firing temperatures will continue to increase in future years, as shown in Fig. 9.2.

Combustion turbines operating on lower grades of fuel encounter corrosion problems and depend heavily on coatings for protection. In recent years the desire for fuel flexibility in turbines has led to the development of alloys for critical hot-gas-path components that constitute a compromise between strength and corrosion resistance and that hence are less susceptible to catastrophic failure. In addition, coatings which are capable of providing longer lifetimes than those intended for aircraft gas turbines, through changes in chemical composition or changes in processing to allow greater thicknesses, have received considerable attention.

Advances in materials and/or coatings for combustion turbines generally have been implemented in two ways: (1) introduction of improved alloys and/or coatings when new turbines are constructed by the manufacturer, and (2) upgrading of materials and/or coatings in existing turbines by the user (often at the suggestion of the manufacturer). Depending on the maintenance philosophy of the particular utility, recommended changes in materials and/or coatings may or may not be adopted. Thus, some utilities operate older turbines with original components, whereas others oper-

ate turbines that are newer or that have been retrofitted with parts made from the latest materials and/or having the most up-to-date coatings. Although few new turbines have been installed by utilities in recent years, new turbine models are being introduced by engine manufacturers with improved materials and coatings. Consequently, a wide variety of alloys and coatings are in use for a given portion of a particular model turbine. The data available on specific alloys are also very limited because of the frequent changes in alloys and coating systems that have been introduced.

Table 9.1 lists compositions of alloys used in turbine-section components. Most of the progress in the area of improved materials for combustion turbines has been made in alloys for blades (or buckets) and vanes (or nozzles) in the turbine section. Materials improvements have been coupled with advances in designs for internal cooling; this has resulted in increased usage of castings rather than forgings. These castings are typically made by lost-wax investment casting under vacuum. Cast parts also are viewed as being more amenable to rejuvenation by heat treatment than are forged parts. Future blades may be made using improved casting procedures such as directional solidification or casting of single

Fig. 9.3. Turbine vanes (Ref 5).

crystals, if these procedures can be made more cost-effective.

Turbine Vanes

The first-stage turbine inlet guide vanes (also sometimes called nozzle partitions) must perform the function of turning and directing the flow of hot gas into the rotating stage of the turbine at the most favorable angle of incidence. They are subjected to impingement of the highest-temperature gases and attain the highest metal temperatures of any component in the turbine. Even though the superalloys used for vanes are capable of creep resistance at temperatures above 925 °C (1700 °F) for short periods in aircraft-engine applications, the desire for component lifetimes of 50,000 to 100,000 h for industrial-turbine vanes means that a high degree of cooling is necessary. Representative designs are shown in Fig. 9.3 and 9.4.

Although there is no centrifugal loading on turbine vanes, the combination of gas bending loads and the thermal gradients caused by vane cooling result in high localized steady-state operating stresses in stationary vanes. Thermal stresses from uneven heating and cooling during start-ups and shutdowns also can cause cracking. The properties a vane alloy must possess include creep strength to resist distortion caused by gas loading and thermal stresses, low-cycle fatigue strength to resist the cyclic thermal strains, and oxidation and sulfidation resistance.

Fig. 9.4. W-501 turbine vane (Ref 5).

Material selection includes alloy strength and material processing as well as requirements of mechanical design and heat transfer. It is common to use the most advanced, highest-strength alloy that also has the other required attributes for the highly cooled first-stage vanes. The design of the later-stage vanes becomes a balance between alloy strength and the amount of cooling. In some cases high-strength alloys and little or no cooling are chosen, whereas in other cases moderate levels of cooling may be used in combination with lower-strength alloys having greater castability and repairability.

Vanes/nozzles are made from cobalt-base superalloys and nickel-base superalloys. They are investment cast individually and then welded to a housing to form a nozzle segment (or stator segment), or are investment cast as segments. Hence, the material must be easily castable into configurations that are large (weighing up to 68 kg, or 150

Table 9.1. Compositions of alloys used in turbine-section components (Ref 3)

Alloy	C	Mn	Si	Cr	Ni	Co	Mo	W	Nb	Ti	Al	Fe	B	Zr	Ta	Other
Blades/buckets																
IN 939 ...... 0.15	0.2(a)	0.2(a)	22.5	Rem	19.0	...	2.0	1.0	3.7	1.9	...	0.009	0.10	1.4	...
IN 792 ...... 0.12	...	...	12.4	Rem	9.0	1.9	3.8	...	4.5	3.1	...	0.02	0.10	3.9	...
IN 738 ...... 0.17	0.2(a)	0.3(a)	16.0	Rem	8.5	1.7	2.6	0.9	3.4	3.4	0.5(a)	0.01	0.10	1.7	...
Inconel 713C ... 0.12	...	...	12.5	Rem	...	4.2	...	2.0	0.8	6.1	0.5(a)	0.012	0.10	...	...
Inconel X-750 ... 0.04	0.5	0.2	15.5	Rem	...	...	...	1.0	2.5	0.7	7.0	...	...	...	0.2 Cu
B-1900(b) 0.10	0.2(a)	0.2(a)	8.0	Rem	10.0	6.0	0.1(a)	0.1(a)	1.0	6.0	0.35(a)	0.015	0.10	4.0	0.015(a) S
GTD-111 ...... 0.10	...	...	14.0	Rem	9.5	1.5	3.8	...	4.9	3.0	...	0.01	...	2.8	...
Udimet 720 ... 0.035	...	...	18.0	Rem	14.7	3.0	1.25	...	5.0	2.5	...	0.033	0.03	...	...
Udimet 710 ... 0.07	0.1(a)	0.2(a)	18.0	Rem	14.7	3.0	1.5	...	5.0	2.5	0.5(a)	0.02	...	...	...
Udimet 700(c) ... 0.07	...	...	15.0	Rem	18.5	5.0	...	...	3.5	4.4	0.5(a)	0.025	...	...	...
René 77(c) 0.07	...	...	15.0	Rem	18.5	5.2	...	...	3.5	4.25	1.0(a)	0.04	...	...	...
Astroloy(c) 0.06	...	...	15.0	Rem	15.0	5.25	...	...	3.5	4.4	...	0.03	...	...	...
Udimet 520 .... 0.05	...	...	19.0	Rem	12.0	6.0	1.0	...	3.0	2.0	...	0.005	...	...	...
Udimet 500 .... 0.08	...	...	19.0	Rem	18.0	4.0	...	...	3.0	3.0	0.5(a)	0.007	...	...	...
Nimonic 105 .... 0.08	...	...	15.0	54.0	20.0	5.0	...	...	1.2	4.7	...	0.005	...	...	...
Nimonic 115 .... 0.20	...	...	15.0	55.0	15.0	4.0	...	...	4.0	5.0	1.0	...	0.04	...	...

Vanes/nozzles

Alloy	C	Mn	Si	Cr	Ni	Co	W	Mo	Cb	Ti	Al	B	Fe	Zr	Ta	Other
X-40(d)	0.50	0.50	0.50	25.0	10.0	Rem	7.5	...	...	...	...	0.010	1.5	...	...	...
X-45	0.25	1.0(a)	...	25.5	10.5	Rem	7.0	...	...	...	...	0.012	2.0(a)	...	...	...
FSX-414	0.25	1.0(a)	1.0(a)	29.5	10.5	Rem	7.0	...	...	...	...	0.01(a)	2.0(a)	...	...	...
ECY-768	0.60	...	...	21.5	10.0	Rem	7.0	...	...	0.2	...	0.010(a)	1.0	...	3.5	...
Mar-M 509	0.60	0.10(a)	0.10(a)	21.5	10.0	Rem	7.0	...	...	0.2	...	0.010(a)	1.0	0.50	3.5	...
Mar-M 302	0.85	0.10	0.20	21.5	...	Rem	10.0	...	...	...	...	0.005	...	0.15	9.0	0.15 N, 0.50(a) Cu
N-155(e)	0.10	1.5	0.5	21.0	20.0	20.0	2.5	3.0	1.0(f)	...	...	...	Rem	...	...	...
WI-52	0.45	0.50(a)	0.50(a)	21.0	1.0(a)	Rem	11.0	...	2.0	...	...	...	2.0	...	(f)	...
Inconel 713C	0.12	...	...	12.5	Rem	...	...	4.2	2.0	0.8	6.1	0.012	0.5(a)	0.10	...	...
IN 738	0.17	0.2(a)	0.3(a)	16.0	Rem	8.5	2.6	1.75	0.9	3.4	3.4	0.01	0.5(a)	0.10	1.75	...
IN 939	0.15	...	...	22.5	Rem	19.0	2.0	...	1.0	3.7	1.9	0.009	...	0.10	1.4	...

Wheels/disks

Alloy	C	Mn	Si	Cr	Ni	Co	W	Mo	Cb	Ti	Al	B	Fe	Zr	Ta	Other
M-152	0.12	...	...	12.0	2.5	...	...	1.7	...	...	...	...	Rem	...	...	0.3 V
V-57	0.08(a)	0.35(a)	0.75(a)	14.8	27.0	...	...	1.25	...	3.0	0.25	0.01	Rem	...	...	0.5(a) V
Incoloy 901	0.10(a)	0.4	0.4	13.5	42.5	...	...	6.0	...	2.7	0.2	...	36.2	...	...	...
Inconel 706	0.03	0.2	0.2	16.0	41.5	...	...	...	2.9(f)	1.75	0.2	...	37.5	...	(f)	0.15(a) Cu
A-286	0.08	1.4	0.5	15.0	26.0	...	...	1.25	...	2.1	0.25	0.003	Rem	...	...	0.3 V
Discaloy	0.04	0.9	0.8	13.5	26.0	...	...	2.75	...	1.75	0.1	...	Rem	...	...	...
Greek Ascoloy	0.15	0.4	0.3	13.0	2.0	...	3.0	...	...	...	...	...	Rem	...	...	0.03(a) S, 0.10(a) Cu
Waspaloy A	0.07	0.5(a)	0.5(a)	19.5	Rem	13.5	...	4.3	...	3.0	1.4	0.006	2.0(a)	0.09	...	...
Cr-Mo-V steel	0.30	...	...	1.0	0.5	...	...	1.25	...	...	...	...	Rem	...	...	0.25 V

(a) Maximum content. (b) Equivalent to PWA 1455. (c) Udimet 700, Astroloy, and René 77 are similar in composition. (d) Equivalent to HS-31. (e) Also referred to as Multimet. (f) Total Cb + Ta content specified.

lb) and complex (containing internal cooling passages). A further requirement is weldability for ease of fabrication (cooling inserts are welded in place) and for repair of service-induced damage. Hollow vanes also may be made from sheet metal in a manner similar to that for titanium-alloy compressor vanes and blades. Some of the older vanes were forged and then welded into segments.

Alloys used for vanes/nozzles typically have greater corrosion resistance but lower elevated-temperature strength compared with those used for blades/buckets for a given stage in the turbine section. More important than strength in vane/nozzle alloys is weld repairability. For example, FSX-414, which is one of the lower-strength alloys in Table 9.1, currently is used in turbines because it is reported to be readily weld-repairable. Vacuum-melted ECY-768 is the latest vane material in some designs, replacing a previous alloy, X-45, because of its higher creep strength. Vacuum-cast alloy Mar-M 509 is also a common material in older engines. The ECY-768 alloy is a modified Mar-M 509 with improved castability.

The cobalt-base alloys for vane applications are solid-solution strengthened by addition of the refractory-metal elements tungsten and tantalum and by formation of carbides of chromium and zirconium. Chromium is also important for the oxidation and corrosion resistance it imparts. These alloys generally are used in the as-cast condition or with an abbreviated heat treatment consisting of a solution treatment followed by an aging treatment to stabilize the carbides. Fabrication processes such as brazing to secure the cooling inserts usually are incorporated in the heat treatment.

Cast nickel-base alloys such as Udimet 500, IN 738, and IN 939 have also been used for some vanes. However, because it is difficult to produce high-quality castings in large multivane segments, the nickel-base alloys have been used for single castings. Shortages in the availability of the strategic metal cobalt could lead to greater use of nickel alloys for vanes. Large three- and four-vane segments cast in N-155 have also

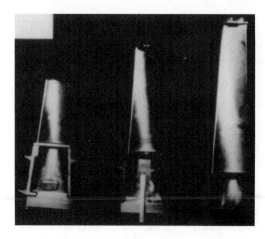

Fig. 9.5. Combustion turbine blades (Ref 5).

been used for some cooler-running last stages (temperatures from 540 to 650 °C, or 1000 to 1200 °F).

Repair of service-run vanes is routinely used to maximize their usefulness. Most of the alloys used are repairable by welding, although the difficulty increases, and the scope of repair becomes more limited, with the strength of the alloy. Occasionally a pre-weld heat treatment is necessary to return the material to a weldable metallurgical condition after extended high-temperature service. Gas tungsten-arc welding using a filler material or similar composition is usually employed, whereas electron beam and plasma welding have been employed to a lesser degree. The scope of repair for vane segments is now being extended to include vacuum brazing to fill minor thermal-fatigue and corrosion-related cracks.

Turbine Blades

Typical turbine blades (buckets) are shown in Fig. 9.5 and 9.6. The blades represent the most difficult materials application in the combustion turbine, with exposure to high temperatures and high centrifugal loadings. The latter are especially significant in the last-row blades in some of the larger turbines, which have airfoil sections approaching 635 mm (25 in.) in length and whose tips may be nearly 1.3 m (50 in.) from the axis of rotation. In addition, there are low-level vibrations, thermal stresses from cooling,

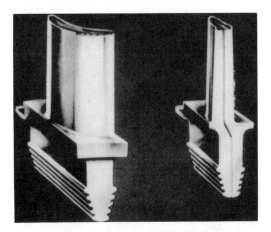

Fig. 9.6. Air-cooled combustion turbine blades (Ref 5; original source, Westinghouse Electric Corp.).

and thermal fatigue associated with starts and stops.

All of the above concerns apply for both aircraft and industrial combustion turbines. As a result, many of the same nickel-base alloys were used for both applications until it became apparent that there were unique problems associated with industrial-turbine blading. In particular, these problems involve component size and resistance to hot corrosion from the action of condensed salts. Alloys such as B-1900, Mar-M 200, and Udimet 700, in part because of their relatively high aluminum and low chromium levels, possess excellent high-temperature strength and oxidation resistance. Consequently, they have been used extensively for blades in aircraft turbine engines. However, the low chromium contents also result in rather poor corrosion resistance at temperatures in the range of about 650 to 925 °C (1200 to 1700 °F), which is precisely the region where industrial turbine blades are expected to operate, often in potentially corrosive combustion environments. To answer this need, a new family of higher-chromium nickel-base alloys has been developed and used successfully for industrial turbine blading. Among the more extensively used materials are the forging alloys Udimet 520, Udimet 710, Udimet 720, and Inconel X-750, and the cast alloys Udimet 500, IN 738, and GTD-111. Advances

made in turbine bucket alloys since 1950 have resulted in increases in allowable metal temperatures totaling approximately 160 °C (285 °F), or about 5.5 °C (10 °F) per year. Although this increase does not appear very large at first glance, an increase of 55 °C (100 °F) in turbine firing temperature corresponds to an increase in attainable output of between 10 and 13% and an improvement in efficiency of between 2 and 4% (Ref 4). Thus, the development of new alloys, although time-consuming (three to four years) and expensive, finds significant rewards in the reduced dollar-per-kilowatt cost of turbines and in the reduced cost of turbine operation — particularly the overriding fuel cost. Figure 9.7 illustrates the increases in rupture strength that have been achieved in blade and vane alloys over the years, permitting higher firing temperatures (Ref 3).

As turbine inlet temperatures have increased and higher-temperature alloys have been employed in the first stage, alloys previously used in first-stage blading have found use in the latter stages. Thus, alloys that are used in first-stage blading in older turbines may be in use in second- or third-stage blading in other, newer turbines (for example, Udimet 500), and no distinction among turbine stages is made in Table 9.1. Most heavy-duty turbines currently in operation use first-stage blades made from investment-cast IN 738, IN 792, or forged Udimet 710. First-stage turbine blades in older aircraft-derivative turbines were made primarily from B-1900, which has encountered numerous corrosion problems in utility service; more recent aircraft-derivative turbines utilize coated Udimet 700 or IN 792 in the first turbine stage. The latest first-stage bucket/blade alloys being used for large industrial turbines are GTD-111 by General Electric and Udimet 720 by Westinghouse. Due to recent problems with Udimet 720, efforts are being made to replace it with a more tough and thermal-fatigue-resistant alloy such as Udimet 520.

The processing methods used to fashion a turbine blade from a nickel-base superalloy have a large influence on the properties

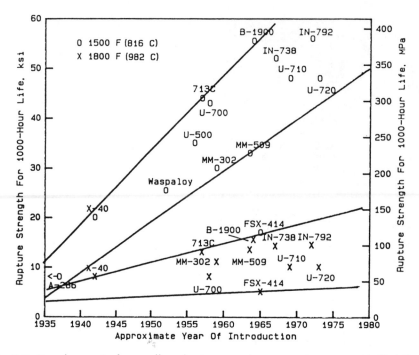

Fig. 9.7. Development of new alloys for increased creep-rupture strength (Ref 3).

of the final product. Processing begins with vacuum melting, vacuum-arc or electroslag remelting, and vacuum casting of the product for conversion to forging bar or for making the melting stock intended for vacuum investment casting. These processes permit tight compositional control, especially of the easily oxidized but critical strengthening ingredients aluminum and titanium, and maintenance of very low gas contents. Precision forging and casting methods are utilized to produce the blades in near-finished sizes.

In the past, precision forging has been used to produce turbine blades in sizes ranging from 100-mm (4-in.) first-row blades to last-row blades approaching 760 mm (30 in.). This method is still used today for its ability to produce a high-quality, low-defect, dimensionally accurate, and stable part. Blades produced by precision forging in alloys such as Udimet 720 have creep strengths nearly on a par with those of cast alloys and somewhat superior high-cycle fatigue resistance. Cooling passages are introduced into forged blades by drilling or electrochemical machining (ECM) of radial holes that pass from the base of the blade to the tip.

With the use of complex ceramic cores, great flexibility in the design of cooling passages is facilitated by investment casting. Developments in casting and solidification control over the last 10 to 15 years have made it possible to produce cast blades with dimensional accuracy approaching that of forged blades. Cast alloys permit greater alloying than forged alloys and can achieve greater creep strength due to that reason and due to other microstructural factors. Unfortunately, highly alloyed cast materials such as IN 738 can contain casting defects (e.g., porosity) that cause a variability in properties in excess of the spread normally anticipated for wrought materials. Hot isostatic pressing (i.e., the simultaneous application of isostatic pressure and temperature) is generally used to close porosity and reduce the property variability. The mechanical properties of cast alloys are also sensitive to the cross section of the part, because it has a strong effect on cooling rate and therefore on microstructure. Care must be taken to design parts using cast-alloy properties derived from specimens that are representative of the dimensions of the blade being considered.

In many precipitation-hardened nickel-base blade alloys, overaging (coarsening) of the gamma-prime precipitates often causes a reduction in creep strength. The aging is accelerated by stress. Alloy compositions are balanced to retard this process as much as possible so that long service lives can be obtained without serious loss of strength. In most alloys, reapplication of the full heat treatment has been shown to return the service-aged and weakened microstructure approximately to its initial condition and strength. Alloy composition is also adjusted to minimize the possibility of formation of brittle phases during service exposure in certain temperature ranges. The embrittling sigma phase has been observed in some nickel-base alloys after service in the range 845 to 925 °C (1550 to 1700 °F). Research has shown that several elements — notably, chromium, tungsten, and molybdenum — can promote sigma-phase formation, as can the application of stress. It is obvious that indiscriminately adding chromium to maximize hot-corrosion resistance could result in an alloy that is rapidly embrittled. Careful balancing with elements that retard sigma formation is hence necessary. Turbine blade alloys are made to a controlled composition, and the analysis of individual heats is compared with a standard that is not sigma-prone.

The most highly alloyed, highest-strength nickel-base alloys are used for the hotter first- and second-stage blades. Later-stage blades, although cooler, are considerably larger and present different problems in alloy selection. Along with rupture and creep strength, these blades must have high tensile strength to resist stresses in the blade root caused by centrifugal loading and must also have good resistance to high-cycle fatigue. Forged blades made from Udimet 520 and Inconel X-750 or cast blades of Udimet 500 are used for this application.

Damage Mechanisms

Materials problems encountered in the turbine sections of heavy-duty gas turbines can be broadly categorized as mechanical-property-related or corrosion-related, whereas some problems belong to both categories or are attributable to site- or machine-specific causes. In general, machines that operate only on clean fuels encounter few corrosion problems, so that operational time between overhauls is limited by stresses imposed by the duty cycle. Base-loaded machines typically suffer creep damage to first-stage blades/buckets, whereas peaking machines are limited by thermal fatigue of components such as first-stage vanes/nozzles, blades/buckets, and combustor components. In addition, in-service microstructural degradation phenomena can render the blade materials increasingly vulnerable to damage by creep, fatigue, or brittle-failure mechanisms. In view of its importance, service-induced damage will be treated as a separate category for purposes of this review.

Mechanical-Property-Related Damage

From a purely mechanical point of view, the damage mechanisms of interest include creep, fatigue, and thermal fatigue.

Creep. Creep failures can occur in hot-section components as a result of continuous exposure to high temperatures and stresses during operation. Unacceptable dimensional changes, creep rupture, and local failure by creep-crack growth constitute failure. Blades and vanes are designed on the basis of a stress to cause rupture or to cause a given amount of elongation in a specified period of time. Hence, they have a specific design life. Premature failures are, however, caused by unanticipated excursions in temperature or stress and by contributing corrosion factors. Inaccuracies in extrapolation of short-time data to estimate the long-time performance at the design stage are also major contributing factors in creep failures. Such extrapolations are often optimistic and do not take into account microstructural-degradation phenomena such as coarsening of the γ' precipitates, grain-boundary carbides, and formation of deleterious phases such as sigma (Ref 6). The effect of such degradation on creep properties will be treated in a separate section.

At the stresses and temperatures encountered in the hot sections of gas turbines, creep failure occurs by the accumulation of damage along grain boundaries in the form of microscopic voids. Such voids can be detected by optical microscopy. Linking up of these voids to form a critical-size crack leads to eventual failure. The accumulation and growth of voids may occur uniformly in the component or in a localized region ahead of a notch or a propagating crack. Stress concentrations, local hot spots, and stress and temperature gradients promote localization of creep damage. Creep cracks are intergranular. Except in cases where damage is severely localized, creep damage is evidenced by multiple cracks parallel with the main crack.

A survey by Cialone *et al* (Ref 3) identified some instances of unacceptable creep of blades in base-load machines with resulting interference between blades and vanes. One suspected cause was higher-than-expected firing temperatures resulting from thermocouple problems such as mis-siting, shorts, or failure. Approaches to dealing with creep problems vary among engine users, but there is a general concern about overheating problems, which may stem from actions such as re-siting of thermocouples or changing of cooling-air flow, which may be the result of partial upgrading of machines. Maintenance budgets for heavy-duty gas turbines often are such that upgrading can be practiced only on a sporadic basis. One approach to reuse of components such as vanes which have experienced creep is to machine them to the original dimensions one time; if creep occurs again, the vane is replaced. Another philosophy is periodic application of heat treatment to remove creep voids and to redistribute carbides. For FSX-414 and X-45 vanes, re-solution annealing is reportedly practiced at 16,000-to-18,000-h intervals to maintain ductility and improve weldability. First-stage blades of IN 738 are reheat treated every 18,000 to 27,000 h, and Udimet 520 blades every 16,000 to 18,000 h (Ref 3); these intervals are considered to be too short. Rejuvenation by hot isostatic pressing is under consideration, and will be weighed against replacement costs. Changes in nozzle design may be required for use in locations where creep is experienced repeatedly.

In applications where creep is a factor, there is concern about the effects of the thermal cycles employed in coating or recoating processes on alloy microstructures, and hence on mechanical properties. Changes such as coarsening of strengthening phases or redistribution of phases can degrade high-temperature creep strength, result in embrittlement, or render the alloy more susceptible to localized corrosion. Obviously, different alloys have different sensitivities to such changes; it is suspected that Udimet 710 is one of the alloys more susceptible to this form of property degradation. Efforts are being made to replace Udimet 710 with Udimet 520 in some engines. As engine firing temperatures are increased in the search for higher efficiencies, problems of creep and of refurbishing and reusing creep-damaged components can be expected to become more common.

Stress-rupture properties, plotted in terms of the Larson-Miller parameter for several blade and vane alloys, are shown in Fig. 9.8 and 9.9. Among the blading alloys, GTD-111 appears to have the best rupture strength, followed by IN 738, IN 939, Udimet 710, and Udimet 720, all of which are comparable, then followed by Udimet 500 and Udimet 520. Among the vane alloys, Mar-M 509 and ECY-768 have the highest rupture strengths, followed by X-40 and then by X-45 and FSX-414. These plots are based on data in Ref 4, 5, and 7.

A very important property that cannot be overlooked in connection with creep strength is creep ductility. Unfortunately, there is very little creep-ductility data from long-time rupture tests. Decreases in rupture ductility may also result from in-service exposure (Ref 8). Values of rupture elongation below 5% can lead to notch-sensitive creep-rupture failures.

Fatigue. Fatigue failures occur as the result of application of repetitive or fluctuating stresses at levels generally much lower than the (single-load) tensile strength of the material. Unlike creep, fatigue is not strongly influenced by temperature. Thus,

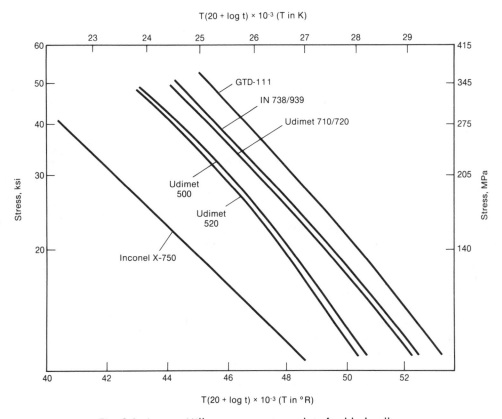

Fig. 9.8. Larson-Miller stress-rupture plots for blade alloys.

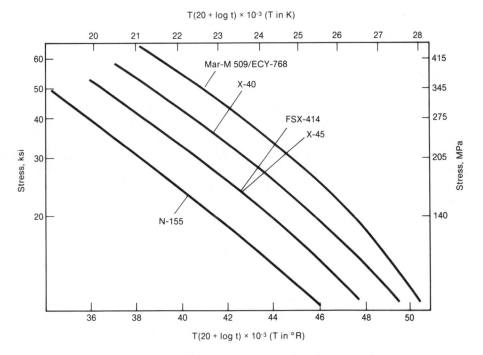

Fig. 9.9. Larson-Miller stress-rupture plots for vane alloys.

fatigue can occur in both hot- and cold-section components.

Fatigue failures are generally categorized as occurring by either high-cycle or low-cycle mechanisms, depending on the frequency of the loading. Typically, low-cycle fatigue occurs as the result of stresses applied once per engine operation cycle, such as those resulting from differential thermal expansion of gas-path components during start-up and shutdown. In such cases, the stress levels required to initiate and propagate a crack are generally on the order of the yield stress. High-cycle fatigue occurs at much higher frequencies associated with resonant vibrations, and the fluctuating stresses are much lower. The resonance of the component coupled with the applied steady and low-frequency transient cyclic stresses becomes an important factor, because it acts to increase the maximum stresses exerted on the part. As would be expected, the differences in the loading conditions for high- and low-cycle fatigue result in different crack-growth mechanisms and fracture appearances.

Depending on the temperature range, low-cycle-fatigue damage can accumulate primarily by plastic deformation or by the combination of plastic and creep deformation at higher temperatures. High-temperature low-cycle-fatigue cracks are, therefore, very similar in appearance to creep cracks, featuring both an intergranular fracture path and the presence of intergranular voids. They can be distinguished from creep cracks both by their locations in areas where steady-state stresses are low and by the typically larger and more numerous voids present. In a number of failures, it is likely that the combined effects of steady-state creep stresses and superimposed low-cycle-fatigue stresses act to propagate the crack. At lower temperatures and higher strains where plastic deformation dominates, transgranular crack propagation occurs.

In high-cycle fatigue, elastic deformation predominates in all temperature regimes. The cracks, therefore, generally initiate and propagate along a transgranular path, giving the fracture a characteristically smooth appearance. A notable feature of many fatigue fractures is the presence of clam-shape "beach marks" which mark the progress of the crack at various stages of its life. These marks are formed by changes in the level of oxidation due to temporary halts in the crack growth as loading or fatigue excitation changes, thus forming a concentric pattern around the crack-initiation point. SEM examination of fatigue-fracture surfaces often reveals the presence of fine striations running perpendicular to the crack direction. Each of these striations marks the advancement of the crack during one loading cycle. Thus, the spacing of striations can be used, in certain instances, to estimate the rate of crack growth and the cyclic stress that caused it.

Typical high-cycle-fatigue data for two commonly used superalloys are shown in Fig. 9.10 (Ref 9). The effect of mean stress (R ratio) can be taken into account in a Smith-type diagram, as shown in Fig. 9.11 (Ref 9). High-cycle-fatigue data on IN 738 and IN 939 alloys may also be found in Ref 7 and 10.

Comparison of the low-cycle-fatigue behavior of different alloys is rendered difficult by a number of factors. Although considerable data on IN 738 are available, published data on the other alloys are very limited. Test temperatures and frequencies employed vary from one investigator to another. Data presentation also varies between investigations. Some plot the number of cycles to specimen failure whereas others plot the number of cycles to crack initiation, which is arbitrarily defined. Sometimes the total strain range is reported; at other times only the inelastic strain range is reported. A limited compilation of data at 850 °C (1560 °F) that could be placed on a comparable basis is shown in Fig. 9.12. For IN 738, the data from Martens, Rosslet, and Walser (Ref 11) and from Nazmy (Ref 12) define the upper-bound curve. Both studies were conducted at 850 °C, the first on as-cast IN 738 and the second on fully heat treated IN 738. The lower-bound behavior

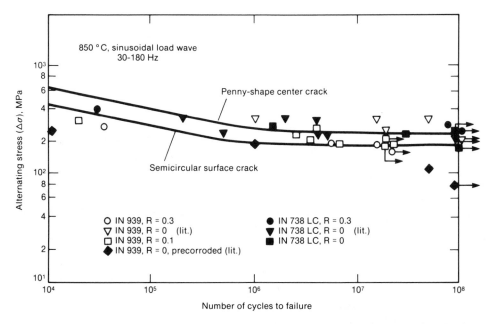

Fig. 9.10. High-cycle-fatigue curves for IN 738 LC and IN 939 at 850 °C (1560 °F) (Ref 9).

for IN 738 is based on the data of Strang (Ref 13) and of Thomas and Day (Ref 14). Since the Strang data are reported for 600 and 750 °C (1110 and 1380 °F), they were extrapolated to estimate the behavior at 850 °C. Similarly, the data of Strang on FSX-414 at 600, 700, and 900 °C (1110, 1290, and 1650 °F) were interpolated to estimate the behavior at 850 °C (Ref 13). A similar interpolation of the data on Udimet 710 was also necessary to put all the data on a constant-temperature basis (Ref 15). Errors from such interpolations and extrapolations are deemed to be minor considering the normally large scatter associated with such data, as clearly evidenced by the behavior of IN 738. In terms of number of cycles to crack initiation, data are also available for Udimet 700 (Ref 16 and 17), IN 738, and GTD-111 (Ref 18). The number of cycles to failure as a function of inelastic strain range has been analyzed by Remy *et al* (Ref 19) for alloy Mar-M 509.

It is important to note that the data presented in Fig. 9.12 are applicable only to uncoated base material tested under continuous cycling in air. The presence of coatings, corrosion processes, in-service degradation, and application of hold times can modify the fatigue behavior substantially, as will be discussed in later sections.

Creep-Fatigue Interaction. Pure low-cycle fatigue under continuous cycling as described earlier is the exception rather than the rule, because during operation machine components are frequently held at high temperatures between cycles. If the temperature of operation or holding is below the creep range, a pure low-cycle-fatigue situation may apply. In the creep range of temperatures, however, an additional damage contribution due to stress relaxation by creep has to be taken into account. Furthermore, actual cycles in machines involve independent variations of strain and temperature, unlike in the isothermal low-cycle-fatigue tests conducted in the laboratory. In view of the ease of testing and data analysis, however, most laboratory studies have focused on isothermal fatigue tests, assuming that such tests conducted at the maximum temperature of operation can define the worst behavior for the material. These aspects were discussed elaborately in Chapter 4.

Data on creep-fatigue interactions pertaining to commonly used blade and vane

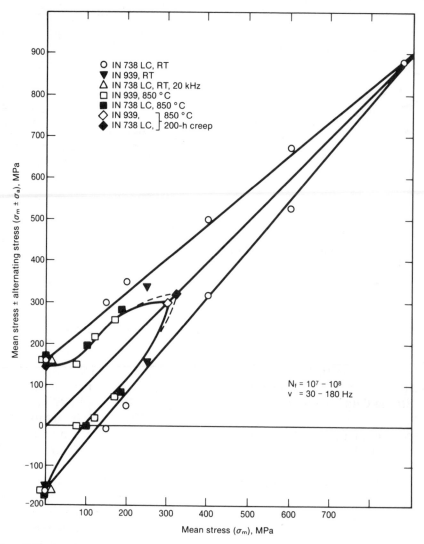

Fig. 9.11. Smith-type diagram for IN 738 LC and IN 939 at room temperature and at 850 °C (1560 °F) (Ref 9).

alloys for land-based turbines are extremely meager and self-inconsistent. These data were reviewed in Chapter 4 but will be briefly summarized here for the sake of continuity.

The most extensive body of data is available with respect to IN 738 LC. Some of the waveforms used (including hold-time effects in tension or compression) and results from the work of Nazmy are illustrated in Fig. 9.13 (Ref 12 and 20). The total strain range versus number of cycles to failure for the various cycles is shown in Fig. 9.13(b). The total strain broken down into the various in-

elastic strains and the Coffin-Manson relationship pertinent to each are illustrated in Fig. 4.23. Based on these results, Nazmy concluded that the CP and CC cycles were more damaging than the PP and PC cycles. In other words, in cycles in which the tension-going part of the cycle was slower or where a hold time was imposed in tension, the damage was found to be more severe. Results of Marchionni, Ranucci, and Picco indicate contrary results on the same alloy, showing the PC cycles in which hold time is imposed during the compressive part of the cycle to be more damaging than

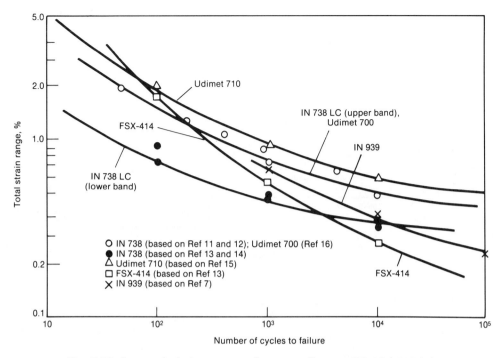

Fig. 9.12. Low-cycle-fatigue curves for superalloys at 850 °C (1560 °F).

the CP cycles (hold time in tension) (Ref 21). Both investigations were conducted at 850 °C (1560 °F), and the reasons for the apparent contradiction are not clear. Detrimental effects due to compressive holds have also been reported for IN 738 by Ostergren (Ref 22). Wells and Sullivan have reported on the damaging effects of compressive holds in Udimet 700 (Ref 23 and 24). In Mar-M 509, both tensile and compressive holds of 2 min at 900 °C (1650 °F) reduced the fatigue life by a factor of 2 compared with the pure fatigue tests (Ref 19). In Udimet 710, a tensile hold for 5 h at 790 °C (1450 °F) was found to reduce the fatigue life by a factor of 5 (Ref 15). The effect of hold time was, however, found to vary with temperature, reaching the greatest severity at the intermediate temperature of 850 °C (1560 °F), as can be seen in Fig. 4.19 and 4.20. In conclusion, the limited amount of data available shows conflicting results regarding the effects of hold time on the fatigue lives of superalloys. Because hold-time effects have been found to depend on numerous variables, such as strain

range, test temperature, material ductility, and environment, considerably more data will be needed before the various effects can be sorted out.

A different form of creep-fatigue interaction in which prior creep damage can lead to a reduction of life subsequently in fatigue has been investigated by Embly and Kallianpur (Ref 10). Their results (see Fig. 9.14) showed a severe reduction in fatigue life due to prior creep exposure. The severity of life reduction increased with increasing temperature of prior exposure and increasing creep strain. Surface cracking by creep was postulated to be the mechanism of fatigue-life reduction. It was also suggested that the surface cracking was influenced by environmental effects, particularly grain-boundary embrittlement phenomena.

The hot-gas-path components of a turbine experience a complex thermal and mechanical history during a typical cycle of operation, consisting of start-up, steady-state operation, and shutdown. A typical thermomechanical cycle for a first-stage blade is shown in Fig. 9.15, where temper-

Notation	Hysteresis loop	Description	Symbol	Wave form
$\Delta\varepsilon_{pp}$		Tension plastic Compression plastic	○	
$\Delta\varepsilon_{pc}$		Tension plastic Compression creep	●	
		Tension plastic Compression creep (relaxation)	□	
$\Delta\varepsilon_{cp}$		Tension creep Compression plastic	◆	
		Tension creep (relaxation) Compression plastic	△	
$\Delta\varepsilon_{cc}$		Tension creep Compression creep	◇	
		Tension creep (relaxation) Compression creep (relaxation)	▲	

(a)

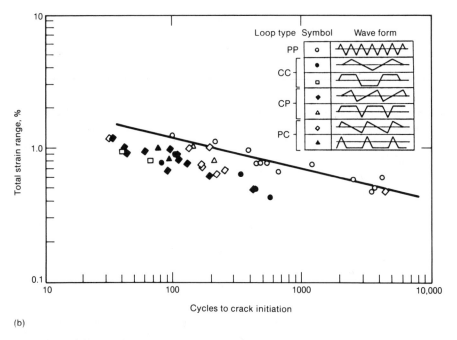

(b)

Fig. 9.13. (a) Typical fatigue cycles employed for, and (b) results of low-cycle-fatigue tests on, cast IN 738 LC (Ref 12 and 20).

ature and mechanical strain variations are plotted for a normal start-up/shutdown cycle. Temperature gradients and mechanical constraints during such complex cycles give rise to cyclic thermal stresses, which in turn can give rise to fatigue damage and eventual failure. It can also be seen from the figure that peak strains, both tensile and compressive, occur at low and intermediate temper-

atures and not necessarily at the peak temperatures in the cycle. Life-prediction models of the past were based on the incorrect premise that the damage processes occurring during thermomechanical fatigue (TMF) can be simulated by isothermal low-cycle-fatigue tests (LCF) conducted at the peak temperature. A more accurate life-prediction and management system (LMS)

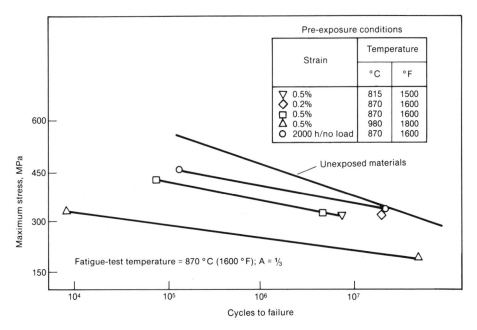

Fig. 9.14. Effect of prior creep exposure to various strain levels in air on the high-cycle-fatigue life of IN 738 at 870 °C (1600 °F) (Ref 10).

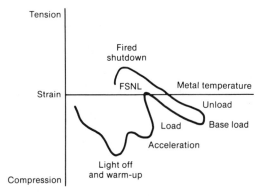

Fig. 9.15. Typical thermomechanical cycle for a first-stage blade, showing leading-edge strain and temperature variations for normal start-up and shutdown (Ref 18 and 25).

that will take into account the material behavior under the complex TMF cycles normally encountered in turbine operation is being developed (Ref 25).

The results of variable-temperature testing in the range 500 to 800 °C (930 to 1470 °F), with and without hold times up to 15 min, using in-phase and out-of-phase cycles on IN 738 LC have been reported by Thomas, Bressers, and Rayner (Ref 26). By compar-

ing the plots of inelastic strain range versus cycles to failure, they concluded that thermomechanical fatigue may be more damaging than thermal fatigue, but that the TMF data were still within the scatterband of the LCF data. No conclusions were drawn regarding the effect of the type of cycle or hold time.

A limited amount of thermomechanical test results on blading alloy Udimet 710 are available from the work of Viswanathan, Beck, and Johnson (Ref 15) (see Fig. 9.16). Increasing the temperature and the hold time in isothermal tests was found to decrease the life. Out-of-phase (OP) thermomechanical fatigue tests in which temperature was cycled between 815 and 980 °C (1500 and 1800 °F) (max tensile strain at 815 °C), with superimposed hold times of 0.23 and 5 h, resulted in even further reductions in fatigue life.

The most comprehensive set of thermomechanical fatigue (TMF) test data generated has been that of GE investigators using coated and uncoated IN 738 and FSX-414 material (Ref 25). Their TMF test facility included several closed-loop servo-hydraulic machines under computer control

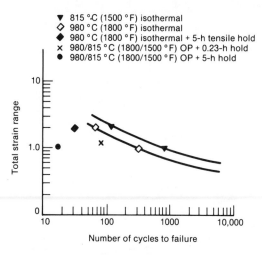

▼ 815 °C (1500 °F) isothermal
◇ 980 °C (1800 °F) isothermal
◆ 980 °C (1800 °F) isothermal + 5-h tensile hold
✕ 980/815 °C (1800/1500 °F) OP + 0.23-h hold
● 980/815 °C (1800/1500 °F) OP + 5-h hold

Fig. 9.16. Variation of number of cycles to failure with strain range for Udimet 710 (Ref 15).

which subjected a test specimen to a specified thermomechanical cycle. Initial testing, using simple TMF cycles such as those shown in Fig. 9.17, have been completed and are providing the basic data on which to build the predictive methodology. Results shown in Fig. 9.18(a) indicated that, at lower total strain ranges, life can vary by a factor of almost 100, depending on the thermomechanical cycle. The shortest lives were associated with the 427 ↔ 982 °C OP cycles, with or without a 2-min hold. The longest lives were associated with diamond cycles, for which strain at maximum temperature was zero. In the figure, 2T denotes a 2-min hold in tension; similarly, 15C means a 15-min hold in compression.

Among the results not immediately explainable by the total-strain correlation is the fact that the isothermal 871 °C results with a 2-min tensile hold show longer lives than the 427 ↔ 871 °C IP cycle. Similarly, there is no obvious reason why the diamond cycles should be better than low-temperature (427 °C) isothermal cycles. As discussed in Ref 25, correlations based on plastic strain or on the net hysteresis energy model were no more useful in gaining an understanding of the data. On the other hand, correlations based on the maximum tensile stress for each cycle provide a more rational basis for ranking the different cycles in terms of severity (Ref 25).

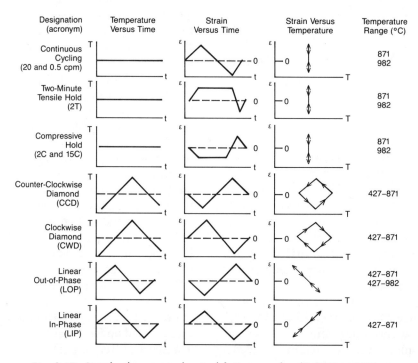

Fig. 9.17. Simple thermomechanical fatigue cycles (Ref 18 and 25).

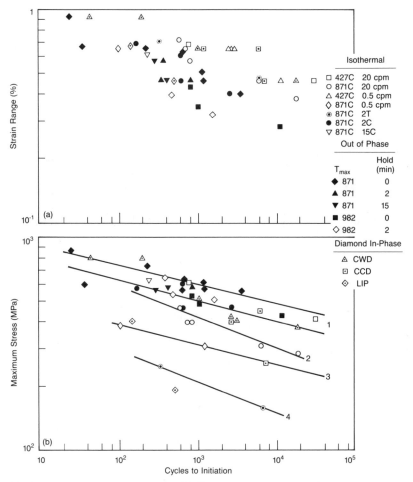

(a) Plot using strain-range criterion. (b) Plot using maximum-tensile-stress criterion.

Fig. 9.18. Fatigue-life data for IN 738 samples tested under thermomechanical fatigue conditions (Ref 18 and 25).

Figure 9.18(b) is a plot of tensile stress versus life in TMF and isothermal tests. The positions of the lines show that several types of cycles can be identified, and that they are ranked according to the temperature at which the maximum tensile stress occurred. The higher the temperature and the longer the hold period, the greater the damage. Based on this criterion, the various types of cycles could be ranked in increasing order of severity as follows:

1. OP cycles and isothermal cycles at 427 °C (least damaging)
2. Diamond cycles, wherein the maximum stress occurred at 649 °C

3. Isothermal cycles at 871 °C (20 cpm and 2-min compressive hold) in which the hold time under tensile stress was minimal
4. IP cycles and isothermal cycles at 871 °C at low frequency (0.5 cpm) wherein longer hold times under high tensile stresses occurred at high temperatures
5. Isothermal cycles at 871 °C with a 2-min tensile hold (most damaging).

GTD-111, another common blading alloy in land-based turbines was also evaluated by Russell *et al* (Ref 18 and 25) using the OP cycle and several cycles simulating actual blade cycles. Depending on the cycle employed, the fatigue life was found to vary by a factor of 100.

Results of thermomechanical fatigue tests on vane alloy FSX-414 are shown in Fig. 9.19(a) and (b), based on the total-strain-range and maximum-tensile-stress criteria. An evaluation of the effect of temperature can be obtained by comparing the results for 2-min holds in cycles having maximum temperatures of 815, 925, and 980 °C (1500, 1700, and 1800 °F) (Fig. 9.19a). It can be seen that the fatigue life decreases with increasing temperature at the highest strain

range. However, the results for the 815 °C cycle appear to converge with the other two sets of data and even suggest a crossover such that the fatigue life for the 815 °C cycle will be lower than for the other temperatures at lower strain ranges. Compared with tests with no hold time, the cycle at 925 °C with 2-min hold resulted in a life reduction of about 20%, whereas the cycle at 980 °C with 2-min hold resulted in a life reduction of approximately 30%. Fatigue life in an isothermal test at 980 °C with a 2-min hold time was comparable to the TMF results for a cycle with a maximum

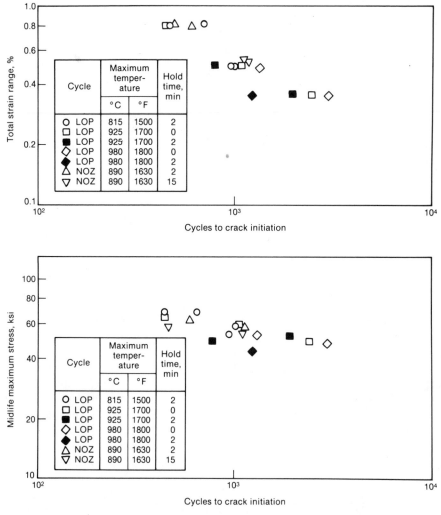

LOP denotes linear out of phase. NOZ denotes an out-of-phase cycle simulative of a nozzle fillet cycle described in Ref 25.

Fig. 9.19. Results of thermomechanical fatigue tests on vane alloy FSX-414 (Ref 25).

temperature of 980 °C and a hold time of 2 min.

Results of Russell *et al* (Ref 18 and 25) and others clearly show the complexity of conducting TMF tests and being able to draw meaningful conclusions. The type of cycle, the maximum temperature in the cycle, the hold time, the strain range, and the presence of coatings—all of these variables affect the fatigue life. The conclusions regarding the relative severity of particular cycles can often be changed if the fatigue lives are plotted in terms of the maximum tensile stress rather than the strain range. Above all, they point out the need for trying to simulate actual component cycles in laboratory tests rather than the usual practice of conducting simplistic isothermal tests.

Hot Corrosion Damage

Blades and vanes located in the hot-gas path in the turbine section are subject to a combined oxidation-sulfidation phenomenon that is commonly referred to as hot corrosion. The phenomenology and mechanisms of hot corrosion have been reviewed in several publications (Ref 27 to 32), and hence only a brief review is required here for the present purpose.

   Types of Corrosion. Three basic types of corrosion attack have been recognized (Fig. 9.20) (Ref 33). In the temperature range 650 to 705 °C (1200 to 1300 °F), a layer-type corrosion, characterized by an uneven scale/metal interface and the absence of subscale sulfides, is observed. At temperatures above 760 °C (1400 °F), a nonlayer-type corrosion (type I) is observed. Type I corrosion is characterized by a smooth scale/metal interface and a continuous, uniform precipitate-depleted zone containing discrete sulfide particles beneath the scale. The transition from one type to the other, which occurs in the range of 705 to 760 °C (1300 to 1400 °F), is characterized by an uneven scale/metal interface containing intermittent pockets of subscale precipitate-depleted zones and sulfides. The layer-type corrosion and the transitional corrosion together are variously referred to as type II

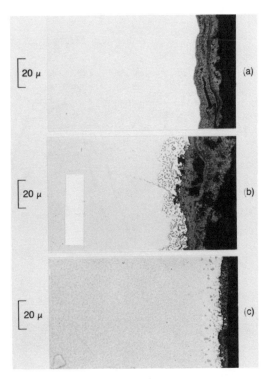

(a) Layer type. (b) Transition type. (c) Nonlayer type.

Fig. 9.20. Three different forms of hot corrosion observed in Udimet 710 (Ref 33).

hot corrosion, low-temperature hot corrosion, and low-power corrosion. The severity of hot corrosion attack as a function of temperature can be schematically depicted, as shown in Fig. 9.21 (Ref 34).

   Figure 9.22 shows the essential features of the corrosion products associated with the three types of corrosion. In the nonlayer-type high-temperature form of hot corrosion (Fig. 9.22a), discrete chromium and titanium sulfide particles are present in a region of the matrix depleted in these elements, adjacent to the base metal. The surface scales consist of protective Cr_2O_3 with some titanium oxides. With decreasing temperature, the chromium and titanium sulfides are increasingly agglomerated into large interconnecting sulfide networks, and the surface scales contain predominantly the oxides of nickel and cobalt (Fig. 9.22b). Complete layer-type corrosion (Fig. 9.22c) is characterized by a continuous layer formed by the chromium and titanium sulfides. The

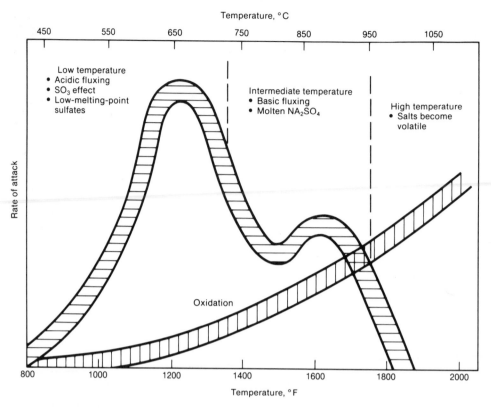

Fig. 9.21. Schematic illustration of the variation in corrosion rate with temperature due to changes in hot-corrosion mechanism (Ref 34).

surface scale in this case contains only the unprotective oxides of nickel and cobalt.

In addition to the corrosion features described above in service-returned blades and in laboratory creep samples, grain-boundary spikes (sharp-pointed cracks) are present in the zone of transition-type corrosion. The spikes usually contain sulfides alone or sulfides followed by oxide penetration. They occur over a narrow region of the blade in a manner suggesting that spike formation is dependent on stress (Ref 33).

The high-temperature form of hot corrosion involves the formation of the hot-gas-path parts of condensed salts that are often molten at the turbine operating temperature. The major components of such salts are Na_2SO_4 and/or K_2SO_4, which apparently are formed in the combustion process from sulfur from the fuel and sodium from the fuel or the ingested air. Because potassium salts act very similarly to sodium salts, specifications limiting alkali content in fuel or air are usually taken to be the sum total of sodium plus potassium.

Very small amounts of sulfur and sodium or of potassium in the fuel and air can produce sufficient Na_2SO_4 in the turbine to cause extensive corrosion problems due to the concentrating effect of the turbine pressure ratio. For example, a threshold level of 0.008 ppm by weight has been suggested for sodium in air; hot corrosion will not occur below this level. Therefore, nonlayer-type hot corrosion is possible even when premium fuels are used. This has been especially true in aircraft-derivative turbines, which have turbine blades made from B-1900 (UNS N13010). Alloy B-1900 has performed well with ultraclean aircraft fuels, but has experienced numerous corrosion problems in land-based service. Other fuel (or air) impurities, such as vanadium, phosphorus, lead, and chlorides, may combine with Na_2SO_4 and form mixed salts having reduced melting temperatures and thus

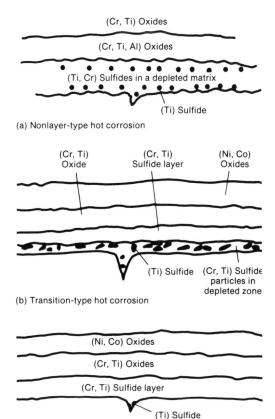

(Cr, Ti) Oxides

(Cr, Ti, Al) Oxides

(Ti, Cr) Sulfides in a depleted matrix

(Ti) Sulfide

(a) Nonlayer-type hot corrosion

(Cr, Ti) Oxide

(Cr, Ti) Sulfide layer

(Ni, Co) Oxides

(Ti) Sulfide   (Cr, Ti) Sulfide particles in depleted zone

(b) Transition-type hot corrosion

(Ni, Co) Oxides

(Cr, Ti) Oxides

(Cr, Ti) Sulfide layer

(Ti) Sulfide

(c) Layer-type hot corrosion

Fig. 9.22. Schematic illustration of corrosion products found at areas typical of the three types of hot corrosion (Ref 33).

broaden the range of conditions over which attack by molten salts can occur. Agents such as unburned carbon can also promote deleterious interactions in the salt deposits.

Research over the past 15 years has led to better definition of the relationships among temperature, pressure, salt concentration, and salt vapor-liquid equilibria so that the location and rate of salt deposition in an engine can be predicted. In addition, it has been demonstrated that a high chromium content is required in an alloy for good resistance to type I hot corrosion. The trend toward lower chromium levels with increasing alloy strength has, therefore, rendered most superalloys inherently susceptible to this type of corrosion. The effects of other alloying additions, such as tungsten, molybdenum, and tantalum, have been documented; their effects of rendering an alloy

more or less susceptible to hot corrosion are known. The near standardization of such alloys as IN 738 and IN 939 for first-stage blades and buckets, as well as FSX-414 for first-stage vanes and nozzles, implies that these are the accepted best compromises between high-temperature strength and hot corrosion resistance. It has also been possible to devise coatings with alloying levels adjusted to resist type I hot corrosion. The use of such coatings is essential for the protection of most modern superalloys intended for duty in first-stage blades or buckets.

The low-temperature form of hot corrosion produces severe pitting and results from the formation of low-melting eutectic mixtures of essentially Na_2SO_4 and cobalt sulfate ($CoSO_4$), a corrosion product resulting from the reaction of the blade surface with SO_3 in the combustion gas. The melting point of the Na_2SO_4-$CoSO_4$ eutectic is 545 °C (1013 °F). Unlike type I hot corrosion, a partial pressure of SO_3 in the gas is critical for the reactions to occur in low-temperature hot corrosion. Knowledge of the relationships between SO_3 partial pressure and temperature inside a turbine allows some prediction of where layer-type hot corrosion can occur. Because first-stage blade metal temperatures in heavy-duty engines range from about 650 to 855 °C (1200 to 1575 °F), all three types of hot corrosion can occur when sulfur and sodium are present in sufficient quantities.

To avoid hot corrosion in land-based combustion turbines, fuel specifications for sulfur, sodium, potassium, and vanadium are typically set at approximately 1% S, 0.2 to 0.6 ppm Na + K, and 0.5 ppm V. Impurity-content limitations can be varied if blade coatings are used, and corrosion inhibitors, such as magnesium, can be added to the fuel. Where the ambient air at the site is contaminated, as in industrial or coastal locations, air filtration is also often practiced.

Problems have been experienced with occasional batches of fuel containing higher-than-specified levels of impurities. A problem that has had to be addressed is the difficulty of accurately measuring low levels of elements such as sodium in fuel oil. Compliance with stringent specifications requires careful supervision and the use of such techniques as centrifuging the oil, which results in increased costs. Impurities from other sources, such as the plum stones (which contain sodium and potassium) used in carboblast cleaning at low engine power, have led to cracking of aluminide coatings and corrosion of blades and vanes. Entrapment of plum-stone fragments in these components places the corrosive species in contact with the alloy surfaces.

Air filtration is not a panacea. It is expensive and requires proper maintenance and monitoring to prevent the periodic release into the engine of material captured on the filters. For example, there are reported instances of collected contaminants being washed off of high-efficiency filters and into engines by sudden heavy storms.

In general, with stringent control of fuel specifications and good air filtration, essentially no unexpected corrosion-related problems are encountered. The life limitation is then the creep-stage blades or vanes. Where such controls cannot be exerted, alloys with some inherent corrosion resistance are used, together with coatings. The alloy used and the type and thickness of the coating are generally the least costly options that correspond to the planned engine maintenance schedules.

Corrosion–Mechanical Properties Interaction. Apart from over-all metal wastage due to hot corrosion, an additional concern has been the degradation of mechanical properties, particularly creep and fatigue resistance. An extensive review of this subject has been published by Grüenling, Keinburg, and Schweitzer (Ref 35). For a detailed list of references to related work, the reader is referred to this review article. Some of the major conclusions from the review are summarized below.

Several investigations have evaluated the degradation of creep-rupture properties of superalloys using different types of simulated as well as accelerated tests. These tests have included creep testing of samples coated with synthetic sulfate deposits, creep testing in a combustion-gas environment, and creep testing of precorroded samples. The presence of chlorides has been one of the variables investigated. The alloys studied have included IN 738 LC, IN 939, Udimet 500, IN 100, and Nimonic 105. In all cases, reductions in rupture life have been reported. In most instances where chlorides were not present, the reduction in rupture life was not substantial (Ref 35 to 41). Grüenling *et al* conclude that under purely sulfidizing/oxidizing conditions, there is no evidence of synergistic effects due to corrosion and creep phenomena and that the decrease in rupture life can be attributed completely to one of the following reasons:

- Losses in cross-sectional area due to corrosive materials wastage
- Alloy weakening in the outer zone due to sulfide formation, depletion of hardening elements, and gamma-prime precipitates
- Notch sensitivity of superalloys at ambient temperatures due to corrosive surface notching
- Superposition of bending stresses due to nonuniform attack on the creep specimens.

The corrosion-induced changes in surface morphology and alloy structure may to some extent facilitate creep-crack initiation (oxide spikes, sulfides, notches) and creep-crack propagation (gamma-denuded areas, sulfidized grain boundaries).

In studies where chlorides have been present in the sulfidizing environment, substantial reductions in rupture life have been demonstrated (Ref 36, 37, and 42 to 44). A dramatic example of the reduction in rupture life for several nickel-base alloys tested at 705 °C (1300 °F) in a 63 wt% Na_2SO_4, 36 wt% $MgSO_4$, 1 wt% NaCl mixture is shown in Fig. 9.23 (Ref 42). The smallest degree of attack was observed in Udimet 720 and coated Udimet 710. These authors invoked a synergistic mechanism, in which corrosion and creep assist each other in a

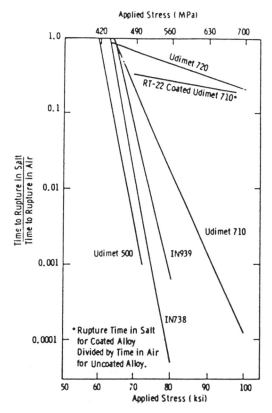

Fig. 9.23. Relative reductions in rupture life due to exposure to sulfate/chloride salt at 705 °C (1300 °F) for several materials (Ref 42).

cyclic sequence, to explain the drastic reduction in life. One possible sequence that was envisioned was as follows: (1) sulfur and/or chlorine penetrate at grain boundaries, assisted by stress, and cause embrittlement; (2) a small crack is initiated along the embrittled boundary, and (3) the embrittler now diffuses to the crack-tip region of triaxial stresses and causes further embrittlement. The process can repeat itself, leading to rapid fracture of the sample. There is also a threshold stress level below which no effects were noted, probably because of stress-assisted diffusion being much slower or inoperative.

The effect of prior sulfidation on the subsequent low-cycle-fatigue behavior of IN 738 LC at 850 °C (1560 °F) has been investigated by Nazmy (Ref 45 and 46). Sulfidation was carried out by exposure to a synthetic ash (mainly sulfates) simulative of the deposit found on blades in an air stream containing SO_2/SO_3 mixtures. In some instances, chlorides were added to the ash deposits. Prior sulfidation at 700 and 750 °C (1290 and 1380 °F) did not affect the fatigue behavior at 850 °C except when chlorides were present. Prior exposure at 850 °C, even in the absence of chlorides, had a pronounced effect on the fatigue behavior at 850 °C (Fig. 9.24). Metallographically observed corrosion effects consisted mainly of formation of chromium-depleted zones and formation of brittle chromium sulfide phases at grain boundaries. Crack-initiation sites always coincided with sulfur-affected areas.

The effect of a chloride-containing salt mixture of sulfate under the influence of type II corrosion on the low-cycle-fatigue behavior of blading alloy Udimet 720 has been investigated at 730 °C (1350 °F) by Whitlow, Johnson, Pridemore, and Allen (Ref 47). The results (Fig. 9.25) show more than an order-of-magnitude reduction in the fatigue life, especially at the higher strain range. The presence of the salt deposits also changed the fracture mode from transgranular to intergranular. Once again, a synergistic mode of failure in which stress-assisted intergranular diffusion of sulfur and chlorine caused embrittlement and fracture, which in turn facilitated further embrittlement ahead of the crack tip, was suggested.

The effect of hot corrosion on high-cycle-fatigue behavior has been investigated by several investigators. Some examples pertaining to observed degradation of fatigue properties in IN 738 (Ref 48 and 49), IN 939 (Ref 48 and 49), and Udimet 720 (Ref 50 and 51) are shown in Fig. 9.26 to 9.28.

From a review of available evidence, Grüenling, Keinburg, and Schweitzer have concluded that there is no evidence of synergistic effects due to pure sulfidation corrosion and fatigue (Ref 35). The observed reductions in fatigue strengths and endurance limits could be attributed mainly to corrosion-induced surface changes such as

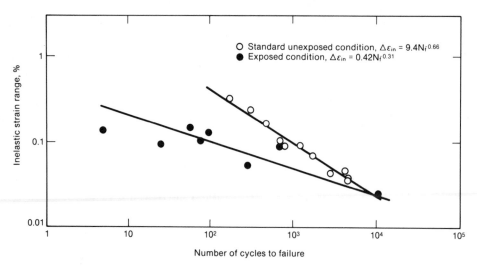

Fig. 9.24. Effect of prior exposure to hot corrosion (without chlorides) on the fatigue life of IN 738 (Ref 45 and 46).

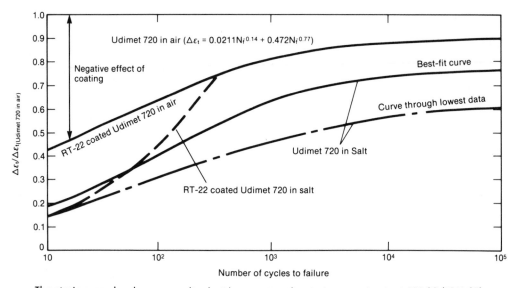

The strain range has been normalized with respect to the strain range in air at 850 °C (1560 °F).

Fig. 9.25. Effect of sulfate/chloride environment on the fatigue life of Udimet 720 at 730 °C (1350 °F) (Ref 47).

pits, notches, precipitate-depleted weak zones, and grain-boundary spikes, which facilitate fatigue-crack initiation.

When chlorides are present, however, the degradation in properties is substantially higher, suggestive of synergistic effects wherein corrosion processes continuously interact with the fracture process, each assisting the other and accelerating the failure. Material wastage and alloy depletion in most cases are not sufficient to explain the effects of corrosion on creep ductilities, stress-rupture lives, and fatigue. The following corrosion phenomena could be attributed to the presence of chlorides (Ref 35):

- Formation of nonprotective porous scales (Ref 43 and 44)
- Severe material loss due to either evaporation or "dusting" processes

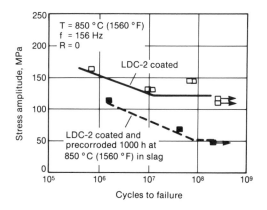

Fig. 9.26. Effect of hot corrosion on high-cycle-fatigue life of IN 738 LC at 850 °C (1560 °F) (Ref 48 and 49).

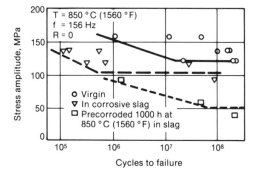

Fig. 9.27. Effect of hot corrosion on high-cycle-fatigue life of IN 939 (Ref 48 and 49).

- Selective corrosion along grain boundaries (Ref 37, 43, and 44)
- Local penetration of protective aluminide coatings and subsequent selective grain-boundary corrosion (Ref 52)
- Intergranular fatigue-crack propagation combined with extensive crack branching (Ref 53 and 54).

Several potential mechanisms by which hot corrosion could influence fatigue-crack growth have been reviewed (see Table 9.2) by Grüenling *et al* (Ref 35). Actual crack-growth data in hot corrosion environments are practically nonexistent. In a study of IN 738 and IN 939 by Hoffelner, crack-growth rates in the Paris-law region were found to be the same in air and in a sulfidizing environment, both being higher than that in vacuum by a factor of 2 (Ref 9). The fatigue threshold stress in air, however, was found to be higher than that in vacuum, presumably because of crack-branching effects.

Coatings. A successful defense against hot corrosion over the years has been the use of coatings. The evolution of coating technology has followed the path shown in Fig. 9.29 (Ref 4).

Historically, the development of corro-

Table 9.2. Possible influences of environment on fatigue-crack propagation in cast nickel-base superalloys at high temperatures (Ref 53 and 54)

Environmentally induced effect	Influence on material	Influence on fatigue-crack-propagation rates(a)
Dissolution of phases	Changes in mechanical properties at the crack tip	↑ ↓
	Weakening of material	↑
	Crack-tip blunting	↓
Crack branching	Reduction of actual ΔK at crack tip	↓
Grain-boundary attack	Weakening of grain boundaries	↑
	Intergranular crack branching	↓
Oxide layer	Prevents rewelding	↑
	Prevents resharpening of crack tip	↓
	Hinders dislocation movement	↓
Crack closure	Reduction of actual ΔK at crack tip	↓

(a) ↑ denotes acceleration of crack propagation; ↓ denotes retardation of crack propagation.

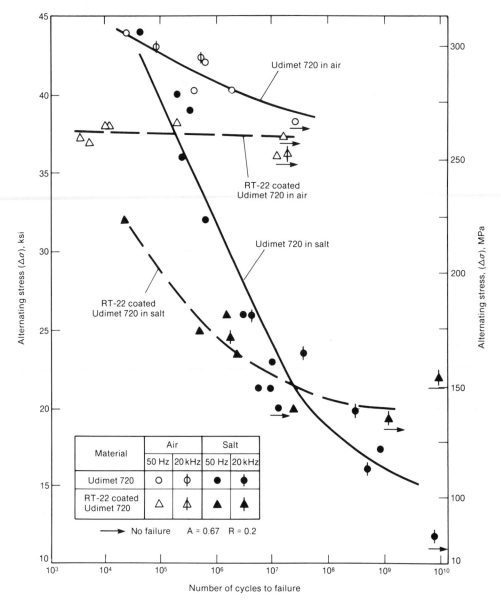

Fig. 9.28. Effect of hot corrosion and coating on the high-cycle-fatigue behavior of Udimet 720 at 705 °C (1300 °F) (Ref 50 and 51).

sion-resistant coatings was aimed at combatting high-temperature hot corrosion. The earliest coatings were the diffusion aluminides. It was found that chromium-modified aluminides offered little additional protection against high-temperature hot corrosion compared with the basic aluminides, whereas the platinum-modified aluminide offered superior protection compared with the basic aluminide. Since then,

it has been found that the chromium-modified aluminides are particularly beneficial against low-temperature hot corrosion, giving results equivalent to those of the platinum-aluminides, with both modified aluminides being better than the basic aluminide.

Although these diffusion aluminides have been successful in reducing hot corrosion, the compositions of these coatings are not

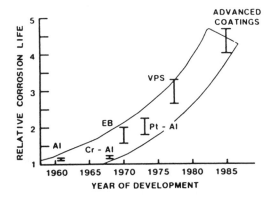

Fig. 9.29. Evolution of coating technology (Ref 4).

readily modified for further improvement in corrosion resistance. Thus, increased attention has been given to the development of overlay coatings, which offer significant compositional flexibility. These generally have the composition MCrAlY, where M can be Ni, Co, or a combination of the two. The actual compositions of these coatings depend on their intended use. Because high-temperature hot corrosion depends on Al_2O_3 for protection, coatings which exhibit the greatest high-temperature protection are generally high in aluminum (11%) and low in chromium (<23%). Low-temperature hot corrosion, on the other hand, depends primarily on Cr_2O_3 for protection, and therefore coatings exhibiting the greatest low-temperature corrosion protection are high in chromium (>30%) and low in aluminum. Other elements, such as silicon, hafnium, tantalum, and platinum, are added to these coatings in an attempt to improve corrosion resistance and spalling resistance. MCrAlY coatings with high-chromium levels have been developed to offer superior low-temperature protection without sacrificing high-temperature protection, because in some cases industrial gas turbines operate under variable loading conditions which could result in exposures to both low- and high-temperature conditions. Coating techniques for overlay coatings have included electron beams (EB), physical vapor deposition (PVD), and vacuum plasma spray (VPS). A comprehensive re-

view of commercially available coatings and their relative performance has been published by McMinn (Ref 55).

In the case of coated components, the integrity of the coating can exert a major influence on component life. Coatings can adversely affect component integrity in two ways. (1) If the heat treatment cycle associated with the coating process is not properly chosen, it may degrade the base-metal mechanical properties. (2) If the coating has low ductility and becomes cracked, the cracks may propagate into the base metal and cause premature failure. In samples of IN 738 coated with a CoCrAlY-type (GT 29) coating (Ref 25) subjected to TMF testing in the laboratory, and in field samples of Udimet 720 blades coated with an aluminide (RT-22) (Ref 56), clear evidence of the association of coating cracks with base-metal cracks has been demonstrated. In the case of aluminum diffusion coating, cracks were found to emanate from the coating surface as well as from the brittle interdiffusion zone between the coating and the base metal, as illustrated in Fig. 9.30. On the basis of engine tests on coated Mar-M 509 vanes, it has been observed that while CoCrAlY coatings ranked as the best from a corrosion-resistance point of view, they were very susceptible to thermal-fatigue cracking. The duty cycle of a turbine must therefore be taken into account in selecting an appropriate coating.*

Obviously, the integrity of a coated blade depends not only on the fatigue resistance of the coating but also on that of the base metal. If the substrate material is sufficiently ductile, the coating cracks may be arrested. It is only when both the coating and the base metal are brittle, such as in the case of the aluminide-coated Udimet 720 blades cited in Ref 56, that rapid crack propagation into the base metal is facilitated.

The effects of coatings on the mechani-

*A. McMinn, R. Viswanathan, and C.L. Knauf, Field Evaluation of Gas Turbine Protective Coatings, *ASME J. Engg. Gas Turbines and Power*, Vol 110, Jan 1988, p 143

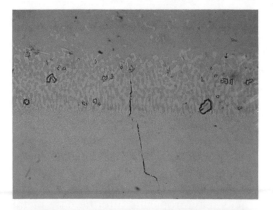

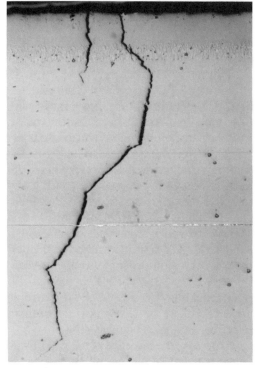

Fig. 9.30. Initiation of thermal-fatigue cracks in the interdiffusional zone (a) and the coating (b) of a Udimet 720 blade coated with aluminide (RT-22) (Ref 56; courtesy of V.P. Swaminathan, South West Research Institute, San Antonio, TX).

cal properties of superalloys have been reviewed in depth by Strang and Lang (Ref 57). The principal conclusions from their paper can be listed as follows:

1. Coatings do not cause degradation of the creep-rupture properties of superalloys provided that postcoating heat treatments are carefully selected and applied and that appropriate stress-correction factors necessary to account for net section change are applied.

2. Coatings can have positive or negative effects on several factors that include surface imperfections, roughness, residual stresses in the coating, particle size within the coating, and fatigue endurance and ductility of the coating. Selection of coatings with good ductility, use of suitable postcoating heat treatments, and optimization of coating thickness consistent with minimum requirements for corrosion protection are some of the ways in which component integrity can be ensured.

3. The most important mechanical property of a coating is its resistance to thermal-fatigue cracking. Most coatings have a ductile-to-brittle transition temperature (DBTT), as shown in Fig. 9.31 (Ref 58 to 61). Selection of a coating on the basis of its ductility/temperature characteristics depends on the maximum strain levels likely to be experienced by the component over its operating temperature range. On this basis, absolute ductility levels as well as DBTT's are important. If operational strain levels are lower than that required to crack the selected coating, the value of its DBTT will not be important. On the other hand, if strain levels are such that cracking of the coating can occur within the operational temperature range, selection of a coating with a low DBTT will be advantageous. In each of these cases, the definition of the DBTT could prove to be important. Lowrie and Boone (Ref 62) define the DBTT as that temperature corresponding to a fracture strain of 0.6%. On this basis, the "best" coatings for general applications could be either the low-aluminum CoCrAlY coatings or any of the low-chromium CoNiCrAlY or NiCrAlY compositions. Selection of pack aluminide and high-aluminum CoCrAlY coatings would necessitate steps to ensure that thermal

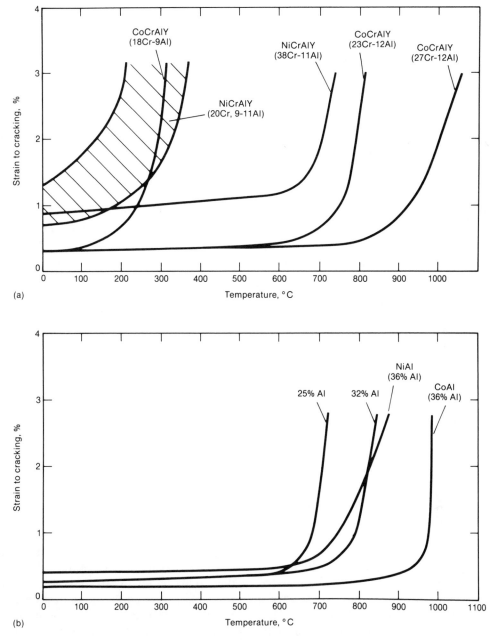

Fig. 9.31. Ductility/temperature characteristics of (a) MCrAlY coatings (Ref 58) and (b) aluminide coatings (Ref 59 to 61).

strains were minimized, particularly at temperatures below the DBTT. In some cases, this may be achieved by careful control of turbine start-up and shutdown cycles. However, in some regions, maximum tensile stresses are primarily due to the reversal of relaxed, steady-state compressive thermal stresses during shutdown and are not strongly affected by start-up/shutdown cycles.

In view of the DBTT behavior exhibited by coatings, it becomes even more critical in evaluating coated components that thermomechanical fatigue tests be performed.

Furthermore, simple TMF cycles in which the maximum tensile strain is made to coincide with the peak temperature (in-phase, IP) or with the lowest temperature (out-of-phase, OP) in the cycle will lead to unrealistic results. In typical blade cycles, as illustrated in Fig. 9.15, both the peak tensile and peak compressive strains occur at intermediate temperatures (not necessarily the same for tensile and compressive) and not at the highest or the lowest temperature in the cycle. Hence, TMF cycles simulative of actual blade cycles must be performed to evaluate the effect of the coating. Such test data on coated components is extremely scarce. Most of the results available in the published literature are based on thermal-fatigue tests using disk-type specimens immersed in fluidized beds; they lead to the qualitative conclusion that coatings, in general, do not degrade the thermal-fatigue resistance of the coated alloys. Although coatings were found to be beneficial from a crack-initiation point of view, they led to higher crack-growth rates. Among the various coatings, the overlay coatings appeared to be better than the aluminides.

The only comprehensive study of the thermomechanical fatigue behavior of coated IN 738 is that reported in Ref 25. These results clearly show reductions in life due to coatings. The coating used was a proprietary CoCrAlY-type vacuum plasma-spray coating known as GT-29. Coating thickness was approximately 0.13 mm (0.005 in.). A comparison of the fatigue curves for the coated and uncoated material, in an out-of-phase cycle with a peak temperature of 870 °C (1600 °F) and no hold, is shown in Fig. 9.32. The fatigue lives of the coated samples were shorter than those of the uncoated samples by a factor of 2. Tests under simulated bucket cycles again showed life reduction by a factor of 3 as a result of the coating. These reductions were due to the fact that the coating material undergoes a ductile-to-brittle transition at an intermediate temperature. For the simple TMF cycle and for the simulated bucket cycles, all of which are out-of-phase cycles, the maximum tensile

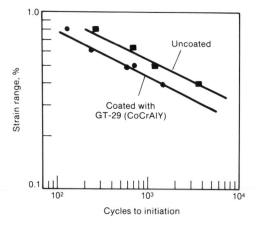

Fig. 9.32. Effect of coating on fatigue life of IN 738 tested in thermomechanical fatigue using linear, out-of-phase cycles with peak temperature of 870 °C (1600 °F) and no hold time (Ref 25).

strains occurred at low temperatures where the coatings were relatively brittle. This facilitated initiation of cracks in the coating, which subsequently propagated into the base metal. Based on these results, it appears that the TMF strength of hardware coated with aluminides might be even worse and may explain the type of field failures reported in Ref 56.

Service-Induced Degradation

One of the major factors affecting the integrity of components and leading to gross errors in life prediction for blades and vanes is the degradation of properties due to service exposure. Design of these components as well as calculational methods of life prediction are based on extrapolation of short-time laboratory data to long times without taking cognizance of such degradation phenomena. An example of this type of error is shown in Fig. 9.33 with respect to creep-life prediction of blade alloy Udimet 700 (Ref 63). The stress-rupture properties show a degradation in rupture life in excess of about 1000 h at 815 °C (1500 °F). Design as well as life prediction based on linear extrapolation of the short-time data (dashed line) would have led to overly optimistic conclusions. Formation of a brittle sigma phase was found to be a cause of the unexpected degradation in this case, and several

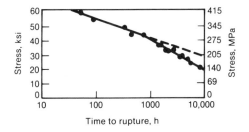

Fig. 9.33. Stress-rupture curve for Udimet 700 at 815 °C (1500 °F), illustrating the risk of life prediction based on linear extrapolation of short-time data (Ref 63).

premature failures of Udimet 700 blades were explained on this basis. Degradation can occur in tensile, creep, fatigue, and impact properties. Characterization studies of such degradations are extremely rare. The limited data available are reviewed below.

Tensile Properties. Susukida *et al* have carried out extensive characterization studies of the degradation of the tensile properties of wrought alloys Udimet 520, Udimet 710, and cast alloys Mar-M 421 and IN 738X (Ref 8 and 64). Results from one of their studies (Ref 64) are shown in Fig. 9.34. With minor exceptions, tensile strength, yield strength, and ductility show decreasing trends both at room temperature and at high temperatures (802 and 871 °C, or 1475 and 1600 °F) as a result of prior exposure

in the range 750 to 950 °C (1380 to 1740 °F). The extent of this reduction increases with increasing exposure time and temperature. It is very unlikely that blades will be exposed in service to 950 °C. On the other hand, blade metal temperatures could conceivably reach 850 °C (1560 °F), although they are usually in the range 790 to 845 °C (1450 to 1550 °F). Because a 10,000-h exposure at 850 °C could correspond to normal service durations of concern in the lower temperature range in terms of producing equivalent damage, the degradations occurring at these conditions are presented in Table 9.3. Substantial reductions (40 to 60%) in tensile and yield strengths and in ductility are observed in some instances. Superimposed stresses in service could further exacerbate the microstructural changes and the tensile-property degradation. Such reductions normally amount to an erosion of the design safety factor in terms of tensile properties, but can also be reflected in terms of reduced fatigue strength, because fatigue strength is related to both tensile strength and ductility. The observed changes in tensile and yield strengths have been attributed mainly to the coarsening of the gamma-prime precipitates (Ref 64).

Impact Toughness. It has been known for some time that the impact toughness of blades can degrade in service, thereby re-

Table 9.3. Degradation of tensile properties for Udimet 520 at 800 °C (1475 °F) and for Udimet 710, Mar-M 421, and IN 738X at 870 °C (1600 °F) due to prior exposure for 10,000 h at 850 °C (1560 °F) (based on Ref 64)

Alloy	Property	Before exposure	After exposure	Decrease, %
Udimet 520	Tensile strength, MPa (ksi)........979 (142)	731 (106)	25
	Yield strength, MPa (ksi) 786 (114)	490 (71)	40
	Reduction in area, % 30	45	(a)
Udimet 710	Tensile strength, MPa (ksi).......689 (100)	586 (85)	15
	Yield strength, MPa (ksi) 441 (64)	345 (50)	2.2
	Reduction in area, % 40	15	60
Mar-M 421	Tensile strength, MPa (ksi).......786 (114)	634 (92)	19
	Yield strength, MPa (ksi) 586 (85)	345 (50)	40
	Reduction in area, % 5	5	0
IN 738X	Tensile strength, MPa (ksi).......689 (100)	634 (92)	20
	Yield strength, MPa (ksi) 510 (74)	345 (50)	33
	Reduction in area, % 2	2	0

(a) 50% increase.

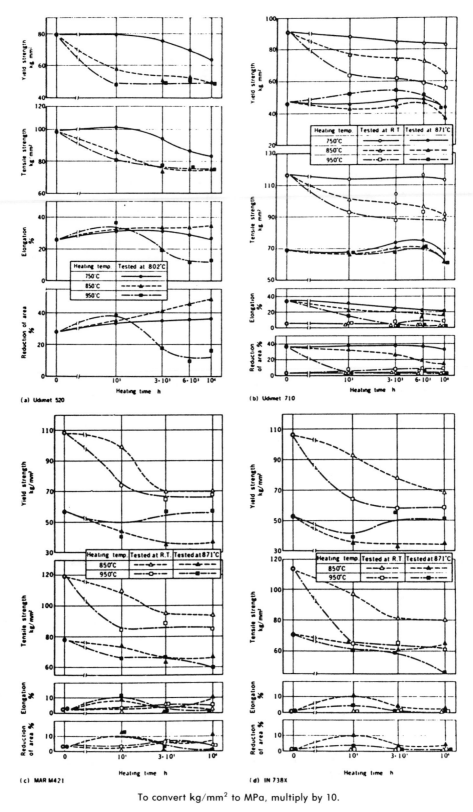

To convert kg/mm² to MPa, multiply by 10.

Fig. 9.34. Effects of prior exposure on the tensile properties of superalloys (Ref 64).

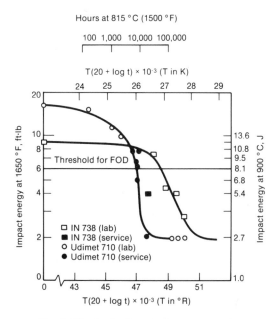

Fig. 9.35. Reduction in impact toughness at 900 °C (1650 °F) for IN 738 and Udimet 710 following laboratory and field exposure in the range 790 to 870 °C (1455 to 1600 °F) (Ref 65 and 66).

ducing their tolerance to defects and resistance to impact damage by foreign objects (FOD). This problem has been particularly acute with blades made of alloys Udimet 710 and Udimet 720, whose service lives had to be curtailed to below 10,000 h. Massive replacement programs have cost the turbine owners millions of dollars.

Figure 9.35 illustrates the decrease in impact toughness measured at 900 °C (1650 °F) with increasing time of prior exposure in the range 790 to 870 °C (1455 to 1600 °F). The time and temperature of exposure have been combined in the form of a Larson-Miller parameter. The usual average exposure temperature of first-row blades of Udimet 710 is about 815 °C (1500 °F). Figure 9.35 shows that a precipitous drop in the impact energy measured at 900 °C occurs due to prior exposure at 815 °C (1500 °F) for 3000 and 20,000 h. It has been reported that 8.13 J (6 ft·lb) at 900 °C as measured in a standard Charpy impact specimen might represent the threshold value of impact energy below which significant FOD might be observed (Ref 65). According to the

data, the threshold value will be reached after 10,000 h of service at 815 °C in the case of Udimet 710.

Figure 9.36 shows the degradation behavior with duration of exposure in the range 790 to 870 °C (1455 to 1600 °F) for three other superalloys, Udimet 520, Udimet 500, and IN 738 (Ref 65). The results show that these alloys also might eventually become vulnerable to FOD. When the data for IN 738 are plotted on the parametric plot in Fig. 9.35, the threshold value for FOD is reached in about 49,000 h at an assumed service temperature of 815 °C (1500 °F). In fact, a first-row blade of IN 738, returned from service after 8242 h, had impact energy as low as 5.15 J (3.8 ft·lb) (Ref 66). Additional results concerning the toughness degradation may also be found in the literature for IN 738 LC (Ref 25), Udimet 520 (Ref 8 and 64), and Udimet 720 (Ref 67). The results of Tsuji on Udimet 720 show that the toughness of the alloy does not degrade appreciably due to prior aging for times up to 10^4 h at 900 °C (1650 °F). These results are difficult to understand in the context of field failures of this alloy related to foreign-object damage (Ref 56). The degradation of toughness in superalloys has been attributed mainly to coarsening of the grain-boundary carbides, which form a continuous network.

Creep-Rupture Strength. It is very difficult to document the degradation in stress-rupture behavior of actual components because the corresponding virgin-material properties are usually not available. It should be possible to use material from the cooler sections of the component (e.g., blade root) as the reference material, but this is seldom done. A more common practice is to compare the stress-rupture life of the service-retrieved component with the minimum required values for virgin material, at a fixed stress/temperature combination typically used for qualification testing (Ref 68). The scatter normally associated with such tests makes it difficult to evaluate accurately the extent of degradation, as will be discussed later.

It is easier to characterize the degradation

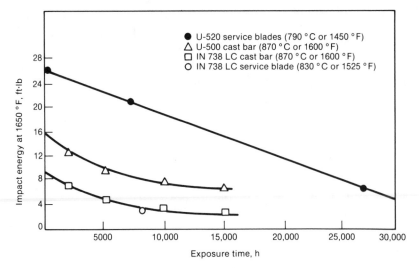

Fig. 9.36. Reduction in impact toughness at 900 °C (1650 °F) for IN 738, Udimet 500, and Udimet 520 following prior exposure in the range 790 to 870 °C (1455 to 1600 °F) (based on Ref 65 and 66).

in laboratory samples where a given heat of material can be characterized in terms of rupture life before and after various aging treatments (with or without stress). An example of results on Udimet 710 is presented in Fig. 9.37 (Ref 8), which shows rupture lives at 845 °C (1555 °F) and a stress of 350 MPa (50 ksi) for specimens aged at 800, 850, and 900 °C (1470, 1560, and 1650 °F) for times up to 10,000 h. Due to the scatter in the data, only a trend curve defining the "minimum" properties is plotted. It can be clearly seen that the rupture life after 10,000 h of aging can be reduced by a factor of 5 compared with virgin material. Results of similar studies on Inconel 700 and Inconel X-750 have been reported by the same authors (Ref 8).

Specimens of alloys Udimet 710, IN 738 X (0.14% C), and Mar-M 421 aged at 850 and 950 °C (1560 and 1740 °F) for times up to 10,000 h have been evaluated at 845 °C (1555 °F) and 350 MPa (50 ksi) by Susukida and coworkers (Ref 8 and 64). Based on their results, a plot of the prior exposure history in terms of a Larson-Miller aging parameter — i.e., $T(°R)(20 + \log t)$ versus rupture life — has been plotted in Fig. 9.38. A sharp drop in life as a result of prior thermal exposure is noted. The data suggest that the rupture lives of these alloys might

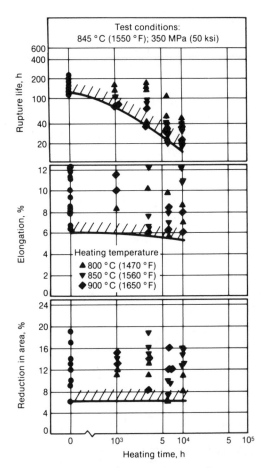

Fig. 9.37. Change in stress-rupture properties of Udimet 710 due to prior exposure (Ref 8).

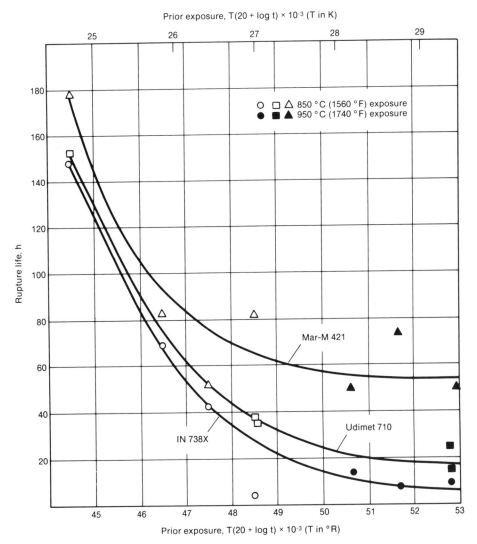

Fig. 9.38. Effect of prior exposure at 850 and 950 °C (1560 and 1740 °F) for times up to 10,000 h on rupture life for superalloy specimens tested at 845 °C (1555 °F) at a stress of 350 MPa (50 ksi) (based on Ref 8 and 64).

be reduced by a factor of 4 to 5 in about 50,000 h under a typical service exposure temperature of 815 °C (1500 °F). Service stresses superimposed on the thermal exposure may further degrade the properties. Observed degradation of creep-rupture properties has been attributed primarily to coarsening of the gamma-prime precipitates.

Thermal-Fatigue Strength. Degradation of tensile strength, ductility, and toughness are all related to fatigue strength and, therefore, would be expected to lead to reduced resistance to fatigue. This is clearly borne out by the results plotted in Fig. 9.39 (Ref 64). An order-of-magnitude reduction in the thermal-fatigue resistance is observed. This reduction was accompanied by and presumably associated with the formation of increasing amounts of sigma phase and grain-boundary carbide coarsening in both Udimet 520 and Udimet 710 (Ref 64).

Degradation of Microstructure. Alloys used for blading in industrial turbines typically derive their creep strength from the precipitate phase known as gamma prime (γ'), which has the composition A_3B where A is nickel and cobalt and B is principally

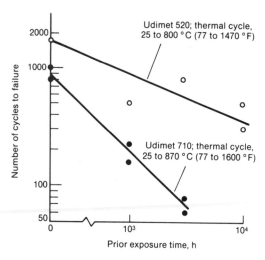

Fig. 9.39. Effect of prior exposure at 850 °C (1560 °F) on thermal-fatigue life for Udimet 520 and Udimet 710 (based on Ref 64).

aluminum with smaller amounts of titanium, niobium, and tantalum. In the case of alloys containing less than about 30 vol% γ', such as Udimet 500, Nimonic 80A, and Nimonic 90, the usual heat treatment cycles result in a uniform size and distribution of fine γ' ranging in size from 0.01 to 0.1 μm (0.4 to 4 μin.). Alloys with higher volume fractions of γ' contain duplex structures with both coarse γ' (0.2 to 2 μm, or 8 to 80 μin.) and fine γ'. Examples of such alloys are IN 738, Udimet 700, Udimet 710, and Nimonic 115. In addition to the γ' precipitates, blocky or scriptlike MC carbides (M = Ta, Ti, Nb) are present in the matrix and $M_{23}C_6$- or M_6C-type (M = Cr, Mo) discrete carbide particles are present at the grain boundaries.

During service exposure, the principal changes in microstructure that have been observed are (1) coarsening of the γ', (2) coarsening of the grain-boundary $M_{23}C_6/M_6C$ carbides, and (3) formation of an additional needlelike phase known as sigma. The first of these changes has been closely related to losses in tensile strength and creep strength, while the other two changes have been related to losses in tensile ductility, creep ductility, rupture strength, impact toughness, and thermal-fatigue resistance. A knowledge of the kinetics of these changes

can be a valuable tool in assessing the condition and the remaining useful life of in-service components.

Gamma-Prime Overaging. Coarsening of the γ' precipitates is generally known to occur by an Ostwald ripening mechanism in which the larger particles grow at the expense of the smaller particles. This mechanism involves a dependence on $t^{1/3}$ (t is time). When the γ' is initially present as uniform fine γ', aging is manifested as an increase in the average γ' size. When a duplex distribution is present initially, the coarse γ' particles grow in size and the volume fraction of the fine γ' diminishes. Figure 9.40 illustrates the γ'-coarsening phenomenon in Udimet 710. The kinetics of γ' coarsening for Udimet 520 and Udimet 710 (Ref 64), IN 738 (0.17% C) (Ref 68 and 69), and IN 939 (Ref 7) are shown in Fig. 9.41. A $t^{1/3}$ rate law is obeyed in all cases. Up to about 800 °C (1475 °F), the kinetics of coarsening are rather slow; at higher temperatures, increasingly rapid coarsening is observed. Comparison of the γ'-coarsening kinetics with the creep-rupture data shown in Fig. 9.38 would lead to the clear conclusion that, while the γ' particles may continue to coarsen with thermal exposure, the rupture strength may initially decrease but then will saturate at some intermediate exposure history. Hence, γ' size alone cannot be used as an index of rupture strength. On the other hand, the γ' size may be used as a good index of thermal history. The usefulness of this will be discussed in a later section.

Carbide Coarsening. The grain-boundary microstructure of nickel-base alloys also changes during service as a result of instabilities in a manner similar to the γ' coarsening. A continuous chain of large $M_{23}C_6$ carbides is known to form from the initially discrete small carbide particles, as shown in Fig. 9.42. These changes often are responsible for degradation in toughness, rupture ductility, and rupture strength. It is, however, difficult to separate the effects of carbide coarsening and those of γ' coarsening because conditions that promote one also promote the other.

Formation of Sigma Phase. The sigma

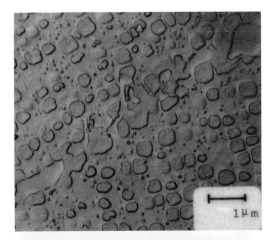

Top: New creep life, 140 h. Bottom: Service, 45,000 h; creep life, 10 h.

Fig. 9.40. Gamma-prime overaging and associated loss of creep strength in Udimet 710 tested at 845 °C and 350 MPa (1555 °F and 50 ksi) (Ref 70; courtesy of P. Lowden, Liburdi Engineering, Ltd., Burlington, Canada).

phase is a brittle, chromium-rich phase that generally forms as platelets (see Fig. 9.43) (Ref 70) and greatly reduces rupture ductility and rupture life in many superalloys. The tendency of any alloy to form this phase depends sensitively on the composition and can vary from heat to heat in the same alloy. Algorithms for calculating the tendency for sigma formation based on composition are generally used to optimize alloy compositions. In spite of this, varying degrees of sigma may still be found in service-exposed parts. The time-temperature fields for sigma formation in alloys Udimet

520 and Udimet 710 have been defined by Susukida, as shown in Fig. 9.44. Results from the study of Moon and Wall have shown that the effect of sigma formation on rupture behavior is not always predictable. They observed a reduction of rupture life due to sigma in Udimet 700 but not in IN 713C, Udimet 500, or Udimet 520 (Ref 63).

Life-Assessment Techniques

The purpose of life assessment for combustion turbine blades and vanes can vary from one context to another. Knowing how much useful life is left in the component is helpful in avoiding forced outages and planning replacement schedules. A more common need is to determine the proper intervals for inspection, repair, and rejuvenation. In addition, assessment techniques also enable the operator to optimize the operating conditions so as to get maximum life from the components. Turbine blades are highly stressed components whose failure can lead to more consequential damage than vane failures can cause. The tolerance of blades to cracks and defects is, therefore, much less than that of vanes. Hence, the failure criterion for blades is in terms of crack-initiation events. Vanes, on the other hand, can usually tolerate large cracks, and hence life-prediction techniques for vanes have emphasized crack-growth-based approaches. The objective of the life-assessment exercise and the failure criterion to be used must be kept clearly in focus in selecting the appropriate technique.

Crack-Initiation Assessment

The techniques for predicting crack initiation include calculations, nondestructive evaluations, and destructive evaluations. The scope and limitations of each were described in Chapter 1. Therefore, the discussion here will focus on unique aspects and problems of life prediction for combustion turbine blades. These problems are again somewhat different based on the assumed damage mechanism—e.g., creep or

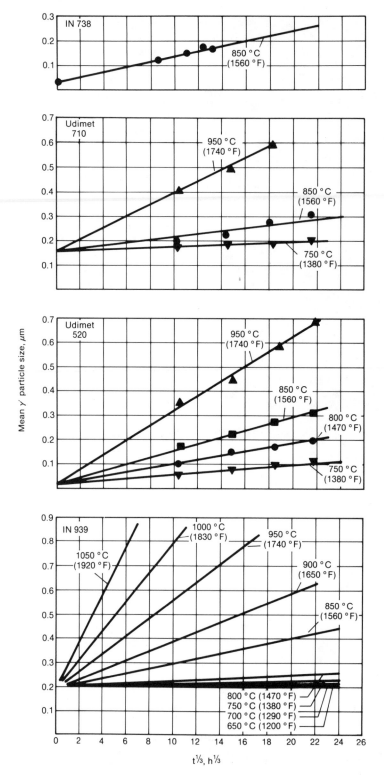

Fig. 9.41. Gamma-prime particle size as a function of $t^{1/3}$ (t is time of thermal exposure) for superalloys (based on Ref 7, 8, 64, and 69).

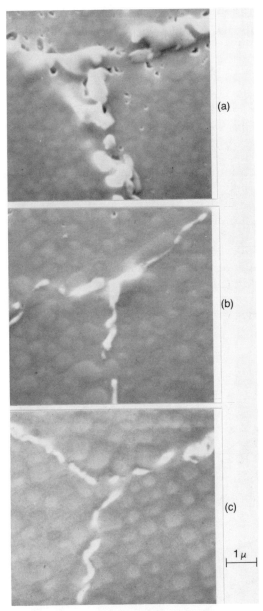

(a) Hot region in airfoil. (b) Cooler regions in shank. (c) Material from airfoil after a rejuvenation heat treatment.

Fig. 9.42. SEM photographs showing coarsening of grain-boundary carbides resulting from 10,000 h of service exposure at 830 °C (1525 °F) in a Udimet 710 blade.

thermal fatigue. Each of these damage mechanisms will be considered separately.

Calculational Techniques. These techniques utilize various damage rules in com-

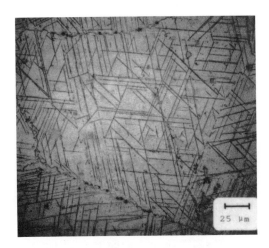

Fig. 9.43. Brittle, platelike sigma phase in Nimonic 115 (Ref 70; courtesy of P. Lowden, Liburdi Engineering, Ltd., Burlington, Canada).

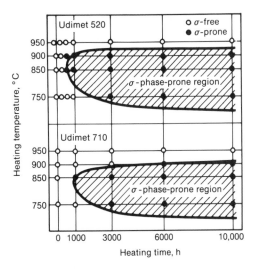

Fig. 9.44. Relationship between sigma-phase formation and exposure conditions for Udimet 520 and Udimet 710 (Ref 64).

bination with service histories and material databases to make estimates of expended life.

Creep. Calculational procedures for creep-life prediction involve linear summation of life fractions consumed under various operating conditions. An estimated temperature history, based on operating records, actual measurements, or microstructural observations, is utilized in combination with the lower-band Larson-Miller-type rupture data for the alloy to calculate the life fractions expended at different temperature-stress combinations. These fractions are added up to calculate the total life expended. The procedure for doing this was illustrated in Chapter 4. Inaccuracies in life prediction arise due to: (1) scatter in virgin-material properties, (2) uncertainties in temperature and stress, (3) effects due to corrosion and coatings, (4) in-service degradation, and (5) errors in the assumed linear damage rule.

Scatter in properties results from heat-to-heat differences in composition and processing. The scatter is expected to be greater in cast material than in forged material. Typical scatterbands for blade alloy IN 738 LC and for vane alloy FSX-414 are shown in

Fig. 9.45. An analysis of the IN 738 LC data under the usual laboratory test conditions of 845 °C and 350 MPa (1550 °F and 50 ksi) shows that the rupture life can vary from 20 to 460 h, or by a factor of about 20. The presence of coatings has not been known to lead to appreciable degradation in rupture life, as can also be seen in Fig. 9.45 (Ref 57).

In the case of uncoated blades, hot corrosion can lead to further uncertainty in rupture life. The magnitude of this uncertainty depends on the alloy, the stress level, the temperature (type I or type II corrosion), and the presence or absence of chlorides. In the absence of chlorides, assumption of a nominal 33% reduction in life (uncertainty factor of 1.5) is reason-

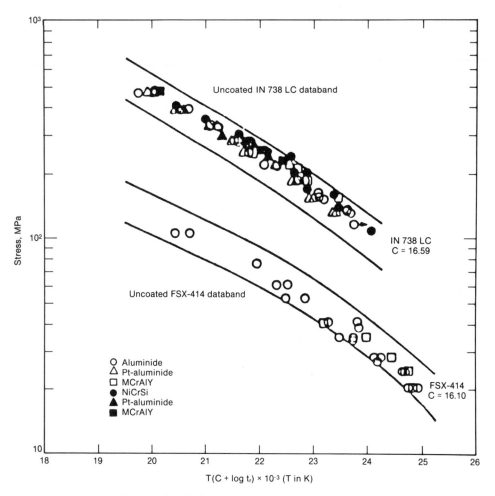

Fig. 9.45. Typical scatterbands for rupture properties observed in IN 738 and FSX-414 (Ref 57).

able. In the presence of chlorides, the reduction will be much greater. Laboratory tests on salt-coated samples in the range 750 to 850 °C (1380 to 1560 °F) have shown that the factor of life reduction is in the range 2 to 10 (Ref 43 and 44). Data of Whitlow *et al* suggest life reduction by as much as a factor of 10^4 for IN 738 tested at 705 °C (1300 °F) (type II corrosion). Under realistic stresses and chloride concentrations that might obtain in industrial turbines in coastal environments, we can assume an uncertainty factor of about 5.

Using the same data in Fig. 9.45, it can be shown that a temperature uncertainty of 10 °C (say 843 versus 853 °C) can lead to an error by a factor of about 1.5 (460 versus 310 h). Uncertainty in stress of 25% (say 350 versus 280 MPa) can increase the estimated rupture life by a factor of about 4 (460 versus 2400 h).

The effects of microstructural changes in causing in-service degradation of rupture life have been described and depicted in Fig. 9.38. To some extent, the microstructural changes are built into the design database provided that the laboratory accelerated tests are carried out at higher temperatures rather than at higher stresses. Because in thermally activated processes, temperature can be "traded" against time, accelerated creep tests at higher temperatures would hopefully simulate the microstructural changes occurring at lower temperatures in longer times. The same thing cannot be said about stress-accelerated tests, because stress cannot be "traded" against time and a high-stress test over a short time would not reproduce the microstructural changes occurring at lower stresses in longer times. Hence, if the design data used for creep design were based on temperature-accelerated tests, then the extrapolated behavior at long times may reasonably coincide with the actual behavior. In practice, however, design data consist of temperature data as well as stress-accelerated test data, and extrapolation of such data is likely to lead to optimistic estimations of alloy performance. It is difficult to judge the effect of service degradation on service life, because the limited data,

as shown in Fig. 9.38, have been generated once again by stress acceleration and indicate a reduction in life by a factor of 10 to 15. At more realistic stresses close to the service stress, however, the rupture lives for the degraded and undegraded conditions are expected to approach each other. In the absence of data, we can assume life to be reduced by a factor of 2 as a result of in-service degradation.

The available data on steels have clearly shown that the life-fraction rule and the linear-damage rule are invalid for tests in which stresses are varied. In the calculational procedure for blades, however, the life-fraction summation is carried out only over a narrow range of conditions. Hence, uncertainties due to this factor can be neglected.

Table 9.4 is a summary of the estimated uncertainty factors arising from the circumstances described above. These estimates are very imprecise and simply reflect the author's judgment based on limited data in the literature. The sole purpose of listing these factors is to indicate the relative importance of the various considerations affecting component life so that priorities can be defined for future research.

Clearly, the major uncertainty factor in life prediction is the scatter in the original properties. Assuming uncertainty factors of 20 due to property scatter, 1.5 due to temperature, 4 due to stress, and 2 due to the unknown degree of service degradation, the cumulative uncertainty in predicted life can be as high as a factor of 240 for the specific circumstances described here. The most obvious way to improve the accuracy of life prediction seems to be "placing" the blade alloy with respect to the scatterband of property data. The only way to do this seems to be to examine the results of the test certificates and the manufacturer's qualification tests for the given heat of the blade alloy. Alternatively, turbine users should institute a standard practice of storing virgin material for later testing if necessary.

An important fact to remember is that neither the stresses nor the temperatures in blades are uniform. Both of these parame-

Table 9.4. Estimated uncertainty factors in life prediction

| | Factor of uncertainty(a) | | | |
| | Creep-rupture life | | Thermal-fatigue life | |
Cause of uncertainty	Factor	Basis of estimate	Factor	Basis of estimate
Scatter in assumed material properties:				
Cast material 20		Fig. 9.45; Ref 57	16	Ref 75
Wrought material.......................... 10		J	10	J
Effect of coatings in air (if present) .... None		Fig. 9.45; Ref 57	3	Fig. 9.32; Ref 25
Effect of hot corrosion:				
Coated material .... None		...	None	...
Uncoated, without chlorides 1.5		Discussion in Ref 35	2 to 5 at low $\Delta\epsilon$	Fig. 9.24
Uncoated, with chlorides 5.0		J	10	Fig. 9.25
10 °C (18 °F) error in assumed temperature....... 1.5		Fig. 9.45	None	...
25% error in:				
Assumed stress............................ 4.0		Fig. 9.45	...	...
Assumed strain range		...	2.0	Fig. 9.46
In-service degradation of:				
Base metal................................. 2.0		J	5 to 15	Fig. 9.39
Coating 2.0		J	2	J
Inaccuracy of damage rules .... None		J	3	J
Stress and thermal gradients in component........ Unknown		Unknown	Unknown	Unknown
Use of isothermal LCF data in lieu of TMF data ..NA		...	3	J

(a) J denotes author's judgment, based on the collective body of information available as reviewed in the text. NA = not applicable.

ters vary along and across the blade, resulting in gradients. Sometimes these gradients are confined to local regions around cooling holes, local hot spots, or other design features. Calculational procedures should be able to take these into account and predict localized failure.

The many uncertainties described above clearly indicate the need for a probabilistic approach to blade-life assessment rather than the deterministic approach currently used. In spite of the many uncertainties described above, calculational procedures using conservative assumptions have a useful and cost-effective role in defining inspection intervals. However, predictions regarding replacement and refurbishment must be calibrated on the basis of inspections and destructive evaluations.

Thermal Fatigue. Prediction of thermal-fatigue life essentially involves a calculation of the life expended using damage rules for a set of assumed or recorded thermal-

history and material data. A variety of damage rules that are commonly used and the procedures for calculating expended life fraction were described in Chapter 4. Only those aspects and problems unique to life prediction of combustion turbine blades are reviewed here.

Most life-prediction studies have simply consisted of evaluating the "fit" of different damage rules to isothermal low-cycle-fatigue data developed in the laboratory. Nazmy and Wuthrich compared the applicabilities of the strain range partitioning (SRP) method, the frequency-modified (FM) Coffin-Manson rule, and the Ostergren damage approaches (ODA) to life prediction for IN 738, and concluded in favor of SRP (Ref 71). For the same alloy, prediction capability within a factor of 2 by SRP has been claimed (Ref 21). Contrary experience indicating the inapplicability of SRP to IN 738 and to René 95 has also been documented (Ref 72 and 73). For a cobalt-base vane alloy, Mar-M 509, the SRP method was found to predict lives within a factor of 3 (Ref 19). For IN 738 LC, Persson *et al* claim better correlation with the Ostergren approach than with the Coffin-Manson relationship (Ref 74).

The most complete comparison of a variety of damage rules as applied to IN 738 LC and Mar-M 509 using isothermal low-cycle-fatigue data has been that of Fischmeister, Danzer, and Buchmayr (Ref 75). Results from this comparison are shown in Table 9.5. The universal slope method could be at once disqualified on the basis of a very large standard error and maximum deviation. The ductility-normalized SRP method was the next poorest. The predictive capabilities of all the other methods were concluded to be about the same, because values of the standard deviation and maximum deviation were about the same. Further evaluations using additional data led these investigators to conclude that the strain-rate-modified accumulation of time-dependent damage (SRM), in which both the creep and fatigue components are converted to an equivalent creep damage and then summed as life fractions, represented a slight improvement over the other methods. Surprisingly, the linear damage summation procedure, similar to ASME Code Case N-47 (see Chapter 4), resulted in predictions

not too far from those of the best method. Creep-fatigue data on IN 738 LC at 850 °C (see Fig. 9.46) (Ref 74) are suggestive of the bilinear damage curves published in Code Case N-47 for other materials. The data plotted in Fig. 9.46 are based on strain-cycling tests at 850 °C (1560 °F) with a total strain range, $\Delta\epsilon_t$, of 0.76%. Hold times of 100 to 2100 s were also superimposed at the maximum tensile strain. The cumulative creep-life fraction expended, $\Sigma t/t_r$, was then plotted as a function of $N_f(hold)/$

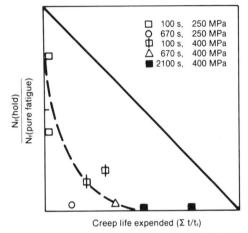

Data were generated using $\Delta\epsilon_t = 0.76\%$, hold times ranging from 100 to 2100 s, and stress levels of 250 and 400 MPa (36 and 58 ksi).

Fig. 9.46. Combined creep-fatigue data at 850 °C (1560 °) for IN 738 LC (Ref 74).

Table 9.5. Performance comparison of life-prediction models (Ref 75)

Method	Descriptor	m	IN 738 LC s	IN 738 LC d_{max}	Mar-M 509 s	Mar-M 509 d_{max}
Universal slopes	$\Delta\epsilon_{tot}$	0	1.08	83	1.71	615
Ductility-normalized SRP	$\Delta\epsilon_{in}$	0	0.41	7.4	0.47	11
Spera model	$\epsilon_{in}(\sigma)$	0	0.37	2.5	0.24	3.1
SRM	$\epsilon_{in}(\sigma)$	1	0.31	4.5	0.35	4.7
Manson-Coffin	$\Delta\epsilon_{in}$	2	0.36	5.6	0.39	7.4
Ostergren, without frequency modification	$\sigma_{max}\Delta\epsilon_{in}$	2	0.31	5.4	0.39	7.4
Manson-Coffin, with frequency modification	$\Delta\epsilon_{in}$	3	0.35	5.2	0.33	3.8
Ostergren, with frequency modification	$\sigma_{max}\Delta\epsilon_{in}$	3	0.32	6.5	0.30	3.3
Analogous to ASME code	$\Delta\epsilon_{in}, \sigma$	4	0.31	4.8	0.32	4.5
SRP–linear damage rule	$\Delta\epsilon_{in}$	8	0.31	3.0	0.33	3.0
SRP–interaction damage rule	$\Delta\epsilon_{in}$	8	0.30	3.3	Not computed	
SRP–general damage rule	$\Delta\epsilon_{in}$	12	0.35	3.1	Not computed	

(a) m = number of fit parameters; s = standard deviation in log N_f; d_{max} = maximum deviation, defined as maximum value of N_f(calculated)/N_f(measured). For further clarification of definitions, see Ref 15.

N_f(pure fatigue), where t is the hold time in a given cycle, t_r is the rupture life, and N_f(hold) is the number of cycles to failure with hold time. As creep-life fraction increases, the fatigue-life ratio decreases, as shown in the figure. Considering the many uncertainties involved in life prediction, the simplicity of the ASME procedure makes it an attractive candidate. The same may be said of the frequency-modified Coffin-Manson approach, which is simple to use and gives reasonably good answers.

It is clear from laboratory studies on low-cycle fatigue that there are divergent views regarding which damage approach provides the best basis for life prediction. Even the best of the methods in a given study predicts life only within a factor of 3. Prediction in the context of laboratory studies has consisted mainly of fitting various damage rules to all the data to determine which rule best describes the data. Independent predictions of life for test conditions outside the envelope used in the study have never been made. In fact, the results of different laboratory studies have even been mutually inconsistent because test conditions are never identical. Several variables, such as test temperature, strain range, frequency, time and type of hold, waveform, ductility of the material, and damage characteristics, have been known to affect fatigue life. The conclusions drawn by each investigator apply only to the envelope of material and test conditions used in that investigator's study. The validity of the damage approaches,

therefore, have to be examined with reference to the material and service conditions relevant to a specific application. Use of broad generalizations based on laboratory tests which may often have no relevance to actual component conditions does not appear to be a productive approach. The limited data available indicate that the fatigue lives of alloys under thermal cycles simulative of machine operation can be appreciably different from those determined in the laboratory under isothermal conditions.

The most extensive evaluation of the thermomechanical fatigue behavior of blading alloys has been that of the GE investigators using coated and uncoated IN 738 LC (Ref 18 and 25). A variety of thermomechanical cycles (see Fig. 9.16) as well as actual blade cycles were utilized in this study. A number of damage parameters were fitted to the data and were compared in terms of which one best described all the data. The results of this study are shown in Table 9.6. Fatigue life (number of cycles to crack initiation) was fitted to each of the first six parameters in the table using the equation

$$\log N = A + B \log \text{(parameter)} \quad \text{(Eq 9.1)}$$

For the most generalized damage parameter (the last one in Table 9.6), the equation used was

$$\log N = A + B \log (\sigma_{max}) + C \log \Delta\epsilon$$

$$\text{(Eq 9.2)}$$

Table 9.6. Comparison of damage parameters (Ref 25)

Parameter(a)	Definition	A	B	C	Standard error	Data spread factor ($\pm 2\sigma$)
Strain range	$\Delta\epsilon$	1.962	−2.83	...	0.293	14.8
Plastic strain range	$\Delta\epsilon_{in}$	1.881	−0.609	...	0.345	24.0
Maximum tensile stress	σ_{max}	15.34	−6.47	...	0.198	6.2
Ostergren parameter	$\sigma_{max}\Delta\epsilon_{in}$	3.66	−0.623	...	0.329	20.8
Leis parameter	$[\sigma_{max}(\Delta\sigma/2)]\Delta\epsilon$	8.47	−1.969	...	0.261	11.0
SWT parameter	$\sigma_{max}\Delta\epsilon$	8.98	−2.35	...	0.230	8.4
General parameter	$\sigma_{max}^B\Delta\epsilon^C$	13.02	−5.41	−0.831	0.189	5.7

(a) Fatigue life (number of cycles to initiation) was fit to each of the first six parameters using the equation log N = A + B log (parameter). For the general parameter, the equation was log N = A + B log (σ_{max}) + C log ($\Delta\epsilon$). The standard error of estimate was determined on log N.

where the constants A, B, and C were determined experimentally and σ_{max} was the maximum tensile stress in the cycle. The standard error was determined with respect to log N. The best description of the data was found to be given by the generalized damage function (GDF) immediately followed by the σ_{max} parameter. Using a generalized damage function, predictions were made for test cycles outside the data set used in developing the parameter and compared with the actual values of N. The GDF was found to predict life within a factor of 3 to 5.

We can now proceed to consider some of the uncertainty factors involved in predicting the thermal lives of components on the basis of laboratory tests. First and foremost is the uncertainty due to scatter in the fatigue-property data. This comprises both heat-to-heat and specimen-to-specimen variations and, in addition, scatter from the measurement procedure. Fischmeister, Danzer, and Buckmeyer have analyzed test data from three data sets on different heats and concluded that for alloy IN 738 LC the standard deviation in log N with respect to the mean is about 0.2. For a $\pm 3\sigma$ (standard deviation), this gives a maximum life/minimum life ratio of about 16. A typical scatterband for IN 738 LC is shown in Fig. 9.47 (Ref 75). The presence of coatings can have a positive or negative effect on thermal-fatigue life. Decreased fatigue life in air due to coatings has been reported for IN 738 (Ref 25) and Udimet 720 (Ref 47). Using data for the base material in the uncoated condition to predict life can result in an uncertainty factor of 2 to 3 (Ref 25). In the presence of corrosive conditions, coatings invariably are helpful and the performance of the coated component in the corrosive medium can be equated with that of the base metal in air. Hence, use of data for the base metal in air presumably does not result in any uncertainties. In the uncoated condition, however, life reduction can result depending on the presence or absence of chlorides. Prior service exposure is known to lead to reduced fatigue life, the degree of which depends on the alloy (Ref

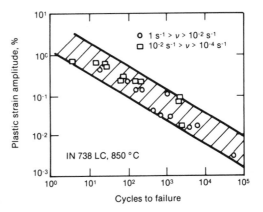

Fig. 9.47. Scatterband for low-cycle-fatigue properties at 850 °C (1560 °F) for IN 738 LC tested at two different frequencies (Ref 75).

64). The uncertainty factor due to this is estimated to be in the range 5 to 15 based on the data in Fig. 9.39. Errors in the assumed damage rules and the use of LCF data in lieu of appropriate TMF data can lead to additional uncertainties, as defined in Table 9.4. The cumulative effect of all these uncertainties is once again a grossly inaccurate prediction. The accuracy of the prediction can be improved by one or more of the following steps: (1) "placing" the particular heat of blade material in the scatterband of virgin-material properties, (2) using TMF data generated under realistic cycles rather than isothermal LCF data, (3) using degraded material and coating properties, and (4) taking the location (coastal versus inland) and history of prior corrosion into account. Methods and data needed to implement many of these steps are sorely lacking at the present time. Additional efforts to improve the damage rules appear to be only marginally fruitful.

Nondestructive Methods. Conventional nondestructive evaluation (NDE) of blades and vanes generally has included visual inspection and dye-penetrant inspection. Eddy-current probes have occasionally been used. These techniques are aimed at detecting the presence of cracks. They are adequate for vanes, which have a high tolerance to cracks. In the case of blades, whose lives are governed by crack initiation, detection of a crack would often constitute grounds for replacement or rejuvenation in selective instances. In view of the fact that significant incipient damage and loss of useful life can occur prior to formation of a manifest crack, development of advanced NDE techniques would be of great value in the case of blades. Currently, such techniques are not available although the grounds for developing them can be identified on the basis of limited data reported in the literature. These data will be reviewed in this section.

Two common indicators of creep-life exhaustion are (1) coarsening of γ' precipitates and (2) formation of cavities at grain boundaries. Qualitative information on how these phenomena progress with respect to life expenditure is available, but a quantitative basis is lacking at the present time. Both of these damage mechanisms lend themselves to detection by nondestructive replication, and hence the subject is of great interest.

Measurement of γ' Size. Correlations between γ' size and rupture strength for specific alloys are not available. Comparison of the γ'-coarsening kinetics (Fig. 9.41) and rupture-strength degradation as a function of prior exposure (Fig. 9.38) indicates that rupture life may degrade rapidly with initial coarsening of γ', but may eventually saturate beyond some critical value of the γ' size. Hence, it is difficult to use γ' size as a direct measure of rupture strength. An indirect but more effective way of using the γ' size is simply to use it as an indicator of the average service temperature at the component location of interest. This information can then be used to calculate the life fraction expended using the standard material-property data. Uncertainty due to lack of knowledge of the local metal temperature and the need to search operating records are thus minimized.

One of the problems in using the current value of γ' size is that it is not only a function of the time-temperature history, but also depends on its initial size prior to service. To estimate the size of the γ' in the virgin condition, samples or replicas can be taken from the cooler regions of the same blade (T < 750 °C or 1380 °F) — e.g., the root section. Once the initial γ' size and the current value are known, the temperature of the blade can be estimated (time is known) by reference to data similar to those in Fig. 9.41.

Creep Cavitation. A second indicator of creep damage in superalloys is the presence of creep cavities at the grain boundaries. Although qualitative observations relating to cavitation have been made by many investigators, quantitative observations are extremely limited. These include the studies of Tipler (Ref 76) and of Stevens and Flewitt (Ref 69) on IN 738 LC and IN 738, of Lindblom on Nimonic 100 (Ref 77), and of Wortmann on Nimonic 108 (Ref 78). Of these studies, specific attempts to use observations on cavitation for life prediction have been made by Wortmann and by Lindblom.

The technique described by Wortmann consists of comparing the degree of creep cavitation in a service-exposed sample with a reference catalog established from laboratory data (Ref 78). The correlation of creep cavitation with the percent of creep life consumed, based on tests on Nimonic 108, is shown in Fig. 9.48.

In the method used by Lindblom, a grading system related to the volume fraction of cavities (relationship not reported) is used as an index of creep life consumed (Ref 77). The grading system used and the correlation between grading and time to failure in Nimonic 100 samples tested at 940 °C and 110 MPa (1725 °F and 16 ksi) are shown in Fig. 9.49. The feature that is immediately striking in the graph in Fig. 9.49 is the steep drop in life observed at very low void ratings, followed by a relative insensitivity of rupture life to cavity ratings beyond 0.5 (see also Fig. 9.49a). This indicates that a very high percent of life may already have been consumed as soon as voids begin to appear. The scatter in void ratings at a given point in life also appears to be very large, as shown by the error bars in Fig. 9.50. In this figure, the black dots indicate average values and the numbers near the dots indicate the numbers of service blades examined.

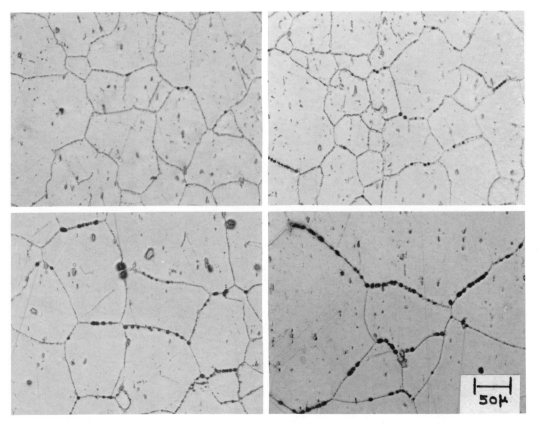

Fig. 9.48. Correlation of degree of creep voiding with percent of creep life consumed. Top left, 40%; top right, 60%; bottom left and right, 80% (Ref 78; courtesy of J. Wortmann, MTU Motoren-und Turbinen-Union, Munich).

The effect of creep voids *per se* on the rupture life of an alloy is somewhat difficult to assess because they are only part of the damage accumulating during service. Other forms of temperature-related microstructural damage are also generated in service, as discussed above. Some estimate of the relative significance of creep voiding and microstructural changes can be made from the results of studies in which service-exposed blade materials have been subjected to reheat treated and hot isostatic pressing (HIP). Both processes restore the alloy microstructure: however, only the HIP process is effective at eliminating creep voids. By comparing the stress-rupture results for the various processes as shown in Fig. 9.51 (Ref 70), the amount of creep-life degradation in the service-exposed material due to creep voiding and microstructural damage may be determined. It should be noted, however,

that this result is representative of only one specific combination of alloy, service stress, and service temperature. For other alloys and service exposures, the relative importance of voiding and microstructure may vary.

At present, there are no nondestructive techniques for detecting incipient fatigue damage (prior to cracking). Correlations between microstructural changes and fatigue-life consumption have not been explored. Correlations between grain-boundary carbide size and impact properties, if available, can provide a reliable way of assessing the susceptibility of blades with respect to foreign-object damage.

Replication. Considerably more research work needs to be done before the microstructure-based techniques can be effectively utilized. Quantitative relationships between microstructural damage index and life con-

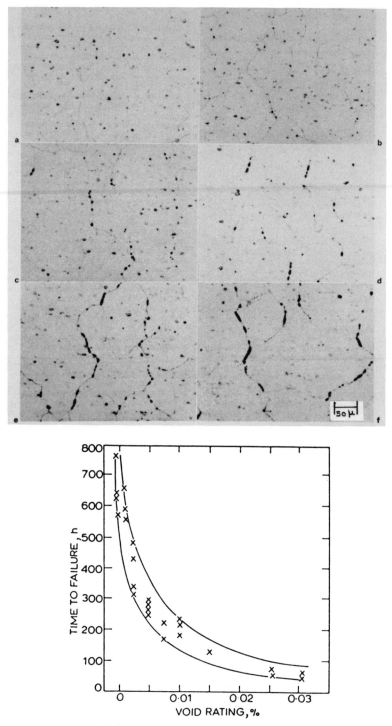

Micrographs show master scale defining creep-void grades: grade a, <0.5; b, 0.5; c, 1.0; d, 1.5; e, 2.0; f, 3.0. Graph shows relationship between creep-void grade and rupture life for a Nimonic 100 sample tested at 940 °C and 110 MPa (1725 °F and 16 ksi).

Fig. 9.49. Decrease in time to failure with creep voiding (Ref 77; courtesy of Prof. Y. Lindblom, F.F.V. Maintenance, Linkoping, Sweden).

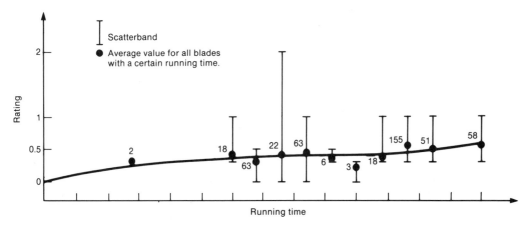

Fig. 9.50. Creep-void rating as a function of service time for Nimonic 100 blades (Ref 77).

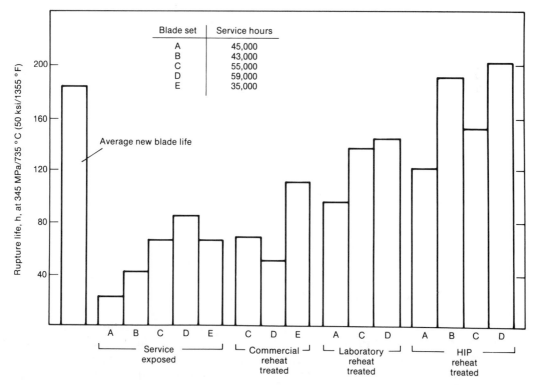

Fig. 9.51. Comparison of stress-rupture life at 345 MPa and 735 °C (50 ksi and 1355 °F) in service-exposed Inconel X-750 blades in the as-received, reheat treated, and HIP reheat treated conditions (Ref 70).

sumption specific to given alloys are needed. Scatter in the results due to heat-to-heat variations and due to test conditions need to be ascertained. Test data needed for developing these correlations must be based on realistic tests approaching the operating conditions as closely as possible. For instance, a given degree of microstructurally observable damage may suggest overly pessimistic predictions of creep life if the correlations are based on stress-accelerated tests rather than temperature-accelerated tests.

Because the relative contributions of damage mechanisms (e.g., γ' coarsening versus cavitation) may change with test conditions, laboratory correlations must be based on tests in which the appropriate damage mechanism is simulated.

Although the replication technique can be readily applied to uncoated blades to detect base-metal damage, coated blades present a problem. The surface preparation associated with replication will cause the coating to be destroyed locally. In such cases, unless a field technique for spot coating is developed, replication may not prove to be viable. Development of remote replication techniques, in which the need for opening up the turbine can be avoided, can render the technique even more useful. Procedures for replication for blades are expected to be similar to those described for steam headers in Chapter 5 and, therefore, will not be repeated here.

Several advantages of replication-based techniques for life assessment can be cited. Although removal and testing of sacrificial blades can provide more accurate answers, the costs of blade removal and testing, of rebalancing the turbine with new blades in place, and of downtime during the laboratory evaluations can sometimes dictate the need for quick answers using replication. Due to scatter in material properties, the extent of creep and microstructural damage can vary from blade to blade, even in a given row of blades in a turbine. Destructive mechanical and metallographic evaluations using one or two sacrificial blades may not, therefore, be adequate to ensure the integrity of the other blades. The plastic replication technique offers a nondestructive method for evaluating the conditions of many blades at many locations. It permits detection and monitoring of creep damage from its very early stages. Because the replicas can be examined at high magnification using a scanning electron microscope (SEM), a very high degree of resolution can be achieved. This technique is capable of detecting damage in selected localized regions, such as airfoils, roots, trailing edges, etc., which is impossible to do through destructive tests. When the replication technique eventually lends itself to a more quantitative interpretation, based on the damage correlations under development, it will be a powerful technique for remaining-life assessment. Some instances of field replication have been documented elsewhere in the literature (Ref 2).

Destructive Techniques. Destructive evaluation of blades provides the most direct and accurate assessment of their current condition. A typical destructive evaluation includes: (1) metallography to identify cracks, cavities, corrosion, coating degradation, size of γ' and grain-boundary carbide particles, and the presence of other detrimental phases; (2) tensile tests to determine if the strength or ductility has been degraded; (3) impact tests to determine the extent of toughness degradation; (4) hardness tests; and (5) accelerated creep or rupture tests to estimate the remaining rupture life. Because deterioration in microstructure or mechanical properties has to be evaluated against the virgin-material condition for the particular batch of material, samples taken from the cooler parts of the blade (e.g., the root) provide a built-in reference condition. Comparison of the condition of the blade in the hot versus the cold sections provides a direct measure of degradation and eliminates many uncertainties associated with the use of literature data as the basis for reference. Another major advantage of destructive tests is that they eliminate the need to know or to estimate the past operating history. By intelligent selection of an accelerated test scheme, the need to assume any type of damage rule is precluded. Destructive tests thus offer the advantage that they directly characterize the current condition of the material, eliminating many uncertainties arising from lack of knowledge of virgin-material properties, operating conditions, and appropriate damage rules.

Metallography. In the case of uncoated blades, metallography can characterize the extent of hot corrosion. The type of hot corrosion found can be an indicator of the local temperature. The widths of the corrosion layer, the alloy-depleted zone, and the remaining cross section give an idea of the load-bearing capability of the component. If deep intergranular penetrations are present, they should be taken into account in calculating the remaining net section. The applicable increase in stress due to loss of cross section should be calculated and compared with the design values to ensure that an adequate safety margin still exists. Metallography can also reveal the presence of cracks that may have been missed by other techniques. The critical crack size that would constitute grounds for immediate blade replacement is a matter of judgment, because "crack initiation" (undefined magnitude) is the governing failure criterion. Internal casting defects and porosity should also be evaluated because they can lead to fatigue failure. The fracture-mechanics basis for such an evaluation is illustrated in the paper by Hoffelner (Ref 9). Other features such as creep cavitation, γ' coarsening, and grain-boundary carbide coarsening can be used to estimate the degradation of rupture strength and impact strength using the procedures described in the previous section. In the case of coated components, loss of coating thickness and coating cracking should be evaluated. If coating degradation is observed without evidence of base-metal cracking or degradation, blade recoating may be sufficient.

Impact tests are carried out on service-exposed blades mainly to characterize their susceptibility to foreign-object damage. Information regarding the type of test that should be done and the minimum acceptable level of impact energy is proprietary to turbine manufacturers. Room-temperature tests using unnotched Charpy specimens (Ref 25), as well as testing at 900 °C (1650 °F) using Charpy V-notch specimens (Ref 65), have been utilized. A minimum acceptable value of 8.1 J (6 ft·lb) for Charpy V-notch energy at 900 °C has been suggested as one criterion to avoid FOD by the work of Crombie *et al* (Ref 65). If the blade has an impact energy at 900 °C that is less than the 8.1-J threshold value, immediate replacement or rejuvenation might be warranted. If the impact energy is higher than

8.1 J, then the future time interval in which the threshold value will be reached will need to be estimated. Using the properties found in the cooler sections of the blade as the starting virgin-material properties, and the hot-section properties as the current values, plots similar to those in Fig. 9.35 and 9.36 can be used to estimate the remaining useful life.

Accelerated Stress-Rupture Tests. In spite of a large body of data in the literature pertaining to postexposure stress-rupture testing of service-returned blades, there has surprisingly been little effort to use the postexposure test as a quantitative tool for remaining-life prediction. The most common practice has been to conduct a single stress-rupture test on a specimen removed from the blade at a given stress and temperature identical to the original qualification test for the blade material. If any significant loss in creep properties has occurred during service, the sample is expected to fail prematurely, below the minimum life specified by the qualification test. Based on such tests on a large number of service-returned blades taken out at different intervals, life-trend diagrams have been developed in which the decrease in life as a function of exposure duration is plotted (Ref 79 to 83). An example of a test-sample-removal procedure and a life-trend diagram for alloy Inconel X-750 is shown in Fig. 9.52 (Ref 82), which highlights the inconsistency of the diagram in picking up the service-exposed degradation effects. Whereas the rupture lives of some of the service-exposed blades are well below the original specification minimum of 100 h at 345 MPa and 730 °C (50 ksi and 1345 °F), several other blades exposed for longer periods fall well within the limits of scatter for the new material. An easy and obvious way to overcome the problem of scatter would have been to test a sample from the cooler part of the blade for direct comparison, although this apparently is generally not done. A major limitation of this procedure is that it employs a high-stress-accelerated test. The degree of degradation defined in this test is likely to be overly pessimistic.

There is considerable evidence in the liter-

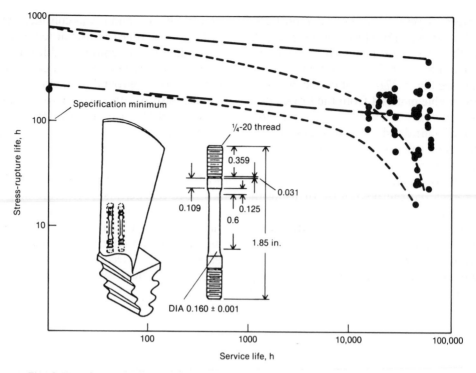

Fig. 9.52. Life-trend diagram for Inconel X-750 tested at 730 °C (1345 °F) (Ref 82).

ature on steels indicating that accelerating the time to rupture by increasing the stress can result in gross overprediction of life consumed. This is illustrated in Table 9.7 using the limited data available on an IN 738 LC blade which had been in service for 14,000 h and had been subsequently tested in the laboratory (Ref 83). In the table, the ratio of the rupture life of the exposed specimen to that of an unexposed specimen tested under identical conditions has been used to compute the remaining life. High-temperature, low-stress testing can be seen to predict a remaining life of 75% of the virgin-material life, whereas high-stress tests at lower temperatures predict remaining lives as low as 5% of the virgin-material life. The superiority of isostress, tempera-ture-accelerated tests over isotemperature, stress-accelerated tests, and the reasons for it, were discussed in detail in the subsection on validity of damage rules in Chapter 3 (see also Fig. 3.33).

Use of a single test, whether accelerated by elevating the temperature or the stress, involves the use of the life-fraction rule to

Table 9.7. Results of accelerated stress-rupture tests on an IN 738 LC blade exposed in service for 14,000 h (Ref 83)

Test temperature		Test stress		Remaining life(a), %
°C	°F	MPa	ksi	
790	1450	620	90	5
815	1500	415	60	10
830	1525	345	50	18
975	1790	90	13	78
990	1810	90	13	75

(a) Remaining life = $1 - [t_r(\text{exposed})/t_r(\text{virgin material})]$, %. Conclusion: accelerating time to rupture by increasing stress can result in gross overprediction of life consumption.

estimate the remaining life under the service conditions. On the other hand, if a mini-mum of two tests can be conducted as a function of temperature and at a constant stress close to the service stress, the log t_r-versus-T data (or 1/T) can be directly ex-trapolated to the service temperature to es-timate the remaining life (see Fig. 3.33). The linear variation of log t_r with T under isostress conditions has been demonstrated for IN 738 and IN 939 by Hoffelner (Ref 84). The life-fraction rule thus can be dismissed altogether. Use of miniature specimens (see Fig. 5.37) can optimize the use of the available material from the service part. Selection of the location of the sample must be made judiciously, on the basis of prior experience on results of nondestructive evaluations.

Creep Tests. Castillo, Koul, and Toscano have explored the possibility of using the second-stage creep rate ($\dot{\epsilon}_s$) to predict rupture life in IN 738 LC using a Monkman-Grant-type relationship (see subsection entitled "Monkman-Grant Correlation" in Chapter 3). They observed that the correla-

tion between $\dot{\epsilon}_s$ and t_r was considerably improved if the time to onset of tertiary creep was taken into account in addition to the time to rupture (Ref 83). Because the two parameters were found to be closely related, a relationship with $\dot{\epsilon}_s$ could be established in terms of t_r alone, as follows:

$$t_r - bt_r^n = K\dot{\epsilon}_s^{-m} \qquad \text{(Eq 9.3)}$$

where b, n, K, and m are experimentally determined coefficients. A logarithmic plot of $(t_r - bt_r^n)$ versus $\dot{\epsilon}_s$, for new as well as service-exposed blade material, resulted in an excellent correlation, as shown in Fig. 9.53. Use of this technique offers the advantage that t_r under service conditions could

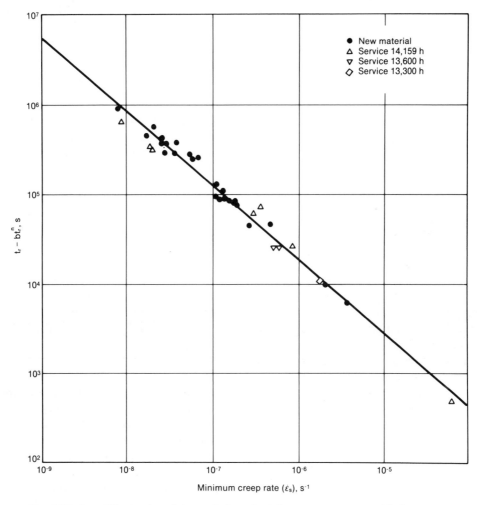

Fig. 9.53. Logarithmic plot of the variation of minimum creep rate $\dot{\epsilon}$ with the parameter $t_r - bt_r^n$ (Ref 83).

be predicted simply from a $\dot{\epsilon}_s$ value determined in a laboratory creep test conducted at the service stress and temperature. Because the $\dot{\epsilon}_s$ value can be determined at much shorter times than t_r, the need for any form of accelerated testing is eliminated. The usefulness of this correlation for life prediction deserves further investigation. Validation of the correlation for additional service-exposed blades of IN 738 LC as well as analysis of available data on other alloys to develop similar correlations seem to be promising avenues.

Thermal-Fatigue Tests. At present, there is no accelerated test procedure that would permit prediction of remaining life (crack initiation) from a thermal-fatigue point of view. In the absence of accelerated tests, the only way to predict incipient fatigue damage (prior to cracking) at present is by calculation. If a suitable accelerated destructive test on the service-degraded material could be developed, many of the uncertainties inherent in the calculational procedure with respect to assumed material properties, prior operating history, and assumed in-service degradation could be eliminated. All available evidence indicates that a thermomechanical fatigue test involving out-of-phase cycles would most closely simulate blade cycles in service. Because most of the fatigue-damage algorithms involve the strain range ($\Delta\epsilon_{total}$ or $\Delta\epsilon_{inelastic}$) in various combinations with the maximum tensile stress (σ_{max}) (see Table 9.6), strain acceleration is one promising approach. In situations where hold-time and frequency effects are not pronounced, simply condensing the hold time (e.g., 2 min in a test cycle versus 5 h in a machine cycle) while holding the strain range at realistic levels also provides a viable accelerated test. This latter type of TMF test has been shown to be successful in ranking service-exposed blades of IN 738 LC in the order of expected remaining life (Ref 25). This test involves a linear out-of-phase cycle, in the temperature range 425 to 870 °C (800 to 1600 °F) and the strain range 0.46%, and a hold time of 2 min in compression. Because microstructural damage

can occur locally in a blade and lead to fatigue cracking from the surface, specialized techniques for preparing and testing samples similar to those described in Ref 25 will also need to be devised.

Crack-Propagation Analysis for Vanes

Because vanes are stationary components, there is no centrifugal stress due to operation. Stresses are primarily thermally induced and arise due to start-stop and other transients. These stresses are sufficiently low to permit large cracks to exist in vanes without failures. Cracks are frequently observed in vanes during inspection. The only question the user often has is how long continued operation of the vane prior to repair can be permitted. Remaining life in this context is the time during which a large observable crack can grow to a critical size that might lead to vane separation. Therefore, what are needed to perform this analysis are crack-growth data under the relevant thermal-fatigue conditions and rate laws governing the crack growth. There is a conspicuous absence of information in either of these areas, because of the noncritical nature of the problem. Even if such information were available, the complexity of vane configurations would make it nearly impossible to calculate the thermal stresses, a knowledge of which is essential for crack-growth analysis. If the stresses were known, some data exist that would permit simplistic analysis of crack growth under pure fatigue and pure creep conditions. Linear-elastic fracture-mechanics methods (see also Chapters 3 and 4) can be applied to these data to estimate the remaining life.

Data on creep-crack-growth rate and fatigue-crack-growth rate are shown in Fig. 9.54 and 9.55, from the work by Hoffelner, for IN 738 and IN 939 (Ref 9 and 84). The creep-crack-growth rate can be expressed in terms of the elastic stress-intensity factor K using the relationship

$$\frac{da}{dt} = CK^m \qquad \text{(Eq 9.4)}$$

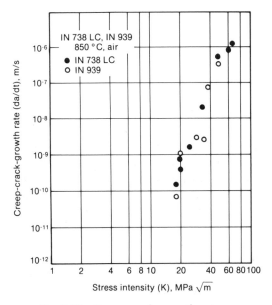

Fig. 9.54. Creep-crack-growth rates as a function of stress-intensity factor for IN 738 LC and IN 939 at 850 °C (1560 °F) in air (Ref 9 and 84).

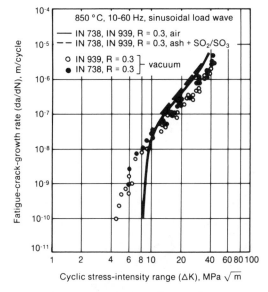

Fig. 9.55. Fatigue-crack-growth rates as a function of cyclic stress-intensity range for IN 738 LC and IN 939 at 850 °C (1560 °F) in different environments (Ref 9 and 84).

where C and m are experimental constants. The remaining life of the vane can be calculated as the time interval t for the crack to grow from its initial size a_i to its final

size a_c (designed arbitrarily). This is given as

$$
t = \int_0^t dt = \int_{a_i}^{a_c} \frac{da}{CK^m}
$$

$$
= \frac{2}{(m-2)CM^{m/2}\sigma^m}
$$

$$
\times \left| \left(\frac{1}{a_i}\right)^{(m-2)/2} - \left(\frac{1}{a_c}\right)^{(m-2)/2} \right|
$$

(Eq 9.5)

where $M = 1.21\pi/Q$ (for a surface crack), σ is stress, and Q is a flaw-shape parameter (given by the quantity in the denominator in the expression for K_I^2 in Fig. 2.5). Because a_i, a_c, σ, C, m, and Q are all known, t can be calculated. The value of a_c can actually be calculated from the fact that the K_{Ic} for the materials is generally about 65 MPa$\sqrt{m}$ (59 ksi$\sqrt{in.}$). The data in Fig. 9.55 also show that the fatigue-crack-growth results obey the Paris law in the intermediate-crack-growth region (Eq 4.44). The number of cycles to failure N_f is determined by the same procedure as above with the difference that K, σ, and t are now replaced by ΔK, $\Delta\sigma$, and N_f, respectively. Other modifications of ΔK may also be necessary to allow for mean-stress effects and crack branching to calculate an effective value of ΔK, as discussed by Hoffelner (Ref 9).

Crack Tolerance of Blades

Combustion turbine blades are subject to high-cycle fatigue. Under fatigue conditions, the factor governing blade life is the threshold value for crack propagation, ΔK_{Th}, rather than the fracture toughness. Although blade materials have large values of K_{Ic} in the range 50 to 80 MPa$\sqrt{m}$ (46 to 73 ksi$\sqrt{in.}$), the ΔK_{Th} values are in the range 4 to 10 MPa$\sqrt{m}$ (3.6 to 9.1 ksi$\sqrt{in.}$) and vary with the R ratio (Ref 85). Using the ΔK_{Th} value of 4 MPa$\sqrt{m}$, corresponding to a high R ratio typical of blade operation, Holdsworth and Hoffelner have illustrated the variation of tolerable crack size with alternating stress, $\Delta\sigma$, at 85 °C (185 °F) for IN 738 and IN 939, as shown in Fig. 9.56 (Ref 85). The presence of corrosive environments can lower the ΔK_{Th} values even further, resulting in even smaller tolerable crack sizes on the order of 0.125 to 0.25 mm (5 to 10 mils). Hence, blades have virtually no tolerance to cracks, and any detectable crack constitutes grounds for repair/replacement.

Integrated Methodology for Life Assessment

The integrated methodology for life assessment of blades includes a three-step process involving calculations followed by non-

destructive evaluations (NDE), and culminating in destructive evaluations. A logic diagram showing the various steps and the decision points is shown in Fig. 9.57.

The calculational procedures should be utilized to set up an inspection schedule and should not be used as a basis for repair/replacement decisions. The various uncertainties delineated in Table 9.4 should be taken into account. NDE evaluations begin with the conventional NDE techniques to detect cracks. If cracks are found, immediate repair/replacement is necessary. If cracks are not detected, replication should be carried out at several locations and on different blades. The purpose of the NDE is primarily to identify the need for, and the critical locations and blades for, further de-

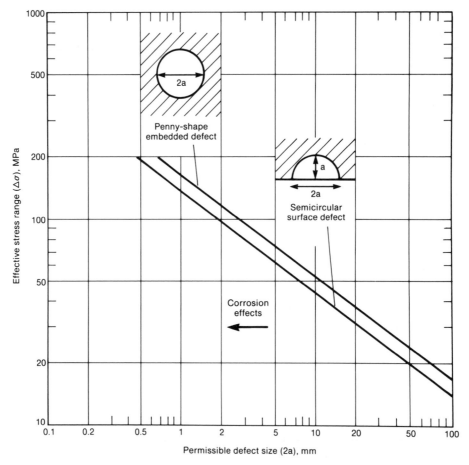

Fig. 9.56. Effect of applied stress on permissible defect size for Nimocast 738 LC blade castings at 850 °C (1560 °F) and R = 0.9 (Ref 85).

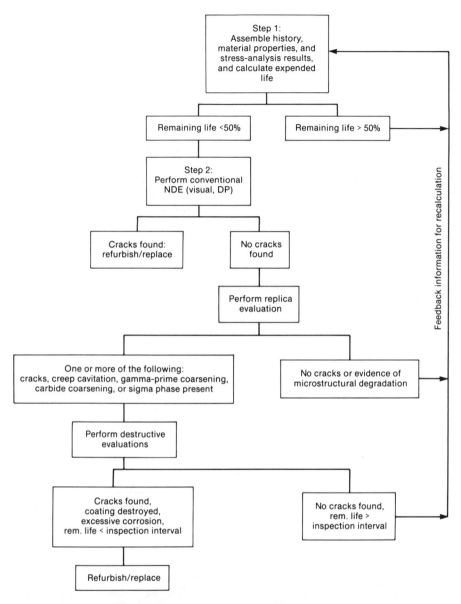

Fig. 9.57. Logic diagram for blade disposition.

structive evaluations. If a comparison of the microstructures at the hot and cold locations shows no evidence of creep cavitation or microstructural changes, further destructive evaluations may not be necessary. If, on the other hand, evidence of such damage is found, sacrificial blades should be removed and subjected to the various destructive evaluations. One or more of the following conditions would constitute grounds for replacement or repair of all the blades:

1. Metallography reveals cracks, severe internal cavitation and microstructural degradation, or unacceptable loss of cross section due to corrosion or loss of coating integrity (if coated).
2. Isostress creep-rupture tests and accelerated TMF tests indicate the remaining life to be less than the inspection interval.
3. Impact tests show the toughness to be below the acceptable level.

If none of these conditions exists, continued operation is deemed to be safe. Information gained from these evaluations should be fed back into the calculational procedure, and the next interval for inspection should be set.

References

1. D.E. Brandt, Heavy Duty Turbo Power: The MS 7001F, *Mech. Engg.*, July 1987, p 28-36
2. R. Viswanathan and A.C. Dolbec, Life Assessment Technology for Combustion Turbine Blades, *J. Engg. Gas Turbines and Power*, Vol 109, Jan 1987, p 115-123
3. H.J. Cialone *et al*, "Combustion Turbine Materials Problems," Report AP-4475, Mar 1986, Electric Power Research Institute, Palo Alto, CA
4. F.P. Lordi, A. Foster, and P.W. Schilke, "Advanced Materials and Coatings," Report GER-3421, G.E. Reference Library, General Electric Co., 1984
5. "Materials for Large Land-Based Gas Turbines," prepared by the National Research Council, Committee on Materials for Large Land-Based Gas Turbines, National Materials Advisory Board, Report AP-4476, Electric Power Research Institute, Palo Alto, CA, Mar 1986
6. D.M. Moon and F.J. Wall, "The Effect of Phase Instability on the High Temperature Stress Rupture Properties of Representative Nickel Base Superalloys," Scientific Paper 68-1D4-STABL-P1, Westinghouse Research Laboratories, Pittsburgh, Aug 13, 1968
7. T.B. Gibbons and R. Stickler, Inco 939: Metallurgy, Properties and Performance, in *Proceedings of the International Conference on High Temperature Alloys for Gas Turbines*, R. Brunetaud *et al*, Ed., Liege, Belgium, Oct 1982, D. Riedel Publishing Co., Dordrecht, 1982, p 369-394
8. H. Susukida, I. Tsuji, H. Kawai, and H. Itoh, "Evaluation of the Properties of Gas Turbine Materials in Long Time Operation," JSME-GTSJ-ASME Joint Gas Turbine Congress, Tokyo, Paper No. 61, 1977
9. W. Hoffelner, High Cycle Fatigue-Life of Cast Nickel-base Superalloys IN 738LC and IN 939, *Met. Trans.*, Vol 13A, 1982, p 1245-1255
10. G.F. Embly and V.V. Kallianpur, "Long-Term Creep Response of Gas Turbine Bucket Alloys," presented at the EPRI Workshop on Life Prediction of High Temperature Gas Turbine Materials, Minnowbrook, Aug 27-30, 1985, Report AP-4477, Apr 1986, V. Weiss and W.T. Bakker, Ed., Electric Power Research Institute, Palo Alto, CA
11. H.J. Martens, A. Rosslet, and B. Walser, Creep-Fatigue Interaction for Two Nickel-base Alloys and a Martensite Heat Resistant Steel, in *High Temperature Alloys for Gas Turbines and Other Applications*, W. Betz *et al*, Ed., Riedel Publishing Co., Dordrecht, Holland, 1986, p 1527-1536
12. M.Y. Nazmy, High Temperature Low Cycle Fatigue of IN 738 and Application of Strain Range Partitioning, *Met. Trans.*, Vol 14A, Mar 1983, p 449-461
13. A. Strang, High Temperature Properties of Coated Superalloys, in *Proceedings of International Conference on Behavior of High Temperature Alloys in Aggressive Environments*, Petten, The Netherlands, Oct 1979, I. Kirman *et al*, Ed., The Metals Society, London, 1979, p 595-612
14. G.B. Thomas and M.F. Day, COST 50/2, UK16, NPL Report DMA(D), 179, National Physical Laboratories, Teddington, United Kingdom, 1979
15. R. Viswanathan, C.G. Beck, and R.L. Johnson, "Low Cycle Fatigue Behavior of Udimet 710 at Elevated Temperatures," Research Report 79-1D4-STABL-R3, Westinghouse Research Laboratories, Pittsburgh, June 19, 1979
16. *Atlas of Fatigue Curves*, H.E. Boyer, Ed., American Society for Metals, Metals Park, OH, 1986
17. C.H. Wells and C.P. Sullivan, *Trans. ASM*, Vol 61, Mar 1968, p 149-155
18. E.S. Russell, "Practical Life Prediction Methods for Thermalmechanical Fatigue of Gas Turbine Buckets," presented at the EPRI Workshop on Life Prediction of High Temperature Gas Turbine Materials, Minnowbrook, Aug 27-30, 1985, Report AP-4477, Apr 1986, V. Weiss and W.T. Bakker, Ed., Electric Power Research Institute, Palo Alto, CA, p 3.1-4.1
19. L. Remy, F. Rezai-Aria, R. Danzer, and W. Hoffelner, Comparison of Life Prediction Methods in MarM509 Under High Temperature Fatigue, in *High Temperature Alloys for Gas Turbines and Other Applications*, W. Betz *et al*, Ed., Riedel Publishing Co., Dordrecht, Holland, 1986, p 1617-1628
20. M.Y. Nazmy, The Applicability of Strain-range Partitioning to High Temperature, Low Cycle Fatigue Life Prediction of 'IN 738' Alloy, *Fatigue Engg. Mater. Struct.*, Vol 4 (No. 3), 1981, p 253-261
21. M. Marchionni, D. Ranucci, and E. Picco, High Temperature Fatigue Life Prediction of a Nickel Superalloy by the Strain Range Partitioning Method, in *High Temperature Alloys for Gas Turbines and Other Applications*, W.

Betz *et al*, Ed., Riedel Publishing Co., Dordrecht, Holland, 1986, p 1629–1638

22. W.J. Ostergren, *ASTM Journal of Testing and Evaluation*, Vol 4, 1976, p 327-339

23. C.H. Wells and C.P. Sullivan, *ASM Trans. Qtrly.*, Vol 60, 1967, p 217-222

24. C.H. Wells and C.P. Sullivan, ASTM STP 459, 1968, p 59

25. "Gas Turbine Life Management System," Final Report on EPRI Project RP2421-2, Electric Power Research Institute, Palo Alto, CA, Feb 1988

26. G.B. Thomas, J. Bressers, and D. Raynor, "Low Cycle Fatigue and Life Prediction Methods," citation of unpublished data of A. Samuelsson, L.E. Larsson, and L. Lundberg, in *High Temperature Alloys for Gas Turbines*, R. Brunetaud, Ed., proceedings of conference held in Liege, Belgium, Oct 1982, D. Riedel Publishing Co., Dordrecht, 1982, p 291-317

27. J. Stringer, Hot Corrosion of High Temperature Alloys, in *Proceedings of the Symposium on Properties of High Temperature Alloys*, Z.A. Faroulis and F.S. Petit, Ed., The Electrochemical Society, Princeton, NJ, 1976, p 513-556

28. S. Mrowec, Attack of High Temperature Alloys in Sulfidizing Gases, in *Proceedings of the Symposium on Properties of High Temperature Alloys*, Z.A. Faroulis and F.S. Petit, Ed., The Electrochemical Society, Princeton, NJ, 1976, p 413-437

29. J. Stringer and D.P. Whittle, High Temperature Corrosion and Coating of Superalloys, in *High Temperature Materials in Gas Turbines*, P.R. Sahm and M.O. Speidel, Ed., Elsevier Publishing Co., Amsterdam, 1974, p 283-312

30. J. Stringer, High Temperature Corrosion of Superalloys, *Mater. Sci. Tech.*, Vol 3, July 1987, p 481

31. F.S. Pettit and C.S. Giggins, in *Hot Corrosion in Superalloys*, Vol II, C.T. Sims, N.S. Stoloff, and W.C. Hagel, Ed., John Wiley & Sons, New York, 1987, p 327-357

32. J.F.G. Conde, E. Erdoes, and A. Rahmel, Mechanisms of Hot Corrosion, in *High Temperature Alloys for Gas Turbines*, R. Brunetaud *et al*, Ed., proceedings of international conference held in Liege, Belgium, Oct 1982, D. Riedel Publishing Co., Dordrecht, 1982, p 99-148

33. R. Viswanathan and A.C. Dolbec, "Life Assessment Technology for Combustion Turbine Blades," International Gas Turbine Conference and Exhibit, Dusseldorf, West Germany, June 1986, ASME Paper 86GT257

34. M. Gell, unpublished work, Pratt and Whitney Aircraft, Ltd., 1987

35. Grüenling, K.H. Keinburg, and K.K. Schweitzer, The Interaction of High Temperature Corrosion and Mechanical Properties of Alloys, in *Proceedings of the International Conference on High Temperature Alloys for Gas Turbines, Liege, Belgium, Oct 1982*, R. Brunetaud *et al*, Ed., D. Riedel Publishing Co., Dordrecht, 1982, p 507-544

36. F. Schmitz, M. Heinrich, and W. Slotty, "Langzeit-Korrosionsversuche mit mechanischer Beanspruchung an hochwarmfesten Legierungen fuer die Schaufeln von ortsfesten Gasturbinen groszer Leistung," Final Report, COST 50/2, Proj. D6-01ZB067, 1980

37. F. Schmitz and K.H. Keienburg, Long Term High Temperature Tests With Simultaneous Mechanical Stress on Hot Corrosion Resistant Materials for Land-Based Gas Turbines, in *Corrosion and Mechanical Stress at High Temperatures*, V. Guttman and M. Merz, Ed., proceedings of international conference, Petten, The Netherlands, May 1980, Applied Science Publishers, London, 1980, p 223-242

38. K.H. Kloos, J. Granacher, and H. Demus, Creep Rupture Strength of Cast Alloys for Gas Turbine Blades Under Hot Gas Corrosion, in *Corrosion and Mechanical Stress at High Temperatures*, V. Guttman and M. Merz, Ed., proceedings of international conference, Petten, The Netherlands, May 1980, Applied Science Publishers, London, 1980, p 243-255

39. K.H. Kloos, J. Granacher, and H. Demus, "Statische und dynamische Festigkeit von Gasturbinenschaufelwerkstoffen unter Hochtemperaturkorrosion," Final Report, DECHEMA-Proj. FE-KKs 4.2/1 und 4.2/1F, 1980

40. F. Schmitz, "Langzeit-Korrosionsversuche mit und ohne mechanische Beanspruchung an hochwarmfesten Legierungen fuer Schaufeln von ortsfesten Gasturbinen groszer Leistung," Final Report, COST 50/2, Proj. D1/6, BCT32, 1980

41. P. Huber and A. Rosslet, "Pruefung der Korrosionsbestaendigkeit von Gasturbinenlegierungen und Schutzschichten," Final Report, COST 50/2, Proj. CH6, 1981

42. G.A. Whitlow, C.G. Beck, R. Viswanathan, and E.A. Crombie, The Effects of a Liquid Sulfate/Chloride Environment on Superalloys, *Met. Trans. A*, Vol 15A, Jan 1984, p 23-28

43. J.C. Galsworthy, The Creep and Stress Rupture Behavior of Superalloys at Elevated Temperatures in Salt Contaminated Environments, in *Environmental Degradation of High Temperature Materials*, The Institute of Metallurgists Conference, Mar 1980, Series 3, No. 13, Vol 2, Paper 2/40, 1301-80-Y

44. J.C. Galsworthy, The Effects of Seasalt on the High Temperature Creep Properties of a Nickel Base Gas Turbine Blade Alloy, in *Corrosion and Mechanical Stress at High Temper-*

atures, V. Guttman and M. Merz, Ed., proceedings of international conference, Petten, The Netherlands, May 1980, Applied Science Publishers, London, 1980, p 223-242

45. M.Y. Nazmy, The Effect of Sulfur Containing Environment on the High Temperature Low Cycle Fatigue of a Cast Nickel Base Alloy, *Scripta Met.*, Vol 16, 1982, p 1329-1332

46. M.Y. Nazmy, H. Wettstein, and A. Wicki, "Experiences With the Material Behavior and High Temperature Low Cycle Fatigue Life Prediction of the In 738 Blading Alloy," North Atlantic Treaty Organization, Advisory Group for Aerospace Research and Development, Conference on Engine Cyclic Durability by Analysis and Testing, 1984

47. G.A. Whitlow, R.L. Johnson, W.H. Pridemore, and J.M. Allen, Intermediate Temperature, Low Cycle Fatigue Behavior of Coated and Uncoated Nickel Base Superalloys in Air and Corrosive Sulfate Environments, *ASME J. Engg. Mater. and Tech.*, Vol 106, Jan 1984, p 43-49

48. K. Schneider, H. Arnim, and H.W. Grüenling, *Thin Solid Films*, Vol 84, 1981, p 29-36

49. H.W. Grüenling, K. Schneider, and H. Arnim, "Influence of Coatings on High Cycle Fatigue Properties of Cast Nickel Base Superalloys," Final Report, COST 50/2, Proj. D2-01ZB067, 1980

50. G.A. Whitlow, J.M. Allen, and E.A. Crombie, Combustion Turbine Blade Design Considerations—Prevention of Corrosion-Assisted Mechanical Failure, *Proceedings of XVth International Congress on Combustion Engines (CIMAC)*, Paris, 1983, p 559-578

51. J.M. Allen and G.A. Whitlow, Observations on the Interaction of High Mean Stress and Type II Hot Corrosion on the Fatigue Behavior of a Nickel Base Superalloy, *ASME J. Engg. Gas Turbines and Power*, Vol 107, Jan 1985, p 220-224

52. K.K. Schweitzer and W. Track, "Festigkeitsverhalten von Turbinenschaufelwerktoffen mit und ohne Schutzschichten unter Heißgaskorrosion," Final Report, COST 50/2, Proj. D13-01ZB097, 1981

53. W. Hoffelner and M.O. Speidel, The Influence of the Environment on the Fatigue Crack Growth of the Nickel Base Superalloys IN 738LC and IN 939 at 850°C, in *Behavior of High Temperature Alloys in Aggressive Environments*, I. Kirman *et al*, Ed., proceedings of international conference held in Petten, The Netherlands, Oct 1979, The Metals Society, London, 1979, p 993-1004

54. W. Hoffelner and M.O. Speidel, Microstructural Aspects of Fatigue Crack Propagation of Cast Nickel Base Superalloys at 850°C in Various Environments, in *Corrosion and Mechanical Stress at High Temperatures*, V. Guttman

and M. Merz, Ed., proceedings of international conference, Petten, The Netherlands, May 1980, Applied Science Publishers, London, 1980, p 275-286

55. A. McMinn, "Coatings Technology for Hot Components of Industrial Combustion Turbines—A Review of the State of the Art," Report AP-5078, Electric Power Research Institute, Palo Alto, CA, Feb 1987

56. G.R. Leverant and A. McMinn, "Analysis of Cracked Udimet 720 Gas Turbine Blades," Report on Project RP2775-4, Electric Power Research Institute, Palo Alto, CA, Oct 30, 1987

57. A. Strang and E. Lang, Effect of Coatings on the Mechanical Properties of Superalloys, in *High Temperature Alloys for Gas Turbines*, R. Brunetaud, Ed., proceedings of conference held in Liege, Belgium, Oct 1982, D. Riedel Publishing Co., Dordrecht, 1982, p 469-506

58. D.H. Boone, Technical Data Sheets, Airco Temescal, Inc., Berkeley, CA, Jan 1976, cited in Ref 57

59. D.H. Boone, "Overlay Coatings for High Temperature Applications," Airco Temescal, Inc., Jan 1976, cited in Ref 57

60. G.W. Goward, *J. Metals*, Oct 1970, p 31-39

61. G.W. Goward, in *Symposium on Properties of High Temperature Alloys*, Las Vegas, Oct 1976, p 806-823, cited in Ref 57

62. R. Lowrie and D.H. Boone, in *International Conference on Metallurgical Coatings*, San Francisco, Apr 1977

63. D.M. Moon and F.J. Wall, "The Effect of Phase Instability on the High Temperature Stress-Rupture Properties of Representative Nickel Base Superalloys," Scientific Paper 68-ID4-STABL-P1, Westinghouse Research Laboratories, Pittsburgh, Aug 1968

64. H. Susukida, Y. Sakumoto, I. Tsuji, and H. Kawai, "Strength and Microstructure of Nickel-base Superalloys After Long-Term Heating," Mitsubishi Technical Bulletin No. 86, Mitsubishi Heavy Industries, Ltd., June 1973

65. E. Crombie, W. McCall, C.G. Beck, and D.M. Moon, Degradation of High Temperature Impact Properties of Nickel-Base Gas Turbine Blade Alloy, *Proceedings of the 12th CEMAC Conference*, Vol C, Tokyo, 1977

66. A.G. Pard, "Long Time Aging Studies of Some Nickel-base Alloys," Final Report on Project RP2775-1, Electric Power Research Institute, Palo Alto, CA, 1988

67. I. Tsuji, "Superalloys for Land Based Gas Turbines in Japan," Japan-U.S. Seminar on Superalloys, 1985, p 259-269

68. R. Castillo and A.K. Koul, "Effects of Microstructural Instability on the Creep and Fracture Behavior of Cast IN 738LC Ni-Base Superalloy," in *High Temperature Alloys for Gas Turbines and Other Applications*, W. Betz *et*

al, Ed., proceedings of international conference held in Liege, Belgium, Oct 1986, Riedel
Publishing Co., Dordrecht, Holland, 1986,
p 1395-1409

69. R.A. Stevens and P.E.J. Flewitt, Intermediate
Regenerative Heat Treatments for Extending
the Creep Life of the Superalloy IN 738,
Mater. Sci. Engg., Vol 50, 1981, p 271-284

70. V.P. Swaminathan, P. Lowden, and J.
Liburdi, "Assessment and Extension of Life of
Combustion Turbine Blades," Report RP2775-
6, Electric Power Research Institute, Palo
Alto, CA, Dec 1988

71. M.Y. Nazmy and C. Wuthrich, The Predictive
Capability of Three High Temperature Low
Cycle Fatigue Models in the Alloy IN 738, in
Mechanical Behavior of Materials IV, J. Carls-
son and N.G. Ohlson, Ed., Pergamon Press,
New York, 1984

72. K. Kuwabara and A. Nitta, *Proc. ICCM3*, Vol
2, 1978

73. M.F. Day and G.B. Thomas, *AGARD Conf.
Proc.*, 1978, p 243

74. P.O. Persson, C. Persson, G. Burman, and Y.
Lindblom, "The Behavior of Nimonic 105 and
Inco 738LC Under Creep and LCF Testing,"
in *High Temperature Alloys for Gas Turbines
and Other Applications*, W. Betz *et al*, Ed.,
proceedings of international conference held in
Liege, Belgium, Oct 1986, Riedel Publishing
Co., Dordrecht, Holland, 1986, p 1501-1516

75. H. Fischmeister, R. Danzer, and B. Buch-
mayr, Lifetime Prediction Models, in *High
Temperature Alloys for Gas Turbines and
Other Applications*, W. Betz *et al*, Ed., pro-
ceedings of international conference held in
Liege, Belgium, Oct 1986, Riedel Publishing
Co., Dordrecht, Holland, 1986, p 495-550

76. H.R. Tipler, Metallographic Aspects of Creep
Fracture in a Cast Ni-Cr Base Alloy, *Advances*

in Fracture Research (Fracture 81), Vol 4, Pergamon Press, Ltd., Cannes, France, Apr 1981,
p 1587-1594

77. Y. Lindblom, "Refurbishing Superalloy Components for Gas Turbines," *Mater. Sci. Tech.*,
Vol 1 (No. 8), Aug 1985, p 636-641

78. J. Wortmann, Improving Reliability and Lifetime of Rejuvenated Turbine Blades, *Mater.
Sci. Tech.*, Vol 1 (No. 8), Aug 1985, p 644-650

79. G. Van Drunen and J. Liburdi, in *6th Turbomachinery Symposium*, Texas A&M University, Houston, Dec 6-8, 1977, Report
H-268, Westinghouse Canada, Inc., 1977

80. G. Van Drunen, J. Liburdi, W. Wallace, and
T. Terada, "Proc. Conf. on Advanced Fabrication Processes, Florence, September 26-28,
1978," in *Conf. Publ. NATO-AGARD-CP-
256*, North Atlantic Treaty Organization, Advisory Group for Aeronautical Research and
Development, 1978, p 13-1

81. W.J. McCall, *Combustion*, Vol 42, 1971, p 27

82. A.K. Koul, R. Castillo, and K. Willett, Creep
Life Predictions in Nickel Base Superalloys,
Mater. Sci. Engg., Vol 66, 1984, p 213-226

83. R. Castillo, A.K. Koul, and E.N. Toscano,
Lifetime Prediction Under Constant Load
Creep Conditions for a Cast Ni-base Superal-
loy, in *International Gas Turbine Conference*,
Dusseldorf, W. Germany, June 8-12, 1986,
ASME Paper 86, GT 241

84. W. Hoffelner, Creep Dominated Damage Processes, Riedel Publishing Co., Dordrecht, Holland, 1986, p 413-440

85. S.R. Holdsworth and W. Hoffelner, Fracture
Mechanics and Crack Growth in Fatigue, in
High Temperature Alloys for Gas Turbines, R.
Brunetaud *et al*, Ed., proceedings of interna-
tional conference held in Liege, Belgium, D.
Riedel Publishing Co., Dordrecht, 1982, p
345-368

Conversion to SI Units

To convert			To convert		
From	To	Multiply by (×) or add (+)	From	To	Multiply by (×) or add (+)
ksi	MPa	×6.894	°F	°R	+460
kg/mm^2	MPa	×9.804	°R	K	×0.556
kg/cm^2	MPa	×0.098	Δ°F	Δ°C	×0.556
bars	MPa	×0.100	inch	cm	×2.54
atmospheres	MPa	×0.100	mils	microns	×25.4
N/mm^2	MPa	×1.0	ft·lb	J	×1.356
MN/m^2	MPa	×1.0	ksi$\sqrt{\text{in.}}$	MPa$\sqrt{\text{m}}$	×1.099
°F	°C	(°F−32)×5/9	MN·m$^{-3/2}$	MPa$\sqrt{\text{m}}$	×1.0
°C	K	+273	calories	J	×4.1868

Index

The symbol (F) or (T) following a page number indicates that information is presented in a figure or a table, respectively. Steel designations comprised of letters and numbers, except for the MAR-M and NASA steels, can be found under the heading *Steels, specific types*.

483